Principles of Chemistry

Ignacio J. Ocasio

Department of Chemistry
Case Western Reserve University

This custom textbook includes materials submitted by the Author for publication by John Wiley & Sons, Inc. The material has not been edited by Wiley and the Author is solely responsible for its content.

Copyright © 1998 by Ignacio J. Ocasio

All rights reserved.

Reproduction or translation of any part of this work beyond that permitted by Sections 107 and 108 of the 1976 United States Copyright Act without the permission of the copyright owner is unlawful. Requests for permission or further information should be addressed to the Permission Department, John Wiley & Sons.

Printed in the United States of America.

ISBN 0-471-32124-9

Principles of Chemistry

Table of Contents.

Chapter 1. <u>Introduction.</u> ... 1

    1.1   Quick History of Chemistry and Some Fundamental Laws. ... 4

    1.2   The Composition of Matter. ... 7

    1.3   International System of Units. ... 9

    1.4   Factor Label Method. ... 12

    1.5   Errors. ... 16

    1.6   Significant Figures. ... 18

    1.7   Significant Figures and Mathematical Operations. ... 20

Chapter 2. <u>Atoms and Moles</u>. ... 27

    2.1   Atoms and Molecules. ... 27

    2.2   Moles. ... 28

    2.3   Atomic and Molecular Weights. ... 31

    2.4   Derivation of Formulas. ... 33

    2.5   Percent Composition. ... 39

    2.6   Chemical Equations. ... 43

    2.7   Stoichiometry. ... 47

    2.8   Limiting Reagents. ... 50

Chapter 3. <u>Solution Stoichiometry</u>. ... 56

    3.1   Molarity ... 56

    3.2   Dilution. ... 59

    3.3   Solution Stoichiometry. ... 61

Chapter 4. <u>Gases</u>. ... 64

    4.1   The Early Laws. ... 64

    4.2   Ideal Gas Law. ... 70

    4.3   Density of Gases. ... 74

Chapter 4: Continuation.

4.4	Gas Stoichiometry.	76
4.5	Partial Pressures.	79
4.6	Kinetic Molecular Theory of Gases.	87
4.7	Molecular Velocities.	90
4.8	Graham's Law of Effusion.	91
4.9	Real Gases.	94

Chapter 5. <u>Thermochemistry</u>. 99

5.1	Energy.	99
5.2	Specific Heats.	100
5.3	Calorimetry.	103
5.4	Enthalpy.	106
5.5	Properties of $\Delta H°$	108
5.6	Hess' Law.	108
5.7	Standard Enthalpies of Formation.	109
5.8	ΔH as a Function of Temperature.	115

Chapter 6. <u>Atomic Structure and Quantum Mechanics</u>. 119

6.1	Early Characterization of the Atom.	119
6.2	Symbols for Atomic Structure.	121
6.3	Isotopes.	122
6.4	Ions.	125
6.5	From Classical To Quantum Physics.	127
6.6	The Photoelectric Effect.	131
6.7	Bohr's Theory.	135
6.8	Wave Mechanics.	144
6.9	Heisenberg's Uncertainty Principle.	146
6.10	The Development of Quantum Mechanics.	148
6.11	The Particle in a Box.	150

Chapter 7. Quantum Numbers and Periodicity. — 156

- 7.1 Quantum Numbers for the Hydrogen Atom. — 156
- 7.2 Multi Electron Atoms. — 165
- 7.2A Energy Levels. — 166
- 7.3 The Aufbau Principle. — 168
- 7.4 The Periodic Table. — 177
- 7.5 Electronic Configuration of Ions. — 182
- 7.6 Magnetic Properties. — 183
- 7.7 Atomic Radii. — 184
- 7.8 Ionization Energies. — 193
- 7.9 Electron Affinities. — 197
- 7.10 Expected Ions from the Periodic Table. — 200

Chapter 8. Bonding I. — 203

- 8.1 Chemical Bonding. — 203
- 8.2 Ionic Bonding. — 205
- 8.3 Covalent Bonding. — 209
- 8.4 The Hydrogen Molecule. — 212
- 8.5 Electronegativity. — 214
- 8.6 Ionic vs. Covalent. — 216
- 8.7 Covalency. — 222
- 8.8 Sigma and Pi Bonds. — 228
- 8.9 Drawing Lewis Structures. — 231
- 8.10 Three Dimensionality of Lewis Structures. — 238
- 8.11 Expected Geometries with Lone Pairs. — 241
- 8.12 Polarity vs. Geometry. — 246

Chapter 9. Bonding II: More Advanced Concepts. — 252

- 9.1 Hybrid Orbitals. — 252
- 9.2 The Molecular Orbital Theory. — 267
- 9.3 Heteronuclear Diatomic Molecules. — 277

Chapter 10. Condensed Phases. — 280

- 10.1 Intermolecular Forces of Attraction. — 280
- 10.2 Solids. — 293
- 10.3 Metals and Closest Packing. — 300
- 10.4 Bonding in Metals. — 307
- 10.5 Ionic Crystals. — 309
- 10.6 Liquids. — 317
- 10.7 Vapor Pressure. — 321
- 10.8 Temperature Dependence of the Vapor Pressure. — 324
- 10.9 Heating Curves. — 328
- 10.10 Phase Diagrams. — 331

Chapter 11. Solutions and Their Properties. — 338

- 11.1 Classification of Solutions. — 338
- 11.2 Henry's Law. — 342
- 11.3 Concentration Units. — 343
- 11.4 Net ionic Equations. — 347
- 11.5 Raoult's Law. — 349
- 11.6 Raoult's Law for Two Volatile Components. — 352
- 11.6A Colligative Properties. — 357
- 11.7 Osmotic Pressure. — 361

Chapter 12. Chemical Kinetics. — 369

- 12.1 Introduction. — 369
- 12.2 Nature of the Reactants. — 370
- 12.3 Concentration of the Reacting Species. — 370
- 12.4 Initial Rate Method. — 374
- 12.5 Solutions to the Differential Rate Equations. — 377
- 12.6 Graphical Method. — 380
- 12.7 Effect of Temperature on the Rate. — 392
- 12.8 Collision Theory. — 393
- 12.9 Temperature Dependence on the Rate Constant. — 396

Chapter 12: Continuation.

    12.10 Transition State Theory.   398

    12.11 Reaction Mechanisms.   402

    12.12 Equilibrium Elementary Steps.   406

Chapter 13. <u>Organic Chemistry</u>.   411

    13.1 Introduction.   411

    13.2 Alkanes.   411

    13.3 Alkenes.   423

    13.4 Stereoisomers I: Geometric Isomers.   424

    13.5 Reactivity of Alkenes.   426

    13.6 Alkynes.   428

    13.7 Aromatics.   429

    13.8 Cycloalkanes.   432

    13.9 Functional Groups.   433

    13.10 Stereoisomers II: Chirality (Asymmetry).   441

    13.11 Nucleophilic Substitution Reactions.   450

    13.12 Second Order Mechanism: The SN2 Mechanism.   451

    13.13 First Order Mechanism: The SN1 Mechanism.   454

Chapter 14. <u>Spectroscopy</u>.   458

    14.1 Experimental Determination of Structure.   458

    14.2 Mass Spectroscopy.   458

    14.3 Infrared Spectroscopy.   471

    14.4 General Diagram for an Infrared Spectrometer.   475

    14.5 Infrared Spectra Analysis.   479

    14.6 Degrees of Unsaturation in Organic Compounds.   486

    14.7 Nuclear Magnetic Resonance Spectroscopy.   487

    14.8 Spin-Spin Coupling.   498

    14.9 NMR Spectra Analysis.   502

    14.10 Additive Effect.   510

    14.11 Organic Reactions Revisited.   521

Chapter 15. Thermodynamics. — 532

15.1	Introduction.	532
15.2	The First Law of Thermodynamics.	534
15.3	Physical and Chemical Transformations.	538
15.4	Expansion Work.	541
15.5	Energy and Enthalpy.	545
15.6	Chemical Transformations.	550
15.7	Entropy.	552
15.8	Entropy Changes for Other Physical Processes.	557
15.9	Second Law of Thermodynamics.	559
15.10	Entropy Changes in Chemical Processes.	563
15.11	Calculation of Absolute Entropies.	566
15.12	Dependence of Free Energy on Pressure.	568
15.13	Equilibrium.	570

Chapter 16. Chemical Equilibrium. — 574

16.1	Introduction.	574
16.2	Le Chatelier's Principle.	574
16.3	Properties of the Equilibrium Constant.	577
16.4	Chemical Equilibrium Problems.	582
16.5	Strategy for Solving Chemical Equilibrium Problems.	589

Chapter 17. Acids and Bases. — 590

17.1	The Brönsted Lowry Acid Base Concept.	590
17.2	Acid Strength and Molecular Geometry.	592
17.3	K_a's and K_b's.	598
17.4	Autoionization of Water.	601
17.5	pH, a Logarithmic Scale.	602
17.6	Hydrolysis.	606
17.7	Common Ion Effect.	608
17.8	Buffers.	609
17.9	The Henderson Hasselbach Equation.	611

Chapter 17: Continuation.

 17.10 Fractions of the Conjugate Forms in a Solution of a Given pH. **615**

 17.11 Acid Base Titrations. **619**

 17.12 Polyprotic Acids. **627**

 17.13 Lewis Acids and Bases. **630**

Chapter 18. <u>Other Aqueous Equilibria: Ksp</u>. **633**

 18.1 Slightly Soluble Salts. **633**

 18.2 Precipitation. **637**

Chapter 19. <u>Electrochemistry</u>. **644**

 19.1 Redox Reactions. **644**

 19.2 Oxidation States. **645**

 19.3 Balancing Redox Reactions by the Ion Electron Method. **646**

 19.4 Electrochemistry: An Introduction. **652**

 19.5 Units. **654**

 19.6 Galvanic Cells. **655**

 19.7 Conventional Notation. **657**

 19.8 Electromotive Force (EMF). **659**

 19.9 Half Cell Potentials. **661**

 19.10 EMF as a Function of Concentration. **667**

 19.11 Calculation of Equilibrium Constants. **671**

 19.12 Electrolysis. **672**

 19.13 Faraday's Laws. **675**

Appendix #1: Thermodynamic Parameters for Some Selected Substances. **678**

Appendix #2: Equilibrium Constants for Some Monoprotic Weak Acids. **683**

    Equilibrium Constants for Some Weak Bases. **683**

    Equilibrium Constants for Some Polyprotic Acids. **684**

Appendix #3: Ksp Values for Some Common Ionic Solids. **685**

Appendix #4: Standard Reduction Potentials for Some Selected Half Cells. **686**

Chapter 1: Introduction.

Why do we study chemistry? This is a question that many of you may be asking yourselves at this point. Others have referred to Chemistry as "the central science"[1] since it is basic to many different disciplines. Just picture the following scenarios: 1. A very old and valuable painting in the Cleveland Museum of Art has many years of grime collected over the original canvas and it looks rather opaque. The curator of this museum has to find someone to clean this painting and restore it to its original luster. What do we need? Someone who has the expertise (and infinite patience) to remove all those years of dirt and grime without removing the actual paint. An intimate knowledge of chemistry is absolutely necessary in order to achieve this. 2. Let's imagine what would happen if someone were to spill copious amounts of water (for example, during a fire) on some of the extremely valuable and irreplaceable texts (collections of complete works by Charles Darwin, Sigmund Freud, and some medieval texts that are also found there) in the Allen Memorial Library[2]. Probably, from our own experience, we would expect for the pages to stick together. An expert with knowledge of chemistry, as well as the forces that are keeping these pages together, would have to step forward to separate the pages without destroying the books. 3. A dead body is recovered somewhere in Cuyahoga county and it is determined that it was a homicide. Examination of the body by a forensic expert would have to determine what chemicals were in this individual's blood stream, the content of the soil in his shoes, etc. Once more, knowledge of chemistry and materials, as well as the geology of the surrounding areas, will be extremely important clues in trying to reconstruct the crime scene and thus, help to solve the murder. 4. A lawyer is presented with a new material by a client who wants to patent it as the best cleansing agent for bathroom tiles. This patent lawyer would benefit immensely from a chemistry degree so that she can understand what the clients are claiming and at the same time, while she is doing all the research in the library, she will be able to understand the science behind this claim. 5. A civil engineer is asked to place a bid for the construction of a bridge along the Cuyahoga river. Knowledge of the chemistry of materials is going to be invaluable in order for this engineer to be able to determine which material would be best in terms of cost, efficiency, safety, environmental issues, etc.

As we can see, chemistry is involved in many different scenarios. You will find that chemistry will be very useful, independent of what your final career goals will be. Of course, scientists, doctors, dentists, etc., have a clear idea of why they need chemistry, but even

---

[1] Brown, LeMay, Bursten, "Chemistry, The Central Science", Seventh Edition, Prentice Hall.
[2] One of the Health Science Libraries here at CWRU.

engineers, lawyers, and media people will find it extremely useful. I hope, therefore, that this will be a course that all of us will enjoy in one way or another.

So where should we start? Let us start at the very beginning. How can we define chemistry? Chemistry is one of the physical sciences that studies the composition, structure and transformation of matter. In order to understand what this definition means, let's concentrate for the time being on a particular object, like a pencil. By the time we finish this course, we should be able to answer questions like: What are the components that make up the pencil? In what proportion are these components combined? How are these components held together? What are the pencil's properties (color, hardness, etc.)? What will happen if you strike the tip of the pencil against the table? Why? Can we still write with this pencil? Why? How "stable" is it? Will it combine spontaneously with other substances to form a new stable substance? For example, if we were to burn it in the presence of oxygen gas, would it burn? If so, what would it form and why? Could we predict its reaction with yet another substance before we actually go into the laboratory to carry it out? How long will it take for these reactions to take place? Is there anything we can do to speed these reactions up? Can we set up these reactions in such a way that as the reaction is taking place, it will do work for us?

Hopefully, we will be able to answer all of these questions and many more once we place all the components of what we call general chemistry together. Keep in mind that in a way, this course is meant to be a <u>survey</u> of many of the different areas that we call chemistry and by no means, will you leave knowing everything there is to know about the subject. This is why we call it *general* chemistry. Our challenge will be to try to make it all fit together by the end of the course. If we succeed in doing this, then we can say that we actually learned something valuable and will have the ability to recall it in the future.

How will we accomplish this? By trying to introduce the types of problems that chemists deal with on a daily basis. We will then proceed to try to understand what the problem is and to analyze it in order to try to reach certain conclusions that will enable us to come to a solution. This will require that we follow some kind of methodology which hopefully we will be able to learn to apply. Another important aspect of this course is problem solving. It is imperative that we apply critical thinking to problem solving, so that we will be able to analyze our answers to see if your numerical answer is what you expected (or at least in the "ballpark"). This is something that we may not have done in high school, but that will not be tolerated at this level. An answer to a problem should make sense, and if we cannot tell if it does, then we certainly do not have a clear idea of what is really going on.

Chapter 1: Introduction

Chemistry is a very modern science and most of us have a misconception of what it is all about. If I were to ask you what comes to your mind when I say chemistry, what would you really say? Probably, you would think of a lab, filled with test tubes with a lot of colored solutions in rather intricate apparatuses, with boiling liquids and some mad scientists working with these. This is a concept that has been portrayed in the movies many times, but it is certainly not accurate nor very modern. It is possible that it may have started that way, but nothing is further from reality these days. The development of computers and lasers has changed chemistry in ways that are hard to imagine. About fifty years ago, the synthesis of a desired compound required many months. Chemists started by doing a literature search[1] to find the most probable synthetic pathway for a similar product and then modified it to get the one they were looking for. There are programs today that will do that for us in a matter of minutes. On top of that problem, once the reaction was carried out came the immense problem of identification of the final compound. Many other reactions were carried out to study this new compound and to finally identify it. Today, there are sophisticated instrumental methods (like infrared spectrometers, mass spectrometers, and nuclear magnetic spectrometers) that will allow us to do the same thing in a matter of minutes.

As one more example, computers can also let us design drugs in a safe environment. There are programs that allow us to see on a computer screen a three dimensional representation of a proposed drug. These programs also allow us to change a given atom or atoms in the molecule to see how its structure changes. Depending on the complexity of the molecule, these changes could completely alter the shape of the molecule. This new shape might fit better to block a site on a given enzyme, for example, if that was the desired property. We can make all these predictions before we even try to synthesize the drug itself. This has tremendously cut the amount of time spent in the actual laboratory synthesis.

We can easily see that chemistry is also central in many engineering disciplines. If you are thinking of engineering as a possible career alternative, you should be interested in materials in one way or another. Whether you are building bridges, computer chips, toys, or anything, a knowledge of materials is going to be extremely advantageous. You should explore the properties of different kinds of materials, the ways in which materials can be shaped, and the uses to which they can be put. Some emphasis should be placed on things like the relationship between internal structure of materials and their properties. Knowledge of these properties and of the ways in which you can shape them is essential if you are even marginally preoccupied

---

[1] For example, the *Journal of the American Chemical Society*, the *Journal of Organic Chemistry*, the *Journal of Inorganic Chemistry*, and the *Journal of Physical Chemistry*.

with the design and manufacture of products.

1.1 Quick History of Chemistry[1] and Some Fundamental Laws.

How did chemistry get started? Probably sometime in high school you learned that Greek philosophers around 400 BC pondered about the possibility that matter was made up of smaller particles. Actually, people like Leucippus and Democritus[2], considered water to possibly be made of smaller particles. Democritus made the analogy that water would be like sand, which although from a distance looks smooth, from up close you can clearly see that it is made up of tiny grains. Democritus proposed that matter is made up of small, indivisible particles that he called "atomos[3]". But he did not stop there. He went on to suggest that these atoms were weightless and with no clear tendency to combine, unless we either forced them to do so, or if they were brought together by chance.

Of course, not everyone agreed. Other philosophers of the time thought differently. Empedocles[4] thought that matter was not made up of smaller particles and was rather continuous. Therefore, there was not such a thing as empty space. He proposed that anything could be made by mixing anyone or all four "elementa": earth, air, fire, and water. Actually, Aristotle[5] believed in Empedocles ideas. The fact that some things could be **transformed** into other things gave birth to a new science: *"Chëmeia."*

People like Epicures (340-271 BC) agreed with Democritus in many ways, but thought that atoms actually had weight. His atoms varied in size, but they were all too small to be seen with the naked eye. One of the things that we have to remember is that in these times, no one tried to test these concepts. A subsequent development of these ideas was that atoms and the laws governing them would determine all things. This certainly created huge problems with the Christian Church, because they regarded this theory as almost heretic.

Later on, many experimental methods were developed to study matter. One of the early precursors of modern chemistry was alchemy and that came forth when Greeks and Egyptians started to experiment with trying to convert cheap metals (like lead) into gold. However, it was the Arabs, after conquering Egypt in 640 AD, that named this science and called it "Alchëmia".

---

[1] Taken in part from: Gray, Harry B., John D. Simon, and William C. Trogler. Braving the Elements. Sausalito: University Science Books. 1995.
[2] Greek philosopher that lived from 460 to 370 BC.
[3] A Greek word that means indivisible, something that cannot be broken into smaller pieces.
[4] Circa 450 BC
[5] 384 - 323 BC

It took a long time (until around 1150 AD) for the Europeans to discover the writings about this science, but once they did, it really took off in Europe during the Middle Ages. Alchemy turned out to be extremely useful, because it lead to the discovery of many new substances like alcohols, metal salts, and acids, and let to the development of techniques such as distillation, sublimation, and crystallization. Certainly, alchemists never achieved their first goal (obtaining gold from inexpensive metals), and there were many who started making fraudulent claims. However, we cannot deny that there were two very important achievements during this period: the development of systematic metallurgy (the extraction of metals from ores) by a German, Georg Bauer, and the medicinal application of minerals by a Swiss alchemist called Paracelsus.

Today, scientists follow what we can call the scientific method. Basically, this involves carrying out various measurements (for example, water freezes at 0°C) and or observations (water is a colorless liquid). Once we collect some of these numbers, we can suggest a **hypothesis**, which is nothing more than a likely explanation to your observations. Of course, you would need to follow this by some experimentation that would test out your hypothesis. These new observations you make, will either be consistent with or force you to alter your hypothesis. We continue doing this over and over, until your hypothesis seems to agree with all observable properties. At this point, we collect all of this and propose a **theory**. This theory will be a collection of various hypotheses which collectively explain a given phenomenon.

A typical misconception is to think that the theory is your final step, answer or solution. Nothing stops at this point. More observations are made with time, new information continually pours in, and you may be forced to improve or clarify your theory. This means that a theory is dynamic, which means that it is constantly being challenged by new observations. Actually, we can test theories even further by asking ourselves related questions that will allow us to make predictions on things that have not been observed up that point. If new experimentation to corroborate this prediction fails, we will be forced to once more refine the theory.

Another term that is often used is a law. Once we observe something over and over in many different scenarios and it always is true, we can establish a law. The law, however, is simply a summary of the observations, whereas the theory attempts to explain these observations.

When did this modern approach start? It wasn't until the 1600's, when Robert Boyle[1] started making real quantitative measurements. His concept was that a substance was an element

---

[1] Irish scientist who lived from 1627 until 1691. His book, *The Sceptical Chemist*, can be seen as the start of modern Chemistry.

unless it could be broken down into two or more substances. Once his idea became more and more universal, the Greek notion that everything was composed of four elements started to disappear. However, Boyle still held to the concept that metals were not elements, so they could theoretically be converted into other metals.

Very soon, experimentation became the norm in science and this lead to many discoveries in the eighteenth century (like the discovery of oxygen gas by Joseph Priestly). Antoine Lavoisier[1] made very careful weighing of reactants and products in chemical reactions that lead to the proposal of the Law of Conservation of Matter. He observed that during a reaction, the total amount of reactants and products (in terms of mass) was constant. He worked with many different types of reactions, but particularly with combustion[2] reactions. His work led others to continue making quantitative observations until Joseph Proust[3] suggested that: a given compound always has the same identical ratio of elements by weight.

These kind of observations led John Dalton[4] to revisit the old Greek thought that matter was composed of small indivisible particles called atoms. He figured that if atoms were the smallest unit, then he could explain Proust's law. Of course, a compound would always contain the same combination of atoms, so the mass ratio would be the same. In 1808, he proposed his modern approach to atomic theory[5]. He proposed the following postulates:

- The smallest unit of matter is the atom.

- All the atoms of a given element are identical in all aspects (mass and properties).

- Atoms of different elements have different atoms (different masses and properties).

- Atoms would be impossible to create or destroy.

- Compounds are formed from the combination of the atoms of different elements in small whole number ratios. Atoms themselves are not changed during these combinations, they are simply rearranged to form molecules.

The smallest unit of a compound is what we call a molecule. This is a group of atoms

---

[1] French scientist who lived from 1743 until 1794 (executed on the guillotine during the French Revolution).
[2] Reactions of materials with oxygen gas. Actually, his measurements proved that oxygen gas was involved in combustion.
[3] Another French scientist who lived from 1754 until 1826.
[4] English scientist who lived from 1766 until 1844.
[5] In a published book, *A New System of Chemical Philosophy*.

that is bound together by forces we will talk about later on. This group of atoms will exist in this united form for a reasonable amount of time, thus we term it as stable. The subscripts we will use to indicate the grouping, will tell us the number of atoms or the relative proportion of atoms in the compound.

The simplest kind of molecules are diatomic molecules. Molecules can also be either elemental (like O_2 or Cl_2) or molecules of compounds (like H_2O or CH_4). In conclusion, when we read the chemical formula for water (H_2O), it tells us at this point in time, that one molecule of water has two hydrogen atoms and one Oxygen atom somehow combined.

1.2 The Composition of Matter.

Before we proceed any further, it is a good idea to define a few terms that we will be using constantly throughout the year.

- Matter – Anything that occupies space and has mass.

- Pure Substance – matter with fixed chemical composition.

- Mixture – Two or more pure substances with variable composition.

- Elements – Cannot be decomposed into simpler substances by chemical means. The periodic table shows about 105 elements (represented by symbols) and about 90 of these are naturally occurring.

I believe that it is interesting to see why some elements have been given the names they have today. Below you can find a few examples[1] that show that the names may have been given for a wide range of reasons, ranging all the way from interesting historical origins to the countries or cities of origin from the scientists who discovered them.

a) Historical origins.

1. **H**ydrogen, is derived from the Greek roots *hydro* and *genes*, which together mean "water forming". This name then makes sense since this element burns in air to form water.
2. **C**obalt (Co) is the English version of the German word *kobold*, which means goblin or evil spirits. It is interesting to note that cobalt occurred as an unwanted metal in what were thought to be copper ores by German miners in the 1800's. Arsenic (As), which

---

[1] Gray, Harry B., John D. Simon, and William C. Trogler. Braving the Elements. Sausalito: University Science Books. 1995.

turned out to be also found in these ores, poisoned some of the workers in the smelting process. Thus, it was thought that cobalt contained evil spirits.

b) <u>Appearance of substance</u>.

1. **I**odine (I) exists as violet crystals that contain I_2 molecules. Its name comes from the Greek *iodes*, which means violet.
2. **C**arbon (C) is derived from the Latin *carbo* (charcoal), and **ca**lcium (Ca) from the Latin *calx* (limestone - a mineral that contains calcium).
3. Gold (Au) comes from the Latin word *Aurum* (shining dawn - from the characteristic brilliant yellow color of gold metal) and the metal lead (Pb) from the Latin word *plumbum* (which means heavy).

c) <u>Names of scientists and cities or countries of origin</u>.

1. **C**urium, **N**obelium, **M**endelevium, **L**awrencium, and **E**insteinium.
2. **Fr**ancium, **Ge**rmanium, **Am**ericium, **Po**lonium, **Ru**thenium (Latin *Ruthenia* or Russia), **Ga**llium (Latin *Gallia* or France), **Hf**nium (Latin *Hafnia* or Copenhagen), **Ho**lmium (Latin *Holmia* or Stockholm), **Lu**tetium (an ancient name for Paris), **Sc**andium, **Y**tterbium (towns in Sweden), **C**alifornium, and **B**erkelium.

- Atom – The smallest particle of an element that can enter into a chemical combination and that has all the properties of that element. Only a few elements truly exist in the atomic form (the noble or inert gases: He, Ne, Ar, Kr, Xe, Rn).

- Compounds – two or more elements combined in a fixed proportion.

- Molecule – The smallest particle of an element or compound that may have a stable, independent existence. A lot of elements consist of molecules formed when two or more atoms are bonded (N_2, O_2, Cl_2, S_8, P_4, ...). In the case of a compound, the smallest unit consists of two or more elements (H_2O, CO_2, CCl_4, ...).

- Homogeneous Mixture – also called solutions. These contain only one phase. Examples of these could be in either of the phases: gaseous (air), liquid (coffee), and solid (14 karat gold, which is a solid made up of 14 parts gold and 10 parts nickel).

- Heterogeneous Mixtures – Two or more phases will be clearly visible. An obvious example is what you obtain when you mix oil and vinegar.

Chapter 1: Introduction

Let's try to interpret all of these things together. Please, look at the diagram below in order to understand the argument that follows. When we hold a chunk of matter in our hands, we basically have one of two things: a pure substance (like pure distilled water[1]) or a mixture (like sea water). If it is a pure substance and we are able to separate it into simpler substances by chemical means, then this pure substance is a compound, whereas if you cannot separate it into anything simpler, then it is an element. On the other hand, if it was a mixture, it will be either a homogeneous mixture (can only see one phase and looks at first glance as a pure substance, but it can be separated into its components by physical means – boiling, etc.) or a heterogeneous mixture.

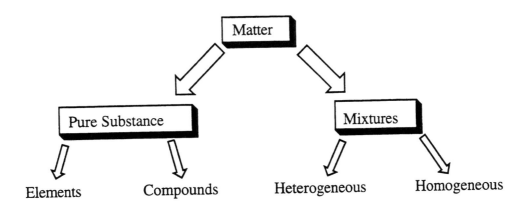

1.3 International System of Units (SI).

Chemistry is an experimental science, which means that it involves making some *observations* and *measurements*. Anytime this happens, we need to agree on the units in which these measurements will be made. We will be using the International System of Units (SI)[2] which is part of the metric system[3]. The three most basic measurable properties are length, mass and time. The SI units for these three are the meter, the kilogram and the second. Of course, there are others of interest to chemists (like electrical current [amperes, A], temperature [Kelvin, K], and amount of substance [moles, mol]), but we will talk about these later.

length	meter	m
mass	kilogram	kg
time	second	s

Since all the units are related by powers of 10, then we can easily use prefixes to make

---

[1] Water that has been deionized and degassed.
[2] From the French *Le Systéme International d'Unités*.
[3] Developed in France in the eighteenth century. In this system, the units are related by powers of 10.

them either larger or smaller. It is imperative that you learn these as soon as possible, since we will be using these throughout the book and in lecture. In one of the units above (the one for mass), we already used one of these prefixes (the **kilo**gram). The prefix kilo means one thousand (1000), therefore, a kilogram is simply 1000 grams.

$$1 \text{ kg} = 1000 \text{ g} = 10^3 \text{ g}$$

Table of Prefixes.

mega	10^6	M
kilo	10^3	k
hecto	10^2	h
centi	10^{-2}	c
milli	10^{-3}	m
micro	10^{-6}	µ
nano	10^{-9}	n
pico	10^{-12}	p
femto	10^{-15}	f

Therefore, $1 \text{ km} = 1 \times 10^3 \text{ m}$

$1 \text{ µg} = 1 \times 10^{-6} \text{ g}$

Some other units that we will be using are referred to as derived units since they come from the ones we have mentioned already. Let's look at three particular examples: area, volume and density.

- Area - Just to make it simple, let's imagine we are referring to the calculation of the area for a rectangular object. In order to determine this, we need to make two measurements: the length (l) and the height (h). The area is then determined as the length times the height. Since these two measurements are technically lengths, they are both in meters, so the SI unit for area will be square meters (m^2).

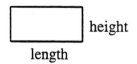

∴ Area = length x height = m x m = m^2. √

- Volume - Again, to make it simple, let's imagine we have a cube. In this scenario, we have a geometrical figure with three identical sides that are mutually perpendicular to one another (at 90° angles). The determination of the volume involves the measurement of the length of one of the sides (in units of meters), and then cubing that quantity. Thus, the SI units for volume (determined as length times height times depth) will be cubic meters (m^3).

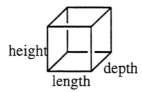

This unit is actually rather large for most of the measurements that we will make in chemistry. Therefore, we will be using another unit: the **liter**, symbolized by the letter **L**. The liter is defined as a small fraction of the actual SI unit, namely, one thousandth of the cubic meter.

$$1\ L = 10^{-3}\ m^3.$$

More so, many times the liter is still too large for our purposes and we will use even a smaller version, the milliliter, mL (approximately the volume occupied by 20 drops of water at room temperature). This unit, the mL, is exactly the same as a cubic centimeter (cm^3), or a cc. You may have seen the show ER[1] in which many times someone says something like "administer 15 cc of saline". This simply is equivalent to 15 mL of a saline solution. Let's prove that 1 mL is exactly equivalent to one cm^3.

$$1\ mL = 10^{-3}\ L = 10^{-3}\ (10^{-3}\ m^3) = 10^{-6}\ m^3.$$

But: $\quad 10^{-6}\ m^3 = (10^{-2})^3\ m^3$

And, $\quad 10^{-2} = centi = c$

$$\therefore 1\ mL = 1\ cm^3.$$

- Density - This property is defined as mass per unit volume. It tells you the mass of a particular volume of a substance.

$$density = \frac{mass}{volume}$$

The actual SI unit would be kg/m^3, but once more, we will be using a smaller unit, namely:

---

[1] Presently seen on NBC on Thursdays at 10 pm.

g/cm³ or g/mL. Density is an intensive property, that is, one that does not depend on the amount of material present. The densities of pure substances depend on temperature, and their values are unique to a given substance. Many times, we can identify a pure metal by just determining its density at room temperature and comparing it to the well known values found in the literature. Below is a table with the densities of some common substances at room temperature. For example, notice that the substance with the highest density from those on the table is gold metal, with a density of 19.32 g/cm³ and the lowest is that of hydrogen gas, with a density of 0.000084 g/cm³. The fact that the gas has a much lower density is expected, since there are a lot less molecules per unit volume.

Densities of Some Common Substances at 20°C.

Substance	Physical State	Density (g/cm³)
Oxygen	Gas	0.00133
Hydrogen	Gas	0.000084
Ethanol	Liquid	0.789
Benzene	Liquid	0.880
Water	Liquid	1.000
Magnesium	Solid	1.74
Salt (Sodium Chloride)	Solid	2.16
Quartz	Solid	2.65
Aluminum	Solid	2.70
Iron	Solid	7.87
Copper	Solid	8.96
Silver	Solid	10.5
Lead	Solid	11.34
Mercury	Liquid	13.6
Gold	Solid	19.32

1.4 **Factor-Label Method.**

This is the method of choice that we will use to solve most of the problems we do in the first few chapters of this book. It is extremely convenient in many ways, since it enables you to keep track of your units as you go along with your calculations. It is almost impossible to make mistakes if you adhere to this method.

And what does this method involve? It uses conversion factors to change (or convert) from one unit to another. Let's take the case of a definition we learned recently.

$$1 kg = 10^3 g$$

$$\therefore \frac{1 kg}{10^3 g} = 1 = \frac{10^3 g}{1 kg}$$

We can see that a kilogram is exactly equivalent to 1000 grams, therefore, if you were to divide one kilogram by 1000 grams (or for that matter 1000 grams by one kilogram), you would get unity since technically these quantities would cancel out. Therefore, multiplying a measured quantity by either conversion factor would not change the quantity itself, but rather just the units, since multiplying the quantity by one does not change its measured value.

In order to convert the initial units (units$_1$) into the desired units (units$_2$), all we need to do is to multiply it by the conversion factor that converts units$_1$ into units$_2$.

$$\#(units_1) \times \frac{\#(units_2)}{\#(units_1)} = \#(units_2)$$

Of course, we must be extremely careful with how we use the conversion factor itself. The initial units must cancel out. If we keep track of the units, we should have absolutely no problems and this will be trivial. For example, if we wanted to convert 10 km into meters, we would have to do the following conversion.

$$10 km \bullet \frac{1000 m}{1 km} = 10,000 m$$

However, if you accidentally inverted the conversion factor, your units would not cancel and you would immediately know you made a mistake.

$$10 km \bullet \frac{1 km}{1000 m} = 0.010 km^2 / m$$

- How many micrometers are there in 35 kilometers?

    In order to solve this problem, we need to first realize that micrometers are much smaller than kilometers. Therefore, we should be aware from the start that our answer has to be a much larger value than the original measurement. The unit conversion should be done in two steps (at least initially, while we are learning how to use this method), mainly, converting kilometers into meters, and then converting the meters into micrometers.

$$35 \text{ km} \times \frac{10^3 \text{ m}}{1 \text{ km}} \times \frac{1 \text{ μm}}{10^{-6} \text{ m}} = 3.5 \times 10^{10} \text{ μm}$$

- How many nanograms are there in 5.0 mg?

Once more, we are going from a given unit (mg) into a much smaller unit (ng), therefore, we should expect the answer to be much larger than the original number itself. The conversion once more involves two conversions: from mg into grams, and then from grams into nanograms.

$$5.0 \text{ mg} \times \frac{10^{-3} \text{ g}}{1 \text{ mg}} \times \frac{1 \text{ ng}}{10^{-9} \text{ g}} = 5.0 \times 10^6 \text{ ng}$$

As a country, we made a commitment to change the English system of units into the metric system in the late 1970's but it has not materialized to date. Therefore, we need to be aware of some conversion factors between these two systems. Although there are many, these are the ones we should know by heart.

$$1 \text{ in} = 2.54 \text{ cm}$$
$$1 \text{ lb} = 454 \text{ g}$$
$$1 \text{ L} = 1.057 \text{ qt}$$
$$1 \text{ gal} = 3.785 \text{ L}$$

Knowing these, we can then use them in the same exact way as we did the ones before (within the metric system).

- A letter weighs 250 g. How much will it cost to mail this letter if the rate is 32 cents per ounce?

As always, we should look for an experimentally measured quantity and start with that number, and then try to convert the units until we get what we desire. The initial number should therefore be the 250 g (the mass of the letter). Of course, the only conversion unit that applies initially is to change from grams to pounds. However, once that is achieved, we can see that the next conversion factor is one that relates the ounces into the actual cost ($0.32 per ounce). This conversion factor is not actually real, since a letter has different rates depending on whether it is the first ounce or the subsequent ounces, but we will assume for the purposes of the questions that they are all rated in the same exact manner. Therefore, we should convert the pounds into ounces before we use the final conversion factor.

$$250g \times \frac{1lb}{454g} \times \frac{16oz}{1lb} \times \frac{\$0.32}{1oz}$$

$$= \$2.82 \quad \sqrt{}$$

- Convert 12.3 lbs/ft³ to g/mL.

    This problem is a bit more complex, because there are different units (cubic feet) and we do not know direct conversion factors between those and the usual metric units. However, we should be able to derive whichever units we need. Once more, we should start with the experimentally determined value (12.3 lbs/ft³). Since we are trying to change the units to grams per mL, this means that we need to change both the numerator and the denominator. That is, the numerator must be changed from pounds to grams (easily done) and the denominator from cubic feet to cubic centimeters (same as mL). Of course, the problem is with the denominator. What is the conversion unit for cubic feet into cubic centimeters? Well, the only thing we do know is that one foot has 12 inches and that one inch is equivalent to 2.54 cm. However, since the feet are cubed, therefore, the conversion factor we already know should also be cubed.

$$\frac{12.3 lbs}{ft^3} \times \frac{454g}{1lb} \times (\frac{1ft}{12in})^3 \times (\frac{1in}{2.54cm})^3 \times (\frac{1cm^3}{1mL})$$

$$= 0.197 \ g/mL \quad \sqrt{}$$

- According to *The Sporting News*, the fastest recorded fastball was thrown by Nolan Ryan and had a velocity of 100.8 miles per hour. Calculate the velocity in meters per second.

    Let's start with the experimentally measured value, the speed: 100.8 miles/hour. All we need to do is to convert the distance from miles into meters (various steps) and then the time from hours into seconds.

$$\frac{100.8 miles}{1hr} \times \frac{5280ft}{1mile} \times \frac{12in}{1ft} \times \frac{2.54cm}{1in} \times \frac{10^{-2}m}{1cm} \times \frac{1hr}{3600s} = 45.1 \ m/s \quad \sqrt{}$$

- A rectangular block has dimensions of 29.0 mm x 35.0 mm x 100.0 mm and has a mass of 289 g. Determine the density of this block in g/cm³.

    The first thing we notice is that we want the density in units of g/cm³. Since the measurements are in mm, then we should convert them into cm before we determine the volume.

∴ Volume = 2.90 cm x 3.50 cm x 10.0 cm = 101.5 cm³.

Finally, the density is given by:

$$Density = \frac{289 g}{101.5 cm^3} = 2.85 g/cm^3$$

1.5 <u>Errors</u>.

We mentioned previously that chemistry is an experimental science, and thus it involves making measurements and observations. These measurements will immediately introduce errors or uncertainties. After working experimentally with numbers, we recognize two different types: <u>exact numbers</u> (whose values we know without a doubt) and <u>inexact numbers</u> (whose values have a certain error on them). The easiest way of knowing whether a number is exact or not, is to ask yourselves if you know for sure what the number is. Let us see one example. You can clearly see that you have five fingers in any one of your hands (there is NO doubt that there are five fingers and you know it is not either 4 nor 6), or that there is exactly two lungs in your body, or the total number of siblings in your family, or the number of hydrogen atoms in a water molecule (two). These are exact numbers because you know exactly their value. We will also apply this exactness to definitions, for example: 1 foot has exactly 12 inches (not 11, not 13).

However, when a number comes from a particular measurement (like the length of you desk, your height, your weight, etc.), then the number is *inexact*. Let's say that two different persons were going to measure the length of the object shown below with the ruler seen right below it.

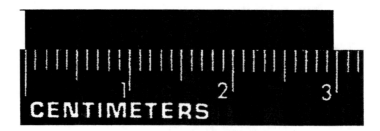

As we can clearly see, it is easy to see that the length is somewhere between 2.9 and 3.0 cm, but when we try to estimate the next decimal place (the next figure), suddenly there is an uncertainty and different people will estimate it differently[1]. This means that no matter how hard we try, a measured value will always have an uncertainty associated with it.

There are two terms that we should clarify at this point. If we make a few determinations

---

[1] I would call it 2.98 cm, but I could not refute a reading of 2.97 cm or 2.99 cm, for example.

of a given measurement we would expect the values to be very close to one another (actually, we probably would think it would always be exactly the same, but we now know that this will not always be the case). *Precision* tells us how close these measured values are to one another. However, *accuracy* will tell us how close a single measurement, or the average of a few measurements, is to the true or real value. We should point out that the **average** of a few measurements can be defined as:

Average = Sum of all the actual values divided by the total number of values.

Often in an experiment, we measure a certain object a number of times and then report its experimental value as the average and its error as the *average deviation*. The average deviation may be defined as the average of the individual deviations from each measurement as compared to its average value.

As an example, let's say that two students are asked to weigh as accurately as possible four pennies minted in 1998. The results they obtain in different balances are shown below.

Penny #	Student #1	Student #2
1	3.2345 g	3.2357 g
2	3.2352 g	3.2353 g
3	3.2349 g	3.2390 g
4	3.2355 g	3.2300 g

The average weight of a 1998 penny for both students can be calculated as follows:

$$Ave_1 = (3.2345 + 3.2352 + 3.2349 + 3.2355)/4 = 3.2350 \text{ g}$$

$$Ave_2 = (3.2357 + 3.2353 + 3.2390 + 3.2300)/4 = 3.2350 \text{ g}$$

We can see right away that these two students, although they have vastly different measurements for the same pennies, have the exact same average results. The easiest way to determine which is more accurate is to determine the average deviation for the average of both students. For the first student:

$$AD_1 = \{|(3.2345 - 3.2350)| + |(3.2352 - 3.2350)| + |(3.2349 - 3.2350)| + |(3.2355 - 3.2350)|\}/4$$

$$AD_1 = \pm 0.0003 \text{ g}$$

and for the second student:

$$AD_2 = \{|(3.2357 - 3.2350)| + (|(3.2355 - 3.2350)| + |(3.2390 - 3.2350)| + |(3.2300 - 3.2350)|\}/4$$

$$AD_2 = \pm 0.003 \text{ g}$$

We can see from those results that the uncertainty on the measurements from student #2 is 10 times larger than that of student #1. In other words, the results of student #1 are <u>more precise</u> than the results of student #2. Thus, the average deviation is an excellent measurement of the degree of precision in a set of experimental values. The accuracy of these values can not be determined with the information given, because we would need to know the actual mass of each penny so that we could see how the student's masses compare to the true values.

1.6 <u>Significant Figures</u>.

To reflect these errors, special care must be taken in writing measurements or numbers to the correct number of *significant figures* (SF). In general, all figures or digits in a written number are significant, the last one being uncertain. In other words, the last SF contains the error for that measurement.

Ex. The number 525 has three SF, the third one is the uncertain one and unless you are told differently, you will assume that the error in the last is ± 1. In conclusion, the true value is somewhere between 524 and 526.

As we can see, determining the number of SF in a quantity is relatively simple. However, we must be careful with the zeros, as they could be misinterpreted. Here are some rules for specific ways in which we should determine whether a zero is significant or not.

Special Cases With Zeros.

- To the left of a number they are <u>not</u> significant.

    One example would be the following number: 0.00038 This number has two SF because the zeroes to the left are not significant. It would be easier to understand if the number was written in scientific notation: 3.8×10^{-4}. In this format, we can see that the only purpose of the zeroes is to indicate the position of the decimal place.

- To the right of a number they are usually significant.

    As an example, let's see the number: 2.5000 This number has five SF. The real value is between 2.4999 and 2.5001. A lot of people believe that it is irrelevant to write these zeros, and they write 2.5 instead. Keep in mind that the number 2.5 has two SF and the implication of such a number would be that the real value would be somewhere between 2.4 and 2.6. This would increase the error by a factor of 1000!!

- Sometimes a zero to the right is not significant and its only purpose is to locate the decimal point.

These cases are a little trickier to see, because it depends on the way the number was determined and the implied error in its measurement. For example, let's say that the population in Athens, Ohio is 30,000 ± 1,000. This number, as measured, has an error in the second digit, which by definition is the last significant figure. Therefore, that population has only two SF (the real number is between 29,000 and 31,000). This situation can be avoided by writing the number in scientific notation (3.0×10^4). Conclusion: the only way of knowing how many significant figures you have in such a number is for us to be told the uncertainty of the number. In the absence of such information, we will assume all figures to be significant!! Example: If you are told that you dissolve a certain amount of a substance in 100 g of water, although information is lacking as to the error for that number, we will assume that all three figures are significant. The only time we will be able to do something differently would be when you are specifically told something to this effect: "...dissolved a certain amount of a substance in 100 g (± 10 g) of water". This would then tell us that we are dealing with only two SF, since there would be an uncertainty in the second digit.

- An exact number contains an infinite number of significant figures.

We have discussed this previously. For example, if we consider a dozen eggs, we would have exactly 12 eggs, not 11 nor 13, therefore the number could be represented as 12.00000... Although we would always write down just the number 12 (and a lot of people could potentially think that there are only 2 SF's), we would immediately know that this number has an infinite number of significant figures.

<u>Rounding-Off</u>.

All of us have a calculator and we often use them to determine the numerical answers to problems. Let's say that we perform a calculation and that the answer that we read from the calculator is: 3.62475923. We will NEVER report an answer to that incredibly large number of significant figures, therefore the answer must be rounded off. We need to do two things: 1) learn how to actually round off, and 2) learn how to decide where to round off a given number to a certain number of significant figures.

Let's determine how to round off first. There are three different cases:

- If the number past the last SF is 4 or less, drop all of them and leave the number unaltered.

    This would be the perfect example if we were to round off the number above to three SF. Since the number that follows is a 4, we would report the answer as: 3.62.

- If the number past the last SF is 6 or higher, drop all of them and add one to the last SF.

    Once more, the perfect example would be the number above rounded off to 4 SF's. Since the fifth figure is a 7, we would drop them and add one to the fourth figure (which would increase from 4 to 5). Therefore, we would report the answer as: 3.625.

- If the number past the last SF is 5, drop them and make sure that the last SF is an even number. This will be achieved in one of two ways: by leaving it that way (if it is even) or by adding one to that last figure (if it is odd).

    The example would be this time rounding off the given number to five SF's. Since the sixth figure is a five, then the decision is for the fifth figure (in this example it is a 7). Since the number 7 is odd, we will add one to it (which makes it an 8). Therefore, we would report the answer as: 3.6248.

1.7 Significant Figures and Mathematical Operations.

This is an approximate method that will be more than sufficient for a lecture course. Of course, in a laboratory course, where actual measurements are made, a more exact method would be needed. We will proceed to describe the approximate method first, and then we will talk about the more exact one: propagation of errors.

When we solve a problem, we often have multiple operations to perform. Some of these involve addition and subtraction, while others involve multiplication and division. The rules for each of these situations are different, so we will discuss them separately. However, we will find that many times, these will be combined in problems, so we need to understand these and apply them for more complex situations.

- Addition and Subtraction - for these kind of operations, we will concentrate in locating the number of decimal places on each of the numbers we are adding (or subtracting). Our answer should have as many decimal places as the number that had the least number of decimal places to start with. For example:

Example: 1.27

1.0025

3.45

0.0038

<u>2.1   </u>

7.8263   =   7.8

The number 2.1 only has one decimal place, therefore, our answer should only have one decimal place. Of course, since we all use a calculator, we are not going to round off each of the numbers to one decimal place; instead we simply add all the numbers to get 7.8263. We should then round-off that answer to one decimal place: 7.8.

- <u>Multiplication and Division</u> - for this kind of operations, we will concentrate on the number of significant figures on each individual number being multiplied or divided. Our answer should have the same number of SF's as that in the quantity with the fewest SF's initially. Let's look at some examples below.

a) (3.0)(4297) / 0.0721

Let's analyze this operation first. We see three numbers: one of them has two SF (the 3.0), one has four SF's (4297), and the other one has 3 SF's (0.0721). This tells us immediately that our answer should have two SF's. The calculator yields: 178,793.342580. Of course, this is a ridiculous number of SF, so we need to round it off to only 2 SF's. That means that we should only keep the first two (the one and the seven). Since the figure that follows is an eight, we should add one to the seven. However, how do we report this number? In scientific notation it would be simple: 1.8×10^5. If we wanted to use regular notation, we would be forced to report it as: 180,000 with the clear understanding that there is an uncertainty on the eight. In other words, in this case, the only reason for the zeroes is to indicate the position of the decimal point.

b) y = 4.4 x (311.8)/(273.1) x (760)/(784 - 3)

Glancing at all the numbers for this operation, we see that there is a number with two SF's (the 4.4), one number with one SF (the 3), two numbers with three SF's (the 760 and the 784), and two numbers with four SF's (the 311.8 and the 273.1). The typical mistake would be to rush into deciding that since there is one number with 1 SF, our answer should have one. This may or may not be the case. This problem involves two different types of operations: subtraction and multiplication/division. Before we decide, we must simplify the expression by solving the subtraction first. This leaves us with the following expression:

$$y = 4.4 \times (311.8)/(273.1) \times (760)/(781)$$

Which is the number with least number of SF's? The number 4.4 (with only 2). Because of this, we can tell that our answer should have only two SF's. The calculator would tell us that the answer is 4.8884328, but we would report it as: <u>4.9</u>. √

c) $(0.057)(265)/2800 + 285/\pi$

In order to solve this problem, once more we must first solve each term separately. It should be clear that the difference between this problem and the last one, is that we would most likely solve each term first (which involves multiplication and division) and THEN we would add them up (different rules). Let's solve each term separately first. The first term will be rounded-off to two SF because 0.057 has only 2 SF. The second term will be rounded-off to 3 SF because the number π will be treated as an exact number. This gives us:

$$= 0.0054 + 90.7$$

Finally, in order to add these two numbers, we simply do so with the calculator, which would yield: 90.7054, but since the rules state that our answer should have the same number of decimal places as the number that had the least (in this case one decimal place as in 90.7), then our answer would be simply: <u>90.7</u> √

<u>Propagated Errors</u> - More exact method.

If you were in a laboratory, the very inexact methods we just described would be inappropriate. Remember that each experimentally determined number has a given error or uncertainty associated with its measurement. In many cases, in order to calculate some quantity, we need to measure various things in the laboratory. Upon multiplying, dividing, adding, or subtracting these experimentally determined values, we will propagate the errors of each of those numbers into the final answer. The meaning of the word propagate is "to continue or spread by generation or successive production".[1] This is exactly what we want to do: we want to spread the errors of all the individual numbers into the answer, because in reality, the more numbers we measure experimentally, the larger the uncertainty of our final answer. Not only that, but each

---

[1] The American Heritage Dictionary of the English Language. 1978.

number may have a different error, that is, some numbers may be more uncertain than others. If for example, we were going to find a density with these two numbers: 15.2234 ± 0.0002 g and 10 ± 1 mL, we can see that the volume has a much higher error than the mass. The error of the volume will therefore have a much larger effect on the final error (that of the density) than the error on the mass.

Before we proceed any further, we should mention that when you report an answer in a laboratory report, this number should be accompanied by its uncertainty. You may notice that the two experimental numbers in the preceding paragraph were reported that way. The experimental uncertainty for a value should only contain one figure. Why is that so? Because anything else is useless. Let's think about it. If our number is 15.2234 g and we know that the last figure (the 4) is really somewhere between 2 and 6 (hence ± 2 on that figure), what is the point of writing something like ± 0.000234? The last SF is the one with the uncertainty and anything else is useless! Conclusion: Any measured number, or for that matter any calculated number from measured values, must be reported to the same decimal place as its uncertainty. Ex: 2.33 ± 0.04; 5.22334 ± 0.00006; 23.5 ± 0.2.

We will use the model proposed by Dr. Terrence J. Swift[1] in his Laboratory Manual[2] for Chemistry 113. We will somewhat paraphrase his treatment of this method in the following paragraphs. Let's assume we have a rectangle with the dimensions shown in the figure below.

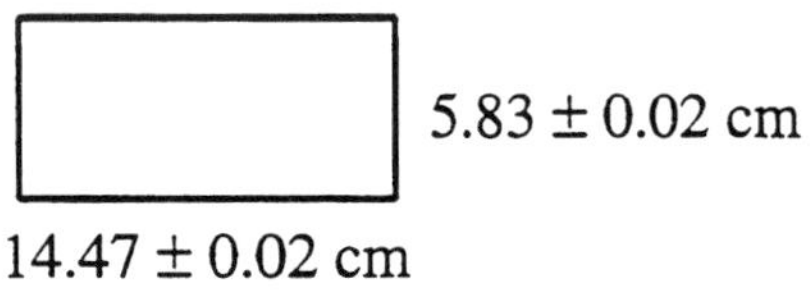

5.83 ± 0.02 cm

14.47 ± 0.02 cm

The most probable are would be: 14.47 cm x 5.83 cm = 84.3601 cm^2, the value we obtain from the experimentally determined values. One quick way to determine the propagated error would be to determine the maximum value of the area (assuming the worst possible scenario for your initial measurement of the experimental numbers) which would give us: 14.49 cm x 5.85 cm = 84.7665 cm^2. If we calculate now the minimum value for the area (once more, picturing the worst possible scenario with the real numbers being on the lower end of the error), we would get: 14.45 cm x 5.81 cm = 83.9545 cm^2. Clearly, we can see that the most probably value could be deviated by 0.4 cm^2 up or down. I suppose that we could report our answer as:

Area = 84.4 ± 0.4 cm^2.

---

[1] Chemistry professor at CWRU, who authored the Chemistry 113 laboratory manual.
[2] Fall 1998 edition.

However, there is one problem. We are assuming the worst of cases, and there is a good chance that some of these two errors could be in opposite directions, which would lead to some cancellation of error in the calculation of the area. The probability of such a cancellation is taken into account by the following formula,

$$\Delta A = \sqrt{(\Delta A_{length})^2 + (\Delta A_{width})^2}$$

where the term ΔA is the calculated uncertainty in the area, and the other two are the uncertainties in the area resulting from the uncertainties in the length and the width that would result if each of those measured quantities were the only uncertain one. Let's determine this for the example in question. We will achieve this by breaking it into various steps.

a. Determine the most probable value (for these purposes, we will keep all the figures from the calculator). We already did this, so the answer is: 84.3601 cm^2.

b. Add the uncertainty to the length to determine the area if the length were the only uncertain figure. This would lead to:

$$A_1 = 14.49 \text{ cm} \times 5.83 \text{ cm} = 84.4767 \text{ cm}^2.$$

c. Do the same thing but with the width this time.

$$A_2 = 14.47 \text{ cm} \times 5.85 \text{ cm} = 84.6495 \text{ cm}^2.$$

d. Now find the two uncertainties (ΔA_{length} and ΔA_{width}) from these numbers by subtracting them from the most probably area.

$$\Delta A_{length} = 84.4767 \text{ cm}^2 - 84.3601 \text{ cm}^2 = 0.1166 \text{ cm}^2.$$

$$\Delta A_{width} = 84.6495 \text{ cm}^2 - 84.3601 \text{ cm}^2 = 0.2894 \text{ cm}^2.$$

e. Calculate the uncertainty in the area with the formula given in the previous page.

$$\Delta A = [(0.1166 \text{ cm}^2)^2 + (0.2894 \text{ cm}^2)^2]^{1/2}$$

$$\Delta A = \pm 0.3 \text{ cm}^2 \quad \surd$$

Of course, the area would be reported as:

$$A = 84.4 \pm 0.3 \text{ cm}^2 \quad \surd$$

Notice that the number we obtained was indeed smaller than the 0.4 cm² we calculated previously, but we expected that since we thought some of the errors would cancel out in the calculation. We should point out that this is NOT an exact method, as there is no such thing. It is just much closer than the previous one, but not as close as the calculus based one. However, it gives us a very good approximation which is why we can use this method in the laboratory.

Let's do one more example.

- Determine the density of a rectangular object that has dimensions of 12.0 ± 0.2 cm, 15.2 ± 0.2 cm, and 1.3 ± 0.2 cm, and a mass of 832.3 g ± 0.3 g.

    In order to determine the density of this object, we need to perform the following operation:

$$density = \frac{mass}{length \times height \times width}$$

$$density = \frac{832.3g}{12.0cm \times 15.2cm \times 1.3cm}$$

Which gives us a most probably density of *3.510037* g/cm³. Now, let's calculate the four density values if only one of the four numbers were contributing to the error.

$Density_1$ = {832.6 / (12.0 x 15.2 x 1.3)} = 3.511302 g/cm³.

$Density_2$ = {832.3 / (11.8 x 15.2 x 1.3)} = 3.569529 g/cm³.

$Density_3$ = {832.3 / (12.0 x 15.0 x 1.3)} = 3.556838 g/cm³.

$Density_4$ = {832.3 / (12.0 x 15.2 x 1.1)} = 4.148226 g/cm³.

The four uncertainties: Δd_1, Δd_2, Δd_3, and Δd_4, are calculated next.

Δd_1 = 3.511302 − 3.510037 = 0.001265 g/cm³.

Δd_2 = 3.569529 − 3.510037 = 0.059492 g/cm³.

Δd_3 = 3.556838 − 3.510037 = 0.046801 g/cm³.

Δd_4 = 4.148226 − 3.510037 = 0.638189 g/cm³.

The error in the density can be determined with the following formula.

$$\Delta d = [(\Delta d_1)^2 + (\Delta d_2)^2 + (\Delta d_3)^2 + (\Delta d_4)^2]^{1/2}$$

$$\Delta d = \pm 0.6 \text{ g/cm}^3.$$

Therefore, the reported density should be:

$$\text{Density} = 3.5 \pm 0.6 \text{ g/cm}^3. \quad \surd$$

Chapter 2: Atoms and Moles.

2.1 Atoms and Molecules.

We have mentioned the existence of atoms, as proposed by Dalton. We will now slowly build up the theoretical background that will allow us to understand the kind of interactions that exist between these atoms as they combine to form molecules. To this point, it will suffice to say that the forces that hold together the atoms in a molecule can be called **chemical bonds**. There are various types of bonds, but we will concentrate for the time being on the ones where the electrons are shared between the atoms. We call these **covalent bonds**. Actually, we could define a **molecule** as the resulting set of atoms in a covalent species.

How can we represent a molecule? The easiest way is to do so is by writing what we call its **chemical formula**. As we saw in the first chapter, we use the symbols of the elements to indicate the types of atoms present and with subscripts to indicate how many of each type of atom we have present in the molecule. In conclusion, when we look at a molecular formula, we see the actual combination of individual atoms that make up a molecule. These number of atoms are exact numbers and can be used as conversion factors, when applicable. Let's look at a couple of examples.

- How many carbon atoms are there in 25 ethanol molecules (C_2H_5OH)?

    There are 25 molecules of this compound. The only thing we know from looking at the molecular formula is that every single molecule contains two carbon atoms, six hydrogen atoms, and one oxygen atom. Therefore,

$$25 \text{ } C_2H_5OH \text{ molecules} \times \frac{2 \text{ C atoms}}{1 \text{ } C_2H_5OH \text{ molecule}} = 50 \text{ C atoms}$$

- How many hydrogen atoms are present in the minimum daily human nutritional requirement of the amino acid lysine, $C_6H_{14}N_2O_2$ if it contains 3.31×10^{22} lysine molecules?

$$3.31 \times 10^{22} \text{ } C_6H_{14}N_2O_2 \text{ molecules} \times \frac{14 \text{ H atoms}}{1 \text{ } C_6H_{14}N_2O_2 \text{ molecule}} = 4.63 \times 10^{23} \text{ H atoms}$$

In this last example we mentioned an extremely large number of atoms. Does that mean that to take this dosage of lysine we must go and start counting those atoms one by one until we get 3.31×10^{22}? Clearly not! If we did and started counting them at a rate of 5 atoms per second, it would take approximately two trillion centuries and that is without ever taking a break! This gives us a fairly good idea as to the minuscule size of atoms.

2.2 <u>Moles</u>.[1]

Since atoms are very small, we need to think of a way in which we can count these incredibly tiny particles. When we deal with macroscopic things, we use units like a dozen and talk about a dozen eggs, for example. Everyone knows exactly what a dozen eggs are and can picture them in their minds. We can see immediately, however, that a dozen molecules would be essentially the same as nothing and we could not link that small number of molecules to a mental picture. We need a unit that will encompass many more molecules; enough that when placed together in a container we could actually see them. A chemist counts atoms in units of 6.02214×10^{23} atoms. This is the unit that we call *one mole of atoms (1 mol)*. This number is also referred to as Avogadro's number (N_0).[2]

Now, how large is that number? Most of us cannot even start to understand how tremendously large this number really is. Let's try to solve a simple problem that will enable us to understand the magnitude of this number, or at the very least, the incredibly small size of atoms. We will achieve this by working not with atoms, but rather with something we can actually relate to: M&M's.

- You can fit 88 regular M&M's in 100 cm^3. Let's say that you want to pile up <u>one mole</u> of M&M's covering the entire surface of the contiguous United States. How high will you have to stack them? The area of the contiguous United States is 3.02×10^6 square miles.

Let's think about the information given. We are told that we have the ability to fit 88 M&M's into a 100 cm^3 container. Although this is not technically telling us the size of an individual candy piece, it is telling us the volume they occupy once they are packed. This is what all we need to know in order to solve the problem. Now, we have one mole of M&M's, therefore we have Avogadro's number of M&M's. We can then use the conversion factor that tells us that we can fit 88 of these into 100 cm^3.

$$6.022 \times 10^{23} \text{ M\&M's} \times \frac{100 \text{ cm}^3}{88 \text{ M\&M's}}$$

---

[1] From the Latin word *moles*, meaning "heap".
[2] Amadeo Avogadro was an Italian scientist who lived from 1776 until 1856. He was a primarily a lawyer, who later became involved with science.

$$= 6.84 \times 10^{23} \text{ cm}^3. \quad \sqrt{}$$

Technically, the volume is equal to the area times the height, that is:

$$\text{Volume} = \text{Area} \times \text{Height}$$

Since we know both the volume and the area, we can easily solve for the height. The only thing we need to be careful with is the fact that these two quantities have different units (the volume is in cubic centimeters, whereas the area is in square miles). Therefore, we need to convert them into similar units. Certainly, it does not matter which one is converted, but we will choose to change the volume into cubic miles, because that way the final answer will be in miles, and we probably have a better conceptual image of miles than of centimeters.

$$6.84 \times 10^{23} \text{ cm}^3 \times \left(\frac{1 \text{ in}}{2.54 \text{ cm}}\right)^3 \times \left(\frac{1 \text{ ft}}{12 \text{ in}}\right)^3 \times \left(\frac{1 \text{ mile}}{5280 \text{ ft}}\right)^3$$

$$= 1.64 \times 10^8 \text{ mile}^3.$$

Now that we have the similar units for the volume and the area, we can solve for the height.

$$\text{Height} = \text{Volume} / \text{Area} = 1.64 \times 10^8 \text{ mile}^3 / 3.02 \times 10^6 \text{ mile}^2$$

$$\text{Height} = \mathit{54.3 \ miles} \quad \sqrt{}$$

This answer tells us that we need to stack the M&M's to a height of over 54 miles covering the entire surface of the continental US in order to account for one mole. If the M&M's were the size of molecules instead, then they would fit into the palm of one of your hands!! Hopefully this problem will illustrate the relative size of atoms as compared to what we usually call small objects like M&M's.

Let's apply this new mole concept to the way we read a chemical formula. For example, when we read the formula of CO_2, we should be able to understand that this compound consists of two elements (C and O) in a mole ratio of 1 : 2. In other words, in order to prepare one mole of carbon dioxide gas, we need to combine one mole of carbon with two moles of oxygen atoms. This new concept can be utilized as a conversion factor. For example:

- How many hydrogen atoms are there in 2.12 moles of methane gas, CH_4?

The first thing we notice is that the actual question refers to something different from the experimentally determined number. Notice that we were given moles of methane gas, whereas we want number of hydrogen atoms. This is something that we have to be very sensitive to, because a mole can apply to anything (marbles, oranges, atoms, molecules, etc.). That is, one mole of methane contains Avogadro's number of molecules because methane is a molecular substance. On the other hand, one mole of hydrogen atoms contains Avogadro's number of atoms. My advice is to convert from the units given to the ones you want <u>while at the mole level</u> , and then change to number of particles using Avogadro's number. Of course, this order in which the units are converted does not matter, and I am only making this suggestion to keep things consistent. Following that advice, we get:

$$2.12 \text{ mol CH}_4 \times \frac{4 \text{ mol H}}{1 \text{ mol CH}_4} \times \frac{6.022 \times 10^{23} \text{ H atoms}}{1 \text{ mol H}} = 5.11 \times 10^{24} \text{ H atoms}$$

Notice that in this case we used the conversion factor of one mole of methane to four moles of hydrogen atoms, which we obtained directly from the chemical formula.

The definition of a mole includes more than just the number of atoms. It is more encompassing and it relates mass to number of particles. Technically, a *mole* of atoms is defined as the number of atoms contained in exactly 12 grams of the most common isotope of carbon (^{12}C). *Isotopes* are atoms that have identical chemical properties but differ in mass.[1] Atomic masses or atomic weights are therefore based on this standard (^{12}C). Thus, the atomic weights of all the elements in the periodic table are standardized to ^{12}C.

Let's define the mass of one ^{12}C atom as twelve (12) atomic mass units (amu)[2]. Thus, we can say that:

1 amu = mass equal to 1/12 of the mass of one ^{12}C atom.

This is an atomic scale and in order to convert this atomic scale to grams, we need to look at the definition of a mole. It states that one mole is the number of atoms in exactly 12 grams of carbon, so if one mole (6.022×10^{23} atoms) weighs 12.0000 g, then one atom must weigh,

$$12.0000 \text{ g} / 6.022 \times 10^{23} \text{ atoms} = 1.99 \times 10^{-23} \text{ g.}[3]$$

---

[1] We will talk more about these in Chapter 6.
[2] Also referred to as 12 daltons.
[3] Don't forget to always check your answers to make it makes sense. In this case, we are looking for the mass of a single atom. Of course, we expect the answer to be a very small number!

This means that one ^{12}C atom weighs 1.99×10^{-23} g. Since that is also equivalent to 12 amu (atomic mass units), this would imply that:

$$1 \text{ amu} = 1.66 \times 10^{-24} \text{ g} \quad \checkmark$$

This represents the conversion factor between atomic mass units and grams.

2.3 <u>Atomic and Molecular Weights.</u>

The atomic weight of an element is a number that is numerically equal to the average mass of an atom of the element (in amu). The atomic weight of carbon (as seen from the periodic table) is 12.01. The reason that it is not exactly 12 is because there is more than one isotope of carbon in nature and that number (12.01) is the weighted average[1] of all the naturally occurring isotopes.

Now, what is the relationship between this number and the mole concept we described earlier? One mole of an element can be obtained by weighing in a balance the atomic weight <u>in grams</u> and that quantity will contain Avogadro's number of atoms. Therefore our definition of a mole contains two parts, one that deals with mass and one that deals with number of particles.

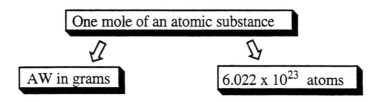

- How many potassium atoms (K) are there in 3.25 g of K?

In order to solve this problem, we should keep in mind that we need to convert the grams of potassium into moles first, and once we have moles we can proceed to convert to number of atoms.

$$3.25 \text{ g K} \times \frac{1 \text{ mol K}}{39.102 \text{ g K}} \times \frac{6.022 \times 10^{23} \text{ K atoms}}{1 \text{ mol K}} = 5.00 \times 10^{22} \text{ K atoms}$$

Don't forget to check your answer to see if at least it is in the right direction. Our answer indicates less than one mole of potassium metal (since it is smaller than Avogadro's number) and that makes sense, since we started with less than one mole of potassium (one mole is

---

[1] Once more, we will talk about this in Chapter 6, but it will suffice to say that if the average is 12.01 (so close to 12), then <u>most</u> of the naturally occurring isotopes of carbon must weigh 12!

about 39 grams and only started with 3.25 g).

As we have seen, we can combine atoms to form molecules. In the latter case, the smallest unit that carries all the characteristics of the substance in the *molecule*. The **molecular weight** is the sum of the atomic weights of all the atoms that constitute the molecule.

Ex. H_2O: MW = 2(1.008) + 16.000 = 18.016 amu.[1]

It would follow that for any molecule, a similar definition for a mole can be made, *i.e.*, a mole contains Avogadro's number of molecules and it can be weighed in a balance by getting the molecular weight in grams. This mass would then contain Avogadro's number of molecules.

$$\text{One mole of an molecular substance} \Rightarrow \begin{cases} \text{MW in grams} \\ 6.022 \times 10^{23} \text{ molecules} \end{cases}$$

A few examples follow.

- How many CO_2 molecules are there in 25.3 grams of CO_2?

    We start with a quick analysis and find out that we are initially given grams of carbon dioxide but we want to end up with carbon dioxide molecules. The problem will be done in two steps: first a conversion to moles of carbon dioxide, then to carbon dioxide molecules.

$$25.3 \text{ g } CO_2 \times \frac{1 \text{ mol } CO_2}{44.011 \text{ g } CO_2} \times \frac{6.022 \times 10^{23} \text{ } CO_2 \text{ molecules}}{1 \text{ mol } CO_2} = 3.46 \times 10^{23} \text{ } CO_2 \text{ molecules}$$

- How many hydrogen atoms (H) are there in 18.0 grams of ethane gas, C_2H_6 ?

    This time, we can see that we are going from grams of ethane to hydrogen atoms. These are two seemingly unrelated units, but we now know they are related through the chemical formula for ethane. Therefore, we will convert the grams of ethane to moles of ethane first. Once we have moles, then we will convert from moles of ethane to moles of hydrogen atoms (a 1:6 mole ratio from the chemical formula), and then finally change the moles of hydrogen into atoms.

---

[1] We can also write that the molecular weight of water is 18.016 g/mol, using the larger scale.

$$18.0 \text{ g C}_2\text{H}_6 \times \frac{1 \text{ mol C}_2\text{H}_6}{30.07 \text{ g C}_2\text{H}_6} \times \frac{6 \text{ mol H}}{1 \text{ mol C}_2\text{H}_6} \times \frac{6.022 \times 10^{23} \text{ H atoms}}{1 \text{ mol H}}$$

$$= 2.16 \times 10^{24} \text{ H atoms.} \quad \surd$$

- How many oxygen (O) atoms are there in 5.55 g of H_3PO_4 ?

We are going from grams of phosphoric acid to oxygen atoms, so once more we need to convert these seeming unrelated units at the mole level.

$$5.55 \text{ g H}_3\text{PO}_4 \times \frac{1 \text{ mol H}_3\text{PO}_4}{97.99 \text{ g H}_3\text{PO}_4} \times \frac{4 \text{ mol O}}{1 \text{ mol H}_3\text{PO}_4} \times \frac{6.022 \times 10^{23} \text{ O atoms}}{1 \text{ mol O}}$$

$$= 1.36 \times 10^{23} \text{ O atoms.} \quad \surd$$

2.4. <u>Derivation of Formulas</u>.

Up to this point, we have been talking about composition of compounds based on the number of constituent atoms. Sometimes it is advantageous to talk about the composition in terms of the masses of the elements that make up the compound. We can go into a laboratory and find out the mass composition of the different elements that make up a compound. As an example, let's say we take an ethane gas sample (not knowing it is ethane, C_2H_6) and burn it in the presence of oxygen gas (O_2). We observe that all of the ethane is gone and it is transformed into two new substances: carbon dioxide (CO_2) and water (H_2O). If we very carefully collect all of the carbon dioxide and all the water formed, we can relate the grams of carbon dioxide formed to the grams of carbon originally present in our sample. We can also relate the grams of water formed to the grams of hydrogen originally present in our sample. This will enable us to determine a relative mass relationship of carbon to hydrogen in our original compound.

The question is, how do we relate these relative masses in order to determine the chemical formula? There are two formulas of interest, mainly the **empirical formula** and the **molecular formula**. The <u>molecular</u> formula is the true formula of a compound, that is, it tells us exactly how many atoms of each constituent element are bonded together in order to make the compound. On the other hand, the <u>empirical</u> formula is the simplest whole number ratio of the atoms that make up a compound.

Examples:

MF: N_2H_4 $B_3N_3H_6$ C_6H_6 H_2O

EF: NH_2 BNH_2 CH H_2O

What information can we get from the relative masses of carbon and hydrogen for the example we were referring to above (ethane)? We will get the empirical formula, which is after all the simplest ratio in which the elements were combined. How can we determine the molecular formula from that information? We would need one more piece of information: the molecular weight. Conclusion: To determine the molecular formula for an unknown compound all we need is the molecular weight and the empirical formula. Let's see a simple example that will show us how this is done.

- Find the molecular formula for a compound whose empirical formula is CH_2 and its molecular weight is 70 amu.

    We were given the two things we need in order to determine the molecular formula. Since the empirical formula is CH_2 and we know that the empirical formula is a multiple of the molecular formula, we know that we will be able to fit the weight of the empirical formula a certain number of times into the molecular weight. CH_2 weighs 14 (twelve from carbon and two from the hydrogens) and 14 fits <u>five</u> times into 70, therefore, the molecular formula must be five times the empirical formula, or:

$$C_5H_{10}.$$

    How do we determine the empirical formula itself? All we need to do is determine experimentally the elemental composition of the unknown compound. Let's start with some simple examples and slowly move into a problem like the one we described at the start of this section.

- What is the empirical formula of a compound that contains 43.6% P and 56.4% O (mass percentages)?

    In order to solve this problem, we first need to evaluate the information given. Remember that formulas are written in terms of moles, not grams. If our compound is 43.6% P, then out of every 100 g of compound, we must have 43.6 g of P and 56.4 g of O. In other words, we can always take the percentages given, assume we have 100 g of compound, and then those percentages may be read as grams instead. Since we said that the formulas are in moles, we need to convert these grams into moles first. Once we achieve that, all we need to do is to find the simplest mole ratio between these elements (empirical formula).

	mass	moles	Simplest Ratio
P	43.6 g	$\frac{43.6}{30.97} = 1.41$	1.41/1.41 = 1
O	56.4 g	$\frac{56.4}{16.0} = 3.53$	3.53/1.41 = 2.5

To find the simplest ratio, we divide the moles of all the components by the smallest number. This usually yields whole numbers. However, in this case we see it does not. Since the empirical formula is the simplest whole number ratio, therefore, we need to multiply these numbers by two, in order to get rid of the fraction. Answer: The simplest ratio is 2 : 5, and the empirical formula must be: P_2O_5, diphosphorus pentoxide.

- An unknown sugar has the following elemental analysis: C (40.00%), H (6.67%), and O (53.3%). If its molecular weight is 240 amu, what would be its molecular formula?

Probably the first thought that crosses our mind is to determine the empirical formula (with the elemental composition, as we did in the previous problem) and then using the empirical formula and the molecular weight, determine the molecular formula. This would be perfectly legitimate, but unnecessary. Let's think about this. Last time we did not know the molecular weight, so all the information we had was the elemental composition, which is why we chose 100 g of compound to decide how many grams of each element we had in the compound. However, this time we do know the molecular weight from the beginning, so it would be easier to simply assume we have one mole of the compound (the molecular weight in grams) and find the percentages of all elements in that one mole. This will give us the actual moles of each element in the molecular formula directly!

	mass	moles
C	0.400 (240 g) = 96.0 g	96.0 / 12.0 = **8**
H	0.0667 (240 g) = 16.01 g	16.01 / 1.01 = **16**
O	0.533 (240 g) = 128 g	128 / 16.0 = **8**

Therefore, the molecular formula must be: $C_8H_{16}O_8$. √

- Caffeine, a stimulant found in coffee, tea, and chocolate, contains 49.48 % C, 5.15 %H, 28.87 % N, and 16.49 % O. Determine the empirical formula. Given that the molecular weight is 194.2, determine its molecular formula.

Since all we want is the empirical formula, we will assume that we initially have 100 g of caffeine, and that way we will directly know how many grams of C, H, N, and O we have in the compound.

	mass	moles	Simplest Ratio
C	49.48 g	4.12	4
H	5.15 g	5.11	5
N	28.87 g	2.06	2
O	16.49 g	1.03	1

Therefore, the empirical formula would be: $C_4H_5N_2O$. √

Finally, the weight of the empirical formula is 97.08, and that is exactly one half of the molecular weight. Therefore, the molecular formula must be twice the empirical formula, that is: $C_8H_{10}N_4O_2$.

There is one more thing we should mention. When we find the simplest ratio, we will never (it would be rare) find an exact number. For example, in the previous problem, 5.11 divided by 1.03 is 4.961165. Of course, we will round off this number to 5, which is what we see on that table. Should we always round off to the nearest whole number? No! We will do so only when it is very close. There are some fractions that we should immediately recognize and therefore, not round off. The example we saw with the P_2O_5 example was 0.500. This fraction is one half, and we should not round that off and instead we multiplied by 2 to get rid of the half. Which other fractions should we recognize? The fractions: 0.200, 0.400, 0.600, and 0.800 are fifths, so multiplying them by 5 will get rid of the fractions. The fractions: 0.250 and 0.750 are quarters, therefore, multiplying them by four will get rid of the fractions. Finally, the fractions of 0.333 and 0.667 are thirds, so multiplying them by three would get rid of the fractions.

- Nicotine is a compound that contains C, H, and N. If a 2.50 g sample of nicotine is burned in O_2, 6.78 g of CO_2, 1.94 g of H_2O, and a certain amount of N_2 gas are formed. What is the empirical formula for the nicotine?

    This problem is a lot more complex than the ones we have done previously, so let's analyze it carefully. We had initially a 2.50 g sample of a compound whose formula we ignore. Upon burning this compound, we find that all of the carbon in the compound is converted into carbon dioxide. Somehow, we will have to figure out how much carbon is trapped in the carbon dioxide. Remember that we cannot do that comparison unless we are at the mole level. From the formula for carbon dioxide, we can see that every mole of carbon dioxide has one mole of carbon. Therefore,

$$6.78 \text{ g } CO_2 \times \frac{1 \text{ mol } CO_2}{44.0 \text{ g } CO_2} \times \frac{1 \text{ mol } C}{1 \text{ mol } CO_2} \times \frac{12.0 \text{ g } C}{1 \text{ mol } C}$$

$$= \underline{1.85 \text{ g of carbon}}.$$

This means that the 2.50 g sample of nicotine contained 1.85 grams of carbon. The rest must have been a combination of hydrogen and nitrogen.

How about the hydrogen? All we know is that all the hydrogen that was originally in the nicotine has been converted into water. Therefore, we need to find out how much hydrogen is in that amount of water. From the molecular formula for water, we can see that there are two moles of hydrogen in every mole of water. Therefore,

$$1.94 \text{ g } H_2O \times \frac{1 \text{ mol } H_2O}{18.0 \text{ g } H_2O} \times \frac{2 \text{ mol } H}{1 \text{ mol } H_2O} \times \frac{1.01 \text{ g } H}{1 \text{ mol } H}$$

$$= \underline{0.218 \text{ g of hydrogen}}.$$

This means that in the 2.50 g sample of nicotine, we must have had 0.218 g of hydrogen. Since we already know that the sample also contained 1.85 g of carbon, the rest must be the nitrogen.

So, grams of N = 2.50 g – 1.85 g – 0.218 g = $\underline{0.43 \text{ g N}}$.

With these numbers (the grams of C, H, and N) we should be able to find the empirical formula.

Chapter 2: Atoms and Moles

	mass	moles	Simplest Ratio
C	1.85 g	0.154	5
H	0.218 g	0.216	7
N	0.43 g	0.0307	1

Therefore, the empirical formula is: C_5H_7N.   √

- A compound contains only carbon, hydrogen, nitrogen, and oxygen. Combustion of 15.70 g of the compound produced 21.4 g of CO_2, 3.11 g of H_2O, and 2.90 g of N_2 gas. Determine the empirical formula for this compound.

    This is a similar problem, so let's go through it quickly. We should be able to get the grams of C from the carbon dioxide, the grams of hydrogen from the water and the grams of nitrogen[1] were actually given. The only way to get the grams of oxygen would be from that data and the mass of the sample (15.70 g). The reason we cannot get the grams of oxygen from the carbon dioxide and from the grams of water is because not all the oxygen in these products came from the compound (some of it did, but the rest came from oxygen gas, an important ingredient for the combustion of any sample of any compound).

$$21.4 \text{ g } CO_2 \times \frac{1 \text{ mol } CO_2}{44.0 \text{ g } CO_2} \times \frac{1 \text{ mol C}}{1 \text{ mol } CO_2} \times \frac{12.0 \text{ g C}}{1 \text{ mol C}} = \underline{5.84 \text{ g C}}$$

$$3.11 \text{ g } H_2O \times \frac{1 \text{ mol } H_2O}{18.0 \text{ g } H_2O} \times \frac{2 \text{ mol H}}{1 \text{ mol } H_2O} \times \frac{1.01 \text{ g H}}{1 \text{ mol H}} = \underline{0.349 \text{ g H}}$$

That leaves us with: $15.70 \text{ g} - 5.84 \text{ g} - 0.349 \text{ g} - 2.90 \text{ g}^2$ = <u>6.61 g of O</u>.

Therefore, we build the table that allows us to calculate the simplest ratios.

---

[1] A 2.90 g sample of N_2 has the same mass if we were to split it into elemental N. Think about it. Let's say that the N_2 is a dollar bill. Of course, to get single N's we would have to split the paper in half. That would yield twice as many pieces of paper (twice as many atoms), but if you weigh them, it would be exactly the same. Therefore, 2.90 g of N_2 = 2.90 g of N.

[2] See the previous footnote for the explanation.

	mass	moles	Simplest Ratio
C	5.84 g	0.486	2.35
H	0.349 g	0.346	1.67
N	2.90 g	0.207	1
O	6.61 g	0.413	2

We recognize the first two as thirds (or very close to it). Therefore, in order to get whole numbers, we will have to multiply all of the simplest ratios by three, which will give us the empirical formula: $C_7H_5N_3O_6$. √

2.5 Percent Composition.

This has been previously discussed, it is called elemental composition. It simply tells us what percentage of the total mass of a given compound corresponds to each element that constitutes that particular compound. If we are asked to determine the percent of any element in a given compound, we should assume that we have <u>one mole</u> of the compound and then it should be very simple.

- What is the % C in Vitamin C, $H_2C_6H_6O_6$?

Since we are assuming that we have one mole of Vitamin C, then to solve this problem we must first determine the molecular weight of Vitamin C.

$$\begin{aligned} 8\,H &= 8.064 \text{ g} \\ 6\,C &= 72.066 \text{ g} \\ 6\,O &= 96.000 \text{ g} \\ \hline &\,176.130 \text{ g} \end{aligned}$$

$$\% \text{ C} = \frac{\text{grams of C in one mole}}{\text{molecular weight}} \times 100$$

$$\frac{72.066 \text{ g}}{176.130 \text{ g}} \times 100 = \underline{\underline{40.92 \,\%}} \quad √$$

- How many grams of Ag are theoretically obtainable from 250 g of an impure silver ore that is 70.0% Ag_2S?

As soon as we read that question, we see two numbers: 250 g of an ore, and 70.0% silver sulfide. The first number we cannot convert into moles, because we ignore how pure is the ore. Therefore, for the time being, this number is useless. However, since we are told that the ore is 70.0% pure, we should be able to immediately determine the grams of silver sulfide in the sample.

$$250 \text{ g of ore} \times \frac{70.0 \text{ g of Ag}_2\text{S}}{100.0 \text{ g of ore}} = 175 \text{ g of Ag}_2\text{S}.$$

Now, how many grams of silver are there in those 175 grams of silver sulfide? We have done calculations like these before, so:

$$175 \text{ g of Ag}_2\text{S} \times \frac{1 \text{ mol Ag}_2\text{S}}{247.9 \text{ g Ag}_2\text{S}} \times \frac{2 \text{ mol Ag}}{1 \text{ mol Ag}_2\text{S}} \times \frac{107.9 \text{ g Ag}}{1 \text{ mol Ag}}$$

$$= \textit{152 g of silver.} \quad \sqrt{}$$

Let's look at a situation that presents a more complex formula. This is the case of **hydrates**. A hydrate is a compound that contains a certain number of moles of water per mole of compound in the crystal lattice[1]. These moles of water molecules become an integral part of the crystal lattice and therefore, become an integral part of its chemical formula. These are normally written as the formula for the substance followed by a dot and the number of moles of water. For example: the formula $MgSO_4 \cdot 5H_2O$ tells us that the solid contains five moles of water imbedded into the crystal lattice per mole of magnesium sulfate. Many times, when we prepare a substance (like magnesium sulfate) using water as a solvent, we obtain a solid that contains water in its lattice. It is hard to tell how many of these waters are in the crystal lattice and how much water is simply making the sample wet. So, experimentally, we dry the sample in an oven to drive off the water that is superficial (not a part of the lattice) and what remains would be the hydrate. It is also true that in many instances, we can dry the sample at a very high temperature and drive off the rest of the water (the one in the lattice). This is very useful, since this information will allow us to determine the number of moles of water in the hydrate itself. Let's look at one example that does exactly that.

- A 4.322 g sample of a hydrated copper(II) sulfate, $CuSO_4 \cdot xH_2O$, was heated to 140°C (temperature at which all the water is removed), cooled down to room temperature and then reweighed. The mass was then measured to be 3.526 g. Determine the value of **x** in the

---

[1] The crystal lattice is a three dimensional arrangement of the atoms and/or molecules in the solid.

formula for this hydrate. The MW of pure $CuSO_4$ is 159.6.

The first thing we see is the number of grams of the sample (4.322 g). Of course, this number is useless initially since we do not know the value of **x** (which must be an integer). However, we are told that after heating up this sample, all the water is removed. This means that the second number given (3.526 g) represents the grams of pure cupric sulfate. These, being grams of a pure substance, can be converted into moles. This was our original goal because we cannot even start to think about comparing different things until we are at the mole stage. By difference, we can figure out how many grams of water we had in the original sample. Once we know the grams of each component (cupric sulfate and water), then we can figure out the formula as we have done before.

grams of $CuSO_4$ = 3.526 g

grams of water = 4.322 - 3.526 = 0.796 g

Now, let's calculate the moles of each.

Moles of $CuSO_4$ = 3.526 / 159.6 = 0.02209 moles of cupric sulfate.

Moles of water = 0.796 / 18.0 = 0.0442 moles of water.

It is evident (from the numbers) that we have twice as many moles of water as we have moles of cupric sulfate, therefore, **x** must be equal to 2, and the formula for this hydrate must be: $CuSO_4 \cdot 2H_2O$. √

- A 6.420 g sample of $ZnSO_4 \cdot xH_2O$ is dissolved in water and the sulfate ion (SO_4^{-2}) is precipitated completely as $BaSO_4$ (MW = 233.4). The mass of pure, dry $BaSO_4$ obtained is 5.559 g. What is the value of **x**?

There is only one number we can use initially, which is the mass of pure barium sulfate. Remember that we can only convert to moles when we start with a pure substance of known chemical formula. Therefore, we can determine the moles of barium sulfate as:

Moles of $BaSO_4$ = 5.559 / 233.4 = 0.02382 mol of $BaSO_4$.

What can we do with this information? How does this number relate to the moles of zinc sulfate? We can see that both substances contain only one mole of sulfate in them. Therefore, there is a one to one mole ratio between them. In other words, since all of the sulfate was precipitated as barium sulfate, we must have had an equal number of moles of zinc sulfate initially. So,

$$\text{moles of ZnSO}_4 = 0.02382 \text{ mol ZnSO}_4.$$

Let's formulate a plan of action. What are we looking for? We want to find **x**. We know that **x** is expressed in terms of moles. There is a whole number relationship between the moles of water and those of zinc sulfate (as seen in the formula: 1 : **x**). Therefore, all we need to do is find the actual moles of each. The problem is that to this point we do not know how many grams of water we had in the original sample. However, we do know the moles of zinc sulfate and we can easily figure out its molecular weight (MW = 161.44). Therefore, we can determine the grams of zinc sulfate in the original sample.

$$\text{grams of ZnSO}_4 = 0.02382 \text{ mol } (161.44 \text{ g/mol}) = \underline{3.8455 \text{ g ZnSO}_4}. \;\checkmark$$

Now that we know the grams of zinc sulfate in the original sample, we can easily determine the grams of water in that same sample (which must be the rest).

$$\text{Grams of water} = 6.420 \text{ g} - 3.8455 \text{ g} = \underline{2.575 \text{ g of water}}.$$

Once this information is gathered, all we need to do is convert them into moles and find the simplest ratio between the two.

	mass	moles	Simplest Ratio
$ZnSO_4$	3.8455 g	0.02382	1
H_2O	2.575 g	0.1429	6

Therefore, **x** is equal to **six (6)**.

Based on what we have learned so far, we can even do problems where we can predict how much product will be formed in a given reaction. As we will see in the next section, these can be worked out in a much simpler way, once we learn how to balance chemical reactions. We can also do them based on the conservation of matter principles we have been applying implicitly.

- The molecular formula for caffeine is $C_8H_{10}N_4O_2$ and its molecular weight is 194. If we completely burn 52.5 g of caffeine, how many grams of carbon dioxide (MW = 44.01) will be formed?

We need to first recognize (or assume) that the every little bit of carbon we had in the original compound (caffeine) was converted to carbon dioxide. What is the relationship between caffeine and carbon dioxide? The carbon itself. There are 8 moles of carbon in caffeine whereas there is only one mole of carbon in carbon dioxide. Therefore, we should see a one to eight mole ratio between the two. Remember that the only way we can compare the caffeine to the carbon dioxide is at the <u>mole</u> level! Therefore,

$$52.5 \text{ g caffeine} \times \frac{1 \text{ mol caffeine}}{194 \text{ g caffeine}} \times \frac{8 \text{ mol C}}{1 \text{ mol caffeine}} \times \frac{1 \text{ mole CO}_2}{1 \text{ mole C}} \times \frac{44.01 \text{ g CO}_2}{1 \text{ mole CO}_2}$$

$$= 95.3 \text{ g of CO}_2. \quad \checkmark$$

- How many grams of water are formed simultaneously in that same reaction?

What is the connection between the caffeine and the water? The hydrogen itself, of course. Caffeine and water both contain hydrogen and every little bit of hydrogen that was initially in the compound will be found eventually in the water (in theory). Therefore, we can compare the caffeine and the water through the hydrogen and the mole level.

$$52.5 \text{ g caffeine} \times \frac{1 \text{ mol caffeine}}{194 \text{ g caffeine}} \times \frac{10 \text{ mol H}}{1 \text{ mol caffeine}} \times \frac{1 \text{ mol H}_2\text{O}}{2 \text{ mol H}} \times \frac{18.02 \text{ g H}_2\text{O}}{1 \text{ mole H}_2\text{O}}$$

$$= 24.4 \text{ g of water.} \quad \checkmark$$

- How many grams of oxygen gas were consumed with the caffeine in the previous reaction?

This question sounds a little bit more complicated, but it actually is simpler. All we need to remember is the principle of conservation of matter. We have formed 95.3 g of carbon dioxide and 24.4 g of water. These are all the products of this reaction. Therefore, the total mass present after the reaction is: 119.7 g of substance. We must have had exactly the same amount of reactants. Therefore, the mass of oxygen gas (the second reagent) must have been: 119.7 g – 52.5 g = <u>67.2 g of oxygen gas</u>. $\checkmark$

2.6. <u>Chemical Equations</u>.

During a chemical transformation, atoms are reorganized from one substance into another. A chemical equation shows exactly this: the reorganization of the atoms. How do these atoms get reorganized? We will observe that some bonds are broken and some bonds are formed during this process. The actual process of balancing a reaction is based on the **law of** conservation of matter, that is, matter can neither be created nor destroyed. What does that mean? Basically, we should have the same number of atoms in the left side (the reactant side) of the equation as we have in the right (the product side). Notice that we are referring to the same number of atoms of a given kind. For example, if we have four carbon atoms on the left side, we **must** have four carbon atoms on the right side. Many times we will indicate the phase (gas, liquid, solid, aqueous solution, etc.) of a given substance as a subscript and in parenthesis.

Ex. $$CH_{4(g)} + 2\,O_{2(g)} \Rightarrow CO_{2(g)} + 2\,H_2O_{(\ell)}$$

What are these coefficients telling us? We can read these equations in various ways. Probably the most evident way is in terms of atoms and molecules. In the example above, we can see that one molecule of methane (CH_4) gas reacts with two molecules of oxygen gas in order to form one molecule of carbon dioxide gas and two molecules of liquid water.

We have already seen that we can read these formulas in terms of moles, and that is probably the most important way in which we should look at them. A properly balanced equation will tell us the relative number of moles of atoms or moles of molecules in which reagents are combined or in which products are formed. Once more, for the methane example above, we can see that every time we completely react one mole of methane gas with two moles of oxygen gas, we will obtain two moles of liquid water and one mole of carbon dioxide gas. Keep in mind that these coefficients must be read in terms of moles, NOT grams. However, we could easily make a mental conversion and say that 16.0 grams of methane (one mole) will combine with 64.0 grams of oxygen gas (two moles), we will get 44.0 grams of carbon dioxide gas (one mol) and 36.0 grams (two moles) of liquid water. Please notice that the mass is conserved (16.0 g + 64.0 g = **80.0 g** = 44.0 g + 36.0 g). This is the main concept behind balancing a chemical equation.

In order to balance a reaction by inspection it is often convenient to follow some simple steps. We should first look for the species that has the most atoms. For example, in the methane reaction above, it would be the methane molecule itself (methane has 5 atoms, oxygen gas has two, carbon dioxide has three and water has also three). Fixing methane assumes that we have one mole of methane gas present initially. The next step would be to analyze the rest of the equation. Are we going to add more carbons later on (with another reagent)? The answer is **no**, since we are only going to add O_2 later (which does not contain C). This means that we have already placed as much carbon (one mole in CH_4) as are ever going to in the reactant side. Therefore, we must have exactly one mole of carbon in the product side. Since the only substance with carbon in the product side is CO_2, which also contains one mole of C, we must

Chapter 2: Atoms and Moles

then fix the amount of carbon dioxide to one mole.

Now, how about the hydrogen? Just like the carbon, all of the H is in the methane. That means that we should fix four moles of H in the product side. Since all of the H will be supplied by the water and there are two H's in every water, the we must add two moles of water. Once we have accomplished this, all of the product side has been fixed. Therefore, we can count how many moles of O have been already fixed in the product side (four O's: two from the carbon dioxide and two from the waters). Therefore, we must have four moles of O in the reactant side. Since there is no O in the methane, all of it must be supplied by the O_2 gas. That is how we decide that we need to fix the oxygen gas to two moles.

Let's do a few more examples. Let's balance the reaction for the combustion of ethane gas (C_2H_6). As long as we have carbon-hydrogen or carbon-hydrogen-oxygen compounds, combustion means its reaction with O_2 gas to form CO_2 gas and liquid water at 25°C. The unbalanced reaction can be seen below.

- Balance the following reaction:

$$\underline{} C_2H_6 + \underline{} O_2 \Rightarrow \underline{} CO_2 + \underline{} H_2O$$

The molecule with the most atoms is the ethane molecule, so we will assume we have one mole of ethane. At this point, our "balanced" reaction looks like this:

$$1\,C_2H_6 + \underline{} O_2 \Rightarrow \underline{} CO_2 + \underline{} H_2O$$

This molecule has two moles of C, therefore, we need to have two moles of carbon in the product side, which means that we need to add two moles of CO_2 gas:

$$1\,C_2H_6 + \underline{} O_2 \Rightarrow 2\,CO_2 + \underline{} H_2O$$

The ethane molecule also has six moles of H, therefore, we need to have six moles of H in the product side. This means that we need to add three moles of water.

$$1\,C_2H_6 + \underline{} O_2 \Rightarrow 2\,CO_2 + 3\,H_2O$$

Now, we can count the total number of O's in the product side (seven O's: four from the carbon dioxide and three from the waters). This means that we need to place seven O's in the reactant side also, but the problem is that they must be added in pairs (O_2). In conclusion, we need to add three and a half moles of O_2.

$$1\,C_2H_6 \;+\; 3.5\,O_2 \;\Rightarrow\; 2\,CO_2 \;+\; 3\,H_2O$$

This is the balanced reaction and there is nothing technically wrong with that. However, we should be able to read the reaction in terms of either moles or molecules. If we read it in terms of moles, we can easily obtain 3.5 moles of oxygen gas. However, we cannot get 3.5 oxygen molecules, because we would have to break one in half. To avoid this problem, we can multiply the entire reaction by two (to get rid of the fraction) and that would lead to the final form for the balanced reaction:

$$2\,C_2H_6 \;+\; 7\,O_2 \;\Rightarrow\; 5\,CO_2 \;+\; 6\,H_2O$$

- Balance the following reaction:

$$\_\_\,KClO_3 \;\Rightarrow\; \_\_\,KClO_4 \;+\; \_\_\,KCl$$

Once more, start by fixing the potassium perchlorate, because it has the most atoms.

$$\_\_\,KClO_3 \;\Rightarrow\; 1\,KClO_4 \;+\; \_\_\,KCl$$

This time, however, we notice immediately that the only element that has been completely fixed in the product side is the O, since we will be adding more K and more Cl later (in the form of KCl). Therefore, we have four O's in the product side. This means that we need four O's in the reactant side also. Since the O's must be added in units of three, that means that we need to add (4/3) moles of potassium chlorate.

$$4/3\,KClO_3 \;\Rightarrow\; 1\,KClO_4 \;+\; \_\_\,KCl$$

Of course, we want to get rid of that fraction immediately, so we will multiply everything we have balanced up to that point by three (to get rid of the fraction). This leads to:

$$4\,KClO_3 \;\Rightarrow\; 3\,KClO_4 \;+\; \_\_\,KCl$$

The only things missing are the K's and the O's (there is one of each missing). Therefore, we need to add one mole of KCl on the product side. This is the final balanced reaction.

$$4\,KClO_3 \;\Rightarrow\; 3\,KClO_4 \;+\; 1\,KCl$$

There are other types of reactions (like simple redox[1] reactions) that we will learn how to

---

[1] Reduction-oxidation reactions.

Chapter 2: Atoms and Moles

balance in Chapter 19.

2.7 Stoichiometry.

We have learned that when we read a balanced chemical equation, we always do so in terms of moles or of molecules, not in terms of grams. However, when you carry out a reaction in a laboratory or in a chemical plant, we need to **weigh** in the reactants and collect the products and weigh them once more. In order to get approximate masses that will give us the amount of product desired, we need to carefully determine how much of the different reactants we should combine. **Stoichiometry**[1] may be defined as the calculation of the quantities of reactants and products involved in a chemical reaction. Since these relationships are read from the balanced equation, they will be in terms of moles, therefore, the general idea will be to convert the grams of the initial substance into its moles, so that we will then be able to convert these moles into the moles of the second substance. Once we know the moles of the desired substance, conversion into grams is trivial (using the molecular weight of the second substance).

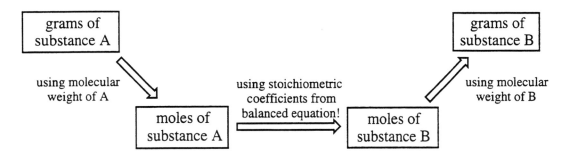

This is probably easier to understand by following a few examples. It should be clear that a stoichiometry problem cannot be solved unless a balanced equation is given, since we need the stoichiometric coefficients from this equation. Therefore, the first step must always involve balancing of the chemical equation.

- How many grams of water are required to react completely with 25.0 g of Fe?

$$3 \, Fe_{(s)} + 4 \, H_2O_{(\ell)} \Rightarrow Fe_3O_{4(s)} + 4 \, H_{2(g)}$$

We can see that the question is trying to relate Fe with water. The balanced equation tells us that there is a 3 to 4 mole ratio between those two substances, that is, for every three moles of iron that get consumed, four moles of water will disappear simultaneously. Since we are starting with grams of iron and the only relationship between these is in moles, the

---

[1] From the Greek *stoicheion* (element) and *metron* (measure).

first thing we will do is convert grams of Fe to moles of Fe. Once we have the moles of iron, we can then convert to moles of water using the stoichiometric coefficients (from the balanced reaction) to give us the moles of water needed to react completely with the initial iron. Of course, once we have moles of water, all we need is the molecular weight of water in order to change it back into grams of water.

$$25.0 \text{ g Fe} \times \frac{1 \text{ mol Fe}}{55.85 \text{ g Fe}} \times \underbrace{\frac{4 \text{ mol H}_2\text{O}}{3 \text{ mol Fe}}}_{\text{from balanced reaction!}} \times \frac{18.0 \text{ g H}_2\text{O}}{1 \text{ mol H}_2\text{O}}$$

$$= 10.7 \text{ g of water.} \quad \checkmark$$

- How many grams of H_2 will be formed in the same reaction?

This time, we are relating the mass of hydrogen gas formed with the initial amount of iron present in the reaction. What is the relationship between Fe and hydrogen gas? The balanced chemical equation tells us that there is a 3 to 4 mole ratio between those two substances, that is, every time three moles of iron disappear in the reaction, we will form four moles of hydrogen gas simultaneously. Once more, we will use the same concept. Since that conversion is at the mole level, we will first change from grams of Fe to moles of Fe, then from moles of Fe to moles of H_2, then finally from moles of H_2 to grams of H_2. Therefore,

$$25.0 \text{ g Fe} \times \frac{1 \text{ mol Fe}}{55.85 \text{ g Fe}} \times \underbrace{\frac{4 \text{ mol H}_2}{3 \text{ mol Fe}}}_{\text{from balanced reaction!}} \times \frac{2.02 \text{ g H}_2}{1 \text{ mol H}_2}$$

$$= 1.21 \text{ g of H}_2 \text{ gas.} \quad \checkmark$$

- How many grams of water will be formed in the complete combustion of 25.0 g of ethanol (C_2H_5OH) ?

The first thing we notice is that a balanced reaction was not given in this problem. Of course, we notice the word combustion, which means that we are burning the sample in the presence of oxygen gas. As we previously learned, this means that it is going to form carbon dioxide gas and liquid water. Balancing the chemical equation, we obtain:

$$C_2H_5OH_{(\ell)} + 3\,O_{2(g)} \rightarrow 2\,CO_{2(g)} + 3\,H_2O_{(\ell)}$$

Since the question deals with ethanol and water, we focus our attention in the fact that there is a 1 : 3 mole ratio between them. Once we realize this, the problem is relatively simple, since we will work it out in exactly the same way as all of the previous ones.

$$25.0 \text{ g C}_2\text{H}_5\text{OH} \times \frac{1 \text{ mol C}_2\text{H}_5\text{OH}}{46.07 \text{ g C}_2\text{H}_5\text{OH}} \times \frac{3 \text{ mol H}_2\text{O}}{1 \text{ mol C}_2\text{H}_5\text{OH}} \times \frac{18.02 \text{ g H}_2\text{O}}{1 \text{ mol H}_2\text{O}}$$

$$= 29.3 \text{ g of water.} \quad \checkmark$$

- Nitric acid is produced commercially in a chemical plant through the Ostwald process. The three steps in this process are shown below.

$$4 \text{ NH}_{3(g)} + 5 \text{ O}_{2(g)} \Rightarrow 4 \text{ NO}_{(g)} + 6 \text{ H}_2\text{O}_{(\ell)}$$

$$2 \text{ NO}_{(g)} + \text{O}_{2(g)} \Rightarrow 2 \text{ NO}_{2(g)}$$

$$3 \text{ NO}_{2(g)} + \text{H}_2\text{O}_{(\ell)} \Rightarrow 2 \text{ HNO}_{3(aq)} + \text{NO}_{(g)}.$$

How many grams of NH_3 must be used in order to form 1.00×10^6 kg of nitric acid by this process?

Let's make sure we understand the process itself. These are **consecutive reactions**, that is, once the product from reaction #1 is recovered, then we proceed with reaction #2, and so on. Therefore, to determine the amount of NH_3 needed, we need to start with the HNO_3 formed (the experimentally determined number) and convert to moles of NO_2 gas from reaction #3. Why NO_2 gas? Because NO_2 is the substance that is common between reactions #2 and #3. Once we know how many moles of NO_2 are needed in the third step, then we relate NO_2 to NO (which is common with step #1). Finally, the NO is related to the NH_3 from step #1. The calculation is show below.

$$1.00 \times 10^9 \text{ g HNO}_3 \times \frac{1 \text{ mol HNO}_3}{63.02 \text{ g HNO}_3} \times \underbrace{\frac{3 \text{ mol NO}_2}{2 \text{ mol HNO}_3}}_{\text{Reaction \#3}} \times \underbrace{\frac{2 \text{ mol NO}}{2 \text{ mol NO}_2}}_{\text{Reaction \#2}} \times \underbrace{\frac{4 \text{ mol NH}_3}{4 \text{ mol NO}}}_{\text{Reaction \#1}}$$

from:

$$= 2.38 \times 10^7 \text{ moles of NH}_3.$$

Therefore,

$$2.38 \times 10^7 \text{ mol NH}_3 \times \frac{17.0 \text{ g NH}_3}{1 \text{ mol NH}_3}$$

$$= 4.05 \times 10^8 \text{ g NH}_3. \quad \checkmark$$

2.8 Limiting Reagents.

All the problems we have worked with so far have involved relatively simple scenarios because in every case, we were given initially only one number: the grams of a given reactant that was going to be consumed entirely. However, when we actually carry out the reaction in the laboratory, we mix different amounts of all the reactants in a given reaction pot. What is left in the reaction vessel once it is all over? We have the products (as expected) but we also have present any excess reactants. The reactant that is consumed 100% is referred to as the limiting reagent (LR). In order to understand how this comes to happen, let's look at the following hypothetical reaction. In this reaction A reacts with B in a one to one mole ratio to form also one mole of C. This makes it a very simple scenario (all coefficients are unity). The only way we would have no reagents left over after the reaction would be if we were to start with equal amounts of A and B. However, in the example seen below, we can see that we have unequal amounts of A and B. Of course, the reaction will come to a full stop once we run out of one of the reagents. Since we have less moles of B than of A, then we are going to run out of B first and we will refer to B as the limiting reagent.

	A	+	B	⇒	C
start:	5 mol		3 mol		0 mol
after:	2 mol		0 mol		3 mol

We can also see that the limiting reagent controlled how many moles of C were formed. Once we run out of B we cannot form any more C. This is why we only obtained three moles of C.

Properties of the Limiting Reagent.

1. It is 100% consumed.

2. It tells us how much product is *formed*.

3. It tells us how much of the other reagents will be *consumed*.

The example above was an extremely simple case because all the coefficients in the reaction were unity. The danger is to think that the only thing we must do is to look for the substance that has the least number of moles in order to decide which is the LR. Let's look at another example to see why this can be extremely dangerous, and why we *must* determine the LR every single time.

Chapter 2: Atoms and Moles

- How many grams of N_2F_4 (MW = 104.0) can theoretically be obtained from the reaction of 5.00 g of NH_3 (MW = 17.0) with 10.0 g of F_2 (MW = 38.0)?

$$2\,NH_3 + 5\,F_2 \Rightarrow N_2F_4 + 6\,HF$$

There are at least two different ways to do a limiting reagent problem.

1. Let's find the LR first. Choose one of the two reagents and *ASSUME* that it is the LR. For example, let's assume that the LR is NH_3. The question then becomes, if NH_3 were the LR, how many grams of F_2 would be needed to react completely with it? *Remember that the LR must be consumed 100%.* Let's determine the grams of F_2 needed to react completely with the 5.00 g of NH_3.

$$5.00\,gNH_3 \times \frac{1\,molNH_3}{17.0\,gNH_3} \times \frac{5\,molF_2}{2\,molNH_3} \times \frac{38.0\,gF_2}{1\,molF_2}$$

$$= 27.9 \text{ g of } F_2 \text{ needed.}$$

Do we have enough F_2 to react with all the NH_3 (all 5.00 grams)? Clearly NOT! We only have 10.0 g of F_2 present, therefore F_2 is the limiting reagent and not the NH_3, as we originally assumed. Now that we know which one is the LR, we can finally calculate the amount of N_2F_4 formed since it is the LR the one that controls the amount of product formed. Therefore, the grams of product formed are:

$$10.0\,gF_2 \times \frac{1\,molF_2}{38.0\,gF_2} \times \frac{1\,molN_2F_4}{5\,molF_2} \times \frac{104.0\,gN_2F_4}{1\,molN_2F_4}$$

$$= 5.47 \text{ g } N_2F_4. \quad \checkmark$$

2. There is an easier way to do this problem.

   a) Let's find the moles of each reagent first:

   $$molNH_3 = \frac{5.00\,g}{17.0\,g/mol} = 0.294\,molNH_3$$

   $$molF_2 = \frac{10.00\,g}{38.0\,g/mol} = 0.263\,molF_2$$

   b) Find the stoichiometric ratio for each reagent. The stoichiometric ratio is defined as the moles of a given reagent divided by the coefficient from the balanced equation. The stoichiometric coefficient for NH_3 is **2** and the stoichiometric coefficient for F_2 is **5**.

$$R_{NH_3} = \frac{0.294}{2} = 0.147$$

$$R_{F_2} = \frac{0.263}{5} = 0.0526$$

c) The smallest of the two ratios corresponds to the one we call the limiting reagent (LR).

Since the ratio for F_2 is smaller than the ratio for NH_3, then the F_2 is the limiting reagent.

d) Use the moles (not the ratio!) of the LR in order to initiate the calculation for the grams of product formed.

$$0.263 \, molF_2 \times \frac{1 \, molN_2F_4}{5 \, molF_2} \times \frac{104.0 \, gN_2F_4}{1 \, molN_2F_4}$$

$$= 5.47 \text{ g of } N_2F_4. \quad \checkmark$$

Every time we carry a reaction, no matter which, we will have a mixture in the reaction vessel that consists of the leftover reagents, the products formed, and any other side products (from other reactions that may have taken place). This means that once we are done, a somewhat difficult task remains: to separate the desired product from the reaction mixture. Many times this leads to getting a lot less product than we were hoping for. The number we calculated above for the previous problem is what we call the theoretical yield, or the amount of product we could obtain as a maximum. Of course, we will never obtain that much, so the actual yield should represent the actual number of grams of the product we are able to recover from the reaction vessel.

- Let's say that we carry out the reaction described in the previous problem but we only obtain 4.50 g of N_2F_4. What would be the percent yield of this reaction?

$$\% \text{ yield} = \frac{\text{actual yield}}{\text{theoretical yield}} \times 100$$

$$\% yield = \frac{4.50g}{5.47g} \times 100$$

$$= 82.3 \% \quad \checkmark$$

Reactions that are carried out in a laboratory should always be reported with a percentage

Chapter 2: Atoms and Moles

yield. This will give an idea of how successful was our synthesis. It bears repeating that no one can prepare a compound to a 100% yield. Many times, a researcher will look for experimental ways of increasing the yield of a reaction, that is, of minimizing the side reactions that will invariably consume some of the reagents to yield a different product and thus give us a lower percentage of the desired product.

- How many grams of water will be formed when you react 10.0 g of $C_2H_{6(g)}$ with 15.0 g of $O_{2(g)}$?

$$2\,C_2H_6 + 7\,O_2 \Rightarrow 4\,CO_2 + 6\,H_2O$$

The first thing we notice is that in this problem, we are given initially two numbers: the grams of both ethane and oxygen gases. Since we do not know which reagent is going to control how much product is formed (the one we call the limiting reagent), we should start by determining this (the LR).

$$mol_{ethane} = \frac{10.0g}{30.07g/mol} = 0.333\,mol_{ethane}$$

$$mol_{oxygen} = \frac{15.0g}{32.0g/mol} = 0.469\,mol_{O_2}$$

Let's calculate the stoichiometric ratios in order to determine which is the limiting reagent, since this cannot be done by comparing the initial moles.

$$R_{C_2H_6} = \frac{0.333}{2} = 0.167$$

$$R_{O_2} = \frac{0.469}{7} = 0.0670$$

Since the ratio for oxygen gas is a lot smaller than the ratio for the ethane gas, then there is no doubt that the limiting reagent is the O_2 gas. We should point out that, at first glance, it seems like the ethane gas should have been the LR because we had not only fewer grams but also fewer moles initially. However, what matters is the stoichiometric ratio! Using the moles of the oxygen gas (the LR) we can now determine the amount of water formed.

$$0.469 \text{ mol } O_2 \times \frac{6 \text{ mol } H_2O}{7 \text{ mol } O_2} \times \frac{18.02 \text{ g } H_2O}{1 \text{ mol } H_2O}$$

$$= \textit{7.24 g water.} \ \sqrt{}$$

- Potassium chromate, a bright yellow solid, can be prepared as follows:

$$4 \text{ FeCr}_2\text{O}_{4(s)} + 8 \text{ K}_2\text{CO}_{3(s)} + 7 \text{ O}_{2(g)} \Rightarrow 8 \text{ K}_2\text{CrO}_{4(s)} + 2 \text{ Fe}_2\text{O}_{3(s)} + 8 \text{ CO}_{2(g)}$$

In a given container, we place 125 g of $FeCr_2O_4$ (MW = 223.84), 175 g of K_2CO_3 (MW = 138.21), and 357 g of O_2 gas. Theoretically, how many grams of K_2CrO_4 (MW = 194.19) are formed in this reaction?

In this example, we see that there are three reagents. Of course, we need to determine the limiting reagent among those three before we try to answer the question. Therefore, let's determine the initial moles of all reagents.

$$mol_{FeCr_2O_4} = \frac{125}{233.84} = 0.535 \, mol$$

$$mol_{K_2CO_3} = \frac{175}{138.21} = 1.27 \, mol$$

$$mol_{O_2} = \frac{357}{32.0} = 11.2 \, mol$$

The stoichiometric ratios for all three should be calculated next.

$$R_{FeCr_2O_4} = \frac{0.535}{4} = 0.134$$

$$R_{K_2CO_3} = \frac{1.27}{8} = 0.159$$

$$R_{O_2} = \frac{11.2}{7} = 1.60$$

Since the ratio for $FeCr_2O_4$ is the smallest, that one is the limiting reagent. Therefore, our final calculation must be started with the moles of the limiting reagent. Remember that the answer we obtain is what we call the theoretical yield.

$$0.535 \, mol_{FeCr_2O_4} \times \frac{8 \, mol_{K_2CrO_4}}{4 \, mol_{FeCr_2O_4}} \times \frac{194.19 \, g_{K_2CrO_4}}{1 \, mol_{K_2CrO_4}}$$

$$= 208 \text{ g of } K_2CrO_4 \text{ are \underline{theoretically} formed.}$$

- What would be the percent yield of the reaction above if only 197 g of potassium chromate are obtained?

We already obtained the theoretical yield for this reaction in the previous part. Remember that the percent yield depends on the ratio of the actual yield to the theoretical yield. Therefore,

$$\% yield = \frac{197g}{208g} \times 100$$

$$= 94.7\% \quad \surd$$

- How many grams of CO_2 gas (MW = 44.01) were <u>actually</u> formed in the reaction above?

The key word here has been underlined. It is asking for the amount of carbon dioxide that was actually formed. Of course, this depends on the % yield, which we already know. Therefore, let's first determine the amount of carbon dioxide that should have been formed theoretically, and then we will determine the actual yield of carbon dioxide.

$$0.535 mol_{FeCr_2O_4} \times \frac{8 mol_{CO_2}}{4 mol_{FeCr_2O_4}} \times \frac{44.01 g_{CO_2}}{1 mol_{CO_2}}$$

= 47.1 g CO_2 formed theoretically.

However, we know that the reaction was only 94.7% efficient. Therefore, we actually obtained:

$$0.947 \ (47.1 \ g) = 44.6 \ g \ of \ CO_2 \ gas.$$

Chapter 3: Solution Stoichiometry.

3.1 Molarity.

In chapter one, we mentioned that solutions were homogeneous mixtures. Homogeneous mixtures have two or more components in a single phase. At first glance, a homogeneous mixture looks like a pure substance, but by simple physical means (like distillation), we are often able to separate its components. Although we mentioned that there are three types of solutions (solid, liquid and gaseous), we will concentrate for the time being on liquid solutions, where the single phase is the liquid phase. As with any mixture, we will be interested in determining the composition of the solution, that is, the relative amounts of the different pure substances that make up the solution. In these solutions, we will separate the components into two different types. The one that is more abundant is referred to as the *solvent*, whereas the other components are called *solutes*. When the solution is said to be **aqueous**, then the solvent is water. We will be dealing primarily with aqueous solutions, as water is the most common solvent of all (and is sometimes referred to as the universal solvent).

The composition of any solution is referred to as its *concentration* – the amount of solute dissolved in a given amount of solvent at a given temperature. Qualitatively, we can use terms like *dilute* or *concentrated*. These terms refer to solutions which have relatively small or large amounts of solute in the solvent (respectively). However, we are more interested in knowing the precise concentration of the solutes in this solution. Therefore, we need to specify the actual amounts of each solute in a given amount of solvent.

One of the simplest ways to measure the concentration of a given solute in a solution is what we call *mass %*. Let's say that a given flask containing an aqueous solution is labeled 40.0 % NaOH. What is the meaning of this label? This means that out of every 100.0 grams of solution we have 40.0 grams of NaOH (which happens to be one mole). Generally speaking, all percentages are always mass percentages and if they are meant to be something else (like mole percent), then you would have to specify that they are a different kind of percentage.

Probably, the most common way to specify the concentration of a solution is using something we call *Molarity (M)*. We can define Molarity as the moles of the solute dissolved in a given volume of the solution, expressed in Liters. Therefore, the units of molarity are moles per liter.

$$Molarity = \frac{moles_{solute}}{Volume_{solution(L)}}$$

Chapter 3: Solution Stoichiometry

How do we actually prepare a solution of a given molarity? The easiest way is by using a volumetric flask. This piece of glassware (let's say it is a 100-mL volumetric flask) is designed in such way that when the bottom of the meniscus[1] (for the solution) is right at the mark we see in the neck of the flask, then such flask contains 100.00 ± 0.02 mL of liquid at room temperature.

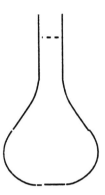

Let's say that we want to prepare a 0.500 M aqueous solution of NaCl (MW = 29.3) using a 100-mL volumetric flask. Immediately we think of the meaning of molarity. If the solution is meant to be 0.500 M, that implies that we need to have 0.500 moles of NaCl in every liter of solution, but since we are only going to prepare one-tenth of that volume (100 mL), then we need to add one tenth of those moles, which would be 0.0500 moles of NaCl. This is equivalent to 1.47 g of NaCl.

$$0.0500 mol_{NaCl} \times 29.3 \frac{grams}{mol} = 1.47 g_{NaCl}$$

Therefore, to prepare this solution we would carefully measure 1.47 g of NaCl and pour it completely (we call this a quantitative transfer) into the volumetric flask. We would then add some water to dissolve the solid. Once it is into one phase, then we would add water carefully until the bottom of the meniscus reaches the mark in the neck of the flask). At this point, we would mix the solution thoroughly to achieve homogeneity. We want to make sure that every drop of the solution has the same concentration. Unless we mix it thoroughly, the solution in the bottom of the flask will be more concentrated than the one in the top, since the last thing we added was water in order to complete the solution. Once we do that, we will have 100 mL of a 0.500 M NaCl. Let's do a few problems with molarity, to see how to manipulate the information in order to calculate concentrations.

- What is the concentration of H_2SO_4 in a solution prepared by dissolving 15.0 g of H_2SO_4

---

[1] The meniscus is the curved edge of the liquid solution. It separates the liquid phase from the gas phase above.

(MW = 98.1) in enough water to form 200 mL of solution?

Since molarity is defined as the moles of sulfuric acid per liter of solution, then we need to determine the moles of sulfuric acid.

$$moles_{H_2SO_4} = \frac{15.0 g_{H_2SO_4}}{98.1 \frac{g_{H_2SO_4}}{mol_{H_2SO_4}}} = 0.153 mol_{H_2SO_4}$$

Therefore,

$$[H_2SO_4] = \frac{0.153 mol}{0.200 L} = 0.765 M$$

- Heme, which is obtained from red blood cells, binds to oxygen (O_2). How many moles of heme are there in 25.0 mL of a 0.00150 M heme solution?

$$M = moles / Volume$$

Therefore, **moles = M x V** √

So,

$$Moles = (0.00150 \, mol/L)(0.0250 \, L) = 3.75 \times 10^{-5} \, mol \, of \, heme.$$

Notice that we have found a new way to determine moles. Up to this point, moles were calculated as the grams of a pure substance divided by its molecular weight. However, if we are dealing with solutions, then the moles of the solute are determined as the molarity of the solute times its volume (in Liters).

- How many grams of NaOH (MW = 40.0) are there in 250 mL of a 0.500 M solution?

Anytime we see the volume and the molarity of the solution written together, the first thought that should cross our minds is the determination of the moles of the solute. Therefore,

$$0.500 \frac{mol_{NaOH}}{L} \times (0.250 L) \times \frac{40.0 g_{NaOH}}{1 mol_{NaOH}}$$

$$= 5.00 \, g \, NaOH \quad √$$

3.2 Dilution.

In the previous section we learned how to prepare a solution by dissolving a certain amount of solute into a volumetric flask and adding enough water until the flask is filled. However, many times we find that it is more convenient to prepare a solution by adding water to a concentrated solution of a given solute until we obtain a lower desired concentration. This process is referred to as dilution. It should be obvious that the addition of water to a certain solution will indeed lower the concentration, but let's think about it anyway. Since molarity is the moles of the solute divided by the volume of the solution and the addition of water has the only effect of increasing the volume of the solution without changing the moles of the solute, then the molarity of the resulting solution would have to decrease (we would be dividing by a larger number).

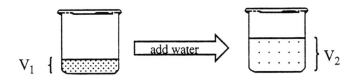

Another situation that involves a dilution is the addition of a solution of a different solute. The principle is the same: the moles of the initial solute remain constant whereas the volume of the final solution increases. Thus, the molarity of each solute will decrease (we can say that each solute was diluted).

Conclusion: In a dilution, *the number of moles of solute does not change as you go from the original solution to the final solution.* ∴ $n_{initial} = n_{final}$.

or,
$$M_1 V_1 = M_2 V_2$$

This equation works always for a dilution problem, but now always on other solution stoichiometry problems, so we should always think carefully before we use it. *If the moles of solute do not change from the initial state to the final state*, then the equation will work. As an example, let look at a problem where we dilute a concentrated solution of NaOH in order to prepare one of lower concentration.

- What volume of 6.00 M NaOH is needed to prepare 250 mL of a 0.500 M NaOH solution?

    Let's first explain how this dilution is achieved. We will take a certain volume of the 6.00 M NaOH solution and transfer it into a 250-mL volumetric flask and then we will dilute to the mark with water. Let's determine the volume of that solution that we should transfer into the volumetric flask.

$$V_2 = \frac{M_1 V_1}{M_2}$$

$$\therefore V_2 = \frac{(0.500M)(250mL)}{6.00M}$$

So, $\qquad V_2 = 20.8 \, mL \quad \sqrt{}$

Notice that we did not convert the volume into liters, even though we were supposed to do so, because in this equation it is a ratio and the two molarities cancel out. The actual unit of molarity is the one that forces us to use Liters for the volume, since it is implicitly moles per liter!

- A certain phosphoric acid solution is 90.0% by mass and has a density of 1.50 g/mL. How many mL of this H_3PO_4 solution are needed to prepare 2.00 L of 0.100 M H_3PO_4 ?

When we read this question we realize that it is a dilution problem. We should solve for the initial volume, just like in the previous problem.

$$V_1 = \frac{M_2 V_2}{M_1}, \text{ but we do not know } M_1 \,!$$

We must solve for the initial molarity before we try to do this problem. How will we accomplish this? Let's think first of the definition of mass percent, which is the only thing we know about the initial solution. What is a 90.0% solution? It is a solution which has 90.0 g of phosphoric acid in every 100.0 g of solution.

However, we need units of moles of H_3PO_4 per liter of solution. Therefore, all we need to do is convert the numerator for moles and the denominator to liters.

$$[H_3PO_4] = \frac{90.0 g_{H_3PO_4}}{100.0 g_{solution}}$$

$$= 13.8 \, M \quad \sqrt{}$$

The calculation of the initial volume is done in exactly the same way as before, therefore:

$$[H_3PO_4]_{initial} = \frac{90.0 \text{ g } H_3PO_4}{100.0 \text{ g solution}} \times \frac{1 \text{ mol } H_3PO_4}{98.0 \text{ g } H_3PO_4} \times \frac{1.50 \text{ g solution}}{1.00 \text{ mL}} \times \frac{1 \text{ mL}}{10^{-3} \text{ L}}$$

$$V_1 = \frac{M_2 V_2}{M_1} = \frac{0.100 \, M \, (2.00 \, L)}{13.8 \, M} = 0.0145 \, L$$

$$\therefore V_1 = 14.5 \, mL \; \checkmark$$

3.3 Solution Stoichiometry.

This unit is simply one more example of stoichiometry, but some (or all) of the substances in the balanced reaction are in solution (indicated by the subscript **aq**). As we learned recently, the moles of these substances can be determined if we know the molarity and the volume (in Liters) for the solute. Remember that: moles of solute = Molarity x Volume (L).

- Consider the following reaction:

$$AlCl_{3(aq)} + 3 \, NaOH_{(aq)} \Rightarrow Al(OH)_{3(s)} + 3 \, NaCl_{(aq)}$$

If you combine 500 mL of a 0.100 M $AlCl_3$ solution with 2.00 L of a 0.100 M NaOH solution, what will be the concentration of all aqueous species after the reaction is over?

We should realize that we were given molarities and volumes for the two aqueous reagents first. Therefore, we will determine the moles[1] of each reagent and decide which of the two is the limiting reagent.

$$n_{AlCl_3} = M \times V = 0.100 \, mol/L \, (0.500 \, L) = 0.0500 \, mol \, AlCl_3.$$

$$n_{NaOH} = M \times V = 0.100 \, mol/L \, (2.00 \, L) = 0.200 \, mol \, NaOH.$$

We will actually determine the limiting reagent with the stoichiometric ratios. Therefore,

$$R_{AlCl_3} = 0.0500 \, / \, 1 = \mathbf{0.0500} \Leftarrow \text{Limiting Reagent!}$$

$$R_{NaOH} = 0.200 \, / \, 3 = 0.0667$$

Now, from these ratios, we can see that the $AlCl_3$ is the limiting reagent. Since the limiting reagent is 100% consumed, we can immediately conclude that the concentration of aluminum chloride will be *equal to zero* once the reaction is complete.

However, the concentration of NaCl must be determined from the limiting reagent.

---
[1] From now on, we will many times use the low case **n** to symbolize moles.

The moles of NaCl formed can be determined from the initial moles of $AlCl_3$, the limiting reagent.

$$n_{NaCl\ formed} = 0.0500\ mol\ AlCl_3 \times (3\ mol\ NaCl\ /\ 1\ mol\ AlCl_3) = 0.150\ mol\ NaCl$$

Since we are trying to determine the molarity of NaCl, we need to divide the moles of NaCl formed by the total volume of the final solution (2.50 L).

$$\therefore [NaCl] = mol\ /\ volume = 0.150\ mol\ NaCl\ /\ 2.50\ L = \mathit{0.0600\ M}.$$

We also want to know the concentration of NaOH after the reaction is over. Since that was not the limiting reagent, some of the NaOH must be left over. The limiting reagent will tell us how much of the other reagents will be consumed. Therefore, let's determine the moles of NaOH that were consumed in the reaction.

$$n_{NaOH\ consumed} = 0.0500\ mol\ AlCl_3 \times 3\ mol\ NaOH\ /\ 1\ mol\ AlCl_3 = 0.150\ mol\ NaOH.$$

However, what we want is not the moles consumed, but rather the moles of NaOH that are still present in solution after the reaction.

$$n_{NaOH\ left\ over} = 0.200\ mol - 0.150\ mol = 0.050\ mol\ NaOH$$

Finally, the concentration of NaOH in solution would be:

$$\therefore [NaOH] = 0.050\ mol\ /\ 2.50\ L = \mathit{0.020\ M}.$$

- Consider the following reaction:

$$H_3A_{(aq)} + 3\ NaOH_{(aq)} \Rightarrow 3\ H_2O_{(\ell)} + Na_3A_{(aq)}.$$

How many mL of a 0.1234 M solution of NaOH are needed in order to react completely with 25.00 mL of a 0.2222 M H_3A solution?

Once more, the first thing we notice is that we were given initially the volume and the molarity of the H_3A solution. We know that this information is sufficient to determine the moles of H_3A present in the solution. These are reacted completely with the NaOH. The only thing we know about the reaction between these two substances is that they do so in a 1 : 3 mole ratio. Therefore, we will start by determining the moles of H_3A with the definition of molarity and then determine the moles of NaOH using this one to three mole ratio in which they reacted. Once we learn how many moles of NaOH must have been added to react with the acid, we will use the definition of molarity once again in order to determine the volume of the NaOH that we must have used.

$$\frac{0.2222 mol_{H_3A}}{L} \times 0.02500 L \times \frac{3 mol_{NaOH}}{1 mol_{H_3A}} = 0.01667 mol_{NaOH}$$

So,

$$[NaOH] = \frac{mol_{NaOH}}{V_{solution}}$$

and finally,

$$0.01667 mol_{NaOH} \times \frac{L}{0.1234 mol_{NaOH}}$$

$$= 0.1350 \text{ L}$$

135.0 mL

Chapter 4: Gases.

4.1 The Early Laws.

We already know that matter exists in three distinct physical phases: gas, liquid, and solid. The gaseous state is by far the simplest to study of the three phases, both experimentally and theoretically. We know that air[1] is a mixture of gases, consisting mainly of nitrogen (N_2) and oxygen (O_2). Most of these gases can be found close to the sea surface, as it has been estimated that 50% of the atmosphere is within 4 miles of sea level, whereas 90% is within 10 miles. This means that as we move up in the atmosphere, it gets thinner and thinner. In 1643, Evangelista Torricelli[2] showed experimentally that the air in the atmosphere exerted **pressure** (P). As we see it today, gases consist of molecules that travel in fast, random motion, colliding with each other and with the walls of the container. It is these collisions with the walls of the container that generate what we call pressure.

Before we proceed any further, we should discuss which units we will be using for pressure. Pressure is defined as force per unit area. Since the units for force are Newtons (N), then:

$$P = \text{Force/Area}$$

$$= N/m^2 \equiv \underline{\text{pascal}} \text{ (Pa)} \qquad \Leftarrow \text{SI unit}$$

Thus, the SI unit for pressure will be the Pascal. Since technically a Newton is a: $kg \, m / s^2$, then the Pascal is a: $kg / m \, s^2$.

<u>Barometer</u> – An instrument that uses the height of a mercury column to measure the pressure of the atmosphere (developed by Evangelista Torricelli). He observed that after filling a tube (that was closed at one end) with mercury and then inverting it into a dish of mercury, the mercury would start to come out (due to gravity) but not completely. Actually, he observed that a column of about 760 mm of Hg would stay in the tube. He attributed this to the pressure exerted by the atmosphere, as seen in the figure below.

---

[1] The four main components of air are: Nitrogen (78.11 %), Oxygen (20.95%), Argon (0.93%), and carbon dioxide (0.034%). Other minor components are: Neon, Helium, methane, Krypton, hydrogen, dinitrogen oxide (laughing gas), and Xenon. These are percentages by volume, not mass.

[2] Italian physicist who lived from 1608 until 1647.

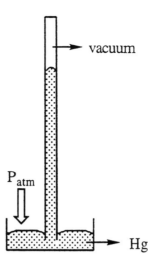

Under normal conditions (sunny day) the atmospheric pressure (the pressure of the atmosphere at sea level) at 0°C corresponds to 760 mm Hg. Therefore,

$$1 \text{ atm} = 760 \text{ mm Hg} = 760 \text{ torr}$$

Notice that the unit torr (corresponding to one mm of mercury only at 0°C) was named in honor of Torricelli. The unit atmosphere, although not technically SI units, will be the one we use most during this chapter. For reference purposes, $1 \text{ atm} = 101.325 \text{ kPa}$.[1]

- The pressure of a certain gas is measure to be 49.5 torr. What is this pressure in atm and Pa?

    These are simple conversion factors, so it will suffice to show how they are calculated below.

$$49.5 torr \times \frac{1 atm}{760 torr} = 0.0651 atm$$

In terms of Pascals, we get:

$$49.5 torr \times \frac{1 atm}{760 torr} \times \frac{101.325 kPa}{1 atm} = 6.60 kPa$$

We should mention at this point, that the pressure of a sample of a gas trapped in a flask or a cylinder could be measured with a simple manometer. This is an instrument that has a U-shaped tube with mercury in it. It is open to the air at one end. The pressure exerted on this end of the mercury is the atmospheric pressure. The other end of the mercury column is in contact

---

[1] For your interest: 1 atm = 14.7 psi (pounds per square inch).

with the trapped gas. If the two sides are equally high, then the gas sample has the same pressure as the atmosphere. On the other hand, if the pressure of the gas is larger than that of the atmosphere, we will observe the case seen in the diagram below, where the gas will push the mercury column away from it, thus creating a height imbalance (Δh). This height difference can be related to the pressure of the trapped gas.

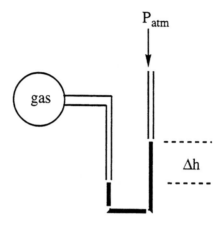

As we mentioned in the first chapter, the first quantitative experiments were performed by the Irish chemist, Robert Boyle, when in 1662 he studied the relationship between the pressure of a gas and its volume. At the time, he maintained the temperature constant (or worked at constant temperature, normally 0°C, since that could be kept using an ice-water bath). He found that the pressure and the volume of a given gas are inversely proportional.

$$V \alpha 1/P$$

$$\therefore V = k/P$$

or: $\underline{PV = k}$ √   (at constant T) ⇒ Boyle's Law

which means that: $P_1V_1 = P_2V_2$.

- An inflated balloon has a volume of 0.500 L at sea level and is allowed to rise to a height of 6.5 km, where the pressure is about 0.400 atm. Assuming that the temperature remains constant (not true at all), what is the final volume of the balloon?

    We recognize that this is Boyle's Law, since the only variables are pressure and volume. Therefore,

$$P_1V_1 = k \qquad \text{and} \qquad P_2V_2 = k$$

and $\therefore P_1V_1 = P_2V_2$.

Therefore, V_2 = (1.00 atm)(0.500 L)/ (0.400 atm) = *1.25 L* √

We should make it clear that there is no need to actually calculate **k** in these Boyle's problems. The reason is simple: **k** depends on the temperature, among other things. Therefore, it is NOT a universal constant and you will not find it in any book. The easiest way to work any problem like this is, as we will see in the next few pages, is to derive an equation for each case: one that deals only with the variables in question as they change from the initial to the final conditions. In the case of Boyle's law it was: $P_1V_1 = P_2V_2$.

Before we look at another of the early laws, we should mention something about temperature scales. The two widely used temperature scales in the world today are the Celsius scale (°C) and the Fahrenheit scale (°F). The Fahrenheit scale is based on the melting point (32°F) and the boiling point (212°F) of water. There are 180°F between these two temperatures. The Celsius scale attempts to make it "metric" by dividing those two same temperature extremes into 100 degrees, calling the freezing point 0°C and the boiling point 100°C. Of course, that makes the size of a degree Fahrenheit smaller (it is only 5/9 of a Celsius) than the size of a degree Celsius. These are the conversion factors from one temperature scale to the other.

$$T_{°C} = (T_{°F} - 32)(\frac{5}{9})$$

$$T_{°F} = [(\frac{9}{5}) \times T_{°C}] + 32$$

- At the time this page was being typed, the temperature was 89°F, a rather warm summer day. If we would have been in most any other country, the temperature would have been reported in °C. Determine that temperature in °C.

    This is a simple conversion factor. Therefore,

    T (°C) = (89 – 32)(5/9) = *32°C*   √

In the century that followed, scientists started to study the properties of gases. Jacques Charles[1] (1746-1823) was the first person to fill a balloon with hydrogen gas and make a solo balloon flight. He found in 1787 that the volume of a gas at constant pressure increases linearly with the temperature of the gas. This linearity does not make it directly proportional, as can be

---

[1] Jacques Charles was a French physicist who lived from 1746 until 1823.

seen below.

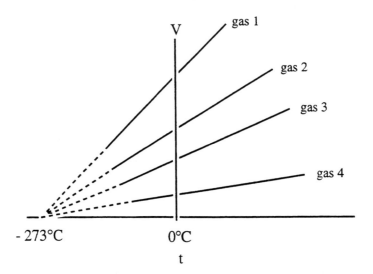

However, he did notice that all gases at a given pressure extrapolate to the same point: at a Volume of zero, the temperature was -273.15°C. Later on, Lord **Kelvin** (1848) identified this temperature as theoretically the lowest attainable temperature, called it the *absolute zero,* thus setting an **absolute temperature scale.** This temperature has been approximated in the laboratory today (to about 0.00001 K) but has not been reached. Since gases cannot have negative volumes, we can understand why the temperature (absolute) cannot be negative.

This absolute temperature scale of Kelvin is very similar to the Celsius scale. The size of the degrees is identical but the scales are shifted by 273.15°C:

$$\therefore T(K) = t(°C) + 273.15°$$

Charles found that the volume is directly proportional to the absolute temperature.

$$V \propto T \quad \text{(not t)}$$
$$\therefore V = k'T$$
$$\text{or} \quad \underline{V/T = k'} \quad \sqrt{} \quad \Leftarrow \text{Charles' Law (at constant P)}$$

Once more, the law can and should be written as follows, since there is no need to actually calculate the numerical value for k'.

$$\frac{V_1}{T_1} = \frac{V_2}{T_2}$$

This law applies as long as the pressure is maintained constant.

Chapter 4: Gases

- A sample of a gas has a volume of 79.5 mL at 45°C. What volume will it occupy at 0°C if the pressure is held constant?

    Let's start this problem by collecting the information given. We cannot forget that the only way that Charles' law applies is if the temperature is in Kelvin. Therefore, as we collect the data, we will also convert it into the appropriate units: V_1 = 79.5 mL, T_1 = 45° + 273 = 318 K, and T_2 = 273 K.

$$\text{but: } \frac{V_1}{T_1} = \frac{V_2}{T_2} \implies \therefore V_2 = \frac{V_1 T_2}{T_1}$$

$$V_2 = \frac{19.5 \text{ mL } (273 \text{ K})}{318 \text{ K}} = \underline{\underline{68.2 \text{ mL}}} \ \sqrt{}$$

- A certain sample of a gas has a volume of 525 mL at an unknown temperature. When the volume of that same sample is changed to 800 mL, its temperature is observed to be 27°C. Determine the initial temperature of this sample, if the pressure was maintained constant throughout the experiment.

    Once more, we need to use Charles' law. Gathering the information, we have: V_1 = 525 mL, V_2 = 800 mL, and T_2 = 300 K. Therefore,

$$T_1 = \frac{V_1 T_2}{V_2} = \frac{(525 mL)(300 K)}{800 mL}$$

$$= 197 K \quad \sqrt{}$$

In 1811, Amadeo Avogadro postulated that equal volumes of gases at the same temperature and pressure contained the same number of molecules. Thus, the volume is directly proportional to the number of particles of the number of moles (n). This is what we call Avogadro's Law.

$$\therefore V \alpha \ n$$

$$\text{or } V = k" n$$

$$\text{so, } \underline{V/n = k"} \quad \sqrt{} \quad \Leftarrow \text{ Avogadro's Law (constant T \& P)}$$

and once more, in order to get rid of the constant (k"), we can write the following equation:

$$\frac{V_1}{n_1} = \frac{V_2}{n_2}$$

This equation works only if the pressure and the temperature are kept constant.

4.2 Ideal Gas Law.

Let's combine all the laws we have seen so far. Mathematically, if A is proportional to B and also proportional to C, then A is proportional to the product of A and B. In the previous section, we saw that the volume is related to the following variables:

$$V \propto 1/P$$

$$V \propto T$$

$$V \propto n$$

and combining all three, we get: $\quad V \propto (1/P)(T)(n)$

To get rid of the proportionality constant, we need to multiply by a constant. In this case, we will symbolize this constant with the capital letter R. This yields the following equation:

$$V = (1/P)\, T\, n\, R$$

Rearranging the equation in a form that we probably are more familiar with, we get:

PV = nRT √ $\qquad$ equation of state for an *ideal gas*.

This is the equation that is referred to as the ideal gas law. Why do we use the term ideal gas? What is the meaning of an ideal gas? We could define it in various ways, but for the time being, it will suffice to say that an ideal gas is one that obeys that equation.

Of course, there are no ideal gases. However, real gases approach ideal behavior under certain conditions, which are mainly *high temperature* (like room temperature) and *low pressure* (like the normal atmospheric pressure). These were the conditions under which most gases were studied in the early days, so apparently (or to the experimental precision of those days) the real gases they worked with (whether it was oxygen, nitrogen, carbon dioxide or air in general) obeyed this equation perfectly.

There is still one thing we ignore. What is this constant **R**? Is it a constant just like the Boyle's law or Charles' law constants? Do we really need to know its value? The answer is yes, we need to know its value. R is a universal constant because the equation we are referring to is an equation of state, that is, one that describes all the variables for a given substance. In other words, if we know all the parameters for that equation, we know everything there is to know about this substance.

In honor of the early pioneers that worked with gases (Boyle, Charles, Gay-Lussac, Avogadro, Amonton, and many others), we will use a very special set of standards for temperature and pressure. We will refer to it as **STP** (which stands for Standard **T**emperature and **P**ressure). Remember that the only "easy" way to establish a constant temperature in those days was using an ice/water bath (which maintains the temperature constant at 0°C) and the pressure could easily be kept constant as long as it was equal to the atmospheric pressure). Therefore, the term STP refers to the following parameters:

$$T = 0°C = 273.15 \, K$$

$$P = 1.000 \, atm$$

Under these conditions most gases behave ideally (that is a fairly high temperature and low pressure). It is also true that under these conditions, one mole of any gas occupies 22.414 L. That is,

$$n = 1.000 \, mol$$

$$V = 22.4 \, L \quad \} \; \textbf{ONLY AT STP!}$$

Let's utilize these conditions in order to determine the numerical value for gas constant, R[1].

$$R = PV/nT = (1.000 \text{ atm})(22.414 \text{ L})/(1.000 \text{ mol})(273.15 \text{ K})$$

$$\therefore \; \boxed{R = 0.08206 \text{ L atm/K mol}} \; \surd$$

- What is the temperature of a 0.100 mol sample of an ideal gas (IG) if it occupies a volume of 2500 mL and it exerts a pressure of 550 torr?

    In order to solve this problem, we should realize that this is a sample of an IG, under a very specific set of conditions. If it is an IG, then it must obey the ideal gas law: PV =

---

[1] Please, notice that R is a constant but every time you use the numerical value of 0.0821, you must express the pressure in atmospheres, the volume in liters, and the temperature in Kelvin. R could also be given (or calculated) in other units (like 8.314 L kPa/K mol), but if we were to use another value for R, then we would have to stay with the units specified by its value.

nRT. Remember that using this equation forces us to use a specific set of units.

$$T = \frac{PV}{nR} = \frac{(\frac{550 torr}{760 torr/atm})(2.500 L)}{(0.100 mol)(0.0821 \frac{L atm}{K mol})}$$

$$= 220 \ K \quad \sqrt{}$$

- A gas filled balloon, having a volume of 250 mL at 912 torr and 25°C, is allowed to rise to a point where the temperature and pressure are –23°C and 2.28 torr, respectively. Calculate the final volume of the balloon under these new conditions.

    The first thing we notice is that this problem differs from the previous one in a major way: this is not a single state, but rather a change in state. That is, we start under one set of conditions and end up in a new set of conditions. We could do this problem in a round about way by using the ideal gas law twice. With the initial conditions, we can determine the moles of the gas (the only unknown would be the initial moles). Once you know the moles of the gas, then we could determine the final volume (since the number of moles is not going to change unless there is a leak).

    However, there is a much easier (shorter) way to do this problem. We should derive our own equation for this particular problem, one that only leaves us with the variables for the problem and gets rid of the ones that stay constant. There are two things that remain constant throughout the problem: the moles (n) and the universal gas constant (R). We did something very similar when we were deriving the equations for Boyle, Charles and Avogadro. In this case, we see that:

$$\mathbf{P\ V} = n\ R\ \mathbf{T}$$

Rearranging all the variables to the left, we get:

$$\frac{PV}{T} = nR = \text{constants}$$

We set the initial conditions equal to the final conditions, and get the following expression:

$$\frac{P_1 V_1}{T_1} = \frac{P_2 V_2}{T_2}$$

$$\therefore V_2 = \frac{P_1 V_1 T_2}{P_2 T_1} = \frac{(912 torr)(250 mL)(250 K)}{(2.28 torr)(298 K)}$$

$$= 83{,}900 \text{ mL} = \textit{83.9 L} \quad \checkmark$$

Notice that we were not forced to use liters, or atmospheres in the equation we derived, because we were not using the numerical value for R. It is this value with its units that forces those units. However, no matter what we try, we must always substitute the temperature in Kelvin, because that is the only time that the volume is proportional to the temperature (as shown previously when we talked about Charles' Law).

- An ideal gas at a given temperature is placed in a cylinder with a movable piston. The initial pressure of this gas is 0.800 atm, when the height of the cylinder was 10.0 inches from the base. What will be the height (in inches) of the piston when the pressure of the gas is recorder as 1.35 atm? The volume is given by $\pi r^2 h$, where r = radius of base and h = height.

This time, we do not even have the option of using PV = nRT, since we are not told the temperature, the moles of the gas, nor the radius of the base (of the cylinder). Therefore, we are forced to derive our own equation in order to solve the problem. Once more, let's start with the ideal gas law and see which are the variables, so that we can separate them on one side of the equation. From the problem, we see that there are only two variables, the pressure and the height of the piston.

$$\mathbf{P} (\pi r^2 \mathbf{h}) = n R T$$

Of course, if we leave the pressure and the height on the left, we can see that the pressure times the height is equal to a lot of constants. Therefore, we can set the initial set of these variables equal to the final set, thus generating our own equation.

$$P \ h = \text{constants.}$$

Or: $\quad P_1 \ h_1 = P_2 \ h_2.$

$$h_2 = \frac{(0.800 atm)(10.0 in)}{1.35 atm}$$

$$\therefore h_2 = \textit{5.93 inches} \quad \checkmark$$

Once more, we were not forced to use SI units simply because the equation we derived did not contain R.

- A bubble of an ideal gas is created under the surface of a pond. The gas is originally in the shape of a sphere of radius 3.21 x 10⁻² m, under a pressure of 5.00 atm, and a temperature of 10°C. As it rises to the surface, it will warm up. Right below the surface its pressure is 1.00 atm and its temperature is 22°C. Determine the new radius for the bubble near the surface.

    Since we are dealing with a change of state, we must derive our own equation. In order to do this, we should start with the ideal gas law and look for the variables. Remember that the volume of a sphere is given by $(4/3)\pi r^3$. Therefore,

$$P(4/3 \pi r^3) = nRT$$

This means that:

$$\frac{Pr^3}{T} = \text{constants}$$

Now, we can set the initial set for these variables equal to the final set, or:

$$\frac{P_1 r_1^3}{T_1} = \frac{P_2 r_2^3}{T_2}$$

And therefore, the final radius is calculated as:

$$r_2 = \sqrt[3]{\frac{P_1 r_1^3 T_2}{T_1 P_2}} = \sqrt[3]{\frac{(5.00 atm)(3.21 \times 10^{-2} m)^3 (295 K)}{(283 K)(1.00 atm)}}$$

$$r_2 = 0.0557 m \quad \sqrt{}$$

4.3 Density of Gases.

We should expect the density of a gas to be a lot smaller than for liquids or solids. This is exactly true. After all, in the initial concept we have of gases, the molecules are far apart (as compared to the condensed phases) and traveling at a high speed. Since we have a lot less molecules per unit volume, we expect the density to be smaller. One of the main properties of a gas that we can obtain from its measured density is its molecular weight. Let's derive an equation that relates the density of a gas to its molecular weight. Since we are dealing with ideal gases, we should start with the ideal gas law.

$$PV = nRT$$

Chapter 4: Gases

But moles (n) is simply the mass divided by the molecular weight (MW), therefore:

$$PV = \frac{mRT}{MW}$$

However, what is density? Density is mass per unit volume (that is, m/V), so:

$$density = d = \frac{P(MW)}{RT}$$

Notice that the density equation is directly proportional to the molecular weight, therefore, if we were to compare different gases under identical conditions (same temperature and pressure), the gas with the highest molecular weight would be the most dense.

- Which gas is more dense at STP, CO_2 or CH_4?

    Under these conditions (273 K and 1 atm), the gas with the highest molecular weight is the most dense. Since the molecular weight of carbon dioxide (44.01) is larger than that of methane gas (16.02), then CO_2 is the denser gas.

- Determine the density of ammonia (NH_3) gas in g/L at 500 torr and 37°C.

    We need to know the molecular weight of ammonia and its pressure (in atm) and temperature (in K). The reason these must be in those units is because the equation uses R.

$$d = \frac{P(MW)}{RT}$$

$$d = \frac{(500 torr \times \frac{1 atm}{760 torr})(17.02 g/mol)}{(0.0821 \frac{L atm}{K mol})(310 K)}$$

$$= 0.440 \text{ g/L} \quad \sqrt{}$$

- Determine the temperature of a sample of oxygen gas that has a pressure of 654 torr and a density of 0.876 g/L.

    Oxygen gas is diatomic, which means that its molecular weight is 32.0 g/mol. Since its pressure is also given, the determination of the temperature should be trivial at this point.

$$T = \frac{P(MW)}{R(d)} = \frac{(654 torr \times \frac{1 atm}{760 torr})(32.0 g/mol)}{(0.0821 \frac{L atm}{K mol})(0.876 g/L)}$$

$$\therefore T = 383 \ K \quad \sqrt{}$$

4.4 <u>Gas Stoichiometry</u>.

In this section, we will look over the same type of problems we have been dealing with for a while now, except we will concentrate on problems that have gases, rather than solids or substances in solution. This means that we will need to determine the moles of a gaseous substance with the ideal gas law. In other words, the moles of a gas can always be determined (as long as the gas is ideal) with the following equation.

$$n = \frac{PV}{RT}$$

It bears repeating that up to this point we have learned three different ways to calculate the number of moles of a substance, mainly:

1.   n = g / MW   if grams are given.

2.   n = M x V   if it is a solution.

3.   n = P V / R T   if it is a gaseous substance.

In the following examples, we will see that in order to solve the problems, we will have to use any one of these (sometimes all three) in order to obtain the moles of the reagents so that we can then determine the limiting reagent.

- How many moles of hydrogen gas do we have present in a 10.0 L sample that is kept at STP?

To solve this problem, we could use the ideal gas law as we stated above. However, since the sample is kept at STP, we can hopefully remember that under these conditions, one mole of any gas (as long as it is ideal) will occupy 22.414 L. Therefore,

$$10.0 L_{H_2} \times \frac{1 mol_{H_2}}{22.4 L_{H_2}}$$

$$= 0.446 \ mol \ H_2. \quad \sqrt{}$$

However, if the conditions would not have been STP and they would have been anything else, let's say as an example that it would have been 1.50 atm and 27°C, then:

$$n = \frac{(1.50 atm)(10.0 L)}{(0.0821 \frac{Latm}{Kmol})(300 K)}$$

∴ n = 0.609 mol H_2 gas. √

- How many liters of carbon monoxide gas (CO), measured at STP, are needed to react completely with 1.00 kg of ferric oxide (Fe_2O_3)?

$$Fe_2O_{3(s)} + 3\ CO_{(g)} \Rightarrow 2\ Fe_{(s)} + 3\ CO_{2(g)}$$

From the balanced reaction it is clear that for every mol of Fe_2O_3 that is consumed, we will form three moles of CO gas, which under conditions of STP (1 atm pressure and 273 K) will occupy 22.4 L per mole of gas formed. Therefore,

$$1000 g_{Fe_2O_3} \times \frac{1 mol_{Fe_2O_3}}{159.7 g_{Fe_2O_3}} \times \frac{3 mol_{CO}}{1 mol_{Fe_2O_3}} \times \frac{22.4 L}{1 mol_{CO}}$$

= 421 L of gas √

- How many L of Cl_2 gas can be obtained at 40°C and 787 torr from 10.0 g of $KMnO_4$ and 100 mL of 1.00 M HCl?

$$2\ KMnO_{4(s)} + 16\ HCl_{(aq)} \Rightarrow 8\ H_2O_{(\ell)} + 5\ Cl_{2(g)} + 2\ MnCl_{2(aq)} + 2\ KCl_{(aq)}$$

Let's calculate the moles of each reagent first. One of the reagents is a solid, whereas the other one is a solution. Afterwards, let's determine the stoichiometric ratios.

n_{KMnO4} = (10.0 g / 158.04 g/mol) = 0.0633 mol $KMnO_4$.

n_{HCl} = 0.100 L (1.00 mol/L) = 0.100 mol HCl.

$$R_{KMnO_4} = \frac{0.0633}{2} = 0.0316$$

$$R_{HCl} = \frac{0.100}{16} = 0.00625$$

Since the stoichiometric ratio for HCl is smaller than that for $KMnO_4$, the HCl must be the limiting reagent.

Chapter 4: Gases

$$0.100 \text{ mol HCl} \times \frac{5 \text{ mol Cl}_2}{16 \text{ mol HCl}} = 0.0313 \text{ mol Cl}_2 \text{ formed.}$$

Using the ideal gas law, we get:

$$V = \frac{0.0313 \text{ mol } (0.0821 \text{ Latm/Kmol}) (313 \text{ K})}{1.036 \text{ atm}}$$

$$= 0.776 \text{ L} \quad \checkmark$$

- A 0.400 g sample of $NaN_{3(s)}$ is heated and decomposes according to:

$$2 \text{ NaN}_{3(s)} \Rightarrow 2 \text{ Na}_{(s)} + 3 \text{ N}_{2(g)}$$

What volume of nitrogen gas, measured at 25°C and 0.980 atm, is obtained?

The experimentally determined number (0.400 g) is a mass of a pure substance. Therefore, we should change it to moles first. Then,

$$0.400 g_{NaN_3} \times \frac{1 mol_{NaN_3}}{64.99 g_{NaN_3}} \times \frac{3 mol_{N_2}}{2 mol_{NaN_3}}$$

$$= 0.00923 \text{ mol N}_2 \text{ gas.}$$

The reason we stopped at moles is that we are not working at STP. If we were, we could have used the conversion factor that states that one mole of any gas occupies 22.4 L, but since the conditions are not at STP, we must stop at moles and then use $PV = nRT$. Thus,

$$V = \frac{0.00923 mol_{N_2} (0.0821 \frac{Latm}{Kmol})(298K)}{0.980 atm}$$

$$= 0.230 \text{ L} \quad \checkmark$$

- A 2.80 L sample of CH_4 gas (measured at 25°C and 1.65 atm) was mixed with an excess of oxygen gas and then ignited. Determine the volume of CO_2 gas formed, if it is measured at a pressure of 2.50 atm and 125°C.

The first thing we need is the balanced chemical equation. Therefore,

$$CH_{4(g)} + 2 O_{2(g)} \Rightarrow CO_{2(g)} + 2 H_2O_{(\ell)}.$$

Since we are told that there is an excess of oxygen gas, we already know that the limiting reagent is the methane gas.

$$n = \frac{(1.65 atm)(2.80 L)}{(0.0821 \frac{Latm}{Kmol})(298K)} = 0.189 mol_{CH_4}$$

Once we know the moles of methane gas, we can proceed to determine the moles of carbon dioxide gas formed, using the coefficients from the balanced reaction. Since in this particular example it happens to be one to one, then the initial moles of methane gas will be converted into an exact amount of carbon dioxide moles. Therefore,

$$\text{moles of } CO_2 = 0.189 \text{ mol} \quad \sqrt{}$$

Finally, to determine the volume occupied by these 0.189 moles of CO_2 gas, we only need to check the conditions at which we are measuring that volume (2.50 atm and 125°C). Therefore,

$$V = \frac{nRT}{P}$$

$$V = \frac{(0.189 mol)(0.0821 \frac{Latm}{Kmol})(398K)}{2.50 atm}$$

$$\therefore \quad V = 2.47 L \quad \sqrt{}$$

4.5 Partial Pressures.

Back on Chapter 1, we mentioned that John Dalton proposed the atomic theory. Some of the experiments that led to this proposal were his studies on mixtures of gases. He observed that the total pressure exerted by a mixture of gases was the same as the sum of the pressure that each gas would exert if it were alone in the container. Basically, this is the statement we know as Dalton's Law of Partial Pressures[1]. If we had three different gases in a mixture, then the total pressure of these gases would be given by:

$$P_{total} = P_a + P_b + P_c.$$

If we consider these gases to be ideal (normally a great assumption), then

---

[1] This statement was postulated in 1803.

$$P_{tot} = n_a RT/V_{tot} + n_b RT/V_{tot} + n_c RT/V_{tot}$$

$$= (n_a + n_b + n_c) RT/V_{tot}$$

$$P_{tot} = n_{tot} RT/V_{tot} \quad \surd$$

Let's think a little bit about what we have said so far. Each gas occupies the entire volume of the container. According to our early description of a gas, this should make sense. A gas consists of fast moving molecules that collide with one another and with the walls of the container. This means that our gas molecules have no restrictions as to where they will go and will travel within the boundaries of your container. This law means that even in a mixture, the molecules of each gas will behave as if they were alone in the container. They will still collide with one another whenever they come in contact, but their collisions will not bring forth any changes. This implies that the collisions will be elastic (the total energy will be conserved). Since the final equation tells us that the total pressure depends on the <u>total number of moles of gases</u> in the mixture, that also implies that the pressure is independent on the actual composition of these substances. In other words, it does not matter whether we have 2 moles of oxygen gas, or instead a mixture of one mole of helium gas and one mole of dimethylether gas (CH_3-O-CH_3), the pressure of these two samples will be exactly the same as long as they are kept in the same container and at the same pressure. The implication of this statement is profound. This means that ideal gases cannot have any kind of interaction between its molecules (attractions nor repulsions). Let try to explain what we mean by that. If a given gas molecule is going to collide with the wall to exert pressure, it is clear that the force with which it will collide will depend on whether this molecule is being held back by an attraction to another molecule or not. Therefore, since we do not observe any difference when we mix the different gases, this means that all gases are exactly the same (a condition that is true for ideal gases). *Conclusion*: In a mixture of gases, the only thing that matters is the total number of gaseous particles present. Actually, there are two implications from this law: 1) that the volume of the individual gas molecules must be unimportant and 2) that the molecules of gases must not exert forces among one another. These two statements are directly related to the definition of an ideal gas!

Since all that matters is how many particles we have of a specific gas, it would be nice to talk about a concentration unit that deals with relative number of particles present in a mixture. We call this the mole fraction and will symbolize it with a capital X.

Mole Fraction (X).

Mole fraction, as implied by its name, is the fraction from the total number of moles in a mixture that belong to one given component (i).

$$X_i = n_i / n_{tot}$$

If we only have two gases in a mixture, (let's call them **a** and **b**), then:

$$X_a = n_a / (n_a + n_b) = n_a / n_{tot}$$

and:

$$X_b = 1 - X_a = n_b / n_{tot}$$

Let's substitute for moles of **a** the ideal gas law ($P_a V/RT$) and for the total moles ($P_{tot} V/RT$) in the expression for mole fraction.

$$X_a = \frac{P_a \, (\cancel{RT/V})}{P_{tot} \, (\cancel{RT/V})} = \frac{P_a}{P_{tot}}$$

We can now express Dalton's law in the following form:

$$P_a = X_a \cdot P_{tot} \quad \sqrt{}$$

Based on what we said previously, this should make a lot of sense. Since the total pressure depends only on the total number of moles of gases present and since the partial pressure of a given gas depends on the relative number of moles of that gas present in the container, the product of the fraction of gas **a** molecules to the total molecules in the container times the total pressure should be equal to the pressure of that particular gas (**a**).

- A mixture of 40.0 g of O_2 gas and 40.0 g of He gas exerts a total pressure of 0.900 atm. What is the partial pressure of O_2 in the mixture?

    There is very little information given: the grams of the two gases (from where we can determine the moles of each) and the total pressure of the mixture. We do not know the temperature nor the volume. Therefore, we must solve it with the following equation:

$$P_{O2} = X_{O2} P_{tot}.$$

Therefore, we need the mole fraction of oxygen gas. This means that we should determine the moles of each component first.

$$n_{O_2} = \frac{40.0 \text{ g}}{32.0 \text{ g/mol}} = 1.25 \text{ mol } O_2 \qquad n_{He} = \frac{40.0 \text{ g}}{4.00 \text{ g/mol}} = 10.0 \text{ mol He}$$

$$\therefore X_{O_2} = \frac{1.25 \text{ mol}}{1.25 \text{ mol} + 10.0 \text{ mol}} = 0.112$$

Finally, $P_{O_2} = 0.112\,(0.900 \text{ atm}) = \mathit{0.101 \text{ atm}}$ √

- How would you calculate the pressure of He in that same flask? Suggest two different ways to do it (once you know the pressure of oxygen gas).

⇒ One of the ways would be to go through the entire process once more and determine the pressure of Helium gas. That is, determine the mole fraction of He in the mixture first and then determine the partial pressure of He (using Dalton's Law of Partial Pressures).

⇒ Another, much simpler way, would be to use the initial statement of Dalton's Law of Partial Pressures, which states that the total pressure is the sum of the partial pressures. Since we know the total pressure and now we know the pressure of oxygen gas, we can determine the pressure of He gas simply by difference.

- To an 11.0 g sample of CO_2 gas, we add 10.5 g of an unknown homonuclear diatomic gas[1], A_2. The original CO_2 sample had a pressure of 2.34 atm and the final mixture has a pressure of 5.85 atm. Determine the identity of the element **A**.

In order to identify an element or a compound, we need to know its molar mass (molecular weight). We are given the partial pressure of the carbon dioxide gas. Remember that the initial pressure of the carbon dioxide will not change in the presence of the new gas. Therefore, in the final mixture, the partial pressure of carbon dioxide is 2.34 atm. However, the partial pressure of carbon dioxide in that final mixture is given by Dalton's Law:

$$P_{CO2} = X_{CO2}\,P_{tot.}$$

So, $2.34 \text{ atm} = X_{CO2}\,(5.85 \text{ atm})$

Therefore, $X_{CO2} = 0.400$

The mole fraction of carbon dioxide is given by the moles of carbon dioxide divided by the total moles of gases present in the mixture. Therefore,

---

[1] The term homonuclear diatomic means that both atoms are of the same element.

$$X_{CO_2} = \frac{n_{CO_2}}{n_{CO_2} + n_{A_2}}$$

$$= \frac{\frac{g_{CO_2}}{MW_{CO_2}}}{\frac{g_{CO_2}}{MW_{CO_2}} + \frac{g_{A_2}}{MW_{A_2}}}$$

Finally, $MW_{A2} = 28.0$ g/mol

The molecular weight of the diatomic molecule is 28.0 g/mol, therefore, the atomic weight of **A** must be half of that number (14.0 g/mol), which identifies the unknown element as **nitrogen (N)**.

- Consider the following reaction which is carried out at a constant temperature of 400 K.

$$2\ PH_{3(g)} + 4\ O_{2(g)} \Rightarrow P_2O_{5(s)} + 3\ H_2O_{(g)}$$

The reaction is carried out in the apparatus shown below. It consists of a 12.5 L bulb (A) which is filled to a pressure of 2.00 atm with PH_3 gas and which is separated from a 10.0 L bulb (B) filled with O_2 gas at a pressure of 3.00 atm. When stopcock C is opened, the gases mix spontaneously and immediately react.

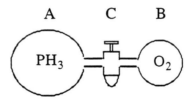

a) Determine the grams of P_2O_5 formed in this reaction.

As always in this type of problem, let's determine the number of moles of all substances given initially (PH_3 and oxygen gas).

$$n_{PH_3} = \frac{P_{PH_3} V}{RT} = \frac{2.00\ \text{atm}\ (12.5\ \text{L})}{(0.0821\ \text{Latm/Kmol})\ (400\ \text{K})} = 0.761\ \text{mol}\ PH_3$$

$$n_{O_2} = \frac{3.00\ \text{atm}\ (10.0\ \text{L})}{(0.0821\ \text{Latm/Kmol})\ (400\ \text{K})} = 0.914\ \text{mol}\ O_2$$

The stoichiometric ratios will tell us which of the two is the limiting reagent.

$$R_{PH_3} = \frac{0.761}{2} = \underline{\underline{0.381}} \quad \text{and} \quad R_{O_2} = \frac{0.914}{4} = \underline{\underline{0.229}} \Bigg\} \therefore O_2 \text{ is LR!}$$

Now that we know that oxygen is the limiting reagent, we can determine how many grams of the product will be formed.

$$0.914 \text{ mol } O_2 \times \frac{1 \text{ mol } P_2O_5}{4 \text{ mol } O_2} \times \frac{142.0 \text{ g } P_2O_5}{1 \text{ mol } P_2O_5}$$

$$= 32.4 \text{ g of } P_2O_5. \quad \surd$$

b) What is the *total* pressure in the system after the reaction is over?

This will depend on the total number of moles of gases present in the flask after the reaction is over.

$$P_{tot} = \underbrace{P_{PH_3}}_{\text{in excess}} + \underbrace{P_{H_2O}}_{\text{formed}} + \underbrace{\cancel{P_{O_2}}^{0}}_{\text{none! (LR)}}$$

Using the limiting reagent, we can determine the moles of phosphine reacted and the moles of water formed.

$$n_{PH_3}^{\text{reacted}} = 0.914 \text{ mol } O_2 \times \frac{2 \text{ mol } PH_3}{4 \text{ mol } O_2} = 0.457 \text{ mol } PH_3 \text{ reacted.}$$

$$\therefore n_{PH_3}^{\text{excess}} = 0.761 \text{ mol} - 0.457 \text{ mol} = 0.304 \text{ mol } PH_3$$

$$n_{H_2O}^{\text{formed}} = 0.914 \text{ mol } O_2 \times \frac{3 \text{ mol } H_2O}{4 \text{ mol } O_2} = 0.686 \text{ mol } H_2O \text{ formed.}$$

The total pressure is then calculated with the total number of moles of gases present and the total volume in which they are found (22.5 L).

$$\therefore P_{tot} = \frac{(0.304 \text{ mol} + 0.686 \text{ mol})(0.0821 \text{ Latm/Kmol})(400 \text{ K})}{(12.5 \text{ L} + 10.0 \text{ L})}$$

$$= 1.45 \text{ atm} \quad \surd$$

- Consider the combustion of butane gas, C_4H_{10}, MW = 58.12.

$$2\ C_4H_{10(g)} + 13\ O_{2(g)} \Rightarrow 8\ CO_{2(g)} + 10\ H_2O_{(\ell)}$$

Into a 25.0 L flask at 27°C, we place an equimolar mixture of four gases: butane, oxygen, carbon dioxide and Neon (an inert[1] gas) at an initial pressure of 10.00 atm.

a) Determine the moles of each gas present in the mixture.

The word equimolar means that we have the same number of moles of each gas in the resulting mixture, therefore, they must all have the exact same partial pressure. Since the total pressure is 10.00 atm, then the partial pressure of each gas initially must be one fourth of that amount, or 2.50 atm. Therefore, the moles of any gas, let's say butane, are given by:

$$n_{butane} = \frac{PV}{RT} = \frac{(2.50 atm)(25.0 L)}{(0.0821 \frac{L atm}{K mol})(300 K)}$$

$$= 2.54\ mol\ of\ each\ gas. \quad \sqrt{}$$

b) What will be the total pressure in the flask (@ 27°C) after the reaction is over?

The question deals with the total pressure in the flask after the reaction is over. We must be very careful to consider only gases for this problem. Let us assume that the only substances that will contribute to the pressure in the flask will be gases.[2] Looking at the balanced equation, we can see that we will keep close attention to the following substances: butane, oxygen, carbon dioxide, and the inert Neon gas (which will always be there contributing to the measured pressure). It is a good idea to construct a table that will show all the gases present in the mixture and that will allow us to keep track of each as the reaction proceeds. Remember that reagents are consumed and that products are formed. This means that we will subtract the moles consumed from the initial amount of reactants (because only the reactants are consumed) and we will add the moles formed to the moles of substances from the reactant side present in the in the initial mixture (remember that things like carbon dioxide can be present in the flask at all times, even before the reaction starts). Finally, we should not forget about the Neon. It is an inert gas, which means that it will neither be consumed nor formed. This means that whatever

---

[1] An inert species is one that will not react, or in other words, will NOT be involved at all in the reaction.
[2] We will see in the future that there in reality some liquids and some solids do contribute with what we call the vapor pressure to the total pressure in the flask.

amount we had initially, will also be present at the end.

Gases in Final Mixture	Initial Moles	Moles Consumed	Moles Formed	Moles Present after Reaction
Butane	2.54 mol	- 0.39 mol	-	= 2.15 mol
Oxygen	2.54 mol	- 2.54 mol	-	= 0 mol
Carbon Dioxide	2.54 mol	-	+ 1.56 mol	= 4.10 mol
Neon	2.54 mol	-	-	2.54 mol

Let's determine the limiting reagent first.

$$R_{but} = \frac{2.54}{2} = 1.27$$

$$R_{O_2} = \frac{2.54}{13} = 0.195$$

Keep in mind that although we have carbon dioxide and neon gases present in the initial mixture, they cannot be considered for limiting reagents since they are not going to react. The Neon will never react (inert) and the carbon dioxide, being a product, will simply be formed as the reaction proceeds, therefore, the moles of carbon dioxide will increase as a function of the progress of the reaction. Anyway, from the stoichiometric ratios we can clearly see that the oxygen is the limiting reagent. What happens to the limiting reagent? It is consumed 100%, so we will write in the column of moles consumed all the moles of oxygen (2.54 moles). Now, the LR will also allow us to determine how many moles of the butane gas will be consumed (0.391 moles of butane consumed), and keeping in mind that we started with 2.54 moles of butane, that leaves us with 2.15 moles of butane present in the final mixture.

$$n_{buta, consumed} = 2.54 mol_{O_2} \times \frac{2 mol_{buta}}{13 mol_{O_2}} = 0.391 mol_{buta}$$

How about the carbon dioxide? We can easily determine the moles of carbon dioxide formed from the limiting reagent.

$$2.54 mol_{O_2} \times \frac{8 mol_{CO_2}}{13 mol_{O_2}} = 1.56 mol_{formed}$$

However, since the carbon dioxide was formed and we had some to start with (2.54 moles), then after the reaction is over we will have present 4.10 moles of CO_2 in the reaction flask.

Let's look at our table. The last column compiles the moles of all gases present in the reaction flask after it is over. The process involves adding the initial moles of a given gas to the moles formed or subtracting the moles of a given gas consumed from the initial moles. Since the total pressure is dependent on the total number of moles of gases, then:

n_{tot} = 2.15 mol + 0 mol + 4.10 mol + 2.54 mol = <u>8.79 mol of gases</u>.

Where do we find all of these 8.79 moles of gas? In the 25.0 L flask. Finally, the total pressure is calculated as:

$$P_{tot} = \frac{n_{tot}RT}{V} = \frac{(8.79 mol)(0.0821 \frac{Latm}{Kmol})(300 K)}{25.0 L}$$

Therefore, $\quad P_{tot}$ = 8.66 atm   √

4.6 <u>Kinetic Molecular Theory of Gases</u>.

Up to this point, we have been talking about the experimentally determined laws that deal with gases. These observations led to the ideal gas law. Not only that, but it should have been clear to this point that all gases behave ideally under ambient conditions. An obvious question should be, WHY? How come this is the case? We should stop for a while to try to construct a model to attempt to explain why gases behave the way they do. Such a model is the Kinetic Molecular Theory of Gases[1] - it attempts to explain the behavior of ideal gases. Below, we can

---

[1] Developed in the 19th century by Clausius, Maxwell and Boltzmann.

see the postulates of the KMT of gases.

1. Gases consist of molecules widely separated[1] in space. The molecules are considered to be "points" in space - have mass but occupy no volume.

2. Gas molecules are in constant random motion. They collide with one another; energy may be transferred, but the total energy of all the molecules remains constant.

3. The average kinetic energy of the molecules of a gas depends on the absolute temperature and it increases as the temperature increases.

$$\therefore KE_{avg} \propto T \quad \text{or:}$$

$$\frac{1}{2}m\overline{(v)^2} \propto T$$

4. Gas molecules exert neither attractive nor repulsive forces on one another.

Certainly, real gases will not "obey" these postulates, but remember that this model is going to try to explain *ideal gases*.

In one of the statements above we mentioned the average kinetic energy of the molecules of a gas. This implies that not all the molecules of a gas at a given temperature have the same KE. However, if you have a large number of molecules (like Avogadro's number), then the *average* of all the individual speeds is always the same at a given T.

Maxwell and Boltzmann determined the molecular energy distribution of a large number of molecules at a given T and found the following. The fraction of molecules {f(v)} that has molecular speeds (relative to the total number of molecules in the container) between v and Δv is given by the expression seen below. A plot of this fraction {f(v)} versus the molecular speed (v) is also shown below. Please, notice that the average molecular speed is a little bit beyond the most probable fraction due to the shape of the curve[2].

$$f(v) = 4\pi \left(\frac{m}{2\pi k_B T} \right)^{3/2} v^2 \exp\left(-\frac{mv^2}{2k_B T} \right)$$

---

[1] Relative to their actual size.
[2] The area under the curve represents the total number of molecules in the sample.

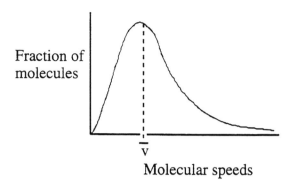

Molecular speeds

At any rate, we can use elementary physics for a collection of gas particles and we can easily derive an equation that is identical with the ideal gas equation (PV = nRT) based on the postulates that we described above. Actually, an expression for pressure can be derived from these postulates[1], mainly:

$$P = \frac{2}{3} \frac{n N_A (\frac{1}{2} m \overline{v^2})}{V}$$

where **P** is the pressure of the gas, **n** is the number of moles of the gas, N_A is Avogadro's number, **m** is the mass of each particle, $\mathbf{v^2}$ (with a bar on top) is the average of the square of the velocities of the particles, and **V** is the volume of the container.

The quantity 1/2 m v^2 (with a bar on top) represents the average kinetic energy of a gas particle. When we multiply it by Avogadro's number, that will give us the average kinetic energy for a mole of gas particles, that is:

$$(KE)_{avg} = N_A (\frac{1}{2} m \overline{v^2})$$

Using this expression, we can rewrite the one for pressure as follows.

$$P = \frac{2}{3}\left[\frac{n (KE)_{avg}}{V}\right] \qquad \text{or:} \qquad \frac{PV}{n} = \frac{2}{3} (KE)_{avg}$$

We should recall that the third postulate of the KMT of gases states that the average kinetic energy of the particles in the gas sample is directly proportional to the absolute temperature. This means that $(KE)_{avg} \propto T$. Therefore,

---

[1] An actual derivation can be seen in any general chemistry book, if you are interested.

$$\frac{PV}{n} = \frac{2}{3} (KE)_{avg} \; \alpha \; T$$

This is the same as saying that: $(PV/T) \; \alpha \; T$

which is a statement of the ideal gas law.

$$(PV/T) = RT \quad \text{or} \quad PV = nRT!$$

From our discussion above, it can be seen that the absolute temperature is a measure of the average kinetic energy of the gas particles. To be more exact, we have just seen in the "derivation" of PV = nRT; the absolute temperature is proportional to 2/3 of the average kinetic energy and that the proportionality constant was the gas constant, R. We can, therefore, write the following important expression which summarizes the meaning of temperature.

$$(KE)_{avg} = \frac{3}{2} (RT)$$

4.7 <u>Molecular Velocities</u>.

If we check the "derivation" above, we may notice that we used an average for the square of the velocities of the gas particles: $(v^2)_{avg}$. The square root of that value, $[(v^2)_{avg}]^{1/2}$, is referred to as the **root mean square velocity** and can be symbolized as v_{rms}. From above, we saw that:

$$(KE)_{avg} = \frac{3}{2} (RT) \quad \text{and} \quad (KE)_{avg} = N_A \left(\frac{1}{2} m \overline{u^2}\right)$$

When we set them equal and then solve for v_{rms}, we get:

$$v_{rms} = \sqrt{\frac{3RT}{N_A m}}$$

and since we know that: $\quad N_A m = MW \quad$ then:

$$v_{rms} = \sqrt{\frac{3RT}{MW}}$$

In this equation, everything must be substituted in absolute SI units in order to get the velocity in m/s. This means that T must be in Kelvin, the molecular weight must be in kg/mol (not grams per mole) and the gas constant, R, cannot be in Latm/Kmol, since those are not pure SI units: it must be in Joules/Kmol. Latm are units of energy and the SI units of energy are Joules (J). One Joule is defined as a kilogram meter squared per second squared (kgm^2/s^2). Using this units for R, we will get that: R = 8.314 J/Kmol.

- Calculate the root mean square velocity for a sample of Oxygen gas at 27°C.

    To do this, we need to know the molecular weight of O_2 in kg/mol (SI units). Other than that, it is direct substitution into that equation above.

$$v_{rms} = \left[\frac{3\,(8.314\ J/Kmol)\,(300\ K)}{0.0320\ kg/mol} \right]^{1/2}$$

$$\therefore v_{rms} = 484\ m/s. \quad \surd$$

What happens when a gas is released in a large volume? It starts to move in a straight, random motion until it collides with another molecule (or the walls). How far does a molecule travel before it collides with another one? This distance is referred to as the **mean free path** and it is typically (ambient conditions) a very small distance, approximately 60 nm. This implies that you have many, many collisions between the molecules under normal conditions. In fact, the molecules collide so often that they exchange kinetic energy constantly and that gives us the wide range of velocities that we discussed earlier (Maxwell-Boltzmann distribution).

4.8 <u>Graham's Law of Effusion</u>.

There is one more phenomenon that we should discuss at this point. It is that of effusion.

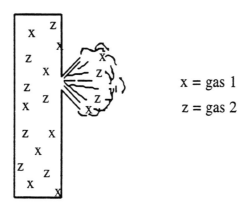

x = gas 1

z = gas 2

Effusion is the term used to describe the passage of a gas through a small orifice from a chamber of high pressure into one of much lower pressure. This is not to be confused with **diffusion**, which simply describes the mixing of gases. The rate of effusion measures the speed at which the gas is transferred into that other chamber (at a lower pressure).

Thomas Graham[1] determined experimentally that the rate of effusion of a gas is inversely proportional to the square root of the mass of its particles. If we take two gases at the same temperature, their relative rates of effusion are:

$$\frac{\text{rate of effusion for gas 1}}{\text{rate of effusion for gas 2}} = \sqrt{\frac{MW_2}{MW_1}}$$

We could "derive" this equation from the KMT of gases. The rate of effusion for a gas is proportional to its root mean square velocity, therefore,

$$\frac{\text{rate of effusion for gas 1}}{\text{rate of effusion for gas 2}} = \frac{v_{rms} \text{ for gas 1}}{v_{rms} \text{ for gas 2}} = \frac{\sqrt{\frac{3RT}{MW_1}}}{\sqrt{\frac{3RT}{MW_2}}} = \sqrt{\frac{MW_2}{MW_1}}$$

One of the main concepts to remember from <u>Graham's Law of Effusion</u>, is that the lighter the gas, the faster it will effuse. We will use it mostly to determine the identity (MW) of an unknown gas, if it is mixed with a gas of known identity.

- Let's say that we have an equimolar mixture of hydrogen gas (H_2) and helium gas (He). Which one will effuse faster?

    When we originally read the question, we realize that it deals with a mixture of two gases trapped in a container with a small leak. This reminds us of Graham's Law.

    $$\frac{\text{rate } H_2}{\text{rate He}} = \sqrt{\frac{4.00 \text{ g/mol}}{2.02 \text{ g/mol}}}$$

    $$= 1.41 \quad \sqrt{}$$

    Therefore, H_2 gas effuses 1.41 times faster than He gas! This is exactly what we expect since Hydrogen gas is lighter than Helium gas.

---

[1] A Scottish chemist who lived from 1805 until 1869.

Let's look at one more example, one in which we utilize the equation to try to determine the identity of an unknown gas mixed with one whose identity we know.

- An equimolar mixture of two gases, Kr and XO, are confined in a container which subsequently develops a leak. It is determined that XO effuses 1.73 times faster than Kr (AW = 83.80). Identify the element X.

    We start by using Graham's Law of Effusion to determine the molecular weight of the unknown gaseous substance (XO). We know that XO is lighter than Kr, because it effuses faster than Kr.

    $$\frac{\text{rate XO}}{\text{rate Kr}} = \sqrt{\frac{83.80 \text{ g/mol}}{MW_{XO}}} = 1.73$$

    Now, solving for the molecular weight, we get:

    $$MW_{XO} = 28.0 \text{ g/mol}$$

    This is not the end of the problem, however. We are supposed to identify element X. Well, one C is bonded to one oxygen and the total molecular weight is 28.0 g/mol. . Clearly, the atomic weight of X can be determined by the difference between the molecular weight and the atomic weight of oxygen.

    $$AW_x = 28.0 \text{ g/mol} - 16.0 \text{ g/mol}$$

    $$= \textit{12.0 g/mol.} \ \sqrt{}$$

    Therefore, the unknown element was *Carbon* (C).

- A certain element (**X**) forms two oxygen compounds that are both gases: XO_2 and XO_3 gases. It is observed that one of them effuses 1.12 times faster than the other. Identify that element (**X**).

    Our first deduction is that the one that effuses faster must be the XO_2, since we already know that the lighter one travels faster. Therefore, Graham's law in this example can be summarized as seen below. Normally, it is easier to write the faster one on top of the left side (or the lightest molecular weight in the denominator of the right side of the equation). Otherwise, this could become a bit confusing.

$$\frac{rate_{XO_2}}{rate_{XO_3}} = \sqrt{\frac{MW_{XO_3}}{MW_{XO_2}}}$$

$$1.12 = \sqrt{\frac{x+3(16.0)}{x+2(16.0)}}$$

and solving for x (the atomic weight of element X) we get:

$$x = 64.0 \text{ g/mol}.$$

Therefore, the element is *sulfur* (S).

Let us consider the behavior of real gases. These are the gases that we actually work with in the laboratory. With modern techniques, we can measure pressures and temperatures for gases far beyond what the early experimentalists were able to do, therefore, we have been able to clearly observe clear deviations from the ideal gas law.

4.9 Real Gases.

So far, we have only considered gases that obey the equation PV = nRT, that is, ideal gases. However, there are no ideal gases, only gases that behave ideally under a certain set of conditions. We have mentioned before that any real gas behaves ideally at low pressures and high temperatures. Why is it that gases do not behave ideally as we deviate away from these conditions? Remember that there were two postulates in the Kinetic molecular theory of gases that were flawed. One stated that the molecules occupy no volume whatsoever, which means that if we were to increase the pressure of a gas sample, we should be able to compress it all the way down to a volume of zero (Charles' Law). We know that this is not true because when we actually do this, we observe that the gas is liquefied first, and subsequently becomes a solid. Further compression does not lead very far and there is a limit as to how far you can compress.[1]

The second statement that is flawed is the one that says that there are no attraction nor repulsion forces in between the molecules. This is also not true, but both statements are very accurate if we work at ambient conditions, because under those conditions (high temperature and low pressure) we have a large volume and the molecules are far apart from one another and their interactions are minimal.

---

[1] For all practical reasons, we can say that condensed phases are incompressible simply because they can only be compressed slightly, particularly as compared to gases. The compressibility of a condensed substance (although very small) is given by its coefficient of thermal expansion (a property of every material).

There is a parameter that we can use to determine how far we are from ideality. We will call this the **compressibility factor**, Z. Let's define Z as:

$$Z = PV/nRT.$$

We should expect this factor to be equal to one if the gas were ideal. If it is not behaving ideally, then this factor will be different from one. In order to determine Z, we need to carefully measure the pressure, temperature, volume and number of moles of a real gas, so that we can then actually calculate Z. Below is a plot of Z versus pressure for various real gases. Notice that all gases approximate ideal behavior (Z = 1) as the pressure goes to zero (low pressures, normally less than five atmospheres).

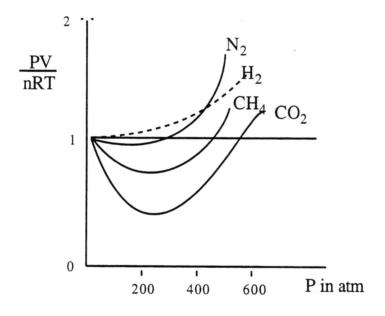

We can also see from the plot above that not all gases behave the same way. Hydrogen gas is the closest to the "ideal" line (through a Z of one) and CO_2 seems to deviate the most. We should point out that all gases have a compressibility factor (Z) larger than one at extremely high pressures. When that happens, PV > nRT, which means that the real gas is more difficult to compress than the ideal gas. This must be due to the fact that under these conditions, repulsive forces are predominant because the molecules are too close to one another.

At intermediate pressures, most gases have a compressibility factor (Z) lower than one. Under these conditions, attractive forces are predominant and it is easier to compress the real gas relative to the ideal one.

We should also point out that the shape of the curves above change with temperature. It has been observed that as the temperature is increased, the shape of the curve becomes more flat and loses some of the initial drop that brings Z to be lower than one. In other words, as the temperature increases, the probability of Z being less than one becomes smaller and smaller. At a certain point, we observe that the curve hugs the Z equals one line for a long time before Z starts to have values that are larger than one. At this temperature, the gas is as ideal as possible. We call this the Boyle's Temperature. Above this temperature, Z is always larger than one. In the example above, we can see that the only gas that is above its Boyle's temperature was hydrogen gas.

All the curves seen below are for a given gas at different temperatures. We can see that the temperature at which this gas is most ideal is 60°C, since Z is very close to one for a large range of pressures. Also, we can notice how Z can be less than one at temperatures lower than 60°C, but not above it.

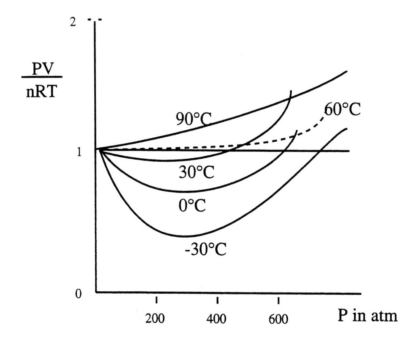

Once we know that a gas does not behave ideally under a given set of conditions, we need to find an equation of state that describes this system. The first such equation to be semi-empirically derived, was proposed by Johannes Van der Waals[1] in 1873. He realized that the problems were the ones we discussed previously (the size of the molecules and the forces of attraction between the molecules). How do these affect the measured (real) pressure and the measured volume of a gas? Well, let's think for a moment of the difference between an ideal molecule as it collides with a wall, and a real one doing the same thing. The ideal molecule

---

[1] A physics professor at the University of Amsterdam who in 1910 received the Nobel prize for his work.

exhibits no intermolecular forces, therefore, nothing stops it from hitting the wall with its full force. However, if we consider a real molecule, we can envision it being "held back" by the attractions to other molecules in the flask, therefore, the force of the impact is less than when it is ideal. This is the same as saying that the real pressure is smaller than the ideal pressure. When Van der Walls realized this concept, he immediately sought to add a correction factor to the real pressure in order to make it ideal.

$$P_{ideal} = P_{real} + \text{correction factor}$$

How can we decide what this correction factor is? This is beyond the scope of this book but it suffices to say that the correction factor ($n^2 a/V^2$, where **a** is a constant that is different for every gas, depends on the strength of the attractions, and it must be determined experimentally) is added to the real pressure in order to make it ideal. So, the ideal pressure is given by:

$$P_{ideal} = P_{real} + \mathbf{a}\,(n/V)^2.$$

The other problem was the volume. The ideal volume is supposed to be compressible to a volume of zero. This is clearly not the case, but he figured that this would be easier to solve if we simply subtract a correction factor from the real volume, mainly the actual volume that the molecules occupy. This volume is referred to as the **excluded volume**, because it actually is the volume that the molecules would occupy if they were in a condensed phase, that is, the volume of the molecules plus the space in between them when they are in direct contact. This leads to the expression:

$$V_{ideal} = V_{real} - n\,\mathbf{b},$$

where **b** is a constant that depends on the size of the molecules. Now that we know an expression of the ideal pressure and one for the ideal volume, we can get multiply them together to get an equation identical to PV = nRT.

$$(P_{real} + \mathbf{a}\,[n/V]^2)(V_{real} - n\,\mathbf{b}) = nRT$$

Notice that this equation only tries to correct real variables in order to make them ideal. It is a semi-empirically derived equation and, although it works better than PV = nRT at conditions that deviate from ideal behavior, it is not the best equation for real gases. There are many more, some totally empirically derived, that mathematically work a lot better for real gases but we will not discuss these equations in this book. Below, we can see a table with the Van der Waals constants for a few gases.

Gas	a, atmL^2/mol^2	b, L/mol
He	0.0338	0.0238
Ne	0.211	0.0171
Kr	2.321	0.0399
H$_2$	0.244	0.0267
O$_2$	1.356	0.0317
CO$_2$	3.588	0.0429
CH$_4$	2.253	0.0437

Chapter 5: Thermochemistry.

5.1 Energy.

Let us start this section with a definition that is rather difficult to put in words: Energy. We will define energy at this point in time as the capacity to do work or to produce heat. We will look exclusively at heat transfers in chemical reactions, and we will spend quite a long time talking about the relationship between heat and work in a subsequent chapter, when we discuss thermodynamics.

Probably the most important characteristic of energy is that it is always conserved. We can refer to this concept as the Law of Conservation of Energy, which states that energy can be converted from one kind to another, but it cannot be created nor destroyed. The implication behind this is that the energy of the universe is constant and that all we can do is convert one type of energy into another.

Chemical energy is a term that we can assign to the form of energy stored in molecules (in bonds and some particular conformations). As a reaction takes place, bonds are broken and formed and the energy trapped in these bonds will either be released or absorbed in the form of heat. It is this type of energy interconversion that we will study in this chapter.

But how can we define heat? Heat (**q**) involves the *transfer* of energy between two bodies due to a temperature difference. Therefore, heat is *not* a property of the system. A body does not have heat, it can only transfer some of its energy to another body IF it is at a higher temperature than the second body. Heat is many times referred to as *thermal energy* and it is associated with the random motion of atoms or molecules in a system. Thermochemistry deals with the study of the heat absorbed or evolved during a chemical or physical process. By convention:

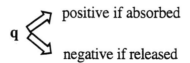

We need to decide what are the SI units for energy before we continue our study of reaction heats. Traditionally, the first unit used for heat (energy) was the calorie[1] (cal). By experience, we know that in order to increase the temperature of a substance, we need to add

---

[1] From the Latin "calor", which means heat.

heat. A calorie is defined as the amount of heat necessary to raise the temperature of one gram of water by one degree Celsius. This is not the SI unit for energy, however. The SI unit is the Joule[1] (J). We can easily follow the meaning of the unit we call Joule by looking at its derivation from another form of energy: mechanical energy (work). How do we define work? Work (W) is done when we apply a force and move an object a certain distance.

$$W = \text{force} \times \text{distance}$$

We also know that force is mass times acceleration[2], therefore,

$$W = \text{mass} \times \text{acceleration} \times \text{distance}$$

But acceleration is change in velocity with respect to time, therefore,

$$W = \text{mass} \times (\Delta v/\Delta t) \times \text{distance}$$

Now, let's derive the units. The SI units for mass are kilograms, for velocity would be meters per second, for time it is second, and for distance it is meters. Therefore,

$$W = \text{kg} \times \text{m/s} \times \text{1/s} \times \text{m}$$

And by definition, one Joule is:

$$1\,J = kg\,m^2/s^2. \quad \surd$$

And what is the relationship between Joules and calories? Experimentally, the amount of heat necessary to increase the temperature of one gram of water by one degree Celsius happens to be 4.184 J. Therefore,

$$1\,\text{cal} = 4.184\,J \quad \surd$$

5.2 Specific Heats.

The specific[3] heat (c) of a substance is by definition the amount of heat necessary to raise the temperature of one gram of a substance by one degree Celsius. This means that the specific heat of water is exactly 1.00 cal/gdeg, or in SI units, 4.184 J/gdeg. This is a property of a substance and from the table that follows, we can see that water is a substance with an

---

[1] Named after James P. Joule, and English physicist who lived from 1818 until 1889.
[2] One of Newton's laws.
[3] The word specific means: per gram.

unusually high specific heat. The amount of heat (q) that a particular object (composed of a pure substance) absorbs as it warms up depends on the mass of the object (which is why we report this property as Joules per gram degree). Therefore, the total amount of heat absorbed would be given by:

$$q = mc\Delta T$$

where   q = heat absorbed or released
m = mass
c = specific heat of substance
ΔT = change in temperature[1]
= final T - initial T

Substance	Specific Heat in J/gdeg
Au	0.130
Cu	0.381
Fe	0.445
Water	4.184
C_2H_5OH	2.462

- How many Joules of heat will be absorbed when 10.0 g of water are warmed up from 25.0°C to 50.0°C?

The first thing we notice is that this is a pure substance (water), therefore, we know its specific heat. With this information, we can easily determine the amount of heat absorbed. Also, remember that ΔT is the final temperature minus the initial temperature.

$$q = mc\Delta T = 10.0 \text{ g } (4.184 \text{ J/gdeg}) (50.0° - 25.0°)$$

$$= 1050 \text{ J} = \textit{1.05 kJ} \quad \sqrt{}$$

---

[1] Notice that ΔT is the same no matter whether you measure it in °C or in K since the size of either degree is identical!

- A 20.0 gram sample of an unknown metal liberates 1250 J as its T is lowered from 69.0°C to 20.0°C. Determine its specific heat (c).

    Once more, this is a pure substance. Therefore, **q = m c ΔT**. The term liberates means releases (given off), which means that its sign is negative by convention. The only unknown is the specific heat, therefore,

$$- 1250 \text{ J} = 20.0 \text{ g (c) } [20.0°C - 69.0°C]$$

Therefore,   **c = 1.28 J/gdeg.**

   Let's try one more problem, but this time we will actually mix two pure substances that have different specific heats and that are originally at different temperatures. Of course, heat will be transferred from the warmer substance to the cooler one until they both reach some intermediate temperature (thermal equilibrium). The implication behind the problem that follows is that there will be no heat loss to the surroundings. This kind of process if referred to as an **adiabatic** process. In other words, all the energy that is lost by the warm object will be gained by the cold one.

- A 50.0 gram sample of Cu (c = 0.38 J/g deg) at 80°C is immersed in 100 g of liquid water (c = 4.184 J/g deg) at 0°C. Determine the final temperature of the system.

    We observe that the copper metal is warmer than the water, therefore, once the process takes place, the copper will lose some energy in the form of heat, which will be absorbed completely by the water. The copper will cool down to a lower temperature (let's call it T_f), and the water will warm up to that same temperature. It is convenient to draw some kind of schematic diagram that shows what we just mentioned.

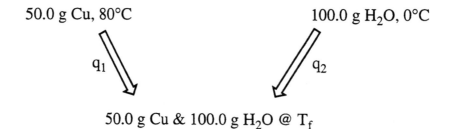

but since there is no heat loss to the surroundings, $q_1 + q_2 = 0$, or: **$q_1 = - q_2$.**

So, $q_1 = m c \Delta T = 50.0 \text{ g } (0.38 \text{ J/gdeg}) (T_f - 80°)$

and: $\qquad q_2 = 100 \text{ g } (4.184 \text{ J/gdeg}) (T_f - 0°)$[1]

Therefore, $\qquad 50.0 (0.38) (T_f - 80) = -100 (4.184) T_f$

$$19 T_f - 1520 = -481.4 T_f$$

$$437.4 T_f = 1520$$

$$\therefore T_f = 3.5°C \ \checkmark$$

b) Why is the final temperature so low?

Because of the fact that the specific heat of water is so much higher than that of the copper. As the copper cools down, it does not give up that much heat, however, it takes a lot of heat to warm up the water.

Another way in which we can measure the amount of heat that must be absorbed by a substance in order for its temperature to increase it what we call its **molar heat capacity** (C_p)[2]. This is exactly the same as the specific heat, except instead of the measurement per gram, it is per mole. We can easily convert one into the other by simply converting the grams into moles. For example, the molar heat capacity for water is:

$$4.184 \frac{J}{gK} \times \frac{18.02 g}{1 mol} = 75.40 \frac{J}{mol K}$$

For example, we can very easily calculate the amount of heat absorbed by 2.50 mol of water as its temperature increases from 25.0°C to 35.0°C from:

$$q = n\, C_p\, \Delta T$$

or: $\qquad q = 2.50 \text{ mol } (75.40 \text{ J/molK})(10.0 \text{ K}) = 1,890 \text{ J} \ \checkmark$

5.3 **Calorimetry**.

As a reaction takes place, some bonds are broken and some bonds are formed. Overall, there is a change in the internal energy of the system and it is reflected as either an absorption or a liberation of energy in the form of heat. Calorimetry is the experimental method through

---

[1] Remember that it is final temperature minus initial temperature.
[2] The reason there is a subscript (p) is to indicate that this value was measured under conditions of constant pressure. It is defined as the amount of heat necessary to increase the temperature of one mole of a substance by one degree Celsius under conditions of constant pressure.

which we will measure this amount of heat that is either released or absorbed. The instrument that measures the heat involved in the reaction is called a calorimeter.

There are many types of calorimeters, but they all operate on the same principle: they measure heat! This basically happens as we allow the actual reaction to take place in some container and the instruments simply records how much heat is given off (or is absorbed) as the reaction takes place. As heat is being given off, if it is trapped in the instrument (a process in which heat cannot escape and which we previously called an adiabatic process), it is then absorbed by its contents. The calorimeter would therefore absorb all that heat and the obvious manifestation of this event would be an increase in its temperature. As long as the calorimeter is adiabatic, then all that heat must have come from the reaction itself.

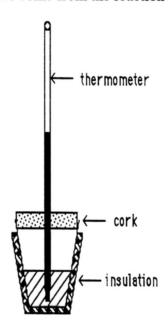

The calorimeter we see above is one used for solution calorimetry. In here we have a Styrofoam coffee cup, which is a good insulator. We know this is true because such a cup maintains coffee warm for a relatively long time. Of course, it is not perfectly adiabatic, so an important pre-requisite is the reactions carried out in this calorimeter should be very quick. This way, anything we measure can be done in a short amount of time (before heat starts to escape). This coffee cup is fitted with a cork and a thermometer. A solution is placed in the cup and is allowed to reach thermal equilibrium (stabilize at a given temperature). It is then mixed with another reagent at the exact same temperature. These react instantaneously and the heat given off will be absorbed by the walls of the container, the water in the solution[1], and the thermometer itself. From all of the substances inside the calorimeter, the one with the highest specific heat is

---

[1] A solution of a given substance is basically all water and a tiny little amount of solute (relative to the solvent).

the water. Therefore, as a first approximation, we can visualize the entire calorimeter as being composed of pure water (in other words, ignore the other components). The temperature increase of the water can then be related to the amount of heat absorbed by that amount of water (as given by: q = m c ΔT). However, a much more accurate determination would have to be made if we were in a laboratory. One way to achieve this would be to use a reaction for which we know exactly how much heat would be released and then measure the temperature rise. That way, we can figure out how much heat was absorbed by the calorimeter and from the equation,

$$q = m_{water} \, c_{water} \, \Delta T$$

determine the mass of water that would have caused this particular temperature increase. Let's look at one very specific example.

Let's say that we want to measure the amount of heat released when we react 50.0 mL of a 1.00 M solution of HCl with 1.60 g of solid pure NaOH (MW = 40.0) {in such proportions, the solid NaOH will be the limiting reagent}. We are aware that these two react according to:

$$HCl_{(aq)} + NaOH_{(s)} \Rightarrow H_2O_{(\ell)} + NaCl_{(aq)}$$

We place the 50.0 mL of the HCl solution in the calorimeter and wait for temperature equilibration. At this point, we add the pre-weighed (1.60 g) solid NaOH pellet's. The reaction will take place very quickly and it will release heat (exothermic). Let us also state that a previous experiment demonstrated that for this type of reaction, any temperature increase in this calorimeter is equivalent to having 60.0 g of water (that would then include everything that absorbs heat in the calorimeter).

The temperature of the water will rise because of the heat absorption. Let's say that experimentally we measure the increase in temperature to be 6.06°C. With that information, we can determine the amount of heat absorbed by the water.

$$q_{water} = m_{water} \, c_{water} \, \Delta T$$

$$q_{water} = (60.0 \text{ g})(4.184 \text{ J/gdeg})(6.06 \text{ deg})$$

$$\therefore q_{water} = 1,520 \text{ J}.$$

where m is the mass of the water (60.0 g in our calorimeter), c is the specific heat of water (4.184 J/gdeg) and ΔT is the change in temperature (6.06°C). Where did all that heat come from? From the reaction! This means that now we know how much heat was **given off** by the mass of NaOH that was consumed.

$$q_{rxn} = -1{,}520 \text{ J.}$$

Certainly, it would be to our advantage to be able to talk about this heat in terms of the reaction itself (as balanced) and not in terms of the actual mass that was reacted. This is accomplished if we talk about the change in enthalpy, ΔH.

5.4 Enthalpy.

The heat released or absorbed during a chemical reaction, measured under conditions of constant pressure, is referred to as the **enthalpy change** (ΔH) of the reaction.

$$\Delta H = H_{products} - H_{reactants}$$

However, the actual enthalpies of the reactants and the products cannot be measured, therefore what we measure directly in the laboratory is the heat released or absorbed during the reaction under conditions of constant pressure, which by definition we call ΔH.

Let's take the case of a reaction $A + B \Rightarrow C$ which releases heat as it proceeds from reactants to products:

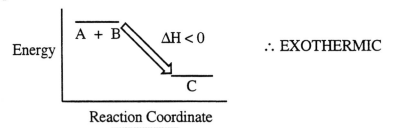

Such a reaction is referred to as an **exothermic** reaction.

If we look at the reverse reaction $C \Rightarrow A + B$, then we will observe that heat is absorbed as we go from reactants to products:

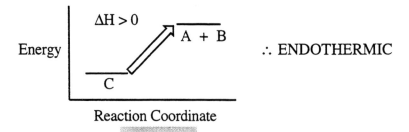

Such a reaction is referred to as an **endothermic** reaction.

ΔH depends on T and P, so it is reported normally in standard conditions at:

$$298 \text{ K } (= 25°C) \text{ and } 1 \text{ atm pressure } \checkmark$$

and in order to indicate the standard conditions we use a ° as a superscript, so we will see it written as ΔH°.

Now, let's go back to the calorimetry example that we were referring to previously. In our example, we saw that the reaction gave off 1,520 J of heat as it took place. Since we are conducting our experiment under constant pressure, then we have just measured the ΔH for the reaction.

$$\Delta H = \frac{- \text{ heat absorbed by water}}{\text{moles of NaOH}} = \frac{-1{,}520 \text{ J}}{\text{moles of NaOH}}$$

$$\Delta H = \frac{-1{,}520 \text{ J}}{\frac{1.60 \text{ g}}{40.0 \text{ g/mol}}} = -38{,}000 \text{ J/mol } \checkmark$$

or: $\quad\quad\quad\quad \Delta H = -38.0 \text{ kJ/mol } \checkmark$

The ΔH will have a negative sign because heat was released by the reaction (the sign indicates the direction of heat flow). The reason that we divided by the moles of NaOH is because the ΔH is an **extensive** property, meaning that it depends on how much mass is involved. Not only that, but it depends on how much of the limiting reagent was used.

- A 50.0 mL sample of 0.100 M Ba^{+2} is mixed with 50.0 mL of 0.100 M SO_4^{-2} in a constant pressure calorimeter. These two ions react to form an insoluble salt, $BaSO_4$. It is observed that the temperature of the calorimeter increases by 0.300°C. If you are told that under these conditions, the calorimeter's temperature increase would be the same as the temperature increase of 115 g of pure water, determine the heat of precipitation of the $BaSO_4$.

$$Ba^{+2}_{(aq)} + SO_4^{-2}_{(aq)} \Rightarrow BaSO_{4(s)}.$$

The first thing we notice is that the actual reaction has a one to one mole ratio between the two ions. Not only that, but we started with identical amounts of each ion (which is 0.00500 moles of each ion as calculated from the molarity times the volume). Therefore, either one of them can be categorized as the limiting reagent.

Let's calculate the heat absorbed by the calorimeter first.

$$q = m\,c\,\Delta T$$

$$= 115\text{ g }(4.184\text{ J/gdeg})(0.300°C)$$

$$q_{water} = 144\text{ J } \sqrt{}$$

If the calorimeter absorbed 144 J of heat, then the reaction must have given off 144 J of heat. Therefore: $q_{reaction}$ = - 144 J. This amount of heat given off by the reaction was for only 0.00500 moles of the limiting reagent. Therefore, the ΔH for the balanced reaction (which has one mole of the LR) would be:

$$\Delta H = -144\text{ J} / 0.00500\text{ mol}$$

$$\Delta H = -28.8\text{ kJ/mol } \sqrt{}$$

5.5 Properties of ΔH°.

There are two important properties for ΔH°. These are a direct consequence of the fact that ΔH° is an extensive property.

1. If the reaction is reversed, the sign of the ΔH is also reversed.

   A + 2 B ⇒ 3 C   ΔH° = -100.0 kJ

   3 C ⇒ A + 2 B   ΔH° = +100.0 kJ

2. The magnitude of ΔH is directly proportional to the quantities of reactants and products in the reaction. If you multiply the coefficients of the balanced reaction by an integer, you must multiply the value of ΔH by the same integer.

   A + 2 B ⇒ 3 C   ΔH° = -100.0 kJ

   2 A + 4 B ⇒ 6 C   ΔH° = -200.0 kJ

5.6 Hess's Law.

Hess's Law is based on the concept that since ΔH° is a **state function**, it only depends on the difference between the reactants and the products and not in the path. In other words, as long as we start with the reactants and end with the products, it does not matter how we get there, whether we do it in one step (direct reaction) or in many steps.

A simple example that illustrates this concept is seen below. Let's say that we want to know the ΔH for the following reaction:

$$2 NO_{2(g)} \Rightarrow N_2O_{4(g)} \qquad \Delta H° = ?$$

However, the only information provided is the following:

1) $\quad 1/2\ N_{2(g)} + O_{2(g)} \Rightarrow NO_{2(g)} \qquad \Delta H° = 34\ kJ$

2) $\quad N_{2(g)} + 2\ O_{2(g)} \Rightarrow N_2O_{4(g)} \qquad \Delta H° = 10\ kJ$

Well, in order to get the ΔH° we want, we must combine the two reactions given in such a way as to obtain the desired one. We can treat the chemical reactions as algebraic equations. Since we want to have two moles of NO_2 in the reactant side, and we observe that the NO_2 appears only in reaction **1**, that means that we need to find a way to use that equation to generate it as two moles of a reagent. The NO_2 appears as one mole and in the product side, therefore, we need to invert that equation and multiply it by two. What effect will that have on the first ΔH°? Its sign will be change from positive to negative (because of the inversion) and it will be twice as large (since it was multiplied by two).

How about the N_2O_4? It is found in reaction **2** and already in the product side as one mole. This is perfect, since that is exactly the way we want to have it in the final equation. Now, let's combine those two equations algebraically in order to obtain the one we want.

$$2 NO_{2(g)} \longrightarrow \cancel{N_{2(g)}} + \cancel{2\ O_{2(g)}} \qquad \Delta H° = -68\ kJ$$

$$\underline{\cancel{N_{2(g)}} + \cancel{2\ O_{2(g)}} \longrightarrow N_2O_{4(g)} \qquad \Delta H° = +10\ kJ}$$

$$2 NO_{2(g)} \longrightarrow N_2O_{4(g)} \qquad \Delta H°_{total} = -58\ kJ$$

5.7 Standard Enthalpy of Formation ($\Delta H_f°$).

Although we already know a way to determine the ΔH° for a reaction by simply measuring it in a calorimeter, the truth is that many times it is not convenient (and sometimes impossible) to determine the heat of a reaction directly. We need to use Hess' Law in order to carry out other reactions, so that by combining these reactions, we can then figure out the ΔH° for the desired reaction. An extension to Hess' Law that will enable us to determine the ΔH° for any reaction we want is through the standard heats of formation.

Chapter 5: Thermochemistry

The **standard enthalpy of formation** can be defined as the heat change when *one mole* of a compound is formed from its elements in their *standard state*. And what do we mean by the standard state? It is the most stable state of a substance at 25°C and under a pressure of 1 atm. This implies that we need to know the standard state of all the elements in the periodic table. These elements can be classified as either solids, liquids, or gases. We should also know whether they are monoatomic species, or diatomic, or polyatomic. Finally, in some instances elements exist in various forms and we need to know which is the stable form (for example: Oxygen is a gas but it can be either diatomic or triatomic). We call these allotropic[1] forms, so we need to know which is the stable allotrope. We should know the following elements from the periodic table. If a given element is not found in this table, it is a solid and we will assume that it is monoatomic.

Standard States:

Stable State	Elements
Gases	H_2, N_2, O_2, F_2, Cl_2, He, Ne, Ar, Kr, Xe and Rn.
Liquids	Br_2 and Hg.
Solids	I_2, $C_{graphite}$, $P_{4\ (white)}$, and $S_{8\ (rhombic)}$.

The standard heats of formation ($\Delta H_f°$) may be obtained from the table on *Appendix #1*. When we get a number from that table we should be able to write the chemical equation that represents that heat of formation. For example, let's say that we look up the heats of formation for liquid water, gaseous HI, methane (CH_4) gas, and gaseous HBr. These are: - 285.8, + 26.5, -74.8, and -36.4 kJ/mol, respectively. The equations representing these four numbers are:

$H_{2(g)} + 1/2\ O_{2(g)} \Rightarrow H_2O_{(\ell)}$ $\Delta H_f° = -285.8$ kJ

$1/2\ H_{2(g)} + 1/2\ I_{2(s)} \Rightarrow HI_{(g)}$ $\Delta H_f° = +26.5$ kJ

$C_{(gr)} + 2\ H_{2(g)} \Rightarrow CH_{4(g)}$ $\Delta H_f° = -74.8$ kJ

$1/2\ H_{2(g)} + 1/2\ Br_{2(\ell)} \Rightarrow HBr_{(g)}$ $\Delta H_f° = -36.4$ kJ

---

[1] Different forms in which an element may exist in nature.

Chapter 5: Thermochemistry

Of course, the only thing we need to be concerned with is knowing which is the standard state of the elements that make up the compound.

- Why is the $\Delta H_f°$ of any element in its standard state equal to zero?

    Because the heats of formation are defined in terms of formation of one mole of a compound from its elements in their standard states. In this case, we are referring to forming one mole of an elements from itself. That is not a reaction, per se, therefore the $\Delta H°$ for such "reaction" would have to be zero.

    Let's use the standard heats of formation in order to get the $\Delta H°$ for the following reaction (which is sometime called the heat of hydrogenation for a given compound – in this case C_2H_4).

$$C_2H_{4(g)} + H_{2(g)} \Rightarrow C_2H_{6(g)} \qquad \Delta H° = ?$$

In order to get the $\Delta H°$ for any reaction all we need is a table of heats of formation and the Hess' Law concept. We do have the table and in there we will find the $\Delta H_f°$ for $C_2H_{4(g)}$ and $C_2H_{6(g)}$, which are -84.68 kJ/mol and $+52.30$ kJ/mol, respectively. We can immediately write down the two reactions which these two numbers represent:

$$2 C_{(gr)} + 3 H_{2(g)} \Rightarrow C_2H_{6(g)} \qquad \Delta H_f° = -84.68 \text{ kJ}$$
$$2 C_{(gr)} + 2 H_{2(g)} \Rightarrow C_2H_{4(g)} \qquad \Delta H_f° = +52.30 \text{ kJ}$$

Let's rearrange these two equations so that, when we add them up, we will get the reaction we want.

$$2 C_{(gr)} + 3 H_{2(g)} \Rightarrow C_2H_{6(g)} \qquad \Delta H_f° = -84.68 \text{ kJ}$$
$$C_2H_{4(g)} \Rightarrow 2 C_{(gr)} + 2 H_{2(g)} \qquad -\Delta H_f° = -52.30 \text{ kJ}$$
$$\overline{C_2H_{4(g)} + H_{2(g)} \Rightarrow C_2H_{6(g)} \qquad \Delta H° = -136.98 \text{ kJ}}$$

It is interesting to note that what we ended up doing was subtracting the heat of formation of the reactants from that of the products. This will always be true, so from now on we do not even need to do this diagram for Hess' Law and instead always perform the following calculation:

$$\Delta H° = \sum \Delta H_f°_{(products)} - \sum \Delta H_f°_{(reactants)} \quad \checkmark$$

- Determine the $\Delta H°$ for the combustion of $CH_{4(g)}$.

$$CH_{4(g)} + 2\, O_{2(g)} \Rightarrow CO_{2(g)} + 2\, H_2O_{(\ell)}$$

In order to determine the heat of combustion of methane gas, we need to find the standard heats of formation for four substances, mainly: methane gas, oxygen gas, carbon dioxide gas, and liquid water. These numbers are readily available from Appendix #1. Once we have the numbers themselves, then we know that the heat of combustion is simply the sum of the heats of formation for the products minus the sum of the heats of formation for the reactants. Keep in mind that the standard heats of formation are for ONE mole of each of these substances, therefore, when we have more than one mole, we need to multiply by the number of moles of that substance. In our example:

$$\Delta H° = \Delta H°_{fCO2} + 2\, \Delta H°_{fH2O} - [\, \Delta H°_{fCH4} + 2\, \Delta H°_{fO2}\,]$$

$$= -393\ kJ + 2\,(-286\ kJ) - (-75\ kJ)$$

$$\therefore\ \Delta H° = -890\ kJ\quad \sqrt{}$$

Since $\Delta H°$ is an extensive property, the only time that the reaction will release 890 kJ of heat is when we burn exactly one mole of methane gas. That is, if we were to burn twice as much methane, we would release twice as much heat. Let's look at one example that proves that we should be very careful about this same point.

- How much heat is released when we burn 2.52 g of methane gas?

In the previous problem, we saw that every time we burn one mole of methane gas, we will release 890 kJ of heat. Therefore,

$$2.52\, g_{CH_4} \times \frac{1\, mol_{CH_4}}{16.04\, g_{CH_4}} \times \frac{-890\, kJ}{1\, mol_{CH_4}}$$

$$= -140\ kJ\quad \sqrt{}$$

Therefore, the reaction will release 140 kJ of heat.

- How many kJ of heat will be released when 1.8 grams of water are formed in the combustion of $CH_{4(g)}$?

Once more, we need to know the $\Delta H°$ for the balanced reaction before we even start thinking about handling this question. However, this time we are told that the reaction in question formed 1.8 grams of water. What is the relationship between the water formed and the heat released? From the reaction, we can see that every time we form two moles of water, 890 kJ of heat will be released.

Therefore,

$$1.8 \text{ g H}_2\text{O} \times \frac{1 \text{ mol H}_2\text{O}}{18.0 \text{ g H}_2\text{O}} \times \frac{-890 \text{ kJ}}{2 \text{ mol H}_2\text{O}} = -44 \text{ kJ}$$

∴ *44 kJ of heat were released!* √

- The amount of heat released by the combustion of 10.0 g of $C_2H_5OH_{(\ell)}$ is absorbed by 8.00 kg of copper metal (c = 0.378 J/g deg) and its temperature increases by 98.5°C. Determine the heat of formation ($\Delta H_f°$) for C_2H_5OH.

We need to realize that there are two different processes going on here. First, we have a certain exothermic reaction taking place. The amount of heat that is given off by this reaction is simply absorbed by a piece of copper metal and the copper will get warm. Since the specific heat of copper is given, we should be able to determine quite easily the amount of heat that the copper absorbed.

$$q_{Cu} = m_{Cu} c_{Cu} \Delta T = 8000 \text{ g} (0.378 \text{ J/gdeg}) (98.5°) = 298,000 \text{ J} \checkmark$$

We have established that the copper absorbed 298 kJ of heat. Where did all that heat come from? From the combustion of 10.0 grams of ethanol. Therefore, the actual burning of the ethanol must have <u>released</u> 298 kJ of heat.

$$q_{rxn} = -q_{Cu} = -298,000 \text{ K} = -298 \text{ kJ} \checkmark$$

To calculate standard heat of combustion, we need to understand that the amount of heat calculated above was only for 10.0 grams of the ethanol. So, in kJ per mole, it would be:

$$\therefore \Delta H° = \frac{-298 \text{ kJ}}{10.0 \text{ g C}_2\text{H}_5\text{OH}} \times \frac{46.07 \text{ g C}_2\text{H}_5\text{OH}}{1 \text{ mol C}_2\text{H}_5\text{OH}}$$

$$= -1370 \text{ kJ/mol C}_2H_5OH \checkmark$$

We just found the heat of combustion for ethanol. Let's write the reaction for which this heat of combustion applies.

$$C_2H_5OH_{(\ell)} + 3\, O_{2(g)} \Rightarrow 2\, CO_{2(g)} + 3\, H_2O_{(\ell)}$$

The heat of combustion (the $\Delta H°$ we just calculated) must be equal to the sum of the heats of formation for the products minus the sum of the heats of formation for the reactants. Therefore,

$$\therefore \Delta H° = 2\, \Delta H°_{fCO2} + 3\, \Delta H°_{fH2O} - \Delta H°_{fC2H5OH}$$

And solving for the heat of formation of ethanol, we get:

$$\therefore \Delta H°_{fC2H5OH} = 2\, \Delta H°_{fCO2} + 3\, \Delta H°_{fH2O} - \Delta H°$$

$$= 2\,(-393\text{ kJ}) + 3\,(-286\text{ kJ}) - (-1370\text{ kJ})$$

$$\therefore \Delta H°_f \text{ ethanol} = -274\text{ kJ} \quad \surd$$

- How many grams of CO_2 gas are formed in the combustion of a certain sample of liquid CH_3COOH, if we observe that the amount of heat released by the reaction is enough to bring 125 g of water, initially at 25.0°C, to 95.0°C.

From the information at the end of the problem, we can determine the amount of heat absorbed by the water, and thus given off by the reaction.

$$q_{water} = m_{water}\, c_{water}\, \Delta T = 125\text{ g }(4.184\text{ J/gdeg})(95.0°C - 25.0°C) = 36{,}600\text{ J}$$

So,
$$q_{reaction} = -q_{water} = -36.6\text{ kJ} \quad \surd$$

Let's determine now the heat of combustion for that liquid. In order to do this, we must balance the reaction first.

$$CH_3COOH_{(\ell)} + 2\, O_{2(g)} \Rightarrow 2\, CO_{2(g)} + 2\, H_2O_{(\ell)}.$$

By definition,
$$\Delta H° = 2\, \Delta H°_{f\,CO2} + 2\, \Delta H°_{f\,H2O} - \Delta H°_{f\,CH3COOH}.$$

$$\therefore \Delta H° = 2\,(-393.51) + 2\,(-285.83) - (-484.5)\text{ kJ} = -874.18\text{ kJ} \quad \surd$$

What is the meaning of that number? The meaning is that when you burn exactly one mole of the liquid, the reaction will release 874.18 kJ of heat. However, the actual sample of the liquid burned must have been a lot smaller than one mole, since only 36.6 kJ of heat was released. Therefore,

$$-36.6 kJ \times \frac{1 mol_{CH_3COOH}}{-874.18 kJ} \times \frac{60.05 g}{1 mol_{CH_3COOH}} = 2.51 g_{CH_3COOH}$$

5.8 ΔH as a Function of Temperature.

Sometimes the ΔH° of a reaction needs to be calculated at other temperatures other than 25°C. That may create a problem, since the table of heats of formation is given at 25°C, so at first glance the table seems to be useless for that particular purpose. The only thing we may be able to obtain from that table is the ΔH° for the reaction at 25°C. However, we can indeed determine the ΔH at any temperature (let's call it temperature **x**) if the heat capacities of all reactants and products are known (assuming they do not change over the temperature range in question)[1]. All that we need to do is construct a temperature cycle where you start with the reactants at the desired temperature (**x**), and then proceed to either cool them or heat them to 25°C. Remember that you CANNOT calculate the ΔH for the reaction at temperature **x**; we can only do this at 25°C. Since this process does not involve a reaction, but rather a physical process (of either cooling or warming up the reactants), the ΔH is simply calculated by the knowledge of their molar heat capacities. Once we have taken the reactants to 25°C, we can then proceed to carry out the reaction at this temperature. The ΔH° for this step is simply given by:

ΔH° = Σ ΔH$_f$°(products) - Σ ΔH$_f$°(reactants)   √

Finally, once we have generated the products at 25°C, all we need to do is either warm them up or cool them down to temperature **x**. Once more, since this step does not constitute a reaction, but rather a physical process, all we need are the molar heat capacities of the products. This completes the cycle, because we started where we wanted (reactants at temperature **x**) and ended up where we wanted also (products at temperature **x**). Now, using Hess's Law, we know that the sum of all these ΔH's will give us the ΔH for the reaction at temperature **x**. Let's follow the example below which illustrates the entire concept.

- Determine the ΔH for the following reaction at 498 K.

$$2 N_{2(g)} + O_{2(g)} \Rightarrow 2 N_2O_{(g)}$$

Since we know the heat of formation of N$_2$O at 25°C only, we can devise a pathway that starts at 498 K with the reactants, cools them down to 298 K, then carry out the reaction, and finally warm up the products to the final temperature (498 K). As we said previously,

---

[1] This is seldom true, but it is normally not a bad assumption, because the temperature dependence of the heat capacity is normally small.

the sum of all these ΔH's will give us the desired ΔH_{498}. I recommend that we start by drawing a temperature diagram (as shown below) that illustrates the process we just described and go from there.

Let's start by building the temperature diagram so that we can see exactly where we are starting and the way in which we should proceed. We want the heat of the reaction at 498 K (top of the diagram), however, the heats of formation in the appendix are useless at that temperature. Therefore, we will do three things. We will cool down the two moles of nitrogen gas and the mole of oxygen gas from 498 K down to 298 K (ΔH_1). Why that temperature? Because that is the temperature at which the table in the back is useful. Now we will actually carry out the reaction ($\Delta H°_2$). Once we generate the products at 298 K, then we will warm up the laughing gas from 298 K to 498 K (ΔH_3). This completes the cycle, because we have ended where we wanted. The sum of the three steps should give us the desired ΔH (according to Hess' Law).

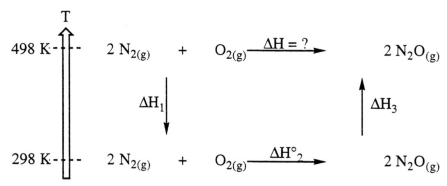

The calculation of $\Delta H°_2$ is simple.

$$\Delta H°_2 = 2\, \Delta H°_f \text{ of } N_2O = 2 \text{ mol } (78.032 \text{ kJ/mol}) = +156,064 \text{ J}$$

To get ΔH_1 and ΔH_3 we need the heat capacities of N_2, O_2 and N_2O gases.

$$\Delta H_1 = \Sigma n\, C_p\, \Delta T = 2 \text{ mol } (29.1 \text{ J/molK})(-200 \text{ K}) + 1 \text{ mol } (26.1 \text{ J/molK})(-200 \text{ K})$$

$$\therefore \Delta H_1 = -16,860 \text{ J}$$

$$\Delta H_3 = 2\, (38.7 \text{ J/molK})(+200 \text{ K}) = +15,480 \text{ J}$$

Finally, the overall ΔH at 498 is the sum of the three which is: *+154,684 J* (or 155 kJ)[1].

---

[1] Please, notice that there is not a very large difference between the DH at 298 K and the DH at 498 K. This is normally the case, but not always.

- Determine the ΔH for the reaction of $C_2H_{6(g)}$ if we start with the reactants at -75°C, and end up with the products at 75°C. Given: C_P for ethane, O_2, CO_2, and water is 70.7, 26.1, 45.7, and 75.3 J/molK, respectively, and assume that they are temperature independent.

This time, we want the ΔH for a reaction that starts at one temperature and ends up at another temperature. Independent of that, we should always start by drawing the temperature cycle that shows what we want and the way we will accomplish it. Just as before, we need to split the calculation into three parts. First of all, we need to warm up the two moles of ethane and the seven moles of oxygen gas from -75°C up to 25°C. Since this is just a physical process, we need only the heat capacities of these substances. Once we have the reagents at 25°C, we can actually carry out the reaction. This is the point where we can use the table in Appendix #1 with the standard heats of formation. Finally, we must warm up the products (the four moles of carbon dioxide and the six moles of liquid water) from 25°C up to 75°C. The sum of the three ΔH's will give us the desired ΔH.

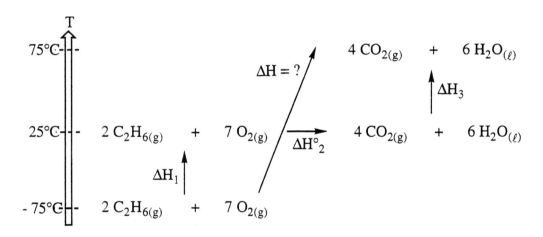

Calculations:

⇒ $\Delta H_1 = n_{ethane} C_{P\,ethane} \Delta T + n_{oxygen} C_{P\,oxygen} \Delta T$

   = 2 mol (70.7 J/molK) (100.0 K) + 7 mol (26.1 J/molK) (100.0 K) = *32,400 J*

⇒ $\Delta H°_2 = 4 \Delta H°_{f\,CO2} + 6 \Delta H°_{f\,H2O} - 2 \Delta H°_{f\,ethane}$

   = 4 (-393.5) + 6 (-285.8) - 2 (-84.7) kJ = *-3,119 kJ*

⇒ $\Delta H_3 = n_{CO2} C_{P\,CO2} \Delta T + n_{water} C_{P\,water} \Delta T$

   = 4 mol (45.7 J/molK) (50.0 K) + 6 mol (75.3 J/molK) (50.0 K) = *31,700 J*.

Finally, the ΔH for the desired reaction (from –75°C up to +75°C) is given by the sum of these three numbers.

$$\Delta H = \Delta H_1 + \Delta H°_2 + \Delta H_3$$

$$= +32.4 \text{ kJ} - 3{,}119 \text{ kJ} + 31.7 \text{ kJ}$$

$$\therefore \Delta H = -3055 \text{ kJ} \quad \sqrt{}$$

Chapter 6: Atomic Structure and Quantum Mechanics.

6.1 Early Characterization of the Atom.

By the end of the 19th century, as people started to accept the concept of atoms, the next logical question became, what are the components that make up an atom, if any? During this period, physicists like J. J. Thomson[1] (who discovered the electron and was awarded the 1906 Nobel prize for the discovery) and Robert Millikan[2] (who was able to determine the mass and the charge of the electron, for which he was awarded the Nobel prize in 1923) were able to advance this notion and prove that there were indeed subatomic particles that were common to all atoms. We recognize the three main sub-atomic particles to be: the electron, the proton and the neutron. The table below compiles the masses and charges of these particles. Keep in mind that these were not all determined at once and that it was through a series of different and sometimes sophisticated experiments that these quantities were determined. For example, the mass and charge of the electron was determined by Millikan's famous oil drop experiment in 1909. In this experiment, he was able to spray a thin mist of tiny oil droplet's in between two metal plates. The droplet's started to fall due to the action of gravity. By shining X-Rays into them, he was able to ionize these droplet's and then by applying a voltage difference between the two metal plates, he was able to suspend them in mid-air. From the charge to mass ratio which was previously determined by J. J. Thomson and the opposing forces that suspended the particles, he was able to determine these quantities. On the other hand, the neutron, which has no charge and thus is much harder to detect, was not discovered until 1932 by James Chadwick[3].

Particle	Actual Mass (kg)	Actual Charge (C)	Relative Mass	Relative Charge
Electron	9.109×10^{-31}	-1.60×10^{-19}	0	-1
Proton	1.673×10^{-27}	$+1.60 \times 10^{-19}$	1	+1
Neutron	1.675×10^{-27}	0	1	0

Let's review the information in the table. The first thing we should point out is that the

---

[1] British physicist who lived from 1856 until 1940.
[2] American physicist who lived from 1868 until 1953.
[3] British physicist who lived from 1891 until 1972.

charges are measured in the SI units of electrical charge, which are Coulombs (C). We also notice that the proton and the electron have exactly the same magnitude for the electrical charge, but opposite sign. This means that they will attract one another and that if we had an equal number of each, the charges would cancel out, leaving a neutral atom. We should also point out that when we state that the relative mass of the electron is zero, it is not because it has no mass, but rather because it is much smaller (about 2,000 times smaller) than that of either the proton or the neutron.

In 1896, Henri Becquerel[1] discovered that certain atoms were unstable enough so that they would actually disintegrate or decompose into other ones with different chemical identity. It was Marie Curie[2] who suggested that this process be called radioactivity. We can identify many different processes by which atoms may decay spontaneously, but three of the most important ones can be seen below.

a. Alpha Rays (α) - consist of two protons and two neutrons, therefore, it is a heavy particle and positively charged. When a substance follows this path of decay, it is observed to radiate these heavy positive particles.

b. Beta Rays (β) - consist of electrons, therefore they are very light and negatively charged. Once more, when a substance follows this path of decay, it is observed to radiate these negatively charged particles we recognize as electrons.

c. Gamma Rays (γ) – is a highly energetic form of light (electromagnetic radiation). Since it is basically light, these rays have no mass and no charge.

In the early 1900's, scientists tried to explain how the electrons and the protons (neutrons were not discovered until the 1930's) were combined to form the atom. One of the first to attempt an explanation was J. J. Thomson (1904) who proposed a "raisin-pudding" atomic model.

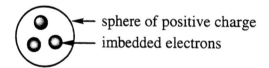

---

[1] French physicist who lived from 1852 until 1908 and discovered radioactivity in Uranium, for which he was awarded the Nobel prize in 1903.

[2] Her maiden name was Marya Sklodowska. She was a Polish chemist and physicist who lived from 1867 until 1934. She is one of the few people who have received more than one Nobel prize in science (one in 1903 in physics and another in 1911 in chemistry).

This model remained until Ernest Rutherford[1] proposed a planetary model in 1909 after his famous gold foil experiment. He directed some alpha (α) particles to a very thin gold foil and observed that the vast majority of these heavy positive particles passed through the foil completely undisturbed. However, a few of them were deflected and very few actually collided with it and bounced right back.

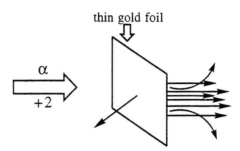

Based on these observations, Rutherford concluded that most of the mass of the atom must be concentrated in a very small volume he called the nucleus and that the electrons must be going around this nucleus in circular orbits thus leaving mostly empty space in an atom.

The simplest view of the atom is that is consists of a tiny nucleus (with a diameter of about 10^{-13} cm) and electrons that move about the nucleus at an average distance of about 10^{-8} cm (1 Å) away from it. As we will see in the near future, the reactivity of elements and compounds depend on the electrons more than anything else: their location and configuration. It is important that you have at least an idea of the relative sizes of atoms since in chemistry we talk about things in the molecular level many times, and later on we have to translate it into a larger level.

6.2 <u>Symbols for Atomic Structure.</u>

An atom is identified by two numbers. These are:

a. Atomic Number (Z) - corresponds to the number of protons in the nucleus. It identifies the element itself. It is the number of protons in the nucleus that makes an atom have the characteristics of a given element.

b. Mass Number (A) - this number represents the total number of protons and neutrons in the nucleus. Therefore: # of neutrons = A - Z.

---

[1] New Zealand physicist who lived from 1871 until 1937. He was awarded the Nobel prize in chemistry in 1908 for his work in atomic structure.

A lot of times the symbol for the element is written with these two numbers included, both on the left: the mass number as a superscript and the atomic number as a subscript.

$$^A_Z El$$

Let's see some examples that illustrate these symbols.

Species	# of protons	# of electrons	# of neutrons
^7Li	3	3	4
$^{16}O^{2-}$	8	10	8
^{80}Br	35	35	45
$^{68}Zn^{2+}$	30	28	38

6.3 <u>Isotopes</u>.

We will define isotopes as atoms having the same atomic number (therefore the same element) but different mass number. The isotopes also have the same properties, therefore the properties of an element depend on the atomic number, not the mass number.

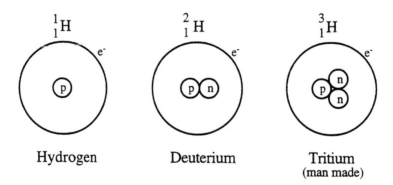

Hydrogen   Deuterium   Tritium (man made)

In the example above, we see three different isotopes for the first element, Hydrogen. Of these three isotopes, the most common is ^1H. In this isotope, we seen the simplest of atoms: a single proton in the nucleus and a single electron somewhere around it. The second isotope is called Deuterium and although rather rare in nature, has many uses in a laboratory, as we will learn at a later time. Deuterium (^2H) has twice the mass of regular hydrogen, because it contains also one neutron in the nucleus. Finally, Tritium (so called because it has three times the mass of regular hydrogen) is radioactive and man-made (that is, it is not normally found in nature).

We mentioned in Chapter one that the most common isotope of carbon is C-12. This is the isotope that we use as a standard for atomic weights (exactly 12 amu). How do we measure the masses of all other isotopes in the world relative to C-12? By using a *mass spectrometer*. When we use this spectrometer, we insert a sample of the substance whose mass we want to know into the ionization chamber. If the substance is a solid, it must be vaporized first (that is normally achieved at low pressures, hence we have an evacuated chamber – a vacuum). The gaseous atoms (or molecules) go through an electron beam, and thus they lose an electron becoming a positive ion. This positive ion is then accelerated through some negative grids with tiny holes in order to generate a beam of fast moving positive ions. This beam of positive ions is then taken through a powerful magnetic field which has the effect of deflecting the beam. The degree of deflection is dependent on the mass of the ions. If we have more than one type of isotope, we will have different degrees of deflection. As the ions collide with a photographic plate (the detector), we will be able to detect two things: the mass of a given ion (the detector has been previously calibrated) and the natural abundance of such ion.

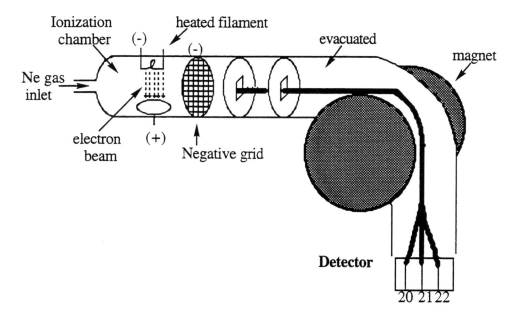

Let's follow the example of Ne (in diagram above), which is a monoatomic gas. Once the Ne has been ionized and later on deflected, it is observed that the beam is resolved into three peaks. This proved the existence in nature of three different isotopes for Ne. The calibration of the machine is done with C-12, which we defined as weighing exactly 12 g. Doing this, we observe the following results. The first thing we see is the what we call the mass spectrum.

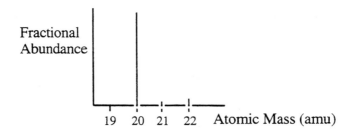

19 20 21 22   Atomic Mass (amu)

However, in modern mass spectrometers, it will also generate a computer printout with the information seen below.

Isotope	mass	% abundance
^{20}Ne	19.9924	90.92%
^{21}Ne	20.9940	0.257%
^{22}Ne	21.9914	8.82%

With this information we should be able to calculate the atomic weight of Ne (the number that appears in the periodic table). The quantity we call atomic weight is a weighted average of all the naturally occurring isotopes. We should learn now how to determine a weighted average.

Since there are three isotopes, we should determine the mass that each one contributes to the atomic weight and then simply add them up. This is achieved by finding the percentage of each of the masses as seen below.

$$\text{AW of Ne} = 0.9092\,(19.9924) = 0.00257\,(20.9940) + 0.0882\,(21.9914)$$

$$= 18.18 + 0.0540 + 1.94$$

$$AW_{Ne} = 20.17 \; \checkmark$$

- The atomic weight of Br is 79.90. The mass spectrum of Br_2 consists of the following three peaks. How should we interpret that data?

Mass	Abundance
157.84	0.2534
159.84	0.5000
161.84	0.2466

We must remember that this is a combination of two atoms. Therefore, having three possible peaks tells us that there are two naturally occurring isotopes. Why? Let's call them **1** and **2**. The possibilities for the combination are: **1-1**, **1-2**, or **2-2**. Of course, the lightest one must be the **1-1**, whereas the heaviest one must be the **2-2**. This means that half of the mass of 157.84 must be the mass of the lighter isotope: 78.92 = **1**. Also, half of the mass of 161.84 must be the mass of the heavier isotope: 80.91 = **2**. It is also clear for all practical purposes, we have equal amounts of both isotopes in nature. We know this for two reasons: first, the atomic weight is essentially half way between the two masses and secondly, the molecule **2-2** is exactly half of all the molecules present in the mixture.

6.4 <u>Ions</u>.

In some of the examples we saw before, we noticed a few species that had a permanent charge. These are called ions. Ions are observed whenever an atom (or a group of atoms) does not have the same number of electrons as it has protons. For example, the sodium ion (Na^{+1}) is one that contains 11 protons but only 10 electrons. The fact that it has one more proton than electrons gives the ion a positive charge. Positive ions are referred to as cations.

An ion may also be negative. An example would be the chloride ion, Cl^{-1}. In this ion we have 17 protons in the nucleus, but there are 18 electrons around it. Therefore, the extra electron gives the ion a negative charge. When a compound is formed between these two ions, the overall charge of the compound will be neutral, therefore, for every mole of sodium ions, we must have a mole of chloride ions, thus the formula: NaCl.

Ions can be polyatomic. Below is a list of common cations and anions. It is convenient to know (memorize) some of these ions. Cations carry the name of the element from which they came. Once more, Na^{+1} is called the sodium ion. However, sometimes there are some elements that form more than one type of cation: example –> Fe^{+2} and Fe^{+3}. These are then called: iron (II) and iron (III), respectively. They also are given common names (ferrous and ferric), where the ending –ous is for the lower charge and the ending –ic is for the higher charge. When you use the common names, the first part of the name is the Latin name (in Latin, iron is called Ferrum). Here are a few cations.

Cation	Name	Cation	Name
Li^+	Lithium	K^+	Potassium
Na^+	Sodium	Mg^{+2}	Magnesium
Ca^{+2}	Calcium	Al^{+3}	Aluminum
Ag^+	Silver	NH_4^+	Ammonium
Fe^{+2}	Ferrous (iron II)	Fe^{+3}	Ferric (iron III)
Ni^{+2}	Nickel	H_3O^+	Hydronium
Cu^{+1}	Cuprous (copper I)	Cu^{+2}	Cupric (copper II)
Pb^{+2}	Plumbous (lead II)	Pb^{+4}	Plumbic (lead IV)

How about the anions? The anions of single atoms carry the ending –ide. For example, the anion formed from chlorine (Cl^{-1}) is called the chloride ion. In polyatomic ions, particularly the ones between an element and oxygen, the endings –ite and –ate represent having the least number and the largest number of oxygens. For example: SO_3^{-2} is the sulfite ion, whereas the SO_4^{-2} is called the sulfate ion. However, in some cases there are too many of these oxyanions. For example, in the case of the anions formed between Cl and O, there are four of them: ClO^-, ClO_2^-, ClO_3^-, and ClO_4^-. These are then called: hypochlorite, chlorite, chlorate, and perchlorate. The prefix hypo- means less than, whereas the prefix per- means more than. Here is a list of some common anions that we should learn.

Anion	Name	Anion	Name
F^-	Fluoride	Cl^-	Chloride
Br^-	Bromide	I^-	Iodide
OH^-	Hydroxide	O^{-2}	Oxide
S^{-2}	Sulfide	CO_3^{-2}	Carbonate
HCO_3^-	Hydrogen carbonate or Bicarbonate (common name)		
NO_2^-	Nitrite	NO_3^-	Nitrate
SO_3^{-2}	Sulfite	SO_4^{-2}	Sulfate
PO_3^{-3}	Phosphite	PO_4^{-3}	Phosphate
CN^-	Cyanide	MnO_4^-	Permanganate
SCN^-	Thiocyanate	$S_2O_3^{-2}$	Thiosulfate
ClO^-	Hypochlorite	ClO_2^-	Chlorite
ClO_3^-	Chlorate	ClO_4^-	Perchlorate
$Cr_2O_7^{-2}$	Dichromate	HPO_4^{-2}	Hydrogen phosphate
$H_2PO_4^-$	Dihydrogen phosphate		

When compounds are made with ions, there are no real connections between the different ions (no real bonds, but rather just attractions). Therefore, the formula is referred to as an empirical formula. As such, all we need to write is the simplest ratio between the two ions. For example, the combination of the aluminum ion (+3) with the sulfate ion (-2) forces us to use two aluminum ions (with a total charge of +6) and three sulfate ions (with a total charge of –6). This yields a neutral compound: $Al_2(SO_4)_3$.

Other examples: $AlCl_3$, $Mg(ClO_4)_2$, MgO, Al_2S_3, and Na_3PO_4.

6.5 From Classical to Quantum Physics.

During the latter part of the 19th century and the early 20th century, while the electrons and protons were being discovered, a series of experiments baffled the scientific community. We must keep in mind that for the longest time (about 200 years) everything that was could be explained with Newtonian physics (also called classical physics or classical mechanics). Actually, in 1894 A. A. Michelson made a speech where he stated something to the effect that from that time on, the only things that would be discovered would be corrections to the previous ones to the sixth decimal place! In other words, if they already had estimated the distance from the sun to the Earth, all they would do from that point on would be to correct that estimation in order to make it more accurate. Just imagine the shock he must have felt when these series of experiments were actually impossible to explain with classical physics. It really took a long time in order for people to accept the fact that the properties of small particles (like atoms and molecules) are NOT governed by the same laws that worked so well for macroscopic objects.

To better understand some of these experiments let's talk for a while about electromagnetic radiation and waves. A *wave* can be described as a vibrating disturbance by which energy is transmitted. Let us see some of the properties of waves.

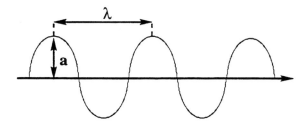

a. Wavelength (λ) - The distance from crest to crest in the wave (see it marked in the diagram above). Being a distance, it is measured in meters. However, since a lot of the waves we will be talking about are rather small, we will sometimes use a very small unit, which is the Angstrom[1] (Å). By definition: $1 \text{ Å} = 10^{-10} \text{ m}$.

b. Amplitude (a) - The distance from the axis of propagation to any point in the wave. Shown above (in the diagram) is the maximum amplitude. The intensity (or brightness) of light is proportional to the square of the amplitude.

c. Velocity of Propagation (c) - A constant and it is referred to as the speed of light. It is the maximum speed attainable for anything. $c = 2.998 \times 10^8 \text{ m/s}$.

d. Frequency (ν) - Defined as the number of waves that pass through a given point in one second. Therefore, the units are: inverse seconds (s⁻) or Hertz (Hz = cycle/s). The frequency of a particular wavelength of electromagnetic radiation is inversely proportional to its wavelength according to the following equation:

$$\nu = c / \lambda \quad \surd$$

Let's define light (or radiation for that matter) as the propagation of energy through space in the form of waves. James Maxwell showed in 1873 proved that visible light consists of electromagnetic waves – waves that have an electric and a magnetic component. He said that they both had the exact wavelength and frequency and the difference was that they were travelling perpendicular to one another (diagram drawn by Daniel Isom[2]).

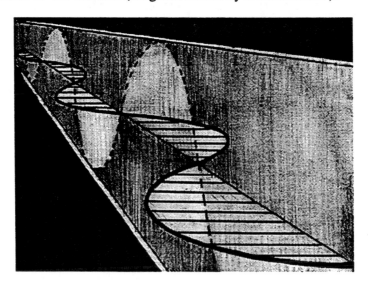

There are many types of electromagnetic radiation, and they differ from one another in

---

[1] In honor of Anders Angstrom, a Swedish physicist who lived from 1814 until 1874.
[2] CWRU undergraduate student from the class of 2000.

wavelength and frequency, but not in the speed of propagation (all propagate at the speed of light).

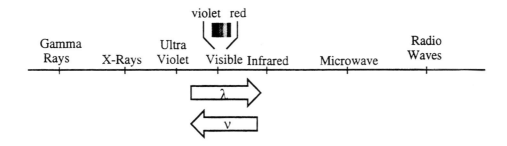

Above is a diagram with some of the different types of electromagnetic radiation. Notice that the wavelength increases from left to right, whereas the frequency increases from right to left. This should be clear since they are inversely proportional. From all of these types, only a very small sample are actually visible (can be detected by your eyes). This region is referred to as the visible region of the electromagnetic spectrum and it goes from Violet (highest frequency and smallest wavelength – about 4000 Å) up to the Red (lowest frequency and largest wavelength – about 7000 Å).

- A certain radiation with a 6000 Å wavelength has an orange color. What is the frequency of this radiation?

$$\nu = c / \lambda = (2.998 \times 10^8 \text{ m/s}) / (6000 \times 10^{-10} \text{ m})$$

$$= 5.00 \times 10^{14} \, s^{-1} \quad \sqrt{}$$

We mentioned that there were a series of experiments in the late 19th century that could not be explained with classical physics. One of the experiments that could not be explained in those days was what we call the **Blackbody Radiation**. When solids are heated, they emit radiation over a wide range of wavelengths. As an example take the case of an electric burner, hot enough to be glowing red. It was observed that the amount of radiant energy emitted depended on the wavelength.

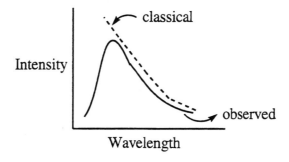

Scientists tried unsuccessfully to explain the results based on the wave theory and Newtonian mechanics but could not. According to classical physics, two things should happen. First, as we mentioned in the introduction to this chapter, the intensity of the emitted light is proportional to the energy, and secondly, the intensity was inversely proportional to the wavelength raised to the fourth power.

$$Intensity \propto \frac{T}{\lambda^4}$$

Let's think about the meaning of that last equation. This tells us that as the wavelength decreases, the intensity should increase. This prediction is seen with the dotted line in the figure above. However, as we can also see, the intensity does increase to a certain point and then it decreases rapidly at low wavelengths. This is what came to be known as **the ultraviolet catastrophe**.

The first person who tackled the problem was a young German physicist by the name of Max Planck[1]. In 1900, Planck studied the black-body radiation from the point of view of thermodynamics. He found that he could account for the experimental curve by assuming that the emitted energy came about from excited atoms that were vibrating in the solid (oscillators). He proposed that the energy of each electromagnetic oscillator is limited to discrete values and cannot be varied arbitrarily. This is a major concept: the Quantization of energy. In particular, he proposed that a given oscillator could only absorb energy of a particular frequency, ν, and that it could only do so by absorbing small chunks of energy he termed **quanta**, or packets. Thus, the energy of the oscillators would be proportional to the frequency, mainly:

$$E \propto \nu$$

Removing the proportionality sign and replacing it with a constant (h), we get:

$$E = h\nu$$

This constant **h** is a fundamental constant that we know today as Planck's constant. It has a value of 6.626×10^{-34} J s.

What is the meaning of this equation? That a given oscillator can only absorb (or emit) energy of $h\nu$. Of course, since we would have more than one oscillator, then we would experimentally observe that only a multiple of hn would be absorbed, depending on how many

---

[1] He lived from 1858 until 1947. He received the Nobel prize in physics in 1918 for his quantum theory.

oscillators get excited. We would therefore observe the absorption of hv, or 2 hv, or 3 hv, but not 1.33 hv. Why? Because a given oscillator would not absorb one third of a packet. If we were to add the wrong amount of energy (let's say 2.50 hv), we would observe the excitation of two oscillators only, and the remaining energy would not be absorbed. In order to account for this principle, we could write this equation in the following form:

$$E = Nh\nu.$$

where: N = number of oscillators.
h = Planck's constant.
ν = frequency

Furthermore, Planck suggested that there were many types of oscillators in the solid (some that absorbed low wavelengths, some that absorbed high wavelengths). He assumed that there were many more of the low wavelength type (which in his assumption above would imply high energy oscillators). His explanation for the ultraviolet catastrophe was that the amount of energy required to excite these high energy oscillators was too high and was not been supplied in sufficient quantities. Let's try to keep in perspective that his proposal was stunning for the time. He was allowing only specific values of energy to be absorbed by the oscillators in the solid, whereas classical physics did not apply any restrictions to the possible energy absorbed. Of course, this was not readily accepted, and it took quite a few years before scientists realized that there was something to this quantization concept after all.

6.6 The Photoelectric Effect.

When a metal in a vacuum is irradiated with monochromatic[1] light of sufficiently high frequency (or low wavelength), electrons are emitted from the surface of the metal. We call these emitted electrons: **photoelectrons**.

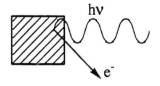

What was the problem with the photoelectric effect? Once more, 19th century people thought that the energy of light would increase with the intensity. Therefore, if light was very intense (high E) it should be absorbed by the metal in order to cause the photoelectron emission.

---
[1] Light of a single wavelength.

That was not the case. The experimental observations can be clearly seen in the plot shown below.

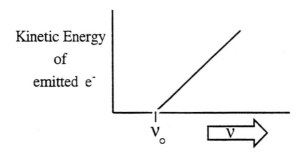

We can see that up to a certain frequency, which we will call the threshold frequency (v_o), no matter what we try, we will not get emission of photoelectrons. Different metals have different threshold frequencies (which at the time was called the metal's work function). Above that frequency, we observe that electrons are indeed emitted with a certain kinetic energy (remember that kinetic energy is given by: $1/2\, m\, v^2$) or a certain velocity. As the frequency is increased, so is the velocity of the emitted photoelectron.

The fact that at a given frequency the emitted electrons always had the same velocity was impossible to explain with classical physics, since in theory, increasing the intensity was supposed to increase the energy and thus increase the velocity of the photoelectrons. However, that was not the case. No matter how intense (or bright) you made the light of a given frequency, the velocity of the emitted electrons did not increase. Instead, what increased was the actual number of electrons emitted.

Albert Einstein[1] explained the photoelectric effect in 1905 using Planck's idea. However, he *quantized light*. He said that light travels through space in small packets of energy he called *photons* with energy : $\underline{E = N h v}$. This is the same exact equation that was proposed initially by Planck. Basically, he is saying that sunlight is a combination of all possible frequencies, and thus is has many packets of different energy (many different photons). Depending on the frequency, different photons have different energies. Applying this concept to the electromagnetic spectrum, we can state that the shorter the wavelength, the higher the energy of the photon. For example, we mentioned that visible light goes from 4000 Å (violet) up to about 7000 Å (red). This would imply that violet light is more energetic than red light.

---

[1] German born American physicist who lived from 1879 until 1955. He received the Nobel prize in physics for his explanation of the photoelectric effect.

- What is the energy of **one** violet photon of light with a wavelength of 4000 Å?

    Since we are asked to determine the energy of a single photon, then the equation above, $E = Nh\nu$ simply becomes: $E = h\nu = hc/\lambda$ (because N = 1). Substituting, we get:

$$E = hc/\lambda = \frac{(6.626 \times 10^{-34} \text{ J s})(2.998 \times 10^8 \text{ m/s})}{4{,}000 \times 10^{-10} \text{ m}}$$

$$\therefore E = 4.966 \times 10^{-19} \text{ J} \quad \sqrt{}$$

Let's try to explain what Einstein said about the photoelectric effect. He stated that the electrons (at the time, the prevailing notion was the raising pudding model for the atom) were stuck to the protons in the atom due to the fact that they had opposing charges (attraction). He went on to mention that in order for one of the electrons in the metal to absorb radiant energy, it would have to absorb a single photon. That is, one electron cannot absorb two or three photons simultaneously. One electron can only absorb one photon of light. Furthermore, in order for us to succeed in separating one of these electrons from the atom, we would have to apply a certain minimum amount of energy (called the threshold energy, E∘). This would immediately explain the fact that below a certain frequency (ν∘), we would not observe the emission of photoelectrons. Above the threshold frequency, the photon would have more than enough energy in order to break the attraction between the electron and the protons. So what would happen to the excess energy (the total energy of the photon minus the threshold energy)? It would be kept by the emitted electron and converted into kinetic energy, and thus take off with a certain velocity as given by: $1/2\, m v^2$. Of course, the higher the frequency of the photon, the higher the energy of the photon, so more energy would be left over to be converted into kinetic energy, therefore, the emitted electron would leave with a much faster velocity.

This was proposed by Einstein just five years after Bohr stunned the world with his proposal for the quantization of the energy for the oscillators (or the atoms) in a solid, which successfully explained the blackbody radiation experiment. Now we have the second of these experiments explained with the same identical equation, except this time it was the light itself that was quantized. In the next few years, one by one all the experiments were successfully explained and in each case, the same equation became the centerpiece of the entire explanation. The last experiment (atomic emission spectra of molecules) that was explained[1] was by far the most baffling, and it proved to be the last straw. The same equation originally proposed by

---

[1] Solved by Niels Bohr in 1913.

Planck ($E = h\nu$) turned out to be the one that defined the emission spectra of atoms. Scientists could no longer ignore that there had to be something behind the quantization concept and a new mechanics had to be developed.

- A certain metal has a threshold frequency of 1.00×10^{-18} J. When a photon of 585Å is strikes the surface of the metal, an electron is ejected. Determine the velocity of this emitted electron.

    Let's determine first the energy of the incident photon of light.

    $$E = hc/\lambda = \frac{(6.626 \times 10^{-34} \text{ J s})(2.998 \times 10^8 \text{ m/s})}{585 \times 10^{-10} \text{ m}}$$

    $$\therefore E = 3.40 \times 10^{-18} \text{ J}.$$

    It can be seen that this photon has more than enough energy to remove the electron from the surface of the metal (you only need 1.00×10^{-18} J in order to remove one electron). The difference in energy is converted into kinetic energy, so that:

    $$E_{photon} - E_{threshold} = 3.40 \times 10^{-18} \text{ J} - 1.00 \times 10^{-18} \text{ J}$$

    $$\therefore KE = 2.40 \times 10^{-18} \text{ J}. \quad \checkmark$$

    $$KE = \frac{1}{2}mv^2$$

    $$2.40 \times 10^{-18} J = \frac{1}{2}(9.11 \times 10^{-31} kg)v^2$$

    mass of an electron

    $$\therefore v = 2.30 \times 10^6 \text{ m/s} \quad \checkmark$$

- When a photon of 254 Å strikes the surface of metal **M**, it is observed that an electron is ejected with a velocity of 3.45×10^6 m/s. Determine the threshold energy of this metal ($E_\circ$).

    $$\frac{1}{2}mv^2 = \frac{hc}{\lambda} - E_\circ$$

    This time the unknown is the threshold energy, while everything else is known. Therefore,

$$\frac{1}{2}(9.11\times 10^{-31} kg)(3.45\times 10^6)^2 = \frac{(6.626\times 10^{-34} Js)(2.998\times 10^8 m/s)}{254\times 10^{-10} m} - E_\circ$$

and: $E_\circ = 2.40 \times 10^{-18} J$ √

6.7 Bohr's Theory.

As we said before, the explanation of the emission spectra of atoms in 1913 by Niels Bohr[1] proved to be the straw that broke the camel's back. We should spend a while trying to understand what is it that Bohr did and how he succeeded in explaining this last experiment.

The emission spectra of an atom is obtained by energizing a sample of the substance, for example by heating it up. This "hot" substance glows in a specific way and if the emitted radiation is passed through a prism, it will break into a characteristic spectrum. The surprising thing is that instead of obtaining a rainbow (which would indicate that the atom absorbed initially any amount of energy, and later emitted it) only a few distinct lines (wavelengths) were observed for a given atom. Yet, every different atom would emit a different spectrum. Actually, we could identify an element due to its emission spectrum.

In 1885, J. J. Balmer looked at the emission spectrum of hydrogen gas and he observed four well defined lines: one was red, another was blue-green, another was violet and the last one was light violet. The specific wavelengths can be seen in the table below.

Wavelength (Å)	Color
6563	Red
4861	Blue-Green
4340	Violet
4102	Light Violet

We can notice right away that the lines keep getting closer and closer (approaching a continuum). There was no explanation for this behavior, because according to classical physics, it should have been a continuum, since a hydrogen atom should be able to possess any amount of energy. A few years later, Rydberg[2] found an empirical mathematical relationship that allowed him to determine Balmer's observed wavelengths (remember that $\nu = c/\lambda$).

---

[1] Danish physicist who lived from 1885 until 1962. He received the Nobel prize in physics in 1922 for his explanation for the emission spectra of atoms.
[2] Johan Rydberg, a Swedish physicist who lived from 1854 until 1919.

$$v = R_H\left(\frac{1}{2^2} - \frac{1}{n^2}\right)$$

where: $R_H = 3.289 \times 10^{15}$ s^{-1}

n = 3, 4, 5, or 6.

Quite a few years went by with no explanation for the emission spectra phenomena. It was not until 1913 when Niels Bohr solved this problem by deriving a very similar equation to that of Rydberg based on the postulates we see below. We will try to derive this equation as closely as possible so as to get an idea of the immensity of this accomplishment.

1. *The hydrogen atom consists of a proton (positive) and an electron (negative). The proton makes up the nucleus and the electron orbits around the nucleus in a circular orbit of radius r.*

    Let's think about this first postulate. What is he saying here? So far all he has done is accept the Rutherford model for the hydrogen atom. Let us call the charge of the electron (which we said it was equal to -1.60×10^{-19} C): **− e**. Of course, the charge of the proton would then be: **+ e**.

From high school physics, we probably remember that when we place two charged particles nearby, a force will be established between them (attractive if they have opposite charges or repulsive if they have similar charges). The potential energy for the attraction between these two particles is given by Coulomb's law.

$$E_{potential} = \frac{q_1 \times q_2}{4\pi\varepsilon_o r}$$

where: q_1 = charge of particle 1
q_2 = charge of particle 2
r = distance between the particles.
ε_o = permittivity of the vacuum =
proportionality constant.
= 8.854×10^{-12} C^2/Jm

Chapter 6: Atomic Structure and Quantum Numbers

The potential energy between the electron and a nucleus in an atom that has a charge of: **Z e** (Z = # of protons in the nucleus) is given by:

$$E_{potential} = \frac{(Ze)(-e)}{4\pi\varepsilon_o r} = -\frac{Ze^2}{4\pi\varepsilon_o r}$$

Of course, the kinetic energy is given by $1/2\, m\, v^2$, therefore, the total energy would be:

$$E_{total} = \frac{1}{2}mv^2 - \frac{Ze^2}{4\pi\varepsilon_o r}$$

Let's look at the same thing in terms of the forces, rather than the energy. There are two forces operating on this electron: an attractive force (between the electron and the nucleus) and a centripetal force. The attractive force is the first derivative of the potential energy with respect to r, therefore, these two forces are given by:

$$F_{attraction} = \frac{Ze^2}{4\pi\varepsilon_o r^2}$$

$$F_{centripetal} = \frac{mv^2}{r}$$

These two forces must exactly offset each other, therefore:

$$\frac{Ze^2}{4\pi\varepsilon_o r^2} = \frac{mv^2}{r}$$

$$r = \frac{Ze^2}{4\pi\varepsilon_o mv^2}$$

At this point in the derivation, we have an equation for the radius of the electron in the atom, and we can see that it can have any possible value, depending on the velocity of the electron. Bohr analyzed the known results and went on to propose the second postulate, which is where quantization is suggested.

2. *Only the orbits which have its angular momentum (defined as the mass times the velocity times the radius: mvr) for the electron equal to **nh/2π** are allowed (where n is an integer).*

This postulate places a restriction as to which orbits are allowed. He did not know why the angular momentum was quantized in integrals of **h/2π**, but using this, everything fit together. According to this statement:

$$mvr = \frac{nh}{2\pi}$$

where: n = integer = 1, 2, 3, ...

and from there:

$$v = \frac{nh}{2\pi mr}$$

Let's substitute this equation for velocity into the last equation from the previous postulate, which gives an equation for the radius in terms of the velocity, and this way get an equation for the radius of the electron in a given orbit (r_n).

$$r = \frac{Ze^2}{4\pi\varepsilon_\circ m \left(\dfrac{nh}{2\pi mr}\right)^2}$$

$$r_n = \frac{\varepsilon_\circ n^2 h^2}{\pi Z e^2 m} = \frac{n^2}{Z} \times \underbrace{\frac{\varepsilon_\circ h^2}{\pi e^2 m}}_{\text{constants}}$$

The last part of the equation has all the constants compiled together. If we were talking specifically about the hydrogen atom (for which Z = 1) and more so, the first orbit (n = 1), then that equation would give us the radius of the first orbit for the hydrogen atom. We shall call this the Bohr's radius and will symbolize it with: a∘. The Bohr radius has a value of 5.29 x 10^{-11} m, or 0.529 Å. This corresponds to the experimentally determined radius. Therefore,

$$r_n = \frac{n^2}{Z} \times a_\circ$$

3. *As long as the electron stays in the orbit, it neither absorbs not emits energy.* This postulate got a lot of resistance because at the time it was thought that as the electron revolved in a circular orbit around the proton, it would lose energy. As soon as it loses some energy, it would move into an orbit with a smaller radius. Of course, it would not take long (a fraction of a second) in order for the electron to collapse into the nucleus.

Look at the equation already derived for the total energy of the electron in the atom. We notice that we can substitute the equation we also derived for the radius for the electron in the atom. However, we would still have an unknown: the velocity of the electron in the atom (v_n).

$$E_{total} = \frac{1}{2}mv_n^2 - \frac{Ze^2}{4\pi\varepsilon_\circ r_n}$$

$$r_n = \frac{n^2}{Z} \times \frac{\varepsilon_\circ h^2}{\pi e^2 m}$$

$$v_n = ???$$

The equation for the velocity can be derived from:

$$v_n = \frac{nh}{2\pi m r_n}$$

Substituting the equation for r_n, we get:

$$v_n = \frac{nh}{2\pi m (\frac{n^2 \varepsilon_\circ h^2}{Z\pi e^2 m})}$$

$$v_n = \frac{Ze^2}{2\varepsilon_\circ nh}$$

And finally, substituting r_n and v_n into the total energy equation, we get:

$$E_{total} = E_n = \frac{-Z^2 e^4 m}{8\varepsilon_\circ^2 h^2 n^2}$$

$$E_n = \frac{Z^2}{n^2} \times (\underbrace{\frac{-e^4 m}{8\varepsilon_\circ^2 h^2}}_{constants})$$

Substituting the values for these constants, we get the equation in the following form:

$$E_n = \frac{Z^2}{n^2} \times (-2.18 \times 10^{-18} J)$$

What is this equation allowing us to determine? The energy of each possible orbit in a hydrogen or a hydrogen-like atom. A hydrogen-like atom is one that has only one electron, just like hydrogen. For example: He^{+1}, Li^{+2}, etc. In these examples, Z would be 2 and 3, respectively. For the hydrogen atom (Z = 1), we could plot the energy of the different orbits and it would look something like the plot below. Notice that the energy levels keep getting closer and closer as the value of **n** increases.

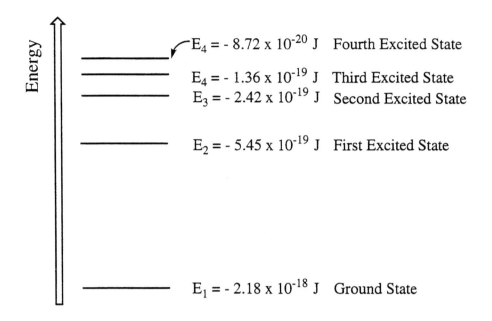

4. *To change from one orbit to another, the electron must emit or absorb a quantity of energy equal to the difference in energy between the two orbits.* In other words, the electron is going to start in a given orbit forever, unless it either absorbs the energy difference between the orbit it is in and the one it is going to. We now have an equation that allows us to determine the energy for the electron in any orbit, therefore, we can figure out the energy in the initial orbit and the energy in the final orbit. If it drops from a higher orbit to a lower orbit (as in the emission spectra), it would release the energy difference as a photon of light (hν).

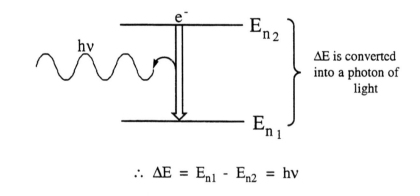

$$\therefore \Delta E = E_{n1} - E_{n2} = h\nu$$

but we already know E_{n1} and E_{n2} (*vide supra* – postulate #3):

$$\Delta E = \frac{Z^2 e^4 m}{8\varepsilon_\circ^2 h^2}\left(\frac{1}{n_1^2} - \frac{1}{n_2^2}\right) = h\nu$$

And solving for the frequency (to get it in the same form as the Rydberg equation from before, we get:

$$v = Z^2 \times \underbrace{\frac{e^4 m}{8\varepsilon_\circ^2 h^3}}_{\text{constants}} \times \left(\frac{1}{n_1^2} - \frac{1}{n_2^2}\right)$$

Substituting the numerical values for these constants, we get the final equation:

$$v = (3.289 \times 10^{15} s^{-1}) \times Z^2 \times \left(\frac{1}{n_1^2} - \frac{1}{n_2^2}\right)$$

This equation is exactly the same equation that Rydberg derived empirically in order to generate the four numbers for the Balmer series in the hydrogen atom, if n_1 is equal to two!!

Let us try to explain the meaning of all of this. Basically, when we placed the hydrogen gas in a tube and then submit it to passage of an electrical current, the hydrogen gas (H_2) will split into two atoms. Of course, we have many hydrogen molecules, let's say Avogadro's number, therefore, we will have zillions and zillions of hydrogen atoms. Normally, a hydrogen atom would be in its lowest possible energy state (n = 1), which we call the ground state. Once the molecule splits, these atoms will be excited to higher n values. Which ones? Since we have so many atoms, you can be convinced that we will have essentially all possible levels: some will be in n = 2, some in n = 3, some in n = 4, etc., plus you will have some in n = ∞. Anytime a system is excited, it will eventually return to the ground state (lowest energy). Therefore, all these atoms will start to emit energy in the form of photons. For example, an atom that starts at n = 25 could emit light and jump to the 10th level (n = 10)[1]. Then it can emit another photon of light as it jumps from the tenth level to the fourth level. It could then jump from the fourth level down to the second level and emit the corresponding photon and finally, it could jump from the second level to the ground state and thus emit one more photon of light. According to Bohr, this pathway would force the atom to emit four photons. From these four photons, only the one that falls into the second orbit will happen to have the right frequency in order to make it "visible". Actually, it would be the blue-green emission. The others would not be visible, since the first two would be a smaller frequency than visible (making it infrared radiation and thus, not visible) and the last one would be a larger frequency than visible (making it ultraviolet and thus, not visible).

From what we just stated, Bohr was able to not only explain the Balmer series for the

---

[1] Of course, this is an imaginary pathway, as it could drop down level by level, or in one big swoop from the 25th to the ground state.

hydrogen atom, but he was able to make predictions as to other emissions that were not visible. This is one of the most amazing things about this derivation: that is was able to explain and predict things to come. The other transitions, the ones that are not visible, were discovered after his predictions and the frequencies of the observed emissions were exactly those predicted by Bohr.

Results.

1. Bohr's theory quantitatively describes the H atom and other H-like atoms (atoms with only one electron).

2. Several other spectral series (other than Balmer's) were predicted and later found

Series Name	n_1	n_2	Spectral Region
Lyman	1	2, 3, ...	Ultraviolet
Balmer	2	3, 4, ...	Visible
Paschen	3	4, 5, ...	Infrared
Prackett	4	5, 6, ...	Far Infrared
Pfund	5	6, 7, ...	Far Infrared

    ```
    n = ∞
    n = 5
    n = 4                                    Prackett
                                             (Far IR)
    n = 3
                              Paschen
                                (IR)
    n = 2
                  Balmer
                   (VIS)


    n = 1
           Lyman
            (UV)
    ```

3. Basically Bohr quantized the energy that the electron might have in an atom. So, **n** can be thought of as a "quantum number" related to energy levels (the allowed ones).

Chapter 6: Atomic Structure and Quantum Numbers

- Determine the wavelength (in Å) for the photon of light emitted when an electron jumps from the second orbit to the ground state (n = 1) in the H-atom.

    We recognize that an electron in a hydrogen atom (Z = 1) is jumping from one orbit to another. Of course, with Bohr's equation we should be able to solve for the frequency.

    $$\nu = (3.289 \times 10^{15} \text{ s}^{-1}) \cdot Z^2 \cdot \left(\frac{1}{n_1^2} - \frac{1}{n_2^2} \right)$$

    $$= 3.289 \times 10^{15} \text{ s}^{-1} (1)^2 \{(1/1)^2 - (1/2)^2\} = 2.467 \times 10^{15} \text{ s}^{-1}.$$

    And what is the relationship between frequency and wavelength?

    $$\nu = c/\lambda.$$

    $$\therefore \lambda = c/\nu = (2.998 \times 10^8 \text{ m/s}) / (2.467 \times 10^{15} \text{ s}^{-1})$$

    $$\lambda = 1.215 \times 10^{-7} \text{ m} \times (1 \text{ Å} / 10^{-10} \text{ m})$$

    $$\therefore \lambda = 1215 \text{ Å}[1] \ \checkmark$$

- Would the wavelength be smaller or larger for the photon emitted in the corresponding transition ($n_2 = 2 \Rightarrow n_1 = 1$) for the Li^{+2} ion?

    From the equation (Bohr's), we can see that the energy of the emitted photon is proportional to ν, which in turn is proportional to Z^2, which in turn is inversely proportional to the wavelength. Therefore, since E α n α Z^2 α $1/\lambda$,

    then the energy will be larger but λ will be smaller!

- Consider the Li^{+2} ion. A certain transition in a corresponding Paschen series, is observed at 1424 Å. Determine the level in which the transition originated.

    This problem deals with a hydrogen-like atom, since this lithium ion is missing two of its three electrons. That means that it is left with only one electron, just like hydrogen. Therefore, Bohr's equation applies where Z = 3 (the atomic number, or the number of protons in the nucleus). Now, what is the Paschen series? This series corresponds to transitions from higher energy levels into the third level (n = 3). Since the only unknown is the initial level for the transition, then we can substitute directly into the Bohr's equation for this system.

---

[1] This corresponds to ultraviolet radiation (less than 4000 Å).

$$\nu = \frac{c}{\lambda} = 3.289 \times 10^{15} s^{-1} \times Z^2 \times (\frac{1}{n_1^2} - \frac{1}{n_2^2})$$

$$\frac{2.998 \times 10^8 m/s}{1424 \times 10^{-10} m} = 3.289 \times 10^{15} s^{-1} (3)^2 (\frac{1}{3^2} - \frac{1}{n_2^2})$$

Finally, solving for n_2, we would get:

$$n_2 = 5 \quad \surd$$

This means that the transition that causes the emission of a photon of light of 1424 Å occurs when the electron jumps from the 5th orbit down to the 3rd orbit.

We have seen that the Bohr theory was very successful in explaining the emission spectra of atoms and that it even succeeded in predicting exactly the wavelengths of emissions that did had not been observed yet. Does that mean that this was the final answer to everything? Did the Bohr theory solve all the problems? No. Many things were not explainable with this theory. For example, it was impossible to tackle the problems for atoms that had more than one electron. Remember that this theory can only handle the attraction of one electron to the nucleus. It was also observed that some lines were more intense than others. For example, in the Balmer series for the hydrogen atom, the red line is very bright, whereas the last violet line is almost impossible to see. This theory can not explain this observation. Furthermore, in the presence of a magnetic field, it was observed that some of these lines split into multiple others, which means that somehow in the presence of a magnetic field, the energy levels split, thus creating other possible transitions. This has been referred to as the Zeeman effect. Of course, Bohr's theory cannot provide the mechanism to explain this either. Even so, Bohr's theory persisted for about 10 years. Different scientists attempted to revise his theory by proposing elliptical orbits, rather than circular, but the problems persisted. Not only that, but keep in mind that this theory was based on an assumption, that there was quantization of the angular momentum. The theory did not explain why there was quantization, it simply assumed that this was the case in order to explain the things that were observed experimentally. This led to the development of a new mechanics: Quantum mechanics.

6.8 <u>Wave Mechanics</u>.

Einstein proposed the famous equation: $E = mc^2$; we can apply it to a photon of light, namely:

$$E = mc^2 = hc/\lambda.$$

And solving for the wavelength, we get:

$$\lambda = \frac{h}{mc}$$

Keep in mind that although a photon does not have a rest mass, but it does have momentum (p), which is defined as the product of the mass times the velocity (mc). Therefore, we could potentially write this equation as:

$$\lambda = \frac{h}{p}$$

Momentum is a property of particles, since it includes the mass. People kept trying to understand why Bohr's theory worked when he proposed quantization of the energy of an electron in an atom. But no matter how hard they tried, there was nothing previously proposed that could allow one to explain such a thing. However, in 1925 Louis De Broglié[1] proposed that just like electromagnetic waves could have particle properties (momentum), maybe particles (like electrons) could also have wave properties. He assumed that electrons could be treated as standing waves (much like the strings in a guitar) and he derived an equation that had the same form as the one above, mainly:

$$\lambda_{electron} = \frac{h}{p} = \frac{h}{(mv)_{electron}}$$

This equation gives particles (or objects) dual properties: particle properties (like mass and momentum) and wave properties (like wavelength). The wave properties of particles becomes significant only when the momentum of the particle (mv) is relatively close to the numerator, which happens to be Planck's constant (h). Since this constant is incredibly small, this means the mass of the particle that will exhibit significant wave properties must be small, therefore, this is only significant for microscopic particles (like electrons). This is why problems did not arise in the past: because everything observable was macroscopic, but as soon as experiments were carried out with microscopic particles (electrons), those five unexplainable experiments showed up.

Of course, this is meaningless unless it can be confirmed experimentally. The very following year, this was confirmed[2] by shining a beam of electrons into a thin gold film and

---

[1] French physicist who lived from 1892 until 1987.
[2] This experiment took place at Bell Laboratories by two American scientists: Davisson *1881 – 1958) and Germer (1896 – 1972).

observing that these were diffracted and the diffraction pattern was exactly the same as one previously observed with X-Rays. Not only that, but the diffraction pattern corresponded to a wave with a wavelength that was exactly the same as the one that was calculated with De Broglié's equation.

- What is the wavelength (in Å) associated with an electron moving at 1.00×10^5 m/s?

    We are talking about an electron, therefore, we should be able to determine the associated wavelength for this electron with De Broglié's equation:

    $$\lambda_{electron} = \frac{h}{mv} = \frac{6.626 \times 10^{-34} Js}{(9.11 \times 10^{-31} kg)(1.00 \times 10^5 m/s)}$$

    $$\therefore \lambda_{electron} = 7.28 \times 10^{-9} \text{ m} = 72.7 \text{ Å} \quad \checkmark$$

Conclusion: *All matter exhibits both particulate and wave properties!*

6.9 <u>Heisenberg's Uncertainty Principle</u> (1926).

In order for us to measure experimentally the velocity and or position of an object, we need to shine light into the object. For example, when a pitcher throws a baseball and we want to know how fast the baseball was coming to the plate, we use a radar gun. Electromagnetic radiation strikes the moving baseball, it is scattered and from this measurement, we can determine the velocity. In order to succeed, the wavelength must be smaller than the object itself. This is not a problem at all, because of the size of the baseball. However, if we try the same thing with an electron, we would be forced to use a photon of very short wavelength. We do know that a short wavelength contains a very high energy. What is the problem? The electron could very well absorb the photon and change its velocity and trajectory. This means that experimentally, we cannot be very accurate as to the velocity, and for that matter, the position of an electron. This immediately changes everything. Up to this point in time, everything is based on observing things. We can tell where is the object we are observing, we can measure its velocity, its trajectory, and all its properties. Now, we can see that when we are dealing with small particles, we cannot predict accurately its position nor its velocity.

Actually, in 1926, Werner Heisenberg proposed that this exact idea. No matter how careful we are in the measurement of an object, we are limited in the precision in which we can know both the position of the particle and its momentum at any given time. Not only that, but he stated that the limitation is of the order of Planck's constant, **h**! Mathematically, his equation

Chapter 6: Atomic Structure and Quantum Numbers

can be written as:

$$\Delta x \cdot \Delta(mv) \geq \frac{h}{4\pi}$$

where: Δx = uncertainty or error in the measurement of the position.

$\Delta(mv)$ = uncertainty in the momentum.

Conclusion: The harder you try to locate the electron, the less you will know about its velocity.

- Determine the uncertainty in the position of an electron and a baseball (mass = 145 g), if both have an uncertainty in their velocities of 0.200 m/s. How do the answers compare to the size of each object (the electron's size is approximately 0.1 nm)?

Remember that we are not talking about the actual position nor the velocities of these particles, but rather the uncertainties in their measurements. In other words, we have not stated what is the velocity of either, but we have rather assumed that someone succeeded in measuring them both to an uncertainty (or error) of ± 0.200 m/s. Let's start with the electron first. For the purpose of this calculation, we will use Heisenberg's uncertainty equation with an equal sign. We will assume that the mass is a constant, therefore, $\Delta(mv)$ is really: mΔv. Therefore,

$$\Delta x \cdot \Delta(mv) = \frac{h}{4\pi}$$

$$\Delta x \cdot (9.11 \times 10^{-31} kg)(0.200 m/s) = \frac{6.626 \times 10^{-34} Js}{4\pi}$$

∴ Δx = 2.89 x 10^{-5} m = 2.63 x 10^{+5} nm.

What is the meaning of this answer? Let's think about it for a moment. The electron is very tiny (in one dimension, we estimate it to be in the order of 0.1 nm). This means that the uncertainty in its position is much larger than the electron itself. Therefore, we have absolutely any idea as to its position. That would be equivalent to saying that a certain person weighs 180 ± 300 pounds. What information did that statement provide? Absolutely NOTHING!

Let's do the calculation for the baseball. This time, the Heisenberg's uncertainty equation looks like this:

$$\Delta x \cdot (0.145 kg)(0.200 m/s) = \frac{6.626 \times 10^{-34} Js}{4\pi}$$

$$\therefore \Delta x = 1.82 \times 10^{-33} \text{ m}$$

And what is the meaning of that number? This uncertainty is so much smaller than the size of a baseball, that we have absolutely no doubt as to the position of the baseball. Therefore, we can see that this limitation is not relevant for macroscopic objects, only for microscopic ones.

- Picture this: Over the Fall break, you meet a graduate student at another university during a party. He tells you that he has measured the position and momentum of an electron traveling in the x-axis and that his numbers were:

$$x = 1.63 \pm 0.08 \text{ Å}$$

$$mv = 1.02 \times 10^{-25} \pm 0.2 \times 10^{-25} \text{ kgm/s}$$

Do you believe him? Why?

The first thing we would consider is Heisenberg Uncertainty Principle, since the student provided the uncertainties. We would immediately mentally calculate the uncertainty in the position times the uncertainty in the momentum.

$$\Delta x \cdot \Delta(mv) = \frac{h}{4\pi} ??$$

$(0.08 \times 10^{-10} \text{ m})(0.2 \times 10^{-25} \text{ kgm/s}) = 1.60 \times 10^{-37} \text{ Js}$ √

Since: $h/4\pi = 5.27 \times 10^{-35}$ Js and the number he reported is smaller than that, it is impossible for this to be true.

6.10 The Development of Quantum Mechanics.

In 1926, both Werner Heisenberg and Erwin Schrödinger independently developed quantum mechanics. Schrödinger used the principles we just discussed (de Broglié and Heisenberg) as a basis to this new mechanics. Quantum mechanics is a probabilistic mechanics, as opposed to classical mechanics, which is exact. When mv is close to the numerical value of

---

[1] Austrian physicist who lived from 1887 until 1961. His presentation of wave mechanics earned him the Nobel prize in physics in 1933.

h, then classical mechanics no longer applies. We cannot measure the position nor the velocity of the electron exactly, thus the exactness of classical mechanics is gone for systems in which mv is close to h (Planck's constant). That is the reason why we calculate only the probability of finding the electron, without even attempting to trace its path!

Schrödinger considered the electron as a wave of energy and described it mathematically (differential equations) as seen below. This is known as Schrödinger equation.

$$\hat{H} \Psi = E \Psi$$

where: $\hat{H}$ = Hamiltonian operator.
= operator of the total energy.
Ψ = wave function for the electron.

This is the fundamental equation of quantum mechanics. An operator is a set of mathematical instructions. In our case, the Hamiltonian is the operator for the total energy, that is, the sum of the kinetic and the potential energy of the particle. In quantum mechanics, this operator takes the following form (in one dimension):

$$\hat{H} = \text{KE operator} + \text{PE operator}$$

$$\hat{H} = -\frac{\hbar^2}{2m}\frac{d^2}{dx^2} + V(x)$$

where: $\hbar = h/2\pi$

$$-\left(\frac{\hbar^2}{2m}\right)\frac{d^2\Psi}{dx^2} + V(x)\Psi = E\Psi$$

where: m = mass of electron
V = potential energy
E = total energy
Ψ = wave function for the electron, which can be separated into three parts and solved separately.

$$\therefore \Psi = R(r)\,\Theta(\theta)\,\Phi(\phi)$$

Solving this equation is the main concern of elementary quantum mechanics. It should be clear that many mathematical functions will not be appropriate functions in quantum mechanics. Why? Because of the restriction that when you apply the Hamiltonian to this wave function, you must be able to regenerate the function itself TIMES a constant. Since the Hamiltonian basically is the second derivative with respect to position, then you can envision a few functions where this may work. Example: sin, cos, or exponential functions. There are many more problems when you solve these equations. A blatant one is that you set up boundary conditions (like saying that the particle cannot be at a certain point) and due to those boundary conditions many times you are able to solve the equation itself. The point is that the solution of this equation does not fix the position of the electron nor does it describe its path. It mainly predicts where the electron is most likely to be found.

It is also important to understand that Ψ does not have a physical interpretation, that is, we cannot describe it pictorially. However, in quantum mechanics, the only function that has a real meaning is Ψ^2, which is a probability function. It is most useful as $4\pi r^2 \Psi^2$, which is referred to as the **radial probability density** - a measure of the probability of finding the electron within the volume of a thin spherical shell of radius r and thickness dr (where dr is an infinitesimal fraction of r).

Before we talk about the solution of this equation for the hydrogen atom and then extend it to multi electron atoms, we should stop for a moment and reflect on the basic principles of quantum mechanics. What better way is there to do this than to actually solve the equation for a very simple system? We will present next the solution to the simplest of all systems: a free particle trapped in a one-dimensional box. The purpose of this derivation is simple: to illustrate how quantum mechanics works and to clearly show how and why quantization arises during the mathematical solution of Schrödinger's equation. This is a special topic that we are placing in this book only to be followed by the people who have taken at least one semester of calculus. If you have not, then ignore this section. It makes quantum mechanics more palatable, but we do not need this section in order to follow the rest of the chapter.

6.11 <u>The Particle in a Box</u>.

Any quantum mechanical problem should start with the description of the system itself. We will define the system as a particle (which can be anything, for example an electron) of mass **m** that is trapped in a one-dimensional box of length **L**. There are two walls of infinite potential energy $\{V(x) = \infty\}$ located at $x = 0$, and at $x = L$. Therefore, the particle can never be physically there. We will also state that within the walls, the potential energy is equal to exactly zero $\{V(x) = 0\}$, which means that the particle is free to come and go as it pleases, with no attractive nor repulsive forces acting on it. Let's look at a diagram that illustrates our system.

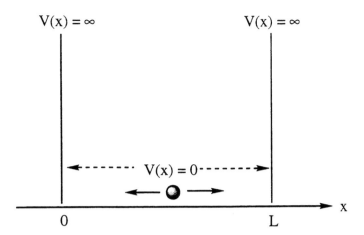

Since inside of the box the potential is zero, the Schrödinger's equation would be:

Now, a possible (mathematically speaking) wave function for this equation would be:

$$\Psi(x) = A \sin(kx)$$

The next step in the solutions of Schrödinger equation is to apply the boundary conditions to our system. Which are the boundary conditions? They were defined by the initial parameters of our system, that is: the particle can never be located at $x = 0$ nor at $x = L$. Therefore, when x is equal to either zero or L, Ψ and Ψ^2 must be equal to zero! Our choice for the wave function was very fortunate, because: $\Psi(0) = 0$, since the sine of zero is also equal to zero.

$$-\frac{h^2}{8\pi^2 m} \cdot \frac{d^2\Psi(x)}{dx^2} = E \cdot \Psi(x)$$

or:

$$\frac{d^2\Psi(x)}{dx^2} = -\frac{8\pi^2 mE}{h^2} \cdot \Psi(x)$$

We must make certain that $\Psi(L)$ is also equal to zero. So,

$$\Psi(L) = A \sin(kL) = 0$$

When is the $\sin(kL) = 0$? This is true anytime kL is equal to either π, or 2π, or 3π, etc. In other words: $\sin(kL) = 0$ only when:

$$kL = n\pi. \qquad n = 1, 2, 3, 4, \ldots$$

$$\therefore k = \frac{n\pi}{L}$$

Substituting into the equation for Ψ, we get:

$$\Psi(x) = A\sin(\frac{n\pi}{L}x)$$

What else must be true? The probability of finding the particle somewhere inside of the box must be 100%, since it cannot penetrate the walls. Therefore, since the probability of finding the particle is given by Ψ², then:

$$\int_0^L \Psi^2(x)dx = 1$$

$$\int_0^L A^2\sin^2(\frac{n\pi}{L}x)dx = 1$$

$$A^2\int_0^L \sin^2(\frac{n\pi}{L}x)dx = 1$$

The integral itself can be easily solved through calculus and it is equal to: L/2. Therefore,

$$A^2(\frac{L}{2}) = 1$$

or:

$$A = \sqrt{\frac{2}{L}}$$

Finally, the wave function has been found to be:

$$\Psi_n = \sqrt{\frac{2}{L}}\sin(\frac{n\pi}{L}x)$$

Now, let's find the energies of all the possible levels, in other words, let's solve the Schrödinger equation for the particle in the box. Since this equation involves knowing the second derivative of Ψ, let's start by doing that.

$$\Psi''(x) = -(\frac{n\pi}{L})^2[\sqrt{\frac{2}{L}}\sin(\frac{n\pi}{L}x)]$$

The quantity in the brackets ([…]) is what we called Ψ, therefore,

$$\Psi''(x) = -(\frac{n\pi}{L})^2\Psi(x)$$

Since we now know the second derivative of Ψ with respect to x, let's solve the Schrödinger equation for the particle in the box.

$$-\frac{h^2}{8\pi^2 m} \cdot \Psi''(x) = E \cdot \Psi(x)$$

$$-\frac{h^2}{8\pi^2 m} \cdot (-[\frac{n\pi}{L}]^2) \cdot \Psi(x) =$$

$$\frac{n^2 h^2}{8mL^2} \cdot \Psi(x) = E \cdot \Psi(x)$$

$$\therefore E_n = \frac{n^2 h^2}{8mL^2}$$

This is the equation we wanted. We have seen how the quantum mechanical solution of the Schrödinger equation gives the energy of the particle. However, the mathematical solution, when applied to the boundary conditions, introduced the quantization process. This brought forth what we call a quantum number (n), which has possible values of 1, 2, 3, 4, ..., ∞. Notice that the equation we derived for the energy of the particle is in terms of **n**, which means that we will obtain a large number of possible energy states. Keep in mind that the energy for this particle cannot have just any value. Its possible values are: $h^2/8mL^2$, or four times that value ($4h^2/8mL^2$), or nine times that value ($9h^2/8mL^2$), or ... Not only that, but we can clearly see that the energy levels are not equally spaced: they are getting further and further apart as **n** gets larger). Below, we can see a diagram that depicts the spacing of the energy levels (in units of $h^2/8mL^2$).

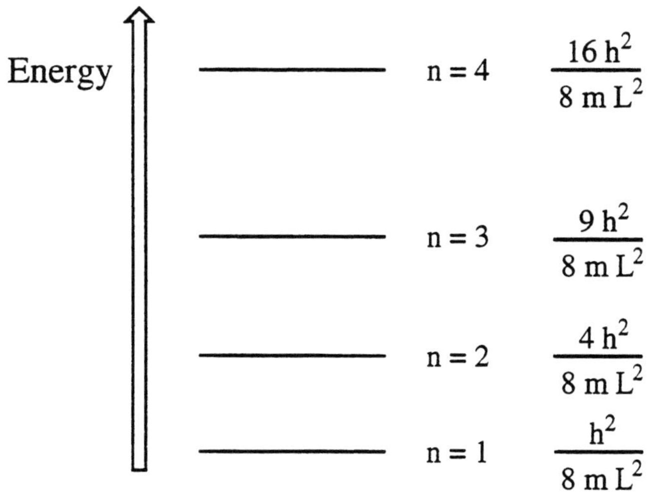

Energy

—————— n = 4 $\frac{16 h^2}{8 m L^2}$

—————— n = 3 $\frac{9 h^2}{8 m L^2}$

—————— n = 2 $\frac{4 h^2}{8 m L^2}$

—————— n = 1 $\frac{h^2}{8 m L^2}$

How about the wave function itself? We already mentioned that the wave function (Ψ) has no physical meaning in quantum mechanics. The quantity that has meaning is the square of the wave function (Ψ^2), which is related to the probability of finding the particle with a given energy. Just to see what we mean, let's make a plot of the square of the wave function and plot it versus the displacement (x) from the wall. We should be able to interpret the plots as probabilities. The diagram below depicts this information. For convenience, we have left the natural separation of energy in between the different values of **n**.

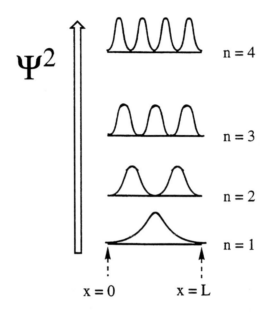

Let's analyze the results. At the lowest energy level (the ground state, **n** = 1), we see that the maximum probability of finding the particle occurs at exactly half of the length for the box, L/2. However, once the particle absorbs the exact amount of energy necessary to jump into the first excited state (**n** = 2), the probability function changes dramatically. We see that at this energy level, the probability of finding the particle is a maximum at: L/4 and at 3L/4. Not only that, but we can clearly see that there is a region of zero probability of finding the particle, which occurs at exactly L/2. We call this region a **node**.

If even more energy is applied and the particle is excited into **n** = 3, then we see that the maximum probability of finding the particle is at: L/6, at L/2, and at 5L/6. At the same time, we can see that there are two nodes at this level: one at L/3 and another at 2L/3. These are regions where there is absolutely no possibility of finding the particle with that much energy.

Most people, when confronted with the notion of a node, are baffled by its implications. For example, in the second level, there is one node at exactly L/2. We saw that the maximum probability for finding the particle is at L/4 and at 3L/4. How does the particle move from one side of the box to the other side, if it can never be at L/2? This is why many times quantum mechanics is called weird mechanics. There are things that cannot be explained with a nice picture and we simply accept it as a mathematical solution, which is all it is, after all.

Finally, we may have noticed that we gave **n** only the possible values of 1, 2, 3, ..., but not the value of zero. Of course, n = 0 would also have been an acceptable solution at first glance, since it would have made the sin (nπx/L) = 0. However, there is a problem. If **n = 0**, then the wave function itself is zero everywhere in the box. In other words, no matter what value you assign to x, the wave function would be zero and then the probability of finding the particle in the box would be zero everywhere. This is an impossibility since the particle has to be somewhere inside of the box. Therefore, even at the absolute zero of temperature, the particle will always have some energy, which we will call the <u>residual energy</u> or the energy of the ground state.

- Determine the wavelength of the photon of energy necessary to excite an electron from the ground state to the second excited state if the electron is trapped in a box of length 1.50 Å.

Let's determine first the energy difference between these two levels: **n** = 1 and **n** = 3. This is given by:

$$\Delta E = \frac{9h^2}{8mL^2} - \frac{h^2}{8mL^2} = \frac{8h^2}{8mL^2}$$

$$\therefore \Delta E = \frac{h^2}{mL^2}$$

So,

$$\Delta E = \frac{(6.626 \times 10^{-34} Js)^2}{(9.11 \times 10^{-31} kg)(1.50 \times 10^{-10} m)^2}$$

And therefore,

$$\Delta E = 2.14 \times 10^{-17} J \quad \surd$$

This amount of energy is converted into a photon of light, therefore, $\Delta E = hc/\lambda$, or:

$$\lambda = \frac{hc}{\Delta E} = \frac{(6.626 \times 10^{-34} Js)(2.998 \times 10^8 m/s)}{2.14 \times 10^{-17} J}$$

$$\therefore \lambda = 9.28 \times 10^{-9} m = 92.8 Å \quad \surd$$

This corresponds to a photon of ultraviolet light (we know this because it is smaller than 4000 Å, which is the smallest visible photon - violet).

Chapter 7: Quantum Numbers and Periodicity.

7.1 Quantum Numbers for the Hydrogen Atom.

The solution to Schrödinger's equation for the hydrogen atom brings forth four quantum numbers, in a very similar way as it did upon solving the particle in the box. For the hydrogen atom, however, we have a very different system than we did with the particle in a one-dimensional box. Here we have an electron attracted to the proton, which is in the nucleus. Thus, we cannot say that the potential energy is zero, since there is an attraction! This complicates the solution immensely. Not only that, but to make it even possible to solve, we must switch to spherical polar coordinates, which split the wave function into three functions: one that deals with a radial part and two others that deal with angular portions. The beauty of this is that we can then solve each equation separately. The mathematical process through which we solve these three equations yield three quantum numbers. Let's look at each equation separately.

1) Principal Quantum Number (**n**).

This first quantum number is derived from the radial part of Ψ, so its solution deals with the radial distribution. When we solve the Schrödinger equation for the hydrogen atom for only the radial part, we get an **energy** term that depends on this first quantum number, **n**. Therefore, the larger **n**, the higher the energy of the electron in the atom. In this equation, the possible values for the principal quantum number happen to be (very much like the quantum number we observed in the particle in the box, and in the Bohr's atom):

$$n = 1, 2, 3, ..., \infty$$

We learned previously the we are not actually interested in describing the actual position of the electron, but rather the probability function that describes the probability of finding the electron with a given energy. This probability function is what we called the radial probability density and it is obtained when we make a plot of $4\pi r^2 \Psi^2$ (radial probability) vs. r (radius). These plots will give us the probability of finding the electron at a certain distance (r) from the nucleus. Remember that $4\pi r^2 \Psi^2$ is just a mathematical equation that depends on **n** and r. Therefore, different values of **n** will yield different probability plots, as we can see in the figure below.

We will interpret the principal quantum number as the energy shell where the electron is most likely to be found. Therefore, when **n** = 1, we can say that the electron is found in the first shell, the one with the lowest energy, or the ground state. Once more, this is very similar to Bohr's interpretation, except he called them circular orbits and we are simply calling them energy shells, without even attempting to describe a path for the electron to follow, since we have proven that to be impossible to do for microscopic particles. The first plot below represents the radial probability density for the first shell in the hydrogen atom (n = 1).

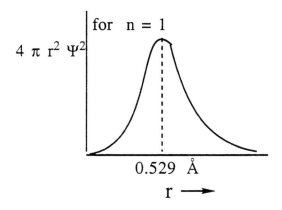

Here we see that the highest probability of finding the electron with that much energy happens to be at 0.529 Å. Once more, this should be a very familiar number. That is the radius for the first Bohr orbit. So what is the difference in the interpretation? Bohr said that the electron was always 0.529 Å away from the nucleus if it was in the ground state. Quantum mechanics states that the electron is most likely to be found at 0.529 Å away from the nucleus, but the probability graph tells us that it can really be anywhere.

If the electron absorbs the exact amount of energy required in order for it to jump to the second shell (**n** = 2), then it should move further away from the nucleus (on the average).

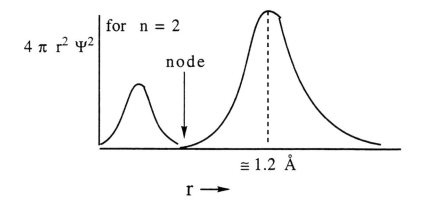

This is what we should expect, as **n** depends on two things: energy and distance from the nucleus. It is also true that if the electron acquires energy, it should have the ability to partially break some of the attraction to the nucleus, therefore, we should expect it to move further away. This is exactly what happens as we can see from the second plot (above).

We now see that the probability density is such that the most probable distance is approximately 1.2 Å (which is larger than 0.529 Å). However, there is one surprise[1]: there is a region (or a distance) where there is no probability of finding the electron in the second shell. Technically, we cannot say where the electron is, as it really can be anywhere, EXCEPT at the nodes. In this node, there is a zero probability of finding the electron. Once more, it is not our concern as to how the electron goes from near the nucleus out to the region where it is more likely to be found. The graph also clearly shows that even though the electron spends most of the time at a distance of around 1.2 Å, it has the ability to penetrate beyond the node and get really close to the nucleus at times. This is a very important property and one that we will revisit because it will help us explain many things about the behavior of atoms and molecules.

In summary, **n** gives us an idea of the energy of the electron in the atom. The higher **n**, the more energy the electron has and the further away from the nucleus it is *most likely* to be found (but it really can be anywhere except at the *nodes*).

Note that when n = 1, there were no nodes and when n = 2, there is one node. The number of nodes is one less than the principal quantum number. Therefore, if we were to plot the radial probability density when n = 5, we would see **four** nodes (that is, four points in space where there is no probability of finding the electron), with the probability density being a maximum relatively far away from the nucleus. Let's check the graph of n = 2 one last time (see the previous page). The maximum probability of finding the electron is around 1.2 Å, but the electron has a measurable probability of being very close to the nucleus at times. We can refer to this as the electron having the ability to penetrate and getting close to the nucleus at times.

2) <u>Angular (or Azimuthal) Momentum Quantum Number (ℓ)</u>.

The second quantum number comes from the solution of one of the angular functions. The solution of this second equation ends up with the quantization of the angular momentum that the electron may have in the atom. It turns out that its solution is also involved in the total energy of the electron in the atom, therefore, the larger ℓ, the larger the energy. This quantum number is related to what we call the energy subshell where the electron is found. Of course, since **n** is also related to the energy, the actual subshell will be represented by a combination of

---

[1] Although only a mild one, since we saw the same thing occur in the particle in the box.

Chapter 7: Quantum Numbers and Periodicity

both quantum numbers and should be represented in parenthesis: (n, ℓ). How is this angular momentum quantum number defined?

$$\ell = 0, 1, 2, ..., (n - 1) \Rightarrow \textit{maximum value for } \ell$$

This definition allows us to see that the second quantum number is not an independent quantum number, as it depends on the numerical value of the first quantum number. In other words, the number of subshells depend on the shell where the electron is found. If the electron is in the first energy shell, $n = 1$, therefore the maximum value of ℓ is zero ($1 - 1 = 0$), which means that ℓ can only have one value (zero). This means that there is only one subshell in the first shell: the (1, 0) subshell.

How about the second shell? In the second shell, $n = 2$, therefore, the maximum value for ℓ is one ($2 - 1 = 1$), which means that then ℓ has two possible values: $\ell = 0$ and $\ell = 1$. This means that we should have two subshells in the second energy shell: the (2, 0) subshell, and the (2, 1) subshell. Both of these subshells are more energetic than the single subshell found in the first shell. However, the one with the most energy would be the (2, 1) subshell since it has the highest value for ℓ.

As one final example, let's look at the third shell ($n = 3$). In this energy shell, we have three possible values for ℓ. The maximum value for ℓ happens to be 2 ($3 - 1 = 2$), therefore, these possible values are: $\ell = 0$, $\ell = 1$, and $\ell = 2$. This yields three distinct subshells in the third energy shell: the (3, 0) subshell, the (3, 1) subshell, and the (3, 2) subshell. Any one of these three subshells has more energy than the first three we have previously mentioned. However, once more, the highest energy is for the (3, 2) subshell, since it has the highest value for ℓ.

The scientists that started the spectroscopy work, did not want to work with all these numbers, so they invented a notation as a substitution for the second quantum number. We should learn this notation immediately.

ℓ	Notation
0	s
1	p
2	d
3	f
4	g
:	alphabetical :

In view of this new notation, let's review what we have seen so far. In the first three energy shells, we found the following subshells:

(1, 0)	⇒	now called the **1s** subshell (the zero stands for s).
(2, 0)	⇒	now called the **2s** subshell.
(2, 1)	⇒	now called the **2p** subshell (the one stands for p).
(3, 0)	⇒	now called the **3s** subshell.
(3, 1)	⇒	now called the **3p** subshell.
(3, 2)	⇒	now called the **3d** subshell (the two stands for d).

Let's predict all the subshells present in the fourth shell (**n** = 4). In this shell, we have four subshells, since ℓ can have the following values: 0, 1, 2, or 3 (maximum value ⇒ one less than 4). Therefore, the possible subshells are the: 4s, 4p, 4d, and 4f.

Summary:

n	(n, ℓ) possible
1	1s
2	2s, 2p
3	3s, 3p, 3d
4	4s, 4p, 4d, 4f
5	5s, 5p, 5d, 5f, 5g

3) <u>Magnetic Quantum Number (m_ℓ)</u>.

The third quantum number comes from the mathematical solution of the second of the angular functions and it turns out to be energy independent. In other words, this number will not have a contribution to the total energy of the electron in the atom. It turns out to simply tell us the possible orientations of the angular momentum in space. Since it is telling us the *orientations* of the angular momentum and that is what the second quantum number tells us (the possible values of the angular momentum), then this quantum number is going to be related to the second quantum number. It is actually defined as follows:

$$\mathbf{m}_\ell = -\ell, ..., 0, ..., +\ell$$

We call this the magnetic quantum number because even though we just mentioned that this quantum number does not contribute to the total energy of the electron in the atom, there are energy contributions if the atom is placed in a magnetic field. Under these conditions, we observe that the different values of this quantum number are no longer equivalent. This will help to explain the Zeeman effect, which the Bohr atom could not explain.

We see that the magnetic quantum number is defined in terms of ℓ. The three quantum numbers together define what we are going to call an orbital.

$$(\mathbf{n}, \ell, \mathbf{m}_\ell) \Rightarrow \text{Representation of an Orbital.}$$

An *orbital* is a region in space where there is a 95% probability of finding an electron with a given energy. Why do we work with orbitals in the hydrogen atom, whereas we did not for the particle in the box? Why are we making the distinction of choosing only a 95% probability of finding the electron as satisfactory whereas in the particle in the box we used 100% probability? Because the limits in the particle in the box were the ends of the box itself. The particle was trapped between zero and L, in a finite space. However, in the case of a hydrogen atom, the boundaries are: the nucleus and the end of the universe. Therefore, if we wanted to use this terminology and mathematically solve the equation that way, the only way to know for sure that we are choosing a volume where the electron is found 100% of the time, we would have to choose the entire universe. Let's get real! There **is** a probability that the electron in a given hydrogen atom is for a fraction of a second in the other end of the universe, just like there is a probability that all the air in the room right now will go to a corner as stay there, while we suffocate to death. However, this probability is so small that it is almost pointless to mention it. How far should we go before we have to be concerned that we are not keeping track of the electron? It turns out that we will only be interested in describing a volume where the electron is going to be found 95% of the time. This chunk of volume in space is what we call an orbital.

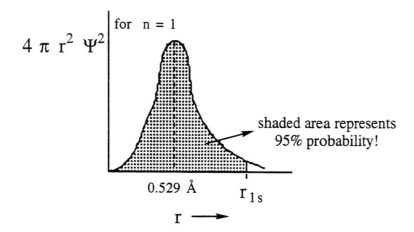

As an example, let's concentrate for a second in the first shell. We already have seen that in this shell there is only one subshell: the 1s subshell (for which $\ell = 0$). However, if $\ell = 0$, then it must follow that the only possible value for the third quantum number must be also zero (0).

Therefore, there is only one orbital in the first shell (or for that matter in the 1sw subshell): the (1, 0, 0) orbital. We have already seen that the most probable distance from the nucleus is about 0.529 Å but we are going to choose a volume which includes 95% of the probability of finding the electron in this first shell (or this 1s subshell). Therefore, we need to go further out until we reach this area. It has been marked as r_{1s} in the graph above. This is what we call the 1s "orbital".

If we were to take a picture of the electron in the 1s orbital (ground state - lowest energy) we could not "freeze" the electron. Instead we would see a spherical "cloud" (blurry). A cross-sectional area of this orbital would look like this:

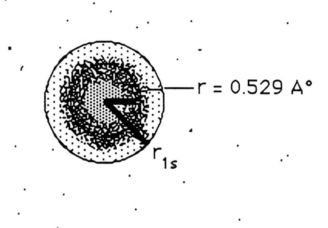

◊ The outer circle is the boundary of the 1s orbital.
◊ The darker inner cloud represents the maximum probability of finding the electron.
◊ The dots on the outside area represent the other 5% probability of finding the electron.

The picture we have seen plotted above tells us that there is a spherical probability around the nucleus in the 1s orbital. It turns out that mathematically this is always the case when we are talking about an s orbital. Conclusion: all s orbitals have spherical symmetry.

Let's talk about the second shell. We have already seen that the second shell contains two subshells: the 2s and the 2p subshells. There is only one value of the magnetic quantum number for the 2s subshell (since $\ell = 0$), therefore, there is only a single 2s orbital. However, in the case of the 2p subshell, $\ell = 1$. Therefore, the possible values for the magnetic quantum number are:

$$m_\ell = -1, 0, \text{ and } +1.$$

Since there are three distinct values for the magnetic quantum number, there are three different orbitals in the 2p subshell, which are: the (2, 1, -1) orbital, the (2, 1, 0) orbital, and the (2, 1, +1) orbital. How do these orbitals differ? Is one more energetic than the others? No. We say that these three orbitals are <u>degenerate</u> in energy (identical energy). They all look alike, but are oriented differently in space. Below we can see one of these p orbitals:

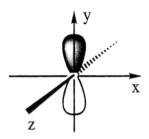

This particular p orbital, like the other two, looks like a dumbbell, and it is oriented along the y axis. Many times it is identified as the p_y orbital. The other two look exactly the same, but they are oriented along the x axis (p_x) and along the z axis (p_z).

- Label all the orbitals in the third shell.

In order to do this, we should be able to generate all the possible values of the magnetic quantum number in the third shell. First of all, we know that there are three subshells in the third shell: the 3s, the 3p, and the 3d. For each of these subshells, the values for ℓ are 0, 1 and 2. Therefore:

ℓ	m_ℓ	Orbitals
0	0	(3, 0, 0)
1	-1, 0, +1	(3, 1, -1), (3, 1, 0), (3, 1, +1)
2	-2, -1, 0, +1, +2	(3, 2, -2), (3, 2, -1), (3, 2, 0), (3, 2, +1) (3, 2, +2)

Therefore, we see a grand total of nine orbitals in the third shell: one labeled s, three labeled p, and five labeled d.

The five <u>degenerate</u> *d* orbitals are shown below. These orbitals are degenerate in energy unless the atom is placed in a magnetic field. Under these conditions, these five orbitals would split energetically. We will talk about this further in a subsequent chapter, when we see the effect of some transition metals on the **d** electrons when they form complex ions (Crystal Field Theory).

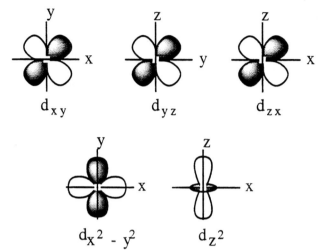

In summary,

Shell (n)	# of Subshells	# of Orbitals
1	1	1
2	2	4
3	3	9
4	4	16
⋮	⋮	⋮
10	10	100

Points of interest:

◊ Number of subshells in a shell = **n**.

◊ Number of orbitals in an entire shell = n^2.

◊ Number of orbitals in a particular subshell (ℓ) = $2\ell + 1$.

4) Spin Quantum Number (m_s).

The solution to the Schrödinger equation for the hydrogen atom initially ended there, with the first three quantum numbers. However, it was observed that if you took a beam of hydrogen atoms and placed them in between the two poles of a magnet, then half of the hydrogens would be attracted to the north pole of the magnet, whereas the other half would be attracted to the south pole of the magnet. This behavior is completely analogous to the one that would be observed if we had two charged particles spinning in their own axis: one spinning up and one spinning down. This led to the introduction of a fourth quantum number: the spin quantum number: m_s.

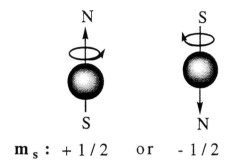

m_s : + 1/2   or   - 1/2

By convention, the two different values of the spin quantum number are set to be: +1/2 and -1/2. It turns out that a particular electron in an atom is defined by a set of four quantum numbers, that is:

$$(n, \ell, m_\ell, \text{ and } m_s)$$

7.2 Multi-Electron Atoms.

We will extend what we have learned for the hydrogen atom to multi-electron atoms. It should be pointed out that in order to solve the Schrödinger equation for multi-electron atoms, some assumptions must be made. These are the subject of a more advanced course called Quantum mechanics, which some of you may take in the future.

Our last statement in the previous section has more information packed in it that it seems at first glance. It tells us that in order to identify an electron in an atom, we need to know four quantum numbers. How do we decide how to assign the fourth quantum number? How do we know how many electrons do we place in each orbital? What are the implications of only having two possible values for the spin quantum number? Many of these questions were answered when Wolfgang Pauli[1] introduced the now famous exclusion principle.

Pauli's Exclusion Principle.

No two electrons in the same atom may have identical sets of all four quantum numbers. Since there are only two possible values for m_s, then the maximum numbers of electrons that can fit in *any* orbital is <u>two</u> and they must have opposite spins (one of them must be +1/2 and the other must be -1/2). If we represent an orbital by a horizontal line, then:

---

[1] Austrian physicist who lived from 1900 until 1958. He won the Nobel prize in physics in 1945.

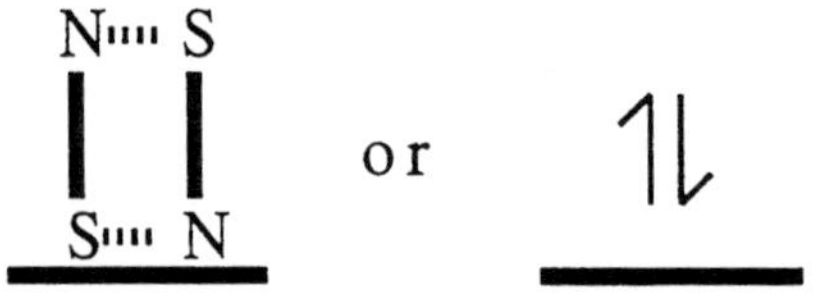

```
    N···· S
    │   │       or    ↑↓
    S···· N
   ───────          ─────
```

This configuration is stabilizing because the opposite poles of the magnets attract, thus helping to minimize the repulsion between the two electrons which reside in the same orbital.

If for some reason, there were three possible values for the spin quantum number, that would imply that every orbital would have the capability of allowing three electrons in, each with a different spin quantum number.

In summary,

Shell (n)	# of Subshells	# of Orbitals	Max. # of Electrons
1	1	1	2
2	2	4	8
3	3	9	18
4	4	16	32
:	:	:	:
10	10	100	200

Of particular interest:

◊ The maximum number of electrons in a shell = $2n^2$.

◊ The maximum number of electrons in a given subshell (ℓ) = $2[2\ell + 1]$.

7.2 Energy Levels.

Recall that **n** and ℓ (the first two quantum numbers) define the total energy of an electron in an atom. Since we know that the first energy level is lower in energy than the second, and that the second is lower than the third, we might be tempted to deduce that the order of the energy levels should be $1s < 2s < 2p < 3s < 3p < 3d < 4s < 4p < 4d < 4f$ ... However, as we can see from the graph below, as **n** gets larger, the energy levels keep getting closer and closer. It actually gets to the point where there is some degree of cross-over (ex. $4s < 3d$). We will see that there is no need to memorize the actual ordering of the subshells because there are quick, easy methods for determining their order.

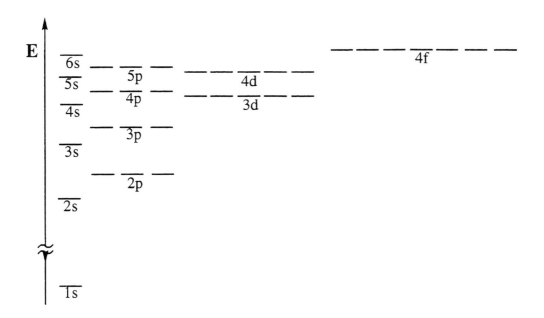

One of the methods to determine the ordering of subshells is the **n** + ℓ rule.

Orbital	n	ℓ	n + ℓ
1s	1	0	1
2s	2	0	2
2p	2	1	3
3s	3	0	3
3p	3	1	4
3d	3	2	5
4s	4	0	4
4p	4	1	5
4d	4	2	6
4f	4	3	7
5s	5	0	5
5p	5	1	6
⋮	⋮	⋮	⋮

Once we have added the values of **n** and ℓ, we need to place them in increasing order. However, we can notice immediately that there are numerous instances where the sums are equal. For example: 2p and 3s both add to three. In these cases, the one with the lower **n** will be lower in energy.

∴ Order: 1s < 2s < 2p < 3s < 3p < 4s < 3d < 4p < 5s < 4d < 5p < 6s < 4f < 5d < 6p < 7s < ...

Also, remember that when we look at orbitals within one particular subshell (like the five different 3d orbitals), they only differ in the value of the magnetic quantum number, but they all have the exact values of **n** and ℓ. Therefore, they are all degenerate in energy.

Another method to remember the ordering of the subshells is the following:

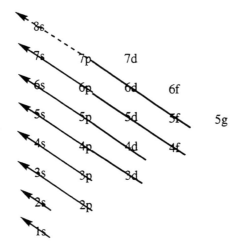

The lowest energetic subshell is the 1s. Start there and follow the arrows. We can see that originally, it seems to go in order (1s, 2s, 2p, 3s, 3p), but once we get past 3p, the chart helps a lot to identify the energetic order of the subsequent subshells. (3p, 4s, 3d, 4p, 5s, 4d, 5p, 6s, 4f, 5d, 6p, 7s, 5f, 6d, 7p, 8s, 5g...). We will now learn how to fill these orbitals with electrons as we look at the different atoms in the periodic table. We will see that the electronic configurations we are going to describe are critically important in determining the reactivity and the properties of different substances.

7.3 The Aufbau Principle.

In this method, we are going to assign each electron in a multi-electron atom to a particular orbital in such a way so as to obtain the lowest energy configuration for this atom. We will call this the ground state electronic configurations. We will follow the rules that we have learned (a maximum of two electrons in any particular orbitals and each with a different spin) and we will start placing these electrons in the lowest possible energy level (1s). We will then steadily go up in energy, orbital by orbital, until all the electrons in the atom are completely assigned.

We will write the ground state electronic configuration of a lot of the elements *by subshells* and *by orbitals*. We will also learn how to assign the four quantum numbers that describe any electron in an atom. Let's start with the simplest example: the hydrogen atom. How many electrons does it have? Only one, of course. Therefore, that single electron must be by itself in the lowest energy orbital, which is the 1s orbital. By subshells, the configuration would be written as: $1s^1$, where the superscript indicates how many electrons are in that particular subshell. On the other hand, if we want to write the configuration by orbitals, we need to draw a horizontal line to indicate the orbital itself, and then draw a half-arrow to indicate the spin of the electron. By convention, the first electron we ever draw in an orbital will be spinning upwards and we will assign it a spin quantum number of + 1/2. Therefore:

1. <u>Hydrogen</u> ($_1$H).

$$1s^1 \quad \Longrightarrow \quad \underline{\uparrow}_{1s} \quad (1, 0, 0, +1/2)$$

The four quantum numbers must be in order: **n**, then ℓ, then $\mathbf{m}_\ell$, then $\mathbf{m_s}$. The first three quantum numbers should be obvious: 1 for **n**, 0 for ℓ (since that is the definition of s) and 0 for the magnetic one (since that is the only possible value (from –0 to +0 ?). Finally, the spin is defined by convention (as we said above) to be +1/2, since it is the first one in the orbital.

2. <u>Helium</u> ($_2$He).

In this atom, we have two electrons: one of them is going to occupy and have the same four quantum numbers as the electron in hydrogen. The question is: where is the second electron? Of course, since there is still a place for an electron in that same orbital (the 1s orbital), it should follow that the second electron will occupy the same orbital, but with the opposite spin.

$$1s^2 \quad \Longrightarrow \quad \underline{\uparrow\downarrow}_{1s} \quad (1, 0, 0, -1/2)$$

What would be the four quantum numbers for that second electron? Since the first three quantum numbers describe the orbital and they are in the same orbital, then the only difference must be in the fourth quantum number, the spin. It must be spinning down with a spin quantum number of -1/2. This illustrates perfectly the Pauli exclusion principle.

3. <u>Lithium</u> ($_3$Li).

Once more, the first two electrons would have exactly the same assignment as those in Helium, however, once the first two electrons are assigned, we should recognize that the

1s orbital is completely full. Not only that, but since there are no other orbitals in the first shell, any subsequent electrons must go into another subshell. The next available (low energy) subshell is the 2s. This one also contains only one orbital. Therefore,

$$1s^2\ 2s^1 \implies \underset{1s}{\underline{\uparrow\downarrow}}\ \underset{2s}{\underline{\uparrow}}\quad (2, 0, 0, +1/2)$$

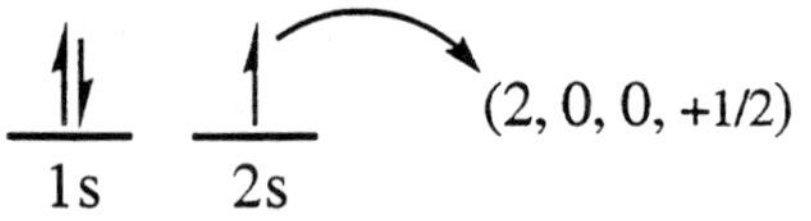

In the configuration by orbitals, please notice that there is a space in between the two horizontal lines. This separation is meant to symbolize an energy difference between the two. Technically, we should write it like this:

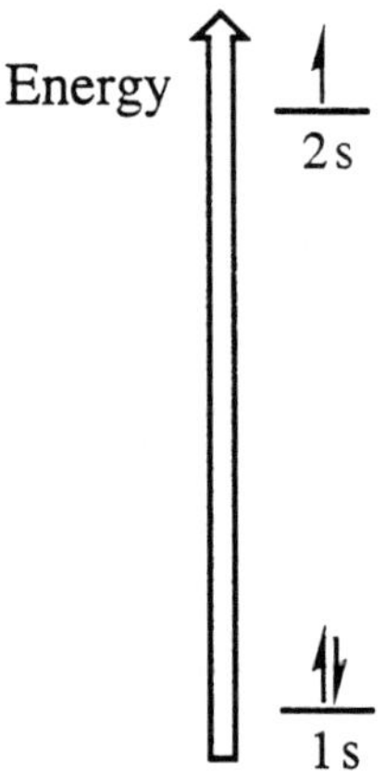

$$\text{Energy} \Bigg\uparrow \begin{array}{c} \underline{\uparrow} \\ 2s \\ \\ \\ \underline{\uparrow\downarrow} \\ 1s \end{array}$$

But we can imagine that pretty soon, we will run out of space! Therefore, from now on, a space in between two horizontal lines will symbolize an increase in energy.

Finally, the first three quantum numbers for the last electron in Lithium are (2, 0, 0). This is because these are the three quantum numbers that describe the 2s orbital. The last one is +1/2 since it is the first electron in that orbital.

4. <u>Beryllium</u> ($_4$Be).

The fourth electron will take its place along the third one, in the 2s orbital with opposite spin.

$$1s^2\ 2s^2 \implies \underset{1s}{\underline{\uparrow\downarrow}}\ \underset{2s}{\underline{\uparrow\downarrow}}\quad (2, 0, 0, -1/2)$$

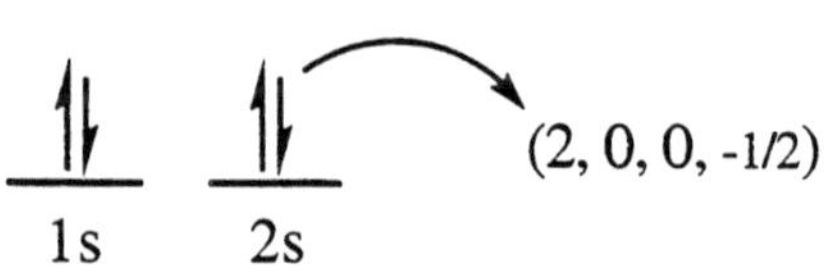

Once more, the first three quantum numbers that describe the last electron in Be start in exactly the same way as those for Li, since they are both in the same orbital. The difference

Chapter 7: Quantum Numbers and Periodicity

between these two electrons is simply the spin quantum number.

5. <u>Boron</u> ($_5$B).

Let's look at the electronic configuration of the Boron. We notice that we fifth electron is going into a new subshell, the 2p subshell. We have three orbitals that are degenerate in energy, therefore, we will be able to introduce up to six electrons in this subshell. This means that the next six elements (from B up to Ne) will be filling the 2p subshell.

$$1s^2\ 2s^2\ 2p^1 \implies \underset{1s}{\underline{\uparrow\downarrow}}\ \underset{2s}{\underline{\uparrow\downarrow}}\ \underset{\underbrace{-1\quad 0\quad +1}_{2p}}{\underline{\uparrow}\ \underline{}\ \underline{}}$$

tie line is used to indicate degeneracy.

Where (in which of the three orbitals) should we place the electron? The electron technically can go into any one of the three orbitals, but we will follow the convention of placing it in the first one on the left. Which is this orbital? Once more, by convention, we will label these orbitals from the most negative value for the magnetic quantum number (in this case −1) to the most positive (+1). Therefore, we will place the fifth electron in the (2, 1, -1) orbital. This means that the four quantum numbers that describe the last electron in the Boron will be: (2, 1, -1, +1/2).

6. <u>Carbon</u> ($_6$C).

The electronic configuration for carbon is: $1s^2\ 2s^2\ 2p^2$. The question is, where do we place this second electron in the 2p subshell? There are three distinct possibilities: either the two electrons are paired up in the −1 orbital, or the second electron goes with the same spin into the 0 orbital, or the electron goes into the 0 orbital with the opposite spin. From these three possibilities, it is the second one that gives the lowest energy, therefore, the one that we observe experimentally. According to **Hund's Rule of Maximum Multiplicity**, electrons stay in degenerate orbitals until each of these orbitals has at least one electron in it. The reason this happens is that this way we can minimize interelectronic repulsion. Therefore, the ground state electronic configuration for carbon is:

$$1s^2\ 2s^2\ 2p^2 \implies \underset{1s}{\underline{\uparrow\downarrow}}\ \underset{2s}{\underline{\uparrow\downarrow}}\ \underset{\underbrace{-1\quad 0\quad +1}_{2p}}{\underline{\uparrow}\ \underline{\uparrow}\ \underline{}} \implies (2, 1, 0, +1/2)$$

Since the last electron is in the (2, 1, 0) orbital, then the four quantum numbers that describe the last electron in the carbon atom would be: (2, 1, 0, +1/2). The next few elements continue filling the 2p in order (from left to right). Once the three orbitals are half-full, then we will go back to fill in completely all the three orbitals. Let's look at these elements (from N to Ne).

7. <u>Nitrogen</u> ($_7$N).

$1s^2\ 2s^2\ 2p^3 \implies$ [1s ↑↓] [2s ↑↓] [2p: ↑(-1) ↑(0) ↑(+1)] ⇒ (2, 1, +1, +1/2)

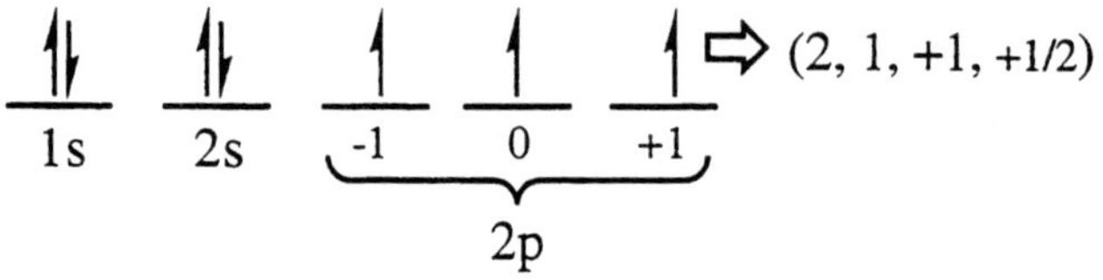

8. <u>Oxygen</u> ($_8$O).

$1s^2\ 2s^2\ 2p^4 \implies$ [1s ↑↓] [2s ↑↓] [2p: ↑↓(-1) ↑(0) ↑(+1)]

(2, 1, -1, -1/2)

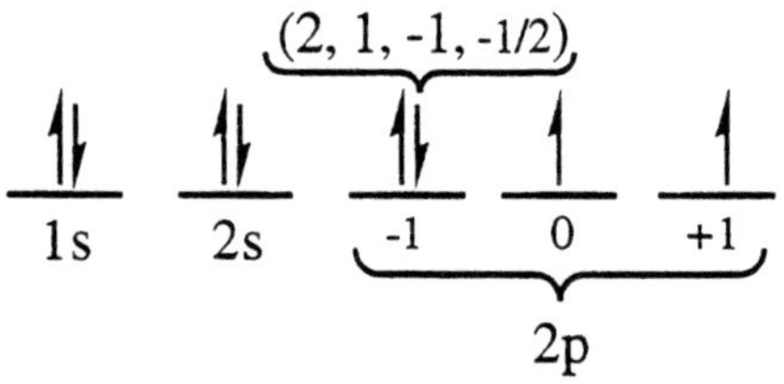

9. <u>Fluorine</u> ($_9$F).

$1s^2\ 2s^2\ 2p^5 \implies$ [1s ↑↓] [2s ↑↓] [2p: ↑↓(-1) ↑↓(0) ↑(+1)]

(2, 1, 0, -1/2)

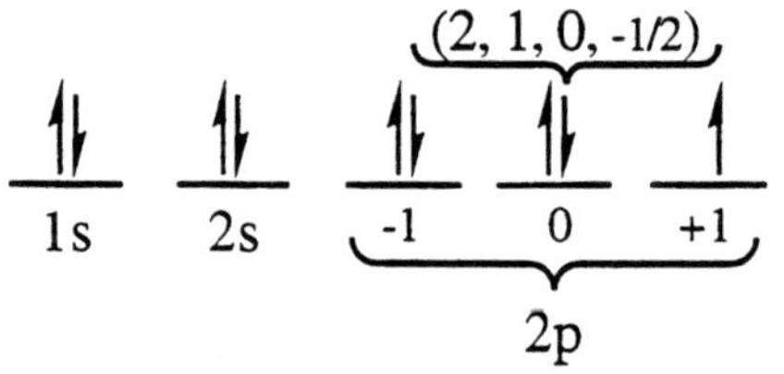

10. <u>Neon</u> ($_{10}$Ne).

$1s^2\ 2s^2\ 2p^6 \implies$ [1s ↑↓] [2s ↑↓] [2p: ↑↓(-1) ↑↓(0) ↑↓(+1)]

(2, 1, +1, -1/2)

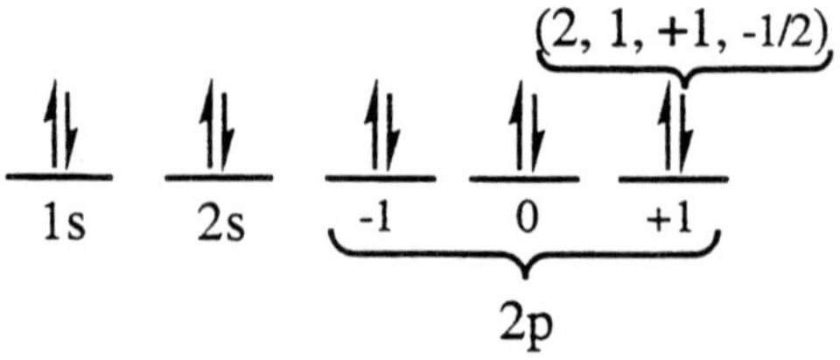

We have completely filled the 2p subshell and the second shell, for that matter. Any more electrons would have to go to another subshell. The next available subshell (lowest energy) would be the 3s. Therefore, the next two elements will be filling up the 3s subshell.

When a chemical reaction takes place, it always involves the ==valence shell electrons==. What do we mean by the valence shell electrons? They are the outermost electrons in an atom, the ones on the outer shell. These always correspond to the ones with the highest principal quantum number (**n**). How are these valence shell electrons involved during a reaction? They are either given to another species, taken away from another species, or shared. We observe that atoms that have a complete octet[1] (**n**s^2 **n**p^6) are unusually stable or unreactive. The other configuration that is unusually stable is that of Helium (a complete first shell: 1s^2). These atoms are extremely unreactive[2] gases and we call them the Noble or Inert gases.

We often write an electronic configuration by simply indicating where the electrons are, past the last noble gas, thus acknowledging the stability of the noble gases. For example, the configuration of oxygen may be written as: [He] 2s^2 2p^4. Let's look at a few other examples.

11. <u>Sodium</u> ($_{11}$Na).

$$[Ne]\ 3s^1 \implies [Ne]\ \underset{3s}{\underline{\uparrow}} \qquad (3, 0, 0, +1/2)$$

12. <u>Aluminum</u> ($_{13}$Al).

$$[Ne]\ 3s^2\ 3p^1 \implies [Ne]\ \underset{3s}{\underline{\uparrow\downarrow}}\ \underset{\underbrace{-1\quad 0\quad +1}_{3p}}{\underline{\uparrow}\ \underline{}\ \underline{}} \quad \Rightarrow (3, 1, -1, +1/2)$$

13. <u>Argon</u> ($_{18}$Ar).

This is another noble vas (with a complete octet: 3s^2 3p^6). Therefore, it is a very stable and unreactive monoatomic gas.

$$[Ne]\ 3s^2\ 3p^6 \implies [Ne]\ \underset{3s}{\underline{\uparrow\downarrow}}\ \underset{\underbrace{-1\quad 0\quad +1}_{3p}}{\underline{\uparrow\downarrow}\ \underline{\uparrow\downarrow}\ \underline{\uparrow\downarrow}} \qquad (3, 1, +1, -1/2)$$

Once we have filled up completely the 3p subshell, we encounter the first dilemma: where do we go next? We have available the 3d subshell, which is still part of the third shell and is therefore closer the nucleus on the average, or the 4s subshell, which would start a new valence shell (the fourth shell, which is a little further away from the nucleus – on the average).

---

[1] An octet is a series of eight valence shell electrons that imparts great stability to a configuration.
[2] They will not react by giving, taking, nor sharing valence shell electrons.

We have already seen that the 4s will be filled out before the 3d. One quick reason would be to say that the 3d is higher in energy because its quantum numbers (3 + 2 = 5) add up to a higher number than the 4s (4 + 0 = 4). This is fine, but what is the real reason for the 4s to be lower in energy than the 3d? After all, the 3d is a little closer to the nucleus than the 4s (4th shell). If it is closer (the 3d), its electrons should interact with the nucleus a little stronger and this have a lower energy. In order to understand this, we should go back once more and think about the radial electron probability density for these two orbitals. We are comparing the 4s and the 3d subshells. The s subshell is in the fourth shell and there were three previous s subshells that are already filled: the 1s, the 2s, and the 3s. However, the probability density of the 4s includes regions of measurable probabilities in these previous regions, separated by nodes. In contrast, the d subshells appear for the first time in the third shell. This means that there is no previous "history" of this type of subshell. Therefore, the probability density of the 3d subshell does not go quite as close to the nucleus as the 4s.

In Section 7.1, we mentioned that the number of nodes in a given shell is one less than the shell number. For example, if we are in the fourth shell, then we will find three nodes, whereas in the third shell we would only have two nodes. We should refine this concept at this point. It turns out that it really depends on which subshell we are referring to at the time. When we refer to a subshell, the question we should ask is: Have we seen this type of subshell previously in this atom? If we are talking about an **s** subshell, we recognize that there are previous **s** subshells (all the way to the first shell). Therefore, if we are talking about the 4**s** subshell, we will see three nodes, each corresponding to the previous shells (1 through 3). If we were talking about a 4**p** subshell, then we would recognize that the first **p** subshell is the 2**p** (remember that there are no **p** subshells in the first shell). Therefore, there would only be two nodes (each corresponding to the previous two **p** subshells: the 2**p** and the 3**p**). In the case of the 3**d** subshell, since that is the first time we would see a **d** subshell, we do not expect to see any nodes.

Let's look at the probability densities for the 3d and the 4s subshells in the figure below. We can see that indeed, the maximum probability of finding the electron in the 3d is closer to the nucleus than that for the 4s. However, due to the ability of the electrons in the 4s subshell to penetrate and get really close to the nucleus, even if it is for very short periods of time, the overall energy of the 4s is ever so slightly lower than that of the 3d.

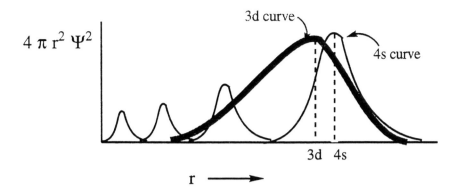

We call this the **penetration effect** and it is observed every time that we have an energy cross over. We could define the penetration effect as the effect of having an enhanced stability[1] of an electron which occupies an orbital that has a probability of being closer to the nucleus at times than orbitals of lower principal quantum number (**n**).

The following is the correct order for penetrating ability. Can you tell why this is the correct order?

Penetrating ability: s > p > d > f

Let's continue with some other examples of ground state electronic configurations.

14. <u>Potassium</u> ($_{19}$K).

$$[Ar]\ 4s^1 \implies [Ar]\ \underline{\uparrow}_{4s} \Rightarrow (4, 0, 0, +1/2)$$

15. <u>Scandium</u> ($_{21}$Sc).

$$[Ar]\ 4s^2 3d^1 \implies [Ar]\ \underline{\uparrow\downarrow}_{4s}\ \underbrace{\underline{\uparrow}_{-2}\ \underline{\ }_{-1}\ \underline{\ }_{0}\ \underline{\ }_{+1}\ \underline{\ }_{+2}}_{3d} \Rightarrow (3, 2, -2, +1/2)$$

16. <u>Vanadium</u> ($_{23}$V).

$$[Ar]\ 4s^2 3d^3 \implies [Ar]\ \underline{\uparrow\downarrow}_{4s}\ \underbrace{\underline{\uparrow}_{-2}\ \underline{\uparrow}_{-1}\ \underline{\uparrow}_{0}\ \underline{\ }_{+1}\ \underline{\ }_{+2}}_{3d} \Rightarrow (3, 2, 0, +1/2)$$

---

[1] The term enhanced stability refers to a stronger attraction between the electron and the nucleus.

Before we go on any further, let's talk about some stable electronic configurations.

a. ***ns² np⁶*** - Noble gas configuration. This is the most stable electronic configuration in the entire periodic table (completely full p subshell). This is the configuration we previously referred to as the complete octet.

b. *Completely Filled subshells* - There is an enhanced stability obtained when a subshell is completely filled. This is the case of the configurations ending in: s^2, p^6, d^{10}, f^{14}, etc. The more degenerate orbitals that we have in a given subshell, the greater the stability obtained when it is completely full. For example, we achieve much greater stabilization by having a completely full d subshell (d^{10}) than by having a completely full s subshell (s^2).

c. *Half Filled subshells* - These are obtained when we have added exactly half of the electrons required to completely fill that shell. For example: s^1, p^3, d^5, f^7, etc. Once more, the more degenerate orbitals we have in a given subshell, the greater the stabilization obtained when we finally we get it half full.

17. <u>Chromium</u> ($_{24}$Cr).

    The expected configuration for Chromium, according to the Aufbau method, is: [Ar] $4s^2$ $3d^4$. However, according to the stabilities mentioned above, it should be (and it is) more stabilizing to "half-fill" the 3d subshell by borrowing one electron from the 4s subshell. Why? Because we would then have a half-full 4s subshell (which is not as stable as completely full, but there is only one orbital, so it is not a dramatic loss), and then also have a half-full 3d subshell (which contains 5 orbitals and this should have a much more dramatic effect. Therefore, the true configuration for Chromium is: [Ar] $4s^1$ $3d^5$.

    [Ar] $4s^1 3d^5$ ⟹ [Ar] ↑(4s) ↑(-2) ↑(-1) ↑(0) ↑(+1) ↑(+2) 3d

18. <u>Manganese</u> ($_{25}$Mn).

    [Ar] $4s^2 3d^5$ ⟹ [Ar] ↑↓(4s) ↑(-2) ↑(-1) ↑(0) ↑(+1) ↑(+2) 3d

19. Copper ($_{29}$Cu).

This is another exception to the Aufbau method. Why is this the case? The expected ground state electronic configuration for copper would be: [Ar] $4s^2\ 3d^9$. In this case, we would have a completely full 4s subshell and an almost full 3d subshell. There are many more orbitals in the 3d than in the 4s, so we would get a much larger stabilization by borrowing an electron from the 4s and bring it into the 3d (and completely filling the 3d subshell).

[Ar] $4s^1 3d^{10}$ ⟹ [Ar] ↑(4s) ↑↓(-2) ↑↓(-1) ↑↓(0) ↑↓(+1) ↑↓(+2) — 3d

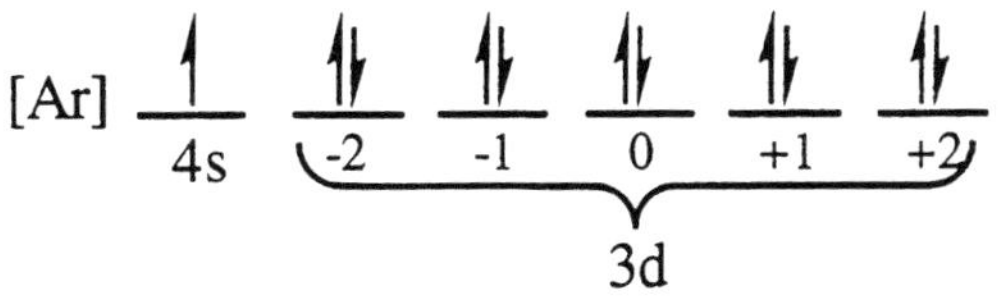

20. Gallium ($_{31}$Ga). Configuration: [Ar] $4s^2\ 3d^{10}\ 4p^1$.

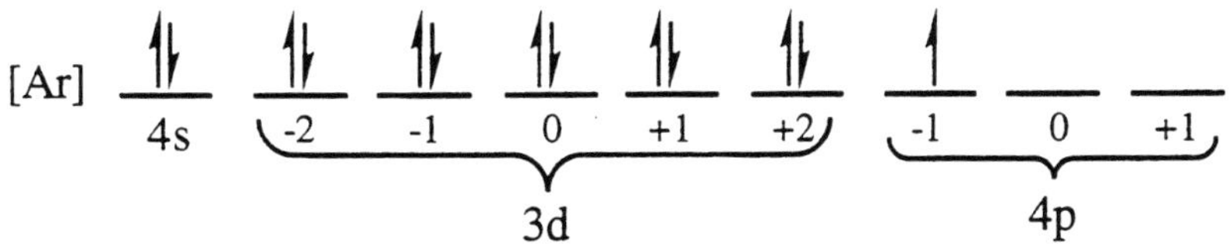

[Ar] ↑↓(4s) ↑↓(-2) ↑↓(-1) ↑↓(0) ↑↓(+1) ↑↓(+2) — 3d ↑(-1) __(0) __(+1) — 4p

7.4. The Periodic Table.

It would be time consuming to have to remember the order of the subshells by either adding the first two quantum numbers (**n** + ℓ) or by the diagonal method, as we have previously seen. However, everything we want to know (including the electronic configuration) can be read directly from the periodic table.

The periodic table is organized in periods and families. The vertical columns are called **families** because the elements that comprise them have similar characteristics and similar electronic configuration.

Examples:

1. The Inert Gases (or Noble Gases).

    - All (except for He) have a configuration of $ns^2\ np^6$ ⟹ Therefore, they all have a complete octet.
    - They are very unreactive and therefore exists in monoatomic form.
    - In many periodic tables, they are labeled as column VIII, or column 8A.
    - Elements: He, Ne, At, Kr, Xe, and Rn (radioactive).

2. The Halogens.

- All have a configuration of: $ns^2\ np^5$. Therefore, they are all missing an electron in order to have a complete octet.
- Very reactive non-metals.
- Elements: F, Cl, Br, I, and At (radioactive).
- In many periodic tables, they are labeled as column VII or column 7A.

3. The Alkali Metals.

- All have a configuration of: ns^1. Therefore, all of them have one more electron than a noble gas (one electron beyond a complete octet).
- Elements: Li, Na, K, Rb, Cs, and Fr (radioactive).
- Very reactive metals. Lithium reacts slowly with water in order to form LiOH and hydrogen gas, whereas Cesium explodes as it reacts with water in order to form CsOH and hydrogen gas.
- In many periodic tables they are labeled as column I, or column 1A.

4. The Alkaline Earth Metals.

- All have a configuration of: ns^2. Therefore, they all have a completely full **ns** subshell, which are two electrons more than a complete octet.
- These metals are not as reactive as the alkali metals. When they react with water, they always form a basic (alkaline) solution, thus their names.
- Elements: Be, Mg, Ca, Sr, Ba, and Ra (radioactive).
- In many periodic tables, they are labeled as column II, or column 2A.

5. The Chalcogens.

- All have the configuration of: $ns^2\ np^4$.
- These non-metals are not quite as reactive as the halogens.
- Elements: O, S, Se, Te, and Po (radioactive).
- In many periodic tables, they are labeled as column VI or column 6A.

Now, we have been mentioning the terms metals and non-metals. What do we mean by those terms? At this point we are making the distinction based on their reactivity and nothing else. **Metals** react by giving away their valence electrons, whereas **non-metals** react by accepting electrons into their valence shell. For example: Sodium metal (Na) has a configuration of: [Ne] $3s^1$. This means that Na has one more electron than the noble gas Neon. Therefore, as the sodium reacts, it will do so by losing its only valence shell electron (the one in the 3s subshell) in order to keep the configuration of the noble gas (Ne). It is true that all alkali metals behave the same way: they react by giving away one electron and thus form a positive ion (+1) with the configuration of a noble gas.

$$Na° \xrightarrow{-1e^-} Na^{+1}$$

configuration: [Ne] $3s^1$ [Ne]

On the other hand, an element like Chlorine (Cl) has a ground state electronic configuration of: [Ne] $3s^2\ 3p^5$. This element is missing one electron in order to acquire the configuration of a noble gas: Argon. Therefore, Cl will react by accepting one electron into its third shell (its valence shell) and thus get the configuration of Ar. All halogens behave in the same way: the react by accepting one electron in order to form a negative ion (-1) with the configuration of a noble gas.

$$Cl \xrightarrow{+1e^-} Cl^{-1}$$

configuration: [Ne] $3s^2 3p^5$ [Ar]

- Let's predict the kind of ions will be formed by the alkaline earth metals. Let's use as an example the metal Calcium.

    Calcium has two more electrons than the noble gas that precedes it: Argon. Therefore, we should expect for Calcium and the others in the family to lose two electrons in order to form +2 ions. For Calcium:

$$Ca \xrightarrow{-2e^-} Ca^{+2}$$

configuration: [Ar] $4s^2$ [Ar]

There is a dark line shaped like a staircase on the periodic table. This line separates the metals (to the left) from the non-metals (to the right). Of course, the separation is not quite as sharp as indicated by the line, and we find quite a few elements that fall in-between. These elements that may behave either way, depending on the situation, are referred to as **metalloids** or **semi-metals**. The metalloids are: Si, Ge, As, Sb, Te, and Po (right around the dark line).

The horizontal lines or rows in the periodic table are referred to as **periods**. We can see seven periods in the table, and the period number corresponds to the principal quantum number, **n**. We observe that the properties of the elements that comprise a given period change dramatically as we move from left to right. This is in marked contrast to the properties of the member of a family. For example, let's look at the third period in the diagram below.

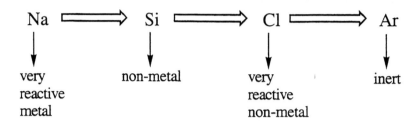

We observe that the first member of this period consists of a very reactive metal (a metal that can rather easily lose its only valence shell electron). The reactivity as metals decrease drastically as we move across this period (through Mg and Al). By the time we get to Silicon we have a non-metal / metalloid. As we move beyond Si, we see that the non-metals become more and more reactive (in their search for electrons) until we get to Cl. Chlorine is a very reactive non-metal. Finally, as in the case in all periods, we end with an inert gas: a totally unreactive non-metal.

The periodic table is also organized in **blocks**. This should be very helpful because we will be able to predict the ground state electronic configuration of most elements in the table by just looking at this organization we are referring to at this point.

The first two families (alkali and alkaline earth metals) are filling up the s subshells with electrons. Therefore, we will classify these first two columns as the **ns** block. On the other hand, the last six columns (ending with the noble gases) are filling up the p subshells with electrons. Therefore, we will classify these as the **np** block. We also can recall from the Aufbau method that after the 4s subshell, we started to fill the 3d subshell (for the next 10 electrons). Looking at the periodic table, this occurs after Ca (element #20), which is in the fourth period. This means that the next 10 elements (from Sc to Zn) we are filling the 3d subshell – an inner subshell since we are already in the fourth shell). We will therefore classify these families (from Sc to Zn and below) as the **(n-1) d** block. Finally, we can also notice that to the right of Lanthanum (La, element #57) is Hafnium (Hf, element #72!). Where are the 14 elements that seem to be missing? They are in a row below (elements 58 to 71). These are filling up the 4f subshell. This means that after the 6s subshell, we will start to fill in the 4f subshell (a deep inner subshell). Therefore, we will classify those rows from below the periodic table as the **(n-2) f** block. This can be summarized in the figure that appears in the following page.

Chapter 7: Quantum Numbers and Periodicity

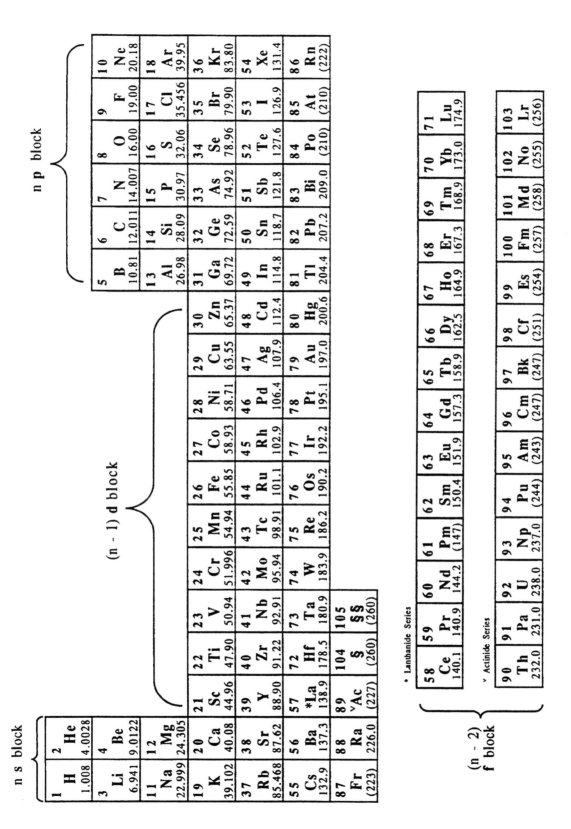

We should finally mention that elements in the **ns** and **np** blocks are often referred to as the **representative elements** and we should be able to tell the ground state electronic configuration of all the representative elements. Elements in the **(n-1) d** block are referred to as the **transition metals**. We should also be able to tell the ground state electronic configuration for the first row of these transition metals. There are many exceptions in these and many more in the **(n-2) f** block, but we are only required to know the first row of the transition metals.

7.5 Electronic Configuration of Ions.

A. Negative Ions.

Negative ions are formed when non-metals accept electrons to try to get a stable configuration, particularly the configuration of a noble gas. For example, Cl has a configuration of [Ne] $3s^2\ 3p^5$, therefore, since it is short one electron from the [Ar] configuration, it would react in such a way as to accept that one electron to form the chloride ion.

$$Cl \xrightarrow{+\ 1e^-} Cl^{-1}$$

configuration: [Ne] $3s^2\ 3p^5$ [Ar]

On the other hand, S has a configuration of [Ne] $3s^2\ 3p^4$, which makes it two electrons shy of a full octet. Therefore, sulfur tends to react in such a way as to accept those two electrons to get the configuration of argon.

$$S \xrightarrow{+\ 2e^-} S^{-2}$$

configuration: [Ne] $3s^2\ 3p^4$ [Ar]

B. Positive Ions.

When we form a positive ion, we must remove electrons from the valence shell. This is the case of metals, which lose their electrons from the valence shell. For example, aluminum has a configuration of: [Ne] $3s^2\ 3p^1$, which gives it three electrons over the configuration of neon. Therefore, Al will react by losing these three electrons (the valence ones) in order to get the configuration of the noble gas (Ne).

$$Al \xrightarrow{-\ 3e^-} Al^{+3}$$

configuration: [Ne] $3s^2\ 3p^1$ [Ne]

Let's look at some cases that are a little different: the transition metals. These metals behave in a very similar fashion: lose their valence shell electrons. However, how many valence shell electrons do these metals have and which are these? Let's take the case of iron (Fe). This is element #26 and its ground state electronic configuration is: [Ar] $4s^2 3d^6$. Of course, we can see that this metal has two immediate valence shell electrons (the ones in the 4s subshell, which is the outermost shell). We should therefore expect Fe to form the Fe^{+2} ion. This is one of the ions that exists naturally for iron. However, it does form one more ion in nature. How can this be? There are apparently no more valence shell electrons. It is possible to remove electrons from the 3d subshell, which is not even completely full. We shall refer to these electrons as being part of a "secondary valence" shell. Remember that the most stable configuration is that of a noble gas, so technically, the iron has eight valence shell electrons: two in the real valence shell and six more in the secondary valence shell. It is impossible to actually remove these many electrons, therefore what we observe is the removal of the real valence shell, and sometimes the removal of one or two more electrons from the 3d subshell. In the case of Fe, it will also form the +3 ion, meaning that we will end up removing the third electron from the 3d subshell. This leads to a configuration of [Ar] $3d^5$ for Fe^{+3}. Actually, the ferric ion is more common in nature than the ferrous ion (Fe^{+2}). This is because its configuration happens to be more stable (half filled 3d subshell) than that of the ferrous ion.

Other examples:

$$Sc \xrightarrow{-3e^-} Sc^{+3}$$
configuration: [Ar] $4s^2 3d^1$ [Ar]

$$Fe \xrightarrow{-2e^-} Fe^{+2} \xrightarrow{-1e^-} Fe^{+3}$$
configuration: [Ar] $4s^2 3d^6$ [Ar] $3d^6$ [Ar] $3d^5$

$$Tl \xrightarrow{-1e^-} Tl^{+1} \xrightarrow{-2e^-} Tl^{+3}$$
configuration: [Xe] $\mathbf{6s^2} 4f^{14} 5d^{10} \mathbf{6p^1}$ [Xe] $\mathbf{6s^2} 4f^{14} 5d^{10}$ [Xe] $4f^{14} 5d^{10}$

7.6 Magnetic Properties.

Most substances do not exhibit magnetic properties unless they are placed in between the two poles of a magnet. Under these circumstances, these substances will either be attracted to the external magnetic field or they will be repelled by the external magnetic field. The substances that end up being attracted to the external magnetic field are referred to as being **paramagnetic**. Paramagnetism occurs anytime we have at least one unpaired electron in the electronic configuration. This is the case of the hydrogen atom, for example, where there is only one electron in the 1s orbital. It would also be the case of the nitrogen atom, where we have three unpaired electrons ($2p^3$). However, this is not the case of the Ne atom. In this case, all of its 10 electrons are paired up. We refer to these atoms as being **diamagnetic**. When these substances that have all their electrons paired up are placed in between the two poles of an external magnetic field, we observe that they will be pushed out of the magnetic field, but the forces they experience are a lot weaker than those experienced by paramagnetic substances.

- Which is more paramagnetic (which has stronger forces towards a magnetic field): Fe^{+3} or Cr metal?

In order to answer this question, we need to know how many unpaired electrons we have in each of these species. Let's figure out the ground state electronic configuration for each of them. Of course, the Fe^{+3} is: [Ar] $3d^5$, whereas the Cr metal is: [Ar] $4s^1\ 3d^5$. Therefore, the Fe^{+3} has 5 unpaired electrons, whereas the Cr° has six unpaired electrons.

Fe^{+3}:   [Ar] __ ↑ ↑ ↑ ↑ ↑
             4s   +1
                    3d

Cr: [Ar] ↑ ↑ ↑ ↑ ↑ ↑
            4s
                 3d

Therefore, the Chromium metal will exhibit stronger magnetic effects than the ferric ion.

7.7 Atomic Radii.

The term atomic radius refers to the *average[1] distance* between the nucleus and the valence shell. These are experimentally measured values, and there are various ways of determining these measurements (depending of the types of compounds these atoms form). Of course, in the case of substances that do not form bonds (like the noble gases), then it is measured as the distance between the atoms in the solid phase. However, for atoms that do form molecules, like H_2, then the atomic radius of the hydrogen atom is simply defined as half the

---

[1] Remember that the electrons are not in a particular place. They are in constant motion and we cannot tell where they are located at any specific time. We do know, however, the more likely place to find them when they are in a given energy shell. That is why we call it the average radius: because the electrons are essentially all over the place.

distance between the two hydrogens in the molecule itself. Remember that distinction when we later compare the actual radii of different atoms. Almost every time there seems to be a slight discrepancy, it turns out that the values were measured differently.

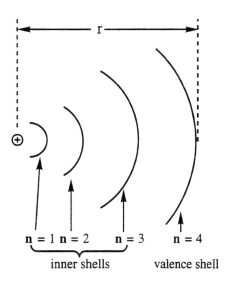

The actual atomic radius will depend on various things. For example, it depends on which shell the electrons are found. The larger the principal quantum number, **n**, the more energy it has and thus it can move further away from the nucleus. We have seen before that as **n** increases, the maximum of the radial probability distribution is shifted towards larger values of r (which means that it is moving further away from the nucleus). Let's look at one example to see if this is true or at least to see if it makes sense. We should expect for sodium (Na) to be larger than lithium (Li) since Li ends in $2s^1$, whereas Na ends in $3s^1$. Since the valence shell for sodium is in the third shell, it should be a larger atom. This is exactly the case, as we can see that the atomic radius for Na is indeed larger than that of Li.

Atom	Configuration	Radius
Li	$1s^2 2s^1$	1.52 Å
Na	$1s^2 2s^2 2p^6 3s^1$	1.86 Å

Conclusion: In general, as you down a family, atoms get larger since their valence shell is in a higher shell (larger **n**).

How about if we were to move across a period? Most people would guess immediately that atoms should also get larger as we move from left to right in the periodic table, since we are adding more and more electrons to the atom. However, we observe exactly the opposite. Let's look at a direct comparison between lithium (Li) and beryllium (Be). We can see that indeed Li is larger than Be.

Atom	Configuration	Radius
Li	$1s^2 2s^1$	1.52 Å
Be	$1s^2 2s^2$	1.11 Å

Let's try to understand why this is the case. We will define the **effective nuclear charge** (ENC) as the nuclear charge **felt** by the valence shell. We have to take into consideration the effect that the inner shell electrons may have on the valence shell. What role do they play? Since they are in between the valence shell and the nucleus, they serve the role of screening part of that nuclear charge from the valence shell itself. In other words, we can visualize the inner shell electrons as blocking or canceling part of the nuclear charge.

Let's think of these two examples: Li and Be. How many inner shell electrons do they have? The Li has two inner shell electrons (the $1s^2$) and one valence shell electron (the $2s^1$). However, the Be still has only two inner shell electrons (the $1s^2$), leaving it with two valence shell electrons (the $2s^2$). In the case of Li, these two inner shell electrons effectively cancel two of the protons from the valence shell, leaving the single electron in the valence shell to feel the attraction of a single proton[1]. Of course, in the case of Be, since it also has only two inner shell electrons, these also cancel out two of the protons from the valence shell, which means that its valence shell now feels the attraction of two protons. Since the attraction of the valence shell for the nucleus is stronger for Be, then it should follow that it must be a smaller atom.

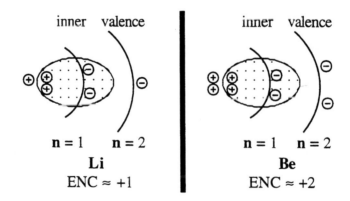

---

[1] Actually, this is a gross exaggeration. The inner shell does not quite cancel two protons. However, the trend is what we are after at this point. The ENC of the Be is indeed larger than that of the Li, so we will "determine" it as we explained in the text, keeping in mind for a future course that this is not quite accurate.

Conclusion: In general, as you move across a period atoms get smaller due to the fact that the ENC keeps increasing, thus the pull on the valence shell is larger and the atom smaller.

We should also point out that not all subshells are equally good screeners. Since we have defined **screeners** as electrons in inner orbitals that have the ability to partially block the nuclear charge, then orbitals that have spherical symmetry (which are the s orbitals) should be able to screen better than those in symmetries that do not completely surround the nucleus. Below we can see that if we were standing from the outside looking in, the s orbital on the left covers the

nucleus extremely well, the p orbital in the center covers the nucleus pretty well, but of course not as well as the s orbital, and the d orbital in the right does the worst job at covering or screening the nuclear charge from the outside. Therefore, the screening ability follows the order seen below:

Screening ability: s > p > d > f

I do not expect you to actually calculate the effective nuclear charge felt by the nucleus, since we have not technically learned how to do this, but instead we should be able to get the general idea of how the effective nuclear changes (and why it does) as we move from left to right in any particular period in the periodic table.

- Which atom would you expect to be smaller, the phosphorus (P) atom or the sulfur (S) atom? Explain your answer in terms of screening ability of the inner shell electrons and the effective nuclear charge felt by the valence shell.

We would expect for the S atom to be smaller since it should have a higher effective nuclear charge. Let's look at the configurations of each atom: P is [Ne] $3s^2\ 3p^3$ whereas the S is [Ne] $3s^2\ 3p^4$. Each atom has the same identical inner shell structure: the neon inner shell. This means that each atom has 10 inner shell electrons. Since S has one more proton (16) than P (15), then the effective nuclear charge for the S has to be larger than that of the P. Remember that this will always be the case in any particular period. For example, in the entire third period, we can see that the atomic radius indeed gets smaller and smaller as we go from sodium metal to argon gas.

Na	Mg	Al	Si	P	S	Cl	Ar
1.86 Å	1.60 Å	1.43 Å	1.17 Å	1.10 Å	1.04 Å	0.99 Å	0.94 Å

<u>Ionic Radii</u>.

How does the radius of ions compare to those of the neutral species? Are ions larger or smaller than the neutral species? As we will see, it depends on whether they are positive or negative ions.

A. <u>Negative Ions</u>.

Negative ions should always be larger than their neutral counterparts, since they contain one or more electrons than the neutral one. These extra electrons are in the same valence shell, therefore the nucleus will be trying to hold on to extra electrons with the same nuclear charge. This makes for a weaker hold and therefore a larger species. This can be seen from the numbers below, where we compare the sizes of the fluorine atom versus the fluoride ion. It should be obvious that nine protons can hold on tighter to nine electrons than they can hold on to ten electrons. Also the added (extra) electron repels the other nine, so the electron-electron repulsion also helps to make the ion larger.

Therefore, $r_{F^-} > r_F$

Conclusion: Negative ions are larger than their neutral counterparts.

Atom	Configuration	Radius
F	$1s^2 2s^2 2p^5$	0.64 Å
F⁻	$1s^2 2s^2 2p^6$	1.36 Å

B. <u>Positive Ions</u>.

How about positive ions? What happens when we form a positive ion? We actually remove electrons from the valence shell. Not only that, but many times we remove all the valence shell electrons, therefore, we end up with an inner shell as the new outer shell. This, by definition, should make the atom smaller. Another way of looking at this is thinking in terms of the attraction between the nucleus and the outermost shell. Once we remove electrons from an atom, the positive ion formed will have less electrons held by the protons in the nucleus, therefore, the attraction should be stronger and the species smaller. Let's look at one particular example: the difference in size between the lithium atom and the lithium ion.

Atom	Configuration	Radius
Li	$1s^2 2s^1$	1.52 Å
Li$^+$	$1s^2$	0.60 Å

The +1 ion will be smaller than Li for the two reasons stated above. First, three protons will hold on tighter to two electrons than to three. Second, the configuration itself makes it evident. Li has an electron in the second shell whereas Li$^+$ has electrons only in the first shell (which is closer to the nucleus).

$$\therefore r_{Li} > r_{Li^+}$$

C. Isoelectronic Species.

The term **isoelectronic** refers to species that have the same number of electrons, therefore they have the same electronic configuration. An example would be the fluoride ion (which has 10 electrons) and the sodium ion (which also has 10 electrons). Of course, 10 electrons would yield a stable configuration: that of the noble gas Neon. We would expect for the fluoride ion to be larger than the sodium ion since we have a smaller nuclear charge (only nine protons in the nucleus as compared to sodium, which has 11 protons) pulling on the same number of electrons (10 electrons). This is exactly what we observe, as can be seen below.

Atom	Configuration	Radius
F^{-1}	$1s^2\ 2s^2\ 2p^6$	1.36 Å
Na^{+1}	$1s^2\ 2s^2\ 2p^6$	0.95 Å

This can be also seen by a much larger representation of an isoelectronic series: from the nitride ion, up to the aluminum ion, all with the neon configuration.

N^{-3}	O^{-2}	F$^-$	Ne	Na$^+$	Mg^{+2}	Al^{+3}
1.71 Å	1.40 Å	1.36 Å	1.12 Å	0.95 Å	0.65 Å	0.50 Å

Radii of the Transition Metals.

How about the transition metals? Do they follow the same pattern, meaning that they get smaller as we move from left to right? In other words, is the effective nuclear charge increasing steadily from left to right as we go from Scandium (Sc) to Zinc (Zn)? Let's look at a very specific example and make a prediction before we try to explain the results. Let's predict the relative size of Ca and Sc. In order to make a prediction, we must first look at the ground state electronic configuration for both atoms.

Ca : [Ar] 4s^2

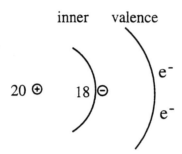

We see that the calcium atom has 20 protons in the nucleus and 20 electrons in the shells. However, 18 of these electrons happen to be in the inner shells (which comprise the argon core), and only two electrons happen to be in the outer or valence shell. According to our past analysis, the 18 inner shell electrons should screen the valence shell from the nucleus, which means that the valence shell should feel the attraction of approximately 2 protons. Now let's analyze Sc in exactly the same way.

Sc : [Ar] 4s^2 3d^1.

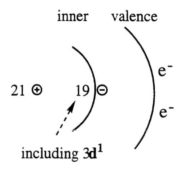

The scandium atom has 21 protons in the nucleus and 21 electrons in the shells. Of these, 19 electrons are inner shell electrons. Remember that the 21st electron (the 3d electron) is also an inner shell electron. Therefore, we see a very similar situation to the case of calcium. That is, there are 19 inner shell electrons that will theoretically screen or block 19 of the 21 protons. Therefore, we should observe an effective nuclear attraction of two protons, just like (or at least very similar) in the case of calcium.

Prediction: Since they both have a very similar effective nuclear charge, we should expect for both to have the same size (or at least very similar). Now let's look at the actual experimental results.

Atom	Configuration	Radius
Ca	[Ar]$4s^2$	1.97 Å
Sc	[Ar]$4s^2\,3d^1$	1.60 Å

What happened here? Why is scandium significantly smaller than the calcium atom? Let's think about it in terms of the differences between the two atoms. What is the main difference between the atoms? The main difference is that Sc is filling an electron into a particular d orbital. This electron, being an inner shell electron, is supposed to block the attraction between the valence shell and the 21st proton in the nucleus. What did we say about d orbitals as screeners? We said that they are not very good screeners. This means that this particular electron in the d orbital is not doing a very good job at screening its proton, therefore the valence shell will feel an attraction for the nucleus stronger than that for the calcium. This should and does make the Sc atom smaller.

As for the rest of the transition metals, we observe that the atoms get increasingly smaller as we go from Sc to Ti and V, but eventually, as we keep using more and more of the d orbitals, the trend is reversed. This is because of interelectronic repulsion between all those inner shell electrons.

Let's look at one more example and use a similar prediction method to decide how the size changes for this triad of elements. Let's compare the atomic sizes for Titanium (Ti), Zirconium (Zr) and Hafnium (Hf). These three elements comprise a small family of transition metals, labeled in many periodic tables as the 2B family. Let's look at the electronic configurations first.

Ti : [Ar] **$4s^2$** $3d^2$;   Zr : [Kr] **$5s^2$** $4d^2$;   Hf : [Xe] **$6s^2$** $4f^{14}\,5d^2$.

The first thing we notice is that the valence shell (in bold face) for Ti is in the fourth shell, for the Zr is in the fifth shell, and for the Hf is in the sixth shell. Based on the idea that the sixth shell is further away from the nucleus than the fifth, and that the fifth is further away from the nucleus than the fourth, we predict that the Hf will be the largest atom, followed by the Zr and then the Ti (which should be the smallest). Now let's look at the actual experimental radii for these three species.

Atom	Configuration	Radius
$_{22}$Ti	[Ar] **4s^{2}3d^2**	1.32 Å
$_{40}$Zr	[Kr] **5s^2** 4d^2	1.45 Å
$_{72}$Hf	[Xe] **6s^2**4f^{14}5d^2	1.44 Å !

How are we going to explain the fact that Hf is **smaller** than Zr? How can this possibly be true? An atom that has its valence shell electrons in the sixth shell is smaller than one with the valence shell electrons in the fifth shell? To answer these questions, we need to once more look at the differences between these atoms. With a very quick glance, we can see that the difference is that the Hf is using the 4f subshell (completely filled) whereas the other two are not. What difference does that make? Well, the f subshells are, by far, the worst screeners yet. They do an extremely poor job at screening the nuclear charge from the valence shell, which means that these 14 electrons which are supposed to block 14 protons are doing such a poor job that the valence shell is feeling a much larger attraction for the nucleus than originally expected. The attraction is so much stronger that the valence shell is pulled much closer to the nucleus, to the point of making the atom a little bit smaller than Zr. We call this reduction in size of atoms beyond the 4f subshell the Lanthanide Contraction. Actually, most of the atoms from Hf to Au are either a bit smaller or the same size as the ones right above them.

- Mixtures are often separated by physical means that rely on the difference in size between the particles (or atoms) that comprise it. A mixture of vanadium (V), niobium (Nb), and tantalum (Ta) is going to be separated by physical means. It is observed that V is easy to separate from the mixture, whereas the Ta is almost impossible to separate from the Nb. Explain this observation.

Once more, we start with the configurations for the three elements.

V : [Ar] **4s^2** 3d^3;   Nb : [Kr] **5s^2** 4d^3;   Ta : [Xe] **6s^2** 4f^{14} 5d^3.

This is due to the fact that the Ta has the 4f subshell completely filled, therefore it is pretty much the same size as the Nb. If they have the same size, that will make it very difficult to separate.

7.8 Ionization Energies.

This is another property of elements that varies periodically. We shall define ionization energy[1] as the energy <u>required</u> to just barely remove an electron from an atom in the gas phase to form an ion in the gas phase. This is equivalent to moving the electron from the ground state (the lowest possible energy state for the electron in that atom) to an energy level where there is absolutely no attraction to the nucleus. This is the energy level we call **n** = ∞. If we were talking about the hydrogen atom, the electron would move from **n** = 1 (ground state) to **n** = ∞ (point of no attraction). The general process can be represented as the loss of an electron by element M. Please, note that the process must be in the gas phase.

$$M_{(g)} \Rightarrow M^{+1}_{(g)} + 1\ e^- \qquad \Delta H = \text{Ionization Energy.}$$

According to the definition, all ionization energies must be positive. How do we know that? Because it is stated that energy is <u>required</u>! This means that we must do work (or the system must absorb energy) in order to completely break the attraction between the electron and the nucleus. If we think about this a little further, we realize that not all elements should require the same amount of work in order to remove an electron. For example, some metals (like Na) react by giving away an electron to form a positive ion with the same electronic configuration of a noble gas whereas some non-metals (like Cl) react by accepting an electron in its valence shell in order to form a negative ion with the configuration of a noble gas. Therefore, when we combine sodium metal with chlorine gas, the compound sodium chloride is formed (NaCl). It should follow that the removal of an electron from Na should be easier than the removal of an electron from the Cl atom. In general, the ionization energies for metals are smaller than the ionization energies for non-metals. Let's look at some tendencies in the periodic table.

As we move down a family in the periodic table, we know that the atoms are getting larger (as we have previously seen). This means that the outermost electron should be easier to remove from a larger element in a family than from a smaller one in the same family. As an example, let look at the ionization energies for the first family: the alkali metals. First of all, they are all in the same family, which means that they all have similar effective nuclear charges (let's call them +1). We can see that the smallest ionization energy is that for Cesium metal, since it is the largest atom.

---

[1] In some texts it is referred to as ionization potential.

Atom	Ionization Energy (kJ/mol)
Li	520
Na	496
K	419
Rb	409
Cs	382

This makes sense, since being larger (Cs), its 6s electron is much further from the nucleus and therefore, its attraction to the positive nucleus must be weaker.

Noble gases have unusually high IE's. This is expected since removing an electron would break its stable configuration. We will see that the ionization energy is sensitive to stable or semi stable configurations.

◊ Lowest ionization energy in periodic table: Cs & Fr
◊ Highest ionization energy in periodic table: He

Conclusion: The ionization energy changes periodically, and the general trend is shown below. We can see that the ionization energy increases as we go up in a family and as we move across a period.

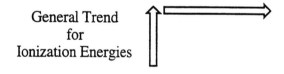

General Trend for Ionization Energies

However, when we look at the actual experimental values for the ionization energies of the element in a period, we observe that the variation is not smooth. For example, let's look at the ionization energies for the elements in the second period in the periodic table. The first thing we observe is that in general, the ionization energies do increase from Li through Ne. However, upon further inspection, we observe that it is not indeed smooth. That is, the ionization energy of Be is larger than that of B and the ionization energy of N is larger than that of O. Why is that so? After all, the effective nuclear charge of B is larger than that of Be, so the attraction should be stronger for the B and it should be harder to ionize. Once more, the effective nuclear charge for the O is larger than that for the N, so it should be harder to remove an electron from the N.

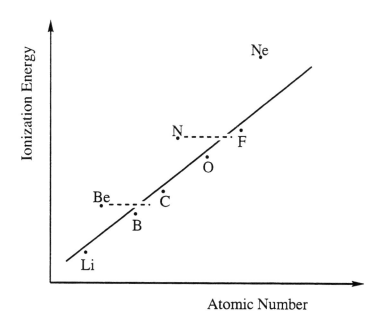

Atomic Number

The first thing we should remember is that the ionization energy, unlike the atomic radius, is a measurement of energy (as the name implies). Energies have a lot to do with stability (or instability). In other words, in order to break an unusually stable system (the meaning of which is a system with lower energy), we would have to do more work or add more energy. This is the case of both Be and N. How is that so? Be has a configuration of: [He] $2s^2$. Therefore, we can see that it has a completely full 2s subshell. Trying to break that configuration in order to form the configuration of: [He] $2s^1$ would require more energy than expected. On the other hand, N has a configuration of: [He] $2s^2\ 2p^3$. This gives the N a half-filled 2p subshell, which imparts a certain stabilization to that particular configuration. If we were to remove an electron from the N, it would form: [He] $2s^2\ 2p^2$. Of course, we can see why it resists it and thus has a larger than expected ionization energy. Notice that in both cases these are not dramatic perturbations. The ionization energy for Be is larger than that for B, but not larger than that for C. Also, the ionization energy for N is larger than that for O, but not larger than that for F. Finally, the ionization energy for the noble gas Ne is unusually high, as expected from its very stable configuration.

Multiple Ionizations.

We have only mentioned the energy required to remove one electron from an atom in order to form a positive ion. However, a lot of positive ions have charges that are larger than +1. This means that for these ions, we would be forced to remove more than one electron. The energy required to remove a second electron from an atom is referred to as the **second ionization energy**. Of course, the third ionization energy would be that amount of energy required to remove the third electron from a particular species. The general processes can be seen below.

$$M^+_{(g)} \Rightarrow M^{+2}_{(g)} + 1\,e^- \qquad \Delta H = \text{Second Ionization Energy}$$

$$M^{+2}_{(g)} \Rightarrow M^{+3}_{(g)} + 1\,e^- \qquad \Delta H = \text{Third Ionization Energy}$$

The second ionization energy for a species is always larger than the first ionization energy. Furthermore, the third ionization energy is always larger than the second ionization energy. This is because each successive electron is more difficult to remove than the previous one. Remember that once we remove an electron the species becomes positive, therefore, the attraction from the positive substance for the remaining electrons becomes much stronger and we would have to do much more work in order to succeed in our efforts to remove another electron.

Let's look at a couple of examples to see how successive ionizations indeed become more difficult. These will be the cases of sodium and calcium metals.

1. Na: [He] $2s^2\,2p^6\,3s^1$.

    $\Rightarrow$ Na$^+$ ([He] $2s^2 2p^6$)   IE$_1$ = 496 kJ/mol
    $\Rightarrow$ Na^{+2} ([He] $2s^2 2p^5$)   IE$_2$ = 4,563 kJ/mol
    $\Rightarrow$ Na^{+3} ([He] $2s^2 2p^4$)   IE$_3$ = 6,913 kJ/mol

2. Ca: [Ne] $3s^2\,3p^6\,4s^2$.

    $\Rightarrow$ Ca$^+$ ([Ne] $3s^2 3p^6 4s^1$)   IE$_1$ = 510 kJ/mol
    $\Rightarrow$ Ca^{+2} ([Ne] $3s^2 3p^6$)   IE$_2$ = 1,146 kJ/mol
    $\Rightarrow$ Ca^{+3} ([Ne] $3s^2 3p^5$)   IE$_3$ = 4,941 kJ/mol

Let's analyze the data. First of all, we do observe that indeed the ionization energy gets larger as we remove more and more electrons. However, upon further analysis, we observe a curious fact. There is a rather large jump from the first to the second ionization energy for sodium metal (approximately a 10 fold increase), whereas the increase from the second to the third ionization energies for Na are not as significant. On the other hand, in the case of Ca we see something quite different. There is an increase from the first to the second ionization energies (a two-fold increase), whereas there is a more dramatic increase from the second to the third ionization energies (a five-fold increase). Why is it different for both metals? It all goes back to the actual configurations. Once you have removed the first electron in the Na, we have achieved the configuration of the noble gas (Ne), therefore, any further ionization will exhibit resistance. On the other hand, it takes two ionizations for Ca in order to reach the stable configuration of the noble gas (Ar), therefore, we observe a more dramatic increase in the ionization energy beyond this point. Once more, we can see that ionization energies are sensitive to stability issues.

7.9 Electron Affinity (EA).

This process is basically the opposite to the one for ionization energy. In here, we will attempt to add one electron to the valence shell of an atom in the gas phase in order to form a negative ion in the gas phase. The energy associated with this process is what we call the electron affinity (EA). The process itself can be seen below.

$$M_{(g)} + 1\,e^- \Rightarrow M^-_{(g)} \qquad \Delta H = \text{Electron Affinity.}$$

In our definition above, we used the term "the energy associated". Please notice this is in contrast to the one we used for ionization energy ("the energy released"). Why is there a difference? This is because the added electron will come into the valence shell and it will form an attraction to the nucleus. The generation of an attraction by itself implies a stabilization, or a lowering of the energy of the system. Therefore, at first glance we should expect all electron affinities to be negative ($\Delta H < 0$). This is true in general:

$$F_{(g)} + 1\,e^- \Rightarrow F^-_{(g)} \qquad \Delta H = -328 \text{ kJ/mol} = EA$$

Not only that, but since the effective nuclear charge increases steadily from left to right, we should expect for the electron affinity to become more negative as we move from left to right. We should also expect the attraction to be stronger if the atom is smaller (since the added electron would be closer to the nucleus). Therefore, we should also expect for the electron affinity to become more negative as we move up a family. The general expectations for the negative character of the electron affinity are depicted below.

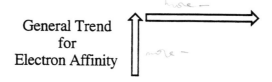

General Trend for Electron Affinity

However, when we look at the experimentally determined values for the electron affinities, we see some blatant exceptions. For example, Beryllium (Be) has a positive electron affinity (as is the case for all the alkaline earth metals). The reason for this is simple in view of what we have stated previously concerning special stabilization of its electronic configuration.

Since these metals have a completely full s subshell, they resist the addition of another electron, because the net result would be to "break" this stable configuration.

Nitrogen and its family follow a very similar pattern and can be explained in the exact same fashion. N has an electron affinity of +7 kJ/mol. This is because the N has a configuration ending in $3p^3$, which has a certain degree of stabilization due to a half-filled subshell. The other members of the family actually have negative electron affinities, but they are a lot smaller than one would expect due to the half-filled subshells. For example:

Silicon (Si):	[Ne] $3s^2\ 3p^2$	EA = -134 kJ/mol.
Phosphorus (P):	[Ne] $3s^2\ 3p^3$	EA = -72 kJ/mol.
Sulfur (S):	[Ne] $3s^2\ 3p^4$	EA = -200 kJ/mol.

Based on the original prediction that the electron affinity gets more and more negative as we move from left to right in a particular period, we would expect the EA for P to be in between those of Si and S. This is not the case, which further reinforces the concept of the partial stabilization caused by the half-filled p subshell.

Finally, we should expect the noble gases to have very positive electron affinities, since they certainly already have a full valence shell and they cannot possibly accept any more electrons into that shell. This is exactly the case. All the noble gases do indeed have positive electron affinities. The addition of one electron to a noble gas will result in the addition to another shell, which will have no attraction to the nucleus, as all the electrons will serve as screeners, therefore, all we have is a huge interelectronic repulsion between the added electron and the other electrons in the noble gas.

- Which element should have a higher (more negative) electron affinity: Oxygen or Fluorine?

    These two elements are in the same period of the table: the second period. Not only that, but they are adjacent to one another. Neither has a particularly stable configuration (O is $1s^2\ 2s^2\ 2p^4$ whereas F is $1s^2\ 2s^2\ 2p^5$), therefore we should observe the addition of an electron to be stabilizing in each case, since the added electron will form an attraction with the nucleus. In the case of F, the added electron will complete the octet plus it will form a stronger attraction with the nucleus due to the fact that the F has a stronger effective nuclear charge. Therefore, we should expect the F to have a more negative electron affinity than the oxygen. This is exactly the case, as we can see from the experimentally determined values:

Oxygen (O): EA = -142 kJ/mol

Fluorine (F): EA = -328 kJ/mol

Noble gases have very positive electron affinities also for similar reasons. In this case it is even worse since the added electron will go to the next shell and will not be attracted to the nucleus at all. Instead all it will feel is a strong repulsion to the electronic cloud.

- Which of the following elements would you expect to have the largest stabilizing effect upon the addition of one electron, that is, which one has the most negative electron affinity? Zinc (Zn) or Gallium (Ga)?

We should start by looking at the electronic configurations for these elements so that we can make an informed decision.

Zn: [Ar] $4s^2\ 3d^{10}$ and Ga: [Ar] $4s^2 3d^{10} 4p^1$

Now, we can immediately see that in the case of Zn, we would be adding one electron to the valence shell, which is the fourth shell. Since the 4s is already full, we would be forced to add this electron into the empty 4p subshell. In the case of Ga we would be adding the extra electron to the 4p subshell, which already contains one electron. It should be easier or more stabilizing to add the electron to the Ga since the Zn has a stable configuration and it should show resistance to the addition. This can be seen from the experimental numbers, since the electron affinities for Zn and Ga are +10 kJ/mol and –29 kJ/mol, respectively.

We should also point out that when we form a negative ion that has a charge larger than –1, then we need to add more than one electron, just like we did with the ionization energies with positive ions. This leads to second, third, and so on, electron affinities. It should be curious that all further electron affinities (beyond the first one) are always positive. This is due to the fact that once we have added one electron, we have created a negatively charged species. Further addition of electrons will simply create an incredible electron-electron repulsion, which no stable configuration will be able to overcome. Let's look at the first and the electron affinities for Oxygen.

Ex: $O_{(g)} + 1\ e^- \Rightarrow O^-_{(g)}$ EA_1 = - 142 kJ/mol

$O^-_{(g)} + 1\ e^- \Rightarrow O^{-2}_{(g)}$ EA_2 = + 844 kJ/mol

From these numbers, it is clear that the overall addition of the two electrons to the oxygen atom does not seem to be stabilizing (since the addition of the two electron affinities is a positive quantity). This means that the fact that the ion is formed at all must be due to something else and not just the fact that the oxide ion has the configuration of a noble gas. This is indeed true and we will see the real reason very shortly when we talk about ionic bonding. We shall see that the real stabilization comes when the ions formed are allowed to interact with one another, and form attractions among themselves. However, before we proceed with bonding, we should summarize the stable configurations that we expect in the periodic table.

7.10 Expected Ions From the Periodic Table.

There are a few configurations that we have already mentioned impart stability to the species. Of these, the most obvious is the noble gas configuration. Elements that do not have initially this configuration will try to react in one way or another in order to either reach this configuration or one of the other stable ones. The most stable are shown below. Knowing this, we should be able to many times predict the ions formed by a given element in the periodic table. Let's look at a few examples.

1. **ns^2np^6**. This is the noble gas configuration. A lot of the elements that are near a noble gas will react by either gaining or losing a few electrons until they reach this particular configuration. These are the easiest to predict.

    Examples: F^{-1}, O^{-2}, Mg^{+2}, Ca^{+2}, Cl^{-1}, Sc^{+3}, ...

2. **$ns^0(n-1)d^{10}$** This is another rather stable configuration for ions in the periodic table. We will call it the Pseudo-noble gas configuration. It is associated to elements that are either at the end of the transition metals, or a little beyond. These elements cannot lose all the electrons (including the d electrons) in order to obtain the noble gas configuration[1]. Therefore, these elements lose only their valence shell electrons and remain with completely full inner shells.

    Examples: Zn^{+2}, Ga^{+3}, In^{+3}, ...

3. **$ns^2(n-2)f^{14}(n-1)d^{10}$**. Maybe we should be more specific and say it is $6s^2\ 4f^{14}\ 5d^{10}$.

---

[1] The amount of energy required to achieve this would be astronomical!

This configuration is especially important for the ions formed by Tl, Pb, and Bi (in the 6th period).

Example: Tl: [Xe] **6s^2** 4f^{14} 5d^{10} **6p^1** (3 valence shell electrons), therefore this element should form the Tl^{+3}, where it loses all three valence shell electrons (the ones that are bold).

There is something curious about this last stable configuration. In the case of Tl mentioned above, we observe that the first electron will be lost easily to form the Tl^{+1} ion, which has the following configuration:

Tl^{+1}: [Xe] **6s^2**4f^{14}5d^{10}.

However, the other two electrons left in the valence shell (the ones in the 6s subshell) are much harder to remove. This is because of the poor screening ability of the 4f subshell. These are very tightly held and a lot of work must be done in order to finally remove these. It is also true that the increased effective nuclear charge due to the poor screening ability of the 4f subshell increases the penetrating ability of the 6s electrons. Because of these reasons, these two electrons in the 6s subshell are often referred to as the **inert pair**[1]. We should point out that although it is harder to form, it does exist in nature. Therefore, the configuration of such ion would be:

Tl^{+3}: [Xe] 4f^{14}5d^{10}.

- Which ions would we expect Lead (Pb) to form? Which would be the most prevalent in nature and why? How about Bismuth (Bi)?

Let's first look at the configuration for lead: [Xe] **6s^2** 4f^{14} 5d^{10} **6p^2**. We can immediately see that lead has four valence shell electrons. Therefore, we should look at the possibility of losing some or all of these electrons. The removal of the 6p electrons should lead to a stable configuration, therefore:

Pb^{+2}: [Xe] **6s^2** 4f^{14} 5d^{10}.

Lead also has the ability to lose the other two electrons: the ones in the 6s subshell. However, these are harder to remove because of the poor screening ability of the 4f subshell, therefore, we should expect for the Pb^{+2} ion to be more prevalent in nature (and it is).

---

[1] Called this only when the 4f is completely full!

Pb^{+4}:   [Xe] $4f^{14}$ $5d^{10}$.

As for Bismuth, its configuration is: [Xe] **$6s^2$** $4f^{14}$ $5d^{10}$ **$6p^3$**. We can see that it has five valence shell electrons, however, it is impossible to remove more than four electrons from an atom to form a stable ion. Therefore, Bi only forms one ion[1]: the Bi^{+3}.

Bi^{+3}:   [Xe] **$6s^2$** $4f^{14}$ $5d^{10}$.

- Explain the following referring to three elements in a given family. Aluminum only forms the +3 ion; Indium (In) can form either the +1 or the +3 ion, and the +3 is the most prevalent one; Thallium (Tl) forms either the +1 or the +3 ion, and the +1 is the most prevalent one.

Once more, we should start by writing the electronic configuration of all three elements with their ions.

Al: [Ne] $3s^2$ $3p^1$. Al^{+3}: [Ne]
In: [Kr] $5s^2$ $4d^{10}$ $4p^1$. In^{+1}: [Kr] $5s^2$ $4d^{10}$. In^{+3}: [Kr] $4d^{10}$.
Tl: [Xe] $6s^2$ $4f^{14}$ $5d^{10}$ $6p^1$. Tl^{+1}: [Xe] $6s^2$ $4f^{14}$ $5d^{10}$. Tl^{+3}: [Xe] $4f^{14}$ $5d^{10}$.

In the case of the Al, we see that there are no **d** nor **f** subshells, which are the poor screeners, therefore, there is nothing to stop us from removing all three electrons in order to reach the configuration of the noble gas. However, in the case of In, we see the presence of **d** electrons. These are not the greatest screeners in the world. Therefore, once we successfully remove the **p** electrons, we find a small resistance to remove the next two (from the **s** subshell). This allows us to form both ions. Since d electrons are not the worst screeners, removal of all three electrons is actually preferred so as to achieve the pseudo-noble gas configuration. Finally, in the case of thallium, we see that this atom has a full 4f subshell. The **f** electrons are the worst screeners, therefore, removal of the 6s subshell is very difficult - so much so that it turns out that the +1 ion is more prevalent in nature than the +3 ion.

---

[1] Bi does not form the –3 ion since it is a metal (metals only form positive ions).

Chapter 8: Bonding I.

8.1 Chemical Bonding.

Up to this point, we have discussed the relationship between the atomic structure (particularly the electronic configuration) and some of the properties of the elements (like ionization energies, atomic sizes, and electron affinities). We also know that atoms combine to form molecules. Why do atoms combine or bond together? Is bonding between atoms critical? Is it important in one way or another?

As an example, let's consider two very basic elements: carbon and silicon. They are members of the same family, and one is right under the other in the periodic table. As we have already seen, this means that they have similar electronic configurations ([He] $2s^2\ 2p^2$ vs. [Ne] $3s^2\ 3p^2$). This would lead us to predict that both of them should form very similar types of compounds. Therefore, their oxides (CO_2 and SiO_2) should be very similar in many ways (reactivities, formulas, properties, etc). We all know that carbon dioxide is a gas which is a product of respiration. How about silicon dioxide? This is the empirical formula for silica[1] which is found in a variety of substances like sand and quartz. Actually, it is one of the most prevalent elements in the Earth's core[2]. Why are they so different? We will see that the carbon forms discrete molecules of carbon dioxide, whereas the silicon forms large networks of silicon and oxygen in a ratio of 1 : 2. We shall see that the entire explanation is bonding.

We shall even see that bonding is important even in simpler substances, like elemental substances. For example, let's compare two common allotropes of carbon: graphite and diamond. These substances are made up exclusively of carbon atoms. Graphite is soft, slippery and conducts electricity. However, diamond is one of the hardest materials we know and it does not conduct electricity. Why are these two materials so different? Why do their properties change so drastically? Of course, once more the answer will be in the way these carbon atoms are bonded to form the solid. Bonding is probably one of the most important topics for anyone who wishes to have some basic understanding of chemistry and the chemistry of materials.

How can we define a bond? It is not an easy task because there are many different types of bonds. We can say that bonding occurs because there are forces that hold a group of atoms together and make the group of atoms function as a unit. Let's say that there are four types of

---

[1] Engineers use the ending –a for the oxides of most elements. For example, soda is the oxide for sodium, boria is the oxide for boron, and silica is the oxide for silicon.
[2] Second only to oxygen.

bonds, mainly ionic bonds, covalent bonds, metallic bonds and the network covalent bonds. The different characteristics of each of these types of bonds can be seen in the table below.

Bonding Types	Examples	Characteristics
Ionic	NaCl	High melting points; they conduct electricity when molten or in solution.
Covalent	H_2O, CH_4	Low melting points; they do not conduct electricity.
Metallic	Na, Cu, Ca	Could have high melting points; always conduct electricity.
Network Covalent	Diamond	Highest melting points; never conduct electricity.

Now that we have seen the different types of bonds found in materials, we should start looking at some of these bonding types and try to understand what actually happens as these atoms come together to form the different types of bonds. We have mentioned that only the valence shell electrons are involved in bonding (in one way or another). It would be to our advantage to have a way to indicate these electrons around a given atom. This is what we call Lewis symbols – a representation of the valence shell electrons drawn around the atomic symbol. For example, let's look at the elements in the second period. Lithium has only one valence shell electron (the $2s^1$) electron. Therefore, the Lewis symbol for Li should just be the letters Li with a dot (representing the electron) around it. The element Beryllium has two valence shell electrons (both in the 2s subshell). Since these are paired up, we should represent it as a pair of dots (electrons) drawn around the symbol Be. Finally, as one more example, the element Boron (B) has three electrons and its configuration is $2s^2\ 2p^1$. Of course, the first two are paired up, whereas the last electron is in a different subshell, so the three dots should show this also.

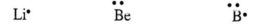

In general, writing these dots is simple. We should divide the symbol for the element into four quadrants.

Chapter 8: Bonding I.

The electrons should be drawn in such way so that they match the electronic configuration. That is, if it is 2s², we should observe a pair of electrons in any one of the quadrants. If it is 2s² 2p¹, we should then observe a pair of electrons in one of the quadrants, and a single dot in any other quadrant. The other elements of this period are shown in the figure below.

$$\overset{..}{\underset{.}{C}}\cdot \qquad \cdot \overset{..}{\underset{.}{N}}\cdot \qquad \cdot \overset{..}{\underset{..}{O}}\cdot \qquad :\overset{..}{\underset{..}{F}}\cdot \qquad :\overset{..}{\underset{..}{Ne}}:$$

There are a couple of things that we should mention. First of all, notice that the carbon has a lone pair[1] plus two unpaired electrons. This corresponds to the ground state electronic configuration of 2s² 2p². Finally, we should notice that the noble gas (Neon) has the complete octet (the stable configuration where both the s and the p subshells are completely full.

We have previously learned that atoms react by either losing some of its valence shell electrons (thus forming positive ions – **cations**), gaining some of these electrons (thus forming negative ions – **anions**), or sharing some of the valence shell electrons in order to try to get a more stable configuration (like the noble gas configuration: ns² np⁶). As an example, let's look at the case of nitrogen. We see that nitrogen is missing three electrons for the complete octet, therefore, N can react by accepting three electrons to form the nitride ion.

$$\cdot \overset{..}{\underset{.}{N}}\cdot \quad + \quad 3\,e^{-} \quad \longrightarrow \quad [:\overset{..}{\underset{..}{N}}:]^{-3}$$

We are not trying to say that all atoms react in such a way as to complete an octet, although many do just that. There are many exceptions to this "octet rule", as we will see later. An obvious one is the case of hydrogen (H), which could react by accepting one electron in order to form the hydride ion, which has the configuration of Helium.

$$H\cdot \quad + \quad 1\,e^{-} \quad \longrightarrow \quad [\,H:\,]^{-1}$$

8.2 Ionic Bonding.

Ionic bonds are formed between <u>metals</u> (low ionization energy) and <u>non-metals</u> (high electron affinity). The metal, with a low ionization energy, will lose the electron rather easily to the non-metal and thus transfer it completely. This is not a "true" bond in the sense of having a connection between the two ions. All we have is a purely electrostatic attraction between the

---

[1] A lone pair is a set of two electrons that are paired up with opposite spins in a given orbital, and therefore are not shared with any other atom.

cation and the anion. We will refer to these compounds from now on as **salts**. Examples of salts: NaCl (table salt), KBr, MgO, Ca(OH)$_2$, etc. The formation of the salt NaCl can be seen below.

$$Na\cdot \; + \; :\!\ddot{\underset{..}{Cl}}\!\cdot \; \longrightarrow \; Na^+ \; [\, :\!\ddot{\underset{..}{Cl}}\!: \,]^{-1}$$

Once we form the ions, the immediate question is: how do these ions interact in order to form a stable solid? In order to answer this question, we need to start by looking at the physics of the interaction between a pair of ions. This can be achieved using Coulomb's law[1].

$$E_{potential} = \frac{q_1 \times q_2}{4\pi\varepsilon_\circ r}$$

Remember that in this equation, the letter q represents the charges of the two ions. Most of these are constants ($\varepsilon_\circ$ is the permittivity of the vacuum = 8.854 x 10^{-12} C^2/Jm). If we substitute the values of all the constants, we get the following expression:

$$E_{potential} = (8.99 \times 10^9 \; J\,m/C^2)(q_1 \, q_2 \, / \, r)$$

As an example, let's take the salt NaCl. This salt consists of two ions: the sodium ions (Na^{+1}) and the chloride ions (Cl^{-1}). When we write these charges of ± 1 for these ions, what are we referring to? The +1 charge in the sodium is telling us that this ion lost an electron, therefore, the charge is that of an electron, but positive. In the case of the chloride, since it has an extra electron, then its charge (-1) is that of an electron. What is the charge of an electron? We have already seen that the charge of an electron is –1.60 x 10^{-19} C. Let's define the letter **e** as that number, that is,

$$e = 1.60 \times 10^{-19} \; C.$$

No matter what the charges for the ions may be, they will always be multiples of **e**. Therefore, the equation may be written as:

$$E_{potential} = (8.99 \times 10^9 \; J\,m/C^2)(q_1 \, q_2 \, / \, r) \, e^2.$$

Substituting the value for **e**, we get:

---

[1] See section 6.7.

$$E_{potential} = 8.99 \times 10^9 \frac{Jm}{C^2} (\frac{q_1 q_2}{r})(1.60 \times 10^{-19} C^2)$$

or

$$E_{potential} = 2.31 \times 10^{-28} Jm (\frac{q_1 q_2}{r})$$

As an example, let's go back to our example of NaCl. The distance between these two ions is 0.276 nm, that is, 2.76 x 10^{-10} m, therefore, substitution into that equation yields:

$E_{potential}$ = 2.31 x 10^{-28} Jm [(+1)(-1) / 2.76 x 10^{-10} m] = -8.37 x 10^{-19} J.

Don't forget that the negative value indicates that we have here an attraction between the sodium and the chloride ion. Of course, in reality we do not have just two ions, but quite a large number. Since our standard to measure things are moles, let's say that we have a mole of these pair of ions. In order to obtain the value of the interactive energy between all of these ions, we need to multiply by Avogadro's number, which will then yield:

$E_{potential}$ = *- 504 kJ/mol.* ✓

We must make clear that the number we just calculated refers to having one mole of each ion interacting in the gas phase. In other words, we took the interaction of two ions by themselves in the gas phase, and then multiplied that number by Avogadro's number. We all know that table salt is a solid, therefore, we must think about the interactions between all of these ions in the solid phase. Therefore, this number should be different, as we will have not only attractions between sodium ions and chloride ions, but we should also have repulsions between sodium ions and repulsions between chloride ions.

In the solid phase, ions are in close proximity, but positioned in such a way so as to maximize the attractions while minimizing the repulsions. In the case of NaCl, we observe that each sodium ion is surrounded by six chloride ions at an equal distance (one above it, one below it, one on each side, one in front, and one in the back). The actual arrangement is dependent, among other things, on the size of the ions and we will talk more about this in a subsequent chapter (solids).

Let's look at the energy aspects of the formation of the salt NaCl. What do we mean by formation? We should recall from Chapter 5, that it is the reaction that forms one mole of a compound from its elements in their standard states. Therefore, we are basically looking for the ΔH for the following reaction:

Chapter 8: Bonding I.

$$Na_{(s)} + 1/2\ Cl_{2(g)} \Rightarrow NaCl_{(s)} \qquad \Delta H = ???$$

What kind of information can we look up in the literature to help us calculate this ΔH using Hess' Law? There are two things we have already talked about: ionization energies and electron affinities. For our substances[1], they are reported as:

Ionization Energy for Na	495 kJ/mol
Electron Affinity for Cl	- 348 kJ/mol

The combination of these two processes can be seen in the scheme below.

$Na_{(g)}$	$\Rightarrow$	$Na^+_{(g)} + 1\ e^-$	$\Delta H = + 495$ kJ/mol
$Cl_{(g)} + 1\ e^-$	$\Rightarrow$	$Cl^-_{(g)}$	$\Delta H = - 348$ kJ/mol
$Na_{(g)} + Cl_{(g)}$	$\Rightarrow$	$Na^+_{(g)} + Cl^-_{(g)}$	$\Delta H = + 147$ kJ/mol

The first thing we notice is that the ΔH for the formation of NaCl is positive. This implies that work must be done in order to form the salt. This is initially very surprising, since we are forming two ions that have the configuration of a noble gas. However, we must also remember that these ions are formed in the gas phase, where the interaction between the particles is minimal. In conclusion: IONS ARE NOT STABLE IN THE GAS PHASE.

What is it then that stabilizes the formation of the salt? It cannot be the configuration of the ions themselves since the $\Delta H > 0$. Once the ions are formed in the gas phase, they will attract electrostatically. As they start to attract, more and more of them will start to get very close to one another to form conglomerates that will keep growing until they result in the solid material. All these attractions that we are forming are accompanied by a large release of energy that over-compensates for the initial input of energy that formed the ions in the gas phase. Actually, the process of crystallizing the ions from the gas phase into the solid releases 786 kJ of heat in the case of NaCl. We will call this the heat of crystallization or lattice energy ($\Delta H_{crys} = - 774$ kJ/mol). Therefore,

$Na_{(g)} + Cl_{(g)}$	$\Rightarrow$	$Na^+_{(g)} + Cl^-_{(g)}$	$\Delta H = + 148$ kJ/mol
$Na^+_{(g)} + Cl^-_{(g)}$	$\Rightarrow$	$NaCl_{(s)}$	$\Delta H = - 774$ kJ/mol
$Na_{(g)} + Cl_{(g)}$	$\Rightarrow$	$NaCl_{(s)}$	$\Delta H = - 626$ kJ/mol

---

[1] Remember that the sodium is losing the electron and the chlorine is accepting it.

Therefore, the crystallization process results in the stabilization of the salt itself.

At this point, we are very close to obtaining the heat of formation for NaCl. The process we have succeeded in obtaining is the one that takes the individual neutral atoms in the gas phase and end with the solid salt. However, the formation process involves starting with the elements in their standard states. The standard state for sodium is a solid metal, whereas the standard state for the chlorine is a diatomic gas. Somehow, we need to convert the solid sodium into sodium gas (a process we call sublimation) and we must break the Cl-Cl bond (a process we call the bond energy[1]). The complete process can be seen below:

$Na_{(s)}$	$\Rightarrow$	$Na_{(g)}$	$\Delta H = + 108$ kJ	(sublimation)
$Na_{(g)}$	$\Rightarrow$	$Na^+_{(g)} + 1e^-$	$\Delta H = + 495$ kJ	(ionization energy)
$1/2\ Cl_{2(g)}$	$\Rightarrow$	$Cl_{(g)}$	$\Delta H = 1/2\ (232)$ kJ[2]	(bond energy)
$Cl_{(g)} + e^-$	$\Rightarrow$	$Cl^-_{(g)}$	$\Delta H = - 348$ kJ	(electron affinity)
$Na^+_{(g)} + Cl^-_{(g)}$	$\Rightarrow$	$NaCl_{(s)}$	$\Delta H = - 774$ kJ	(lattice energy)
$Na_{(s)} + 1/2\ Cl_{2(g)}$	$\Rightarrow$	$NaCl_{(s)}$	$\Delta H = - 412$ kJ	

So, in conclusion, we can say that the heat of formation of $NaCl_{(s)}$ is -412 kJ/mol or it can be said that the formation of one mole of NaCl is accompanied by the release of 412 kJ per mol of salt.

8.3 Covalent Bonding.

We have seen what happens when two atoms with extremely different properties (one with low ionization energy – the metal, and one with a high electron affinity - the non metal). For these, we observe a net transfer of electrons from the metal to the non-metal. But what happens when the two atoms are similar in their properties? Neither one is strong enough to completely let go or completely give away their electrons. This is the case of the interactions between a non-metal with another non-metal. We will observe that instead of transferring electrons, these will be shared by both atoms since neither has a great tendency to lose electrons (they have both high ionization energies and high electron affinities). Let's look at a very simple example: the formation of a single bond between two hydrogen atoms.

---

[1] We can find the bond energy by observing which frequency of light has just enough energy to break the bond. Since the energy of the light is given by hv, then we can determine the energy.

[2] The number 232 kJ/mol refers to the dissociation of the Cl-Cl molecule to form two chlorine atoms. Since we want only one Cl, we divided it by two.

H• + •H ⟶ H:H

We see that two electrons are shared between the two hydrogen atoms. We can (and we will) write a solid line to represent the two shared electrons (H – H), keeping in mind that the line represents two electrons (one from each hydrogen).

Covalent bonds have a directional character. This means that, unlike in ionic bonds, there is a true connection between the two atoms and we can clearly identify the place where the electrons of the bond are most likely to be found. These bonds have a very important characteristic which we call the **bond length**. We will define the bond length as the average distance between the two nuclei. Why did we use the term average distance? Because the atoms that form the bond are in a constant vibration mode, thus the bond is not static. It elongates and shortens all the time, and what we measure as the bond length then turns out to be the average. Why do bonds vibrate? This will be answered in the next section when we look carefully at the energetics of the hydrogen molecule.

Homonuclear diatomic molecules are examples of purely covalent molecules. Both atoms have exactly the same tendency to attract the electrons, therefore the electrons are perfectly or equally shared between the two atoms. Examples:

H:H

:B̈r:B̈r:

As we said before, the bond length is the average distance between the two nuclei. Once you measure these experimentally, then half that distance is estimated to be the atomic radius for the atom. Below we can find a partial table with atomic radii.

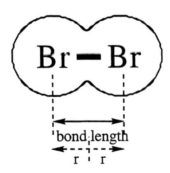

Atomic Radii in Å

H 0.37							He 0.50
Li 1.52	Be 1.11	B 0.88	C 0.77	N 0.70	O 0.66	F 0.64	Ne 0.70
Na 1.86	Mg 1.60	Al 1.43	Si 1.17	P 1.10	S 1.04	Cl 0.99	Ar 0.94
K 2.31	Ca 1.97	Ga 1.22	Ge 1.22	As 1.21	Se 1.17	Br 1.14	Kr 1.09
Rb 2.44	Sr 2.15	In 1.62	Sn 1.40	Sb 1.41	Te 1.37	I 1.33	Xe 1.30
Cs 2.62	Ba 2.17	Tl 1.71	Pb 1.75	Bi 1.46	Po 1.50	At 1.40	Rn 1.40

Once we know the covalent radius of all atoms, then we can predict fairly accurately any covalent bond length as:

$$(\text{bond length})_{x-y} = r_x + r_y$$

- What is the bond length of C-O in CH_3-O-H ? Given: $r_C = 0.77$ Å and $r_O = 0.66$ Å.

This should be as simple as adding the two radii to get the bond length. Therefore, the bond length should be given by:

$$\text{length}_{C-O} = 0.77\text{Å} + 0.66\text{Å} = 1.43\text{Å}$$

Just to prove that this is an accurate estimation of bond lengths, the actual experimentally determined value for the length between the carbon and the oxygen in this molecule is found to be 1.434 Å!

We should point out that some atoms have the ability to form multiple bonds. Carbon happens to be a great example for this, as it is one of the few atoms in nature that has the ability to form multiple bonds and long chains. Carbon can form single, double and triple bonds with itself. These different bonds are not equally long nor equally strong. In general, **multiple bonds are stronger and shorter than single bonds.**

Bond	Bond Length
C – C	1.54 Å
C = C	1.34 Å
C ≡ C	1.21 Å

Therefore, the carbon-carbon triple bond would be the shortest one and the strongest. Why is it the strongest bond? Simply put at this point, because it consists of the most bonds between the two atoms. Let's go back and look now at the energetics of the formation of a covalent bond and for this purpose, let's use the simplest molecule possible: the hydrogen molecule.

8.4 The Hydrogen Molecule.

The formation of one mole of these molecules from two moles of single hydrogen atoms releases 435.1 kJ of heat. This means that the ΔH for the process is:

$$2\,H \Rightarrow H_2 \qquad \Delta H = -435.1 \text{ kJ/mol of } H_2.$$

What is this telling us? A negative ΔH implies that the hydrogen molecule is lower in energy than the individual hydrogen atoms. Therefore, the molecule must be in a lower ground energy state than the atoms. This is expected if each nucleus shares the two electrons in the bond, thereby completing or filling its valence shell ($1s^2$).

What forces are at work in the hydrogen molecule? Technically, there are four main forces that play a role in the stabilization of this molecule.

1) An attractive force between each nucleus and the electron originally belonging to the other nucleus.

2) A repulsive force between the two nuclei.

3) A repulsive force between the two electrons.

4) An attractive force (magnetic) between the two electrons **if** they have opposite spins – they *must* have opposite spins in order to occupy a common orbital = Pauli's Principle.

Chapter 8: Bonding I.

Energy Aspects.

Let's explain what happens when we approach two H atoms (from a point where they do not attract) to form a bond. In order to see it better, let's follow the following graph that depicts how the potential energy changes as a function of the internuclear distance.

- As the atoms approach from infinity the attraction increases, thus the energy starts to decrease. The atoms keep getting closer and closer and as they do so they move faster and faster toward each other. It should be clear that the predominant force at this point is the attraction of one nucleus for the electron of the other atom (and vice-versa). It is also evident that as the atoms get closer the attraction should increase. The word attraction implies a stabilization or a lowering of the potential energy.

- At the point of maximum stability (lowest part of the curve) the atoms have the maximum speed of approach, so they cannot suddenly stop and thus keep getting closer. At this point the <u>repulsion</u> between the two nuclei is getting stronger (∴ energy is going up) and thus the speed of approach slows down until it stops.

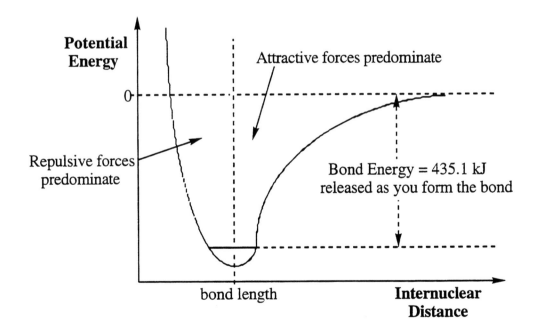

- Since the forces are predominantly repulsive at this point, the nuclei start moving away from one another, gathering speed as they do so. Once more, when they reach the minimum point they are at the maximum speed so they keep going past this point. However, now the predominating attractive forces take over and therefore the atoms will stop and return again.

- This vibration will continue on and on and on "per secula seculorum" (forever).

- What we call the bond length is nothing more than the average length between the maximum and the minimum distance during the vibration. This usually corresponds to the minimum of the curve.

- The bond energy is the amount of energy liberated or released upon the formation of the covalent bond. It corresponds to the energy difference between the atoms in the bond when they are infinitely apart and when they are at their vibrating level (see graph above).

This explains why the atoms vibrate constantly in the bond. We should point out that even at the absolute zero of temperature, there is a minimum energy for this vibration, and we can call this the residual energy, as we did in Section 6.11 when we were talking about the particle in the box.

8.5 Electronegativity.

We have seen how homonuclear diatomic molecules are perfect examples of covalent molecules since both atoms have the same attraction for the electrons in the bond (being identical, they have the same ionization energies and electron affinities). However, most diatomic molecules are heteronuclear. What happens then? Let's take the H-F molecule as an example. The hydrogen and the fluorine do not have the same attraction for the electrons in the bond, since they have different ionization energies and different electron affinities. We have established that since they are both non-metals, they are not different enough to create into an ionic species. However, the molecule will not have an even electronic distribution. The diagram below gives you an idea of how the electrons in the bond are drawn more towards the fluorine than the hydrogen. How does this molecule differ from the hydrogen molecule? This molecule has a permanent dipole, that is, there is a permanent charge separation in this molecule. Since the electrons seem to spend more time around the fluorine than the hydrogen, then the fluorine has a partial negative charge whereas the hydrogen has a permanent partial positive charge.

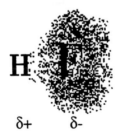

δ+   δ-

From now on, we will use the term polar bond to refer to these bonds that have a permanent charge separation and thus a permanent dipole. What causes this charge separation? Why are these two atoms different in its pull for the electrons in the bond? There are two main reasons as to why the fluorine atom pulls the electrons more tightly than the hydrogen atom. The two reasons are basically size and the effective nuclear charge. Let's think about each one individually. Since we are talking about the attraction of the atom for the electrons in the bond, as smaller atom should have a stronger pull for the electrons since they will be closer to the nucleus. Similarly, the larger the effective nuclear charge, the stronger the pull for the electrons. Let's analyze these two atoms in terms of the two factors.

The hydrogen atom ends up winning the first factor since it is indeed smaller than the fluorine atom (H = 0.37 Å and F = 0.64 Å). However, the fluorine atom wins the second factor in a big way since the effective nuclear charge for the F is much larger than that of the hydrogen (+1). So, the hydrogen atom is small but attracts with only one proton, whereas the fluorine atom is slightly bigger and attracts the electrons with quite a few more protons.

It would be very convenient to have a scale that measures the ability of an atom to attract electrons towards itself in a covalent bond. Such a scale was developed by Linus Pauling and it is called the **electronegativity** scale. He based this scale by comparing the dissociation energies of molecules like H-H, F-F, and H-F, and came up with the scale we see in the next figure. In this electronegativity scale, the atom with the largest electronegativity is the fluorine and is given a value of 4.00 (these numbers have no units). The lowest electronegativity is that of cesium metal with a value of 0.37. In general, electronegativity increases as we move up a family since the atoms are getting smaller, and increases from left to right in the periodic table, since the effective nuclear charge is increasing in that direction. Notice also that oxygen has the second highest electronegativity in the periodic table, followed by chlorine and then nitrogen. Actually, we can generalize that the electronegativity decreases diagonally as we move away from fluorine. The noble gases are not included because, as we said before, electronegativity is derived from bond energy data and these gases do not form bonds (except for Xe).

---

[1] American scientist who lived from 1901 to 1994. He won two Nobel prizes, one for chemistry and one for peace.

H 2.20						
Li 1.00	Be 1.47	B 2.04	C 2.50	N 3.04	O 3.50	F 4.00
Na 0.95	Mg 1.31	Al 1.81	Si 1.90	P 2.19	S 2.55	Cl 3.16
K 0.82	Ca 1.00	Ga 1.80	Ge 1.89	As 2.18	Se 2.45	Br 2.96
Rb 0.79	Sr 0.95	In 1.78	Sn 1.85	Sb 2.05	Te 2.10	I 2.66
Cs 0.74						

Cs 0.74 } lowest Electronegativity

There are other ways to calculate electronegativity values and some (like Robert Mulliken's scale) are related to some sort of average of the ionization energies and the electron affinities of the element in question. The values obtained from this scale are slightly different from Pauling's scale but have been "scaled" to give fluorine a value of 4.00. These can be found in other general chemistry books.

8.6 Ionic vs. Covalent.

Is there a simple way of telling whether a compound is going to be ionic of covalent? Many people have wondered about this and there are many different and sophisticated ways to determine this. One rather elegant way is using what has come to be called bond-type-triangles. These use electronegativity differences and averages to separate compounds into four types: covalent, ionic, metallic and semi-metallic.

We will, however, simplify the process much further. In general, ionic compounds are observed when there is a fairly large difference in electronegativity whereas covalent compounds have much smaller differences in their electronegativities. The fact is that most compounds have a certain ionic character. The exceptions are the homonuclear diatomic molecules, which are purely covalent since the electronegativity difference is zero. If the two atoms have a different electronegativity, we observe that they have an ionic character, even when we consider them covalent. In summary, there are three possibilities.

---

[1] From Spencer, Bodner, and Rickard's general chemistry book published by Wiley.

Chapter 8: Bonding I.

1. Ionic compound – large difference in electronegativity between the atoms.

    a. CsF: ΔEN = 4.00 - 0.74 = 3.26 (largest difference). This is the "most ionic" substance since this is the largest difference in electronegativities from the periodic table.

    b. LiF: ΔEN = 4.00 - 1.00 = 3.00.

2. Pure Covalent compound – they have an electronegativity difference of zero.

    a. F_2: ΔEN = 4.00 - 4.00 = 0.00.

    b. H_2: ΔEN = 2.20 – 2.20 = 0.00.

3. Polar Covalent compounds – These are the intermediate cases and they are observed when the electronegativity difference is not that large.

    a. HF: ΔEN = 4.00 - 2.20 = 1.80.   COMPARED TO

    b. HCl: ΔEN = 3.16 - 2.20 = 0.96.

    c. CO: ΔEN = 3.50 - 2.50 = 1.00.

An obvious question would arise if we tried to quantify these parameters. Where is the cutoff between a large difference and an intermediate one? This is further complicated by the fact that molecules like HF have a considerable ionic character since they have a substantial difference in electronegativity. We will simplify this in a major way by simply stating that ionic compounds are formed between metals and non-metals whereas covalent bonds are formed between non-metallic atoms. This is an oversimplification, but it will be more than adequate for our purposes.

Below we can see a pictorial representation of these bonding types. We can see that we have a whole range of situations, from purely ionic and distorted ions, to polar covalent bonds and perfect covalent bonds and the entire thing is based on the electronegativity difference between the atoms in the molecule.

Chapter 8: Bonding I.

	Name	Example	ΔEN
(+) (−)	Pure Ionic	CsF	Very large
(+) (−)	Distorted Ions	AlCl$_3$	Large
δ+ ⬭ δ−	Polar Bonds	HF	Small
⬭	Pure Covalent	F$_2$	Zero

Now let's go back to the HF molecule and try to understand what we meant when we stated that it had a considerable ionic character. We already know that HF consists of a polar single bond. This means that there is a permanent dipole in the molecule – separate centers of positive and negative charges. All polarized bonds have a dipole moment (μ). This is a vector, so it has both magnitude and direction.

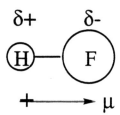

The dipole moment (μ), measured in Debyes (D), will give you an idea of the strength of the electrical field generated by the charge separation in the molecule. It is defined by:

$$\mu = q \times d$$

where: q = magnitude of charge separation ($\delta+$), in C.
d = distance between the two poles in the dipole.

Therefore, the strength of a dipole depends on two things: the charge separation, which sometimes is referred to as the fraction of ionization (δ), and the distance between the two atoms

---

[1] Peter J. Debye was an American chemist and physicist who lived from 1884 until 1966. He won the Nobel prize in Chemistry in 1936.

in the bond (d). A molecule with a high dipole moment is more polar than one with a lower dipole moment, therefore, this is a good measure of the polarity of the molecule.

Technically, the dipole moment is measured in "coulomb meters", but this unit is way too large for molecules. Thus, we use another unit for measuring dipoles: the Debye (D). The relationship between Debyes and SI units is: D = 3.336 x 10^{-30} C m. Let's assume that we have two charges (one +1 and the other –1, thus it would be ionic) at a distance d. Using that equation, let's calculate the distance that would have to separate these two in order to create a dipole of exactly 1.00 Debye. Remember that a charge of +1 is technically +e (that is, +1.60 x 10^{-19} C). Thus,

$$\mu = 1.60 \times 10^{-19} \text{ C} \times \text{distance (in meters)} = 3.336 \times 10^{-30} \text{ C m} = 1.00 \text{ Debye}$$

$$\therefore \ d = 2.085 \times 10^{-11} \text{ m} = 0.2085 \text{ Å} \quad \checkmark$$

This was an extreme scenario: two full ions. However, in a covalent molecule, we do not have full charges and instead, we have partial charges (δ+ and δ-, or more technically: ± δe). Therefore, the charge separation (δ) can be determined from the following equation:

$$\delta = \frac{(0.2085 \text{Å}/D) \times \mu(D)}{d(\text{Å})}$$

What do we need in order to determine the charge separation between the two dipoles in the bond? Two things: the actual dipole moment (which can be measured experimentally by electrical and spectroscopic methods), and the distance between the two atoms (in Å). As an example, let's go back to the HF molecule and calculate δ. Experimentally, the dipole moment for HF is 1.82 D, and the bond length if 0.917 Å[1]).

$$\delta = \frac{(0.2085 \text{ Å}/D) \times 1.82 D}{0.917} = 0.414$$

This delta of 0.414 represents the fraction of ionization for this molecules. That is, if we multiply it by 100, we can say that the molecule is 41.4% ionic, or that is shows 41.4 % ionic character. Another way of looking at this is that the permanent charge on the hydrogen atom is +0.414, whereas the permanent charge on the fluorine is –0.414.

---

[1] This is close to the one calculated with the atomic radii from section 7.7. The difference is due to the fact that the atomic radii were determined from the H-H and the F-F molecules, which are 100% covalent, whereas HF is not and there is an uneven share of electrons, thus a different bond length.

How do we measure µ experimentally? We place the liquid (or the solution) in between two plates that generate an external electrical field. If the molecules are polar, their permanent dipoles will orient themselves with the external magnetic field as seen below.

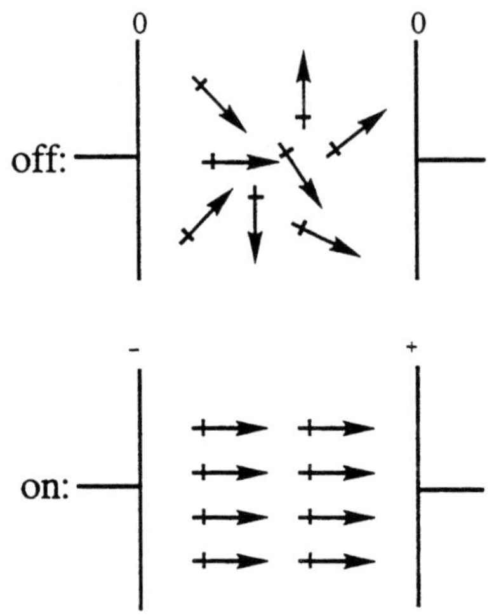

We have seen that the measured dipole moment gives us an insight into the degree of polarization in the molecule. Can we tell anything else about the molecule based on this? It turns out that the answer is yes because it will give us an idea of the geometry of the molecule itself. Of course, if we are talking about a diatomic molecule it is pointless since two atoms define a straight line. But as soon as we have three or more atoms, the possibilities for the geometrical configuration of all the bonds in the molecule could be very large. In molecules that have three or more atoms (referred to as polyatomic molecules), the dipole moment (µ) depends on various things:

1. the electronegativity difference between all the bonded atoms.
2. the directions of the bond dipole moments.
3. the angles between the bonds.
4. the effect of the unshared electron pairs (lone pairs).

What is the meaning of these four points? We need to determine all the individual dipoles (for each bond in the molecule) and then, remembering that they are vectors, determine the vector sum.

Let's look at one particular example: the case of carbon dioxide (CO_2). Knowing that carbon is the central atom, there are two and only two possibilities to draw this molecule. The molecule is either linear or it is not (which would make it angular). At this point in time, how can we tell whether it is linear or angular? We can measure experimentally the dipole moment (μ) for carbon dioxide, and based on the results obtained, we can make an educated guess as to the geometry for the molecule.

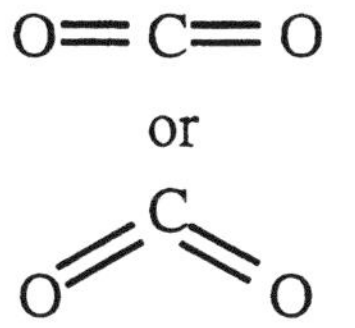

$$O=C=O \quad \text{Linear}$$

or

$$\text{Angular}$$

Why can we make this assertion? Because of the polarity of the C=O bond. The difference in electronegativity between the carbon and the oxygen is (3.50 – 2.50) equal to 1.00. This means that the bond itself is polar. If we wanted to actually calculate the dipole, we would need to know that distance between the two atoms also, but it suffices to know that the bond is polar. If the molecule were linear, then:

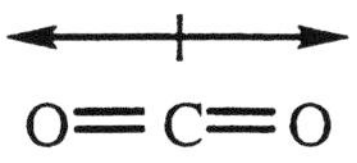

$$O=C=O$$

but:  ←─┤ + ├─→ = 0 D

Therefore, the molecular would be **non-polar**.

However, if the molecule were angular, then it would be a very different situation:

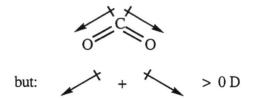

but: ↗ + ↖ > 0 D

Therefore, the molecule would be **polar**.

Experimentally, the dipole moment for the carbon dioxide gas is measured to be exactly equal to zero, therefore, we can immediately conclude that the carbon dioxide molecule must be linear and non-polar.

- BF_3 is a non-polar molecule (μ = 0). The B is the central atom and it is singly bonded to the three fluorine atoms. Predict the geometry of this molecule.

    Let's first explore all the possibilities. What possible geometries could we have with this molecule if all three fluorines are bonded to the boron atom? There are three distinct possibilities, which are shown below. The first possibility could be called a T-shaped geometry (since it actually looks like a T). Possibility **2** is what we call a Trigonal shaped geometry, where all atoms are directly on the plane of the paper at 120° angles. Finally, possibility **3** is called Pyramidal shape, where the four atoms form a small pyramid with the boron on the top. The wedge represents a bond coming out of the plane of the paper, whereas the dotted line represents a bond going into the plane of the paper.

```
      F—B—F                F   F                  B
         |                   \ /              F◣ ╱ ╲ F
         F                    B                   F
                              |
                              F
         1                    2                   3
```

Let's explore the polarity of each of these structures. First of all, it should be clear that the B-F bond is very polar, since there is a substantial difference between the electronegativity of B and that of F. In structure **1**, the B-F bond to the left cancels with the B-F bond to the right. However, the B-F bond going down does not cancel with anything above. Therefore, this structure has a measurable dipole and can be discarded right away. How about structure **2**? If we actually were to add up three identical vectors that happen to be on the same plane 120° apart, we would see that they would cancel out. Therefore, this is a distinct possibility. Finally, structure **3** is clearly polar since the three polar bonds (B-F bonds) do not cancel out. Actually, the resulting dipole would be pointing down, right in between the three fluorines. Conclusion: The molecule must be trigonal planar (structure **2**).

8.7 Covalency.

We have seen that the polarity of molecules depend on the geometry of the molecules. We should concentrate in the next few sections on learning how to draw molecules and therefore, deciding whether they are polar or non-polar. As we will see, this is essential in determining the properties of different materials.

Let's define covalency as the number of bonds normally formed by an atom that completes its octet. This usually corresponds to the number of electrons that an atom is missing in order to complete its octet. For example, Fluorine has seven valence shell electrons, therefore it is missing one electron and we can conclude that fluorine forms one bond. The obvious exception to this statement is hydrogen, since hydrogen will not complete an octet (only two electrons fit into the 1s orbital. This method applies only to some simple molecules, so it will only serve as an introduction.

1. Hydrogen - This element has only one valence shell electrons and since it is in the first shell, it only needs one electron to complete its "pair". We can say that the covalency

Chapter 8: Bonding I.

for hydrogen is one, which means that hydrogen can only form one bond. As we shall see, hydrogen can never form anything else other than a single bond!

H• ⟹ missing one electron

2. Fluorine – This element has seven electrons in its valence shell. Therefore, it needs one electron to complete its octet. We can therefore say that it also has a covalency of one and it forms single bonds (just like hydrogen). Examples: H-F, F-F. The same is true for the other halogens (Cl, Br, and I).

:F̈• ⟹ missing one electron

3. Oxygen – This element has six electrons in its valence shell. Therefore, it needs two electrons to complete its octet. We can therefore say that it has a covalency of two and it will normally form two bonds. As we shall see, these can be two single bonds to two other atoms or a double bond to another atom. Examples: H-O-H and O=O.

:Ö• ⟹ missing two electrons

4. Nitrogen – This element has five electrons in its valence shell. Therefore, it needs three electrons in order to complete its octet. We can therefore say that it has a covalency of three and it will normally form three bonds. As we shall see, these can be three single bonds to three other atoms, a single bond to an atom and a double bond to another, or a triple bond to one other atom. Examples: NH_3 (three single bonds), H-N=O, or N≡N.

:N̈• ⟹ missing three electrons

5. Carbon – This element has four electrons in its valence shell. Therefore, it needs four electrons in order to complete its octet. We can therefore say that it has a covalency of four and it will normally form four bonds. As we shall see, these can be four single bonds to four other atoms, or two single bonds to two other atoms and a double bond to yet another atom, or a single bond to one atom and a triple bond to another atom. Examples: CH_4 (four single bonds), $H_2C=O$, and H-C≡N.

•C̈• ⟹ missing four electrons

We can also assume that the elements below these (at least the non-metals) have the same covalency. For example, sulfur (S) should have a covalency of 2, just like oxygen, and phosphorus (P) should have a covalency of 3, just like nitrogen. With this information, we should be able to determine Lewis structures for some simple molecules. Let's do a few examples.

1. CH_4O.

This is like a puzzle. We know that carbon forms 4 bonds, hydrogen forms one and oxygen forms 2. We should always start with the atoms that have a covalency larger than one and bond them together. This will give us an idea of where to place the ones with the single covalency. The atoms that form more than one bond are the carbon and the oxygen, therefore, we should bond them together and draw the lines that indicate that there are more bonds to come for both the carbon and the oxygen (three more bonds for the carbon and one more bond for the oxygen). Since this corresponds to the number of univalent atoms left (atoms that form only single bonds, like H), then we can conclude the drawing by simply writing the hydrogens along the drawn lines and completing octets for the oxygen (with lone pairs).

$$-\overset{|}{\underset{|}{C}}-O- \longrightarrow H-\overset{H}{\underset{H}{\overset{|}{C}}}-\overset{..}{\underset{..}{O}}-H$$

2. C_2H_6O.

These are the same types of atoms: carbon, oxygen and hydrogen, so we should follow the same directions and bond together the atoms that form more than one bond together first. We should discover immediately that there are two distinct ways to bond these atoms together: C-C-O, or C-O-C. This would therefore yield two different compounds with the same molecular formula. These are referred to as **isomers** – same molecular formula but different order for the bonding of the atoms.

$$H-\overset{H}{\underset{H}{\overset{|}{C}}}-\overset{H}{\underset{H}{\overset{|}{C}}}-\overset{..}{\underset{..}{O}}-H \quad \text{Ethanol (alcohol in beer)} \qquad H-\overset{H}{\underset{H}{\overset{|}{C}}}-\overset{..}{\underset{..}{O}}-\overset{H}{\underset{H}{\overset{|}{C}}}-H \quad \text{Dimethyl Ether}$$

These are two very different compounds, with extremely different properties. Ethanol is a liquid and it is the alcohol in beer and other alcoholic beverages and its consumption will make you drunk. The second one (the ether) is actually a gas and has totally different properties than ethanol. As we shall see in a later chapter, the reason for the difference is all in the bonding. That is, the order in which the atoms are bonded has profound consequences in the properties of materials.

Many students get confused as to what constitutes an isomer. For the time being, it will suffice to say that in order to have a different isomer, we should change the order in which we bond the multivalent atoms (the ones that form more than one bond) while making sure that it is not exactly the same thing backwards. For example, in the case of the ethanol molecule, the first two are not different isomers since the three atoms (the two carbons and the oxygen) are still bonded in the same order. There are interconverted by a simple rotation. This is not clear from the diagram at this point but we should point out that these molecules are not planar, as shown. Once we see the real geometry of this molecule, it would be clear as to why these first two are identical. In the case of the last two, they are the same because the order is exactly backwards (first the oxygen and then the two carbons). This is achieved by flipping the molecule (like lifting it on one end and flipping it).

Conclusion: These are **not** three different isomers! There is just one: they are all the same structure but *viewed* from different angles.

So far, we have seen a couple of examples of molecules in which all the bonds are single. However, we mentioned above that there are many atoms that can form double or triple bonds. How can we account for the formation of multiple bonds?

Multiple Bonds.

When we have multiple bonds we have two atoms sharing more than one pair of electrons. When two electron pairs are shared we get a double bond. Example: O=O.

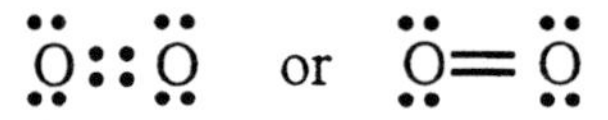

$$\ddot{\text{O}}::\ddot{\text{O}} \quad \text{or} \quad \ddot{\text{O}}=\ddot{\text{O}}$$

When three electron pairs are shared we get a triple bond. Example: N_2.

$$:\text{N}:::\text{N}: \quad \text{or} \quad :\text{N}\equiv\text{N}:$$

Let's do an example where we have isomers that contain multiple bonds with a more complex system (as compared to either O_2 or N_2). Remember that he first thing we will do is connect all the atoms that have a covalency larger than one.

3. Draw all the isomers for C_4H_8.

As we already know, we should start by connecting together all the atoms that form more than one bond (the multivalent atoms). In this case, these are the carbons which form four bonds. This would yield the "picture" seen below.

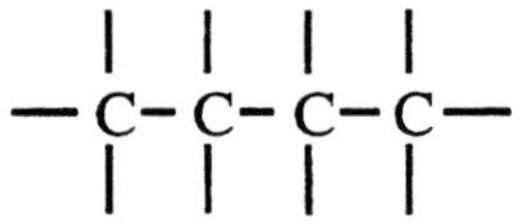

$$-\overset{|}{\underset{|}{C}}-\overset{|}{\underset{|}{C}}-\overset{|}{\underset{|}{C}}-\overset{|}{\underset{|}{C}}-$$

The first thing we notice is that there are still ten "open bonds" to fill out with the remaining atoms (the 8 univalent hydrogens). Of course, this means that we have two extra lines. This means that we must get rid of two of those "potential bonds". ". In order to do this, we need to create an extra connection between two of the carbons. For example:

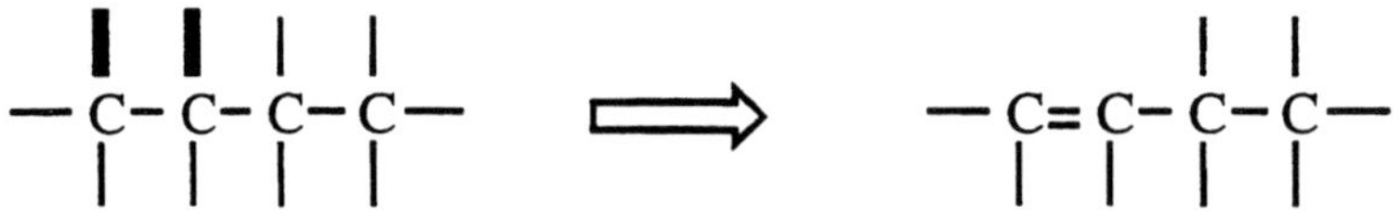

$$-\overset{|}{\underset{|}{C}}-\overset{|}{\underset{|}{C}}-\overset{|}{\underset{|}{C}}-\overset{|}{\underset{|}{C}}- \quad \Longrightarrow \quad -\overset{}{\underset{|}{C}}=\overset{|}{\underset{|}{C}}-\overset{|}{\underset{|}{C}}-\overset{|}{\underset{|}{C}}-$$

Notice that there are two **highlighted** "potential bonds" which we will get rid of and as we do, we create an extra connection between Carbons #1 and #2. Now we have only eight "potential bonds" which are filled out with the remaining H's to get our first isomer.

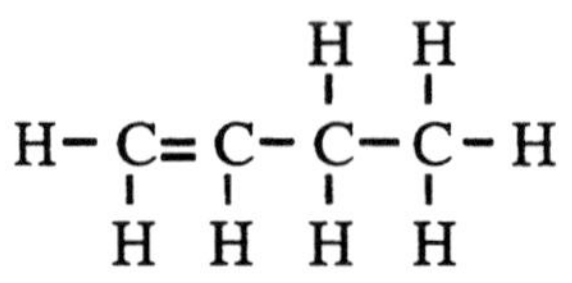

$$\text{H}-\overset{}{\underset{\text{H}}{C}}=\overset{\text{H}}{\underset{\text{H}}{C}}-\overset{\text{H}}{\underset{\text{H}}{C}}-\overset{\text{H}}{\underset{\text{H}}{C}}-\text{H}$$

Another possible isomer has the double bond between Carbons #2 and #3.

```
      H       H
      |       |
  H - C - C = C - C - H
      |   |   |   |
      H   H   H   H
```

We can get a third isomer by doing a different connectivity between the carbons, so that instead of having a chain of four carbons in a row, we have three carbons in a row and the fourth one coming off the middle one. We still have to create the extra bond between two carbons in order to have only 8 H's in the molecule.

```
              H
              |
     H - C = C - C - H
         |   |   |
         H   |   H
             |
         H - C - H
             |
             H
```

There is only one of this type because drawing the double bond between any one of the two carbons will yield the exact molecule (viewed from a different angle).

Are there any more isomers with this formula? Not with a double bond. However, initially we could have chosen to get rid of two "potential bonds" that were not adjacent. This will yield what we will call a **ring**. In order to form a ring, we need to get rid of two "potential bonds" between two atoms, just like the case of a double bond.

```
   |   |   |   |                    |   |
 - C - C - C - C -       ⟹       - C — C -
   |   |   |   |                    |   |
                                    |   |
                                  - C — C -
                                    |   |
```

This yields a ring with four carbons and we have eight "potential bonds" which will be filled out with the eight hydrogen atoms. This is isomer #4.

```
       H    H
       |    |
   H - C —— C - H
       |    |
       |    |
   H - C —— C - H
       |    |
       H    H
```

Finally, we can have one more isomer by making a ring of three carbons (you need *at least* three atoms that have covalency larger than one in order to make a ring) as seen below (isomer #5).

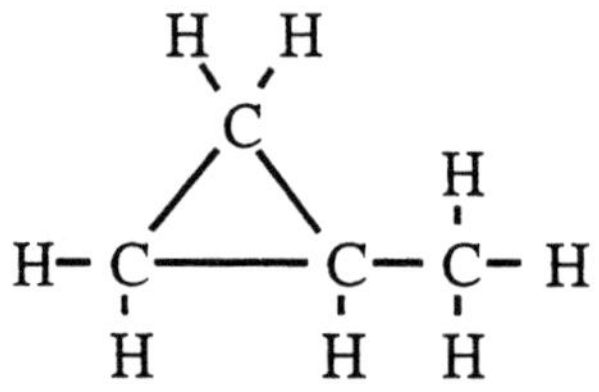

$$\begin{array}{c} HH \\ \diagdown\diagup \\ C \\ \diagup\diagdown \\ H-C\!\!-\!\!-\!\!-\!\!-\!\!C-C-H \\ ||| \\ HHH \end{array}$$

(with an additional H on the right C)

Let's summarize what we have seen so far. When using the covalency method, we should always bond together the multivalent atoms and see how many "potential bonds" are left to be filled with the univalent atoms. If they do not correspond, we will have more "potential bonds" than actual univalent atoms. For every two extra "potential bonds", we have what we call a degree of unsaturation. What is a degree of unsaturation? We have seen that a degree of unsaturation can be obtained one of two ways: with a double bond or with a ring. Before we continue, we should learn how these double bonds are formed and in what types of bonds they are involved.

8.8 Sigma (σ) and Pi (π) Bonds.

So far, we have spent some time talking about bonds and the energy aspects related to their formation, whether they are ionic (which involve pure electrostatic attractions) or they are covalent (in which case the electrons are shared between the two atoms). We saw that covalent bonds are formed basically because upon their formation, energy is going to be released and thus the system is going to be stabilized (energy will be lower for the bonded atoms than for the individual atoms). What must take place between the two atoms that are going to form a bond? As we have seen already, the electrons of one atom must be attracted to the nucleus of the other atom, and vice-versa. These electrons are found in *atomic orbitals*, which are regions in space where there is a large probability of finding the electrons with a given energy. As the atomic orbitals get closer, they will interact or **overlap** with one another to form a volume in space where these electrons will reside "comfortably" (meaning attracted to both nuclei) in between the two atoms. We call this area a **bond**. It should follow that a strong bond will be achieved whenever we get a good overlap between the atomic orbitals. Similarly, a weaker bond will be obtained if the overlap is not as great.

Not all bonds are of the same kind. The classification will depend on what type of overlap we observe between the atoms along the internuclear axis. We will mention at this point

Chapter 8: Bonding I.

two different types of overlap, one that occurs directly along the internuclear axis and one that occurs above and below the internuclear axis.

Let us define a *sigma* (σ) bond as one which contains the maximum electron density **between** the two atoms right along the internuclear axis. A sigma bond may be formed from the interaction of two **s** atomic orbitals (see below), two **p** atomic orbitals (see below), if they approach "head to head", an **s** and a **p** orbital, or any hybridized orbital (next section) and an **s** orbital. At any rate, you should be able to see that the maximum electron density is right in between the two nuclei.

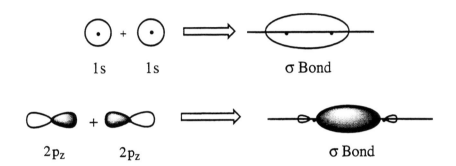

It is important to realize that *all single bonds are sigma* and theoretically allow for free rotation along this bond. So, when you use the covalency method to draw a molecule, if all the bonds depicted are single, you know instantaneously that all of them are sigma bonds. Examples: CH_4, CH_3CH_3, and $CH_3CH_2CH_3$ (practice drawing these). However, you may find that a molecule has multiple bonds (double or triple bonds), as we saw in some of the examples in the previous section. *In any multiple bond there is <u>one sigma bond</u> and the rest are pi (π) bonds.* And what is a π bond? It is a bond in which the electron density is a maximum **above and below** the internuclear axis (NOT along the internuclear axis). A π bond may be formed from the interaction of two **p** atomic orbitals, as seen below.

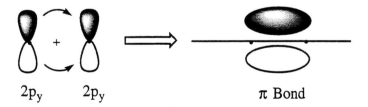

It should be pretty clear from the diagram above that the rotation along the internuclear axis is prevented by the π bond (the electronic cloud above and below the internuclear axis), and the only way in which you could accomplish this would be by breaking the π bond and after the rotation, reforming it (breaking the bond would require too much energy, therefore, it is highly unlikely). This is another one of the differences between a sigma and a pi bond: The sigma bond allows for rotation along the internuclear axis, whereas the pi bond impedes this rotation.

From a structure, we can easily determine the total number of sigma and pi bonds.

- Determine the total number of sigma and pi bonds in the following molecules: CH_3CH_2-OH and C_2H_4O, which are drawn below.

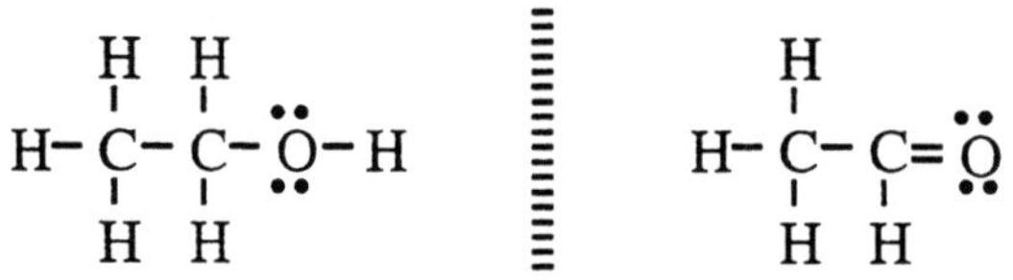

$$
\begin{array}{cc}
\text{H H} & \text{H} \\
\text{| |} & \text{|} \\
\text{H-C-C-}\ddot{\text{O}}\text{-H} & \text{H-C-C=}\ddot{\ddot{\text{O}}} \\
\text{| |} & \text{| |} \\
\text{H H} & \text{H H}
\end{array}
$$

For the molecule on the left, all the bonds are single and therefore sigma in character. Conclusion: It contains 8 sigma bonds. However, the molecule on the right has one double bond, which consists of one sigma bond and one pi bond. Therefore, the molecule on the right has 6 sigma bonds and 1 pi bond.

Now that we are able to recognize sigma and pi bonds, let's revisit our discussion on covalency. We mentioned that when we draw a molecule, we must first connect all the atoms with covalency larger than one together and see if the remaining "potential bonds" are equivalent to the number of univalent atoms. We will still do this but only with one purpose in mind: to determine the number of degrees of unsaturation. A saturated molecule is one that has the maximum number of univalent atoms. An unsaturated molecule has less univalent atoms than it is capable of bonding. In the last example from section 8.7, we saw that there were ten "potential bonds" in the system, but only 8 univalent atoms (eight H's). This implies that we had two "potential bonds" too many. Every two H's in excess leads to one degree of unsaturation. Therefore, in that example, we had one degree of unsaturation. Since all isomers must have exactly the same number of degrees of unsaturation, this means that there are two ways to get one degree of unsaturation: 1) by forming a π bond in the molecule or 2) by forming a ring. This is useful piece of information, because in order for us to draw all the isomers, we simply need to make sure that we have the same number of degrees of unsaturation.

- Draw three different isomers for C_3H_4BrN.

As always, start by connecting the polyvalent atoms (C and N, since H and Br have a covalency of one).

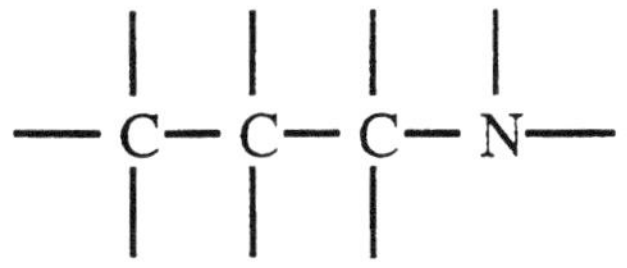

Chapter 8: Bonding I.

$$-\overset{|}{\underset{|}{C}}-\overset{|}{\underset{|}{C}}-\overset{|}{\underset{|}{C}}-\overset{|}{\underset{|}{N}}-$$

We can see that there are nine "potential bonds", but we only have five (4 H's and one Br) univalent atoms. Therefore, we have four "potential bonds" too many, which translates to **two** degrees of unsaturation. How can we get two degrees of unsaturation? 1) two pi bonds, 2) one pi bond and one ring, or 3) two rings (impossible in this case since we only have four polyvalent atoms - remember that you need at least three polyvalent atoms to form one ring). We will now proceed to draw as many different connections between these atoms exhibiting two degrees of unsaturation and then proceed to make sure that each atom obeys covalency.

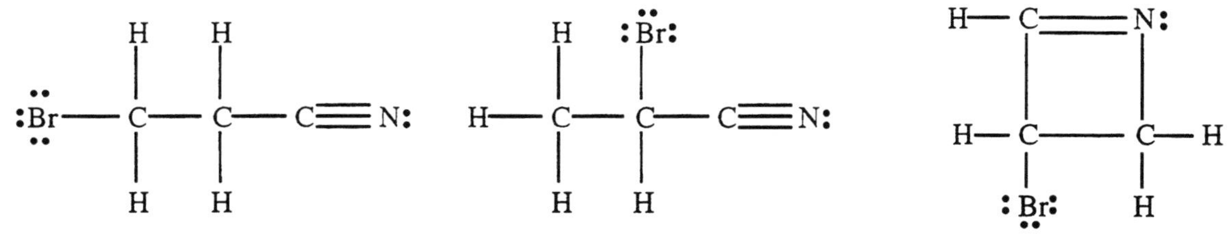

Are these the only three isomers for this molecular formula? Of course not. We chose to draw these three simply because these were the first three that crossed our minds. We can draw any isomer we desire, as long as each isomer has two degrees of unsaturation, we have the correct number of C, N, H, and Br, and we complete the octets on all the atoms (by drawing lone pairs, when necessary). Let's draw three more (of the zillions possible) below.

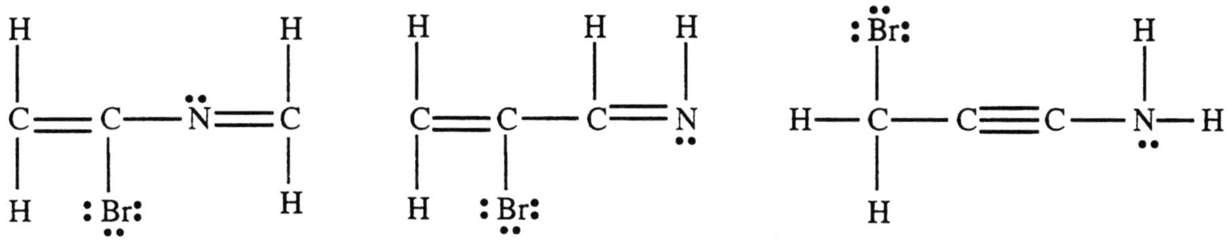

8.9 <u>Drawing Lewis Structures.</u>

At this point, it is convenient that we learn how to draw simple sketches of covalent molecules. In the next few lectures following that, we will also learn how to precisely determine the actual geometry of these sketches and will learn part of the theoretical background for these decisions.

To sketch the general structure of a molecule, we need to keep in mind a couple of things. First, the electrons that are involved in the bonding process are the valence shell electrons. Furthermore, not all the valence shell electrons are involved in bonding, which means that we must decide which are and which are not. A pair of electrons (valence) that is shared between two atoms is referred to as a bond and is drawn with a single line. If two pairs of electrons are shared between two atoms, then we would have a double bond and that would be represented with two lines. The case of a triple bond is completely analogous. An electron pair that is not shared between two atoms (and therefore is part of a specific atom) is referred to as a lone pair and is symbolized as two dots on that atom.

Secondly, these valence shell electrons repel one another, since they have the same charge. As it turns out, the largest repulsion is between lone pairs themselves, then between lone pairs and bond pairs, and finally between bond pairs themselves. Keeping these two things straight, we should be able to sketch some simple molecules.

We will be following the rules below to sketch these molecules.

1. Count the total number of valence shell electrons in the entire molecule.
2. Sketch a simple drawing where the least electronegative atom is singly bonded to the other atoms. In more complicated molecules (where there is no central atom), we have to use a little common sense to decide where to start the drawing, but this will become increasingly easier as we go along.
3. Start trying to complete an octet for every atom bonded to the central atom (since they are more electronegative than the central), at all times keeping track of how many electrons you are using in the process.
4. If when you are done with step #3, you still have some electrons to draw, go and place them in the central atom. This may lead to various situations: If you ran out of electrons, then you have the case of an apparent incomplete octet for the central atom. If you complete the octet and still have electrons to draw, then you must expand the octet for the central atom.
5. Determine the formal[1] charges on the atoms.
6. Based on the formal charges, we may need to establish extra bonds between the atoms bonded to the central atom and the central atom itself, as we will see in some of the following

---

[1] Formal charges are fictitious charges that go along with the electronegativity difference between the atoms.

examples.

Let's simply do a few examples now to see how all these rules apply. Once we become proficient in sketching these molecules, we will go back and learn how to draw them more accurately in three dimensions.

1. NH_3.

In this molecule, we can reason that the nitrogen is the central atom. We mentioned earlier that the least electronegative atom is the central atom, but that is impossible in this case, since hydrogen can only form one bond, therefore, there is no way that it could ever be considered to be a central atom. We start by counting the total number of electrons in this molecule. Nitrogen contributes 5 electrons, whereas each hydrogen contributes one electron, for a total of 8 electrons. Therefore, we must draw eight electrons in the sketch. We immediately proceed to draw the three apparent bonds between the nitrogen and each hydrogen, as follows.

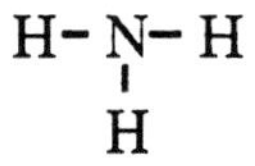

$$\begin{array}{c} H-N-H \\ | \\ H \end{array}$$

This takes care of six of the electrons (two in each bond), leaving us with the task of drawing two more electrons. This means that we must place these two remaining electrons on the N (as a lone pair).

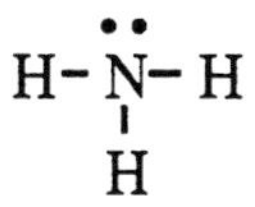

$$\begin{array}{c} \overset{..}{N} \\ H-N-H \\ | \\ H \end{array}$$

As for formal charges, there are none. Let's figure out how to determine these. Go back to each individual atom and check to see how many of the electrons around it belong to that atom. From each bond pair, one electron belongs to one atom, whereas the other electron belongs to the other atom. In a lone pair, both electrons belong to that atom. In the case of each H, they still have only one electron (one from the bond). Since they brought one electron into the bonding situation, they are still neutral (no excesses or deficiencies of electrons). How about the N? Well, from the structure above, we can see that out of the eight electrons around the N, only five belong to N (one from each bond, plus the two from the lone pair). This corresponds with the number of electrons that N brought into the bonding situation, therefore, it carries no formal charges.

2. CO_2.

Once more, start by counting all the valence shell electrons in the molecule. The carbon has four electrons in its valence shell, and each oxygen has six, for a total of 16 electrons. That means that we must draw 16 electrons in the molecule. Since the carbon is less electronegative than oxygen, we will start by drawing a simple structure that shows the carbon singly bonded to each oxygen.

$$O-C-O$$

Now, that takes care of four electrons (two from each bond), which means that we are still missing 12 electrons. Our next task is to complete the octet for each oxygen.

$$:\ddot{O}-C-\ddot{O}:$$

This takes care of all 16 electrons. One thing that we notice immediately is that the carbon does not have a complete octet. Let's determine the formal charges to see if there is anything we can (or should) do about it. Each oxygen has a formal charge of -1, because they contain in the drawing 7 electrons (the 6 from the lone pairs, plus one from the bond). Since oxygen only brought in 6 electrons into the bonding situation, it has one extra electron, thus making it negative. The carbon only has two electrons in this drawing (one from each bond), and since it brought in 4 electrons, that gives it a +2 charge. To satisfy the octet for the carbon, each oxygen (remember that they have extra electrons since they are negatively charged) will share a lone pair with the carbon, thus each forming an extra bond with it (a double bond). Now we have a structure that seems more complete. Each atom has a complete octet and there are no formal charges.

$$:\ddot{O}-C-\ddot{O}: \longrightarrow :\ddot{O}=C=\ddot{O}:$$
$$\phantom{:\ddot{O}}{-1}{+2}\phantom{-\ddot{O}:}{-1}$$

3. SF_4.

Let's start the same way, by determining the total number of electrons in the molecule. The sulfur contributes 6 electrons, and each fluoride contributes 7 electrons, for a total of 34 electrons. Since the sulfur is the least electronegative atom, we will make it the central atom, and all four fluorines bonded to it.

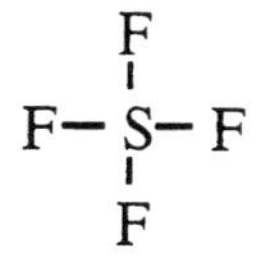

Chapter 8: Bonding I.

```
      F
      |
  F — S — F
      |
      F
```

That takes care of eight electrons, therefore, we still have to draw 26 more. We will first complete the octets for the fluorines.

```
       ..
      :F:
       |
   .. ..  ..
  :F—S—F:
   ..  |  ..
      :F:
       ..
```

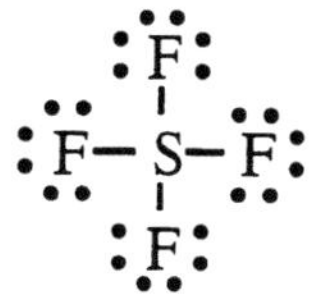

In this example, we can see that after completing the octets for the fluorines we still are left with two more electrons to draw. That means that we will have to expand the octet for the sulfur. Atoms (non metals) in the third period and beyond, like P, S, Cl, etc., have the ability to expand their octets, if necessary. This would yield the structure below, which has the correct total number of electrons for the molecule. Notice that none of the atoms have a formal charge either.

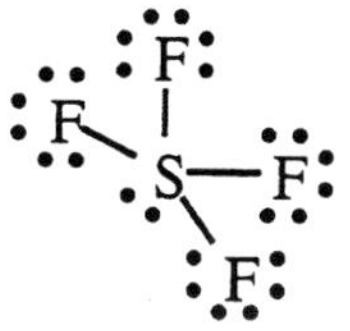

```
        ..
       :F:
    .. \ |   ..
   :F.  S — F:
    ..  / ..
       :F:
        ..
```

4. BF_3.

The total number of electrons in this molecule is 24 (3 from the boron and 21 from the three fluorines). This can be satisfied immediately by the following structure.

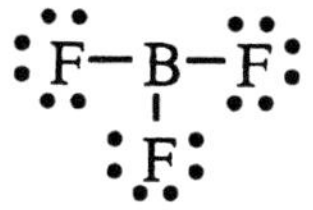

```
   ..       ..
  :F — B — F:
   ..   |   ..
       :F:
        ..
```

We notice right away that the boron does not have a complete octet. However, there are no formal charges in any atom. Therefore, we are not going to form extra bonds with the boron to satisfy its octet. In general, trivalent[1] Boron and Aluminum compounds do not complete octets in their valence shells.

---

[1] The word trivalent means, in this case, that it forms three bonds. Examples: BCl_3, $AlCl_3$, BH_3, etc.

5. Phosphite Ion, PO_3^{-3}.

The total number of electrons in this ion is 26 (5 from the phosphorus, 18 from the oxygens, plus an extra three electrons from the -3 charge). That will lead to the following structure.

$$\left[\ominus \ddot{\underset{..}{O}} - \underset{|}{P} - \ddot{\underset{..}{O}} \ominus \atop :\underset{\ominus}{\ddot{O}}: \right]^{-3}$$

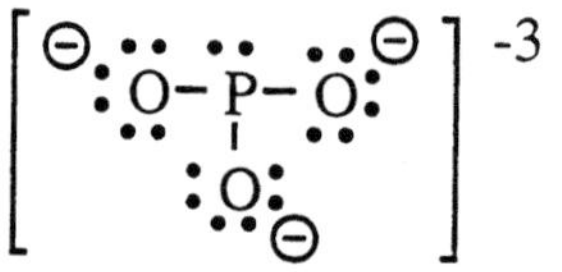

Try to do this by yourselves and corroborate the formal charges.

6. Nitrate ion, NO_3^{-1}.

The total number of electrons in this ion is 24 (5 from the N, 18 from the three O's, plus an extra electron from the negative charge). This leads initially to the following structure.

$$\ominus \ddot{\underset{..}{O}} - \overset{+2}{\underset{|}{N}} - \ddot{\underset{..}{O}}: \ominus \atop :\underset{\ominus}{\ddot{O}}:$$

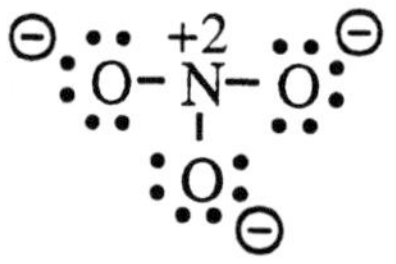

Once more, we see that there is no complete octet on the nitrogen, while the three oxygens are negative (thus with extra electrons). Either[1] one of these three oxygens will be able to form a double bond with the nitrogen to give the following structure.

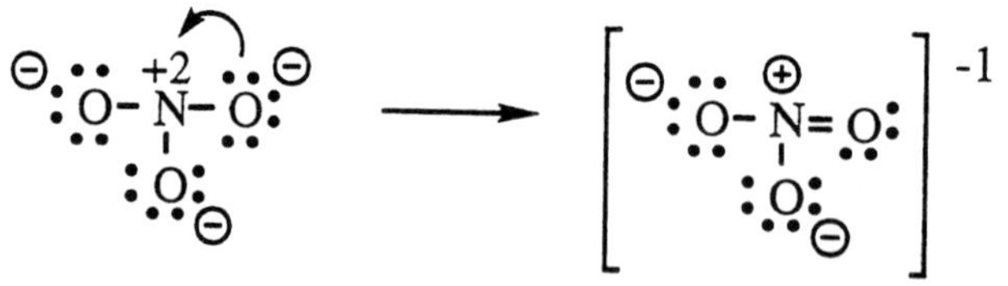

$$\ominus \ddot{O} - \overset{+2}{N} - \ddot{O} \ominus \quad \longrightarrow \quad \left[\ominus \ddot{O} - \overset{\oplus}{N} = \ddot{O}: \atop :\underset{\ominus}{\ddot{O}}: \right]^{-1}$$

and:

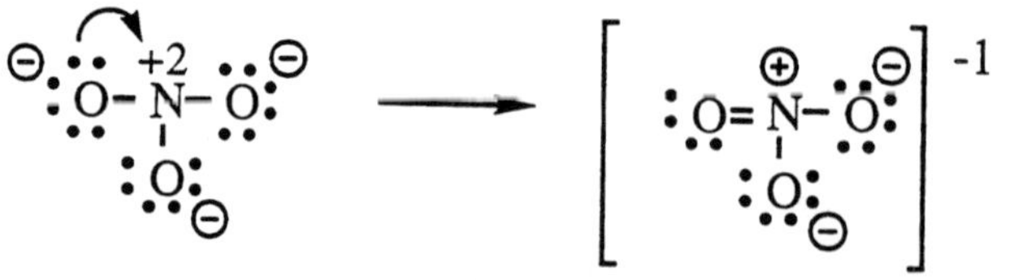

$$\ominus \ddot{O} - \overset{+2}{N} - \ddot{O} \ominus \quad \longrightarrow \quad \left[:O = \overset{\oplus}{N} - \ddot{O}: \ominus \atop :\underset{\ominus}{\ddot{O}}: \right]^{-1}$$

and:

$$\ominus \ddot{O} - \overset{+2}{N} - \ddot{O} \ominus \quad \longrightarrow \quad \left[\ominus \ddot{O} - \overset{\oplus}{N} - \ddot{O} \ominus \atop :\underset{||}{O}: \right]^{-1}$$

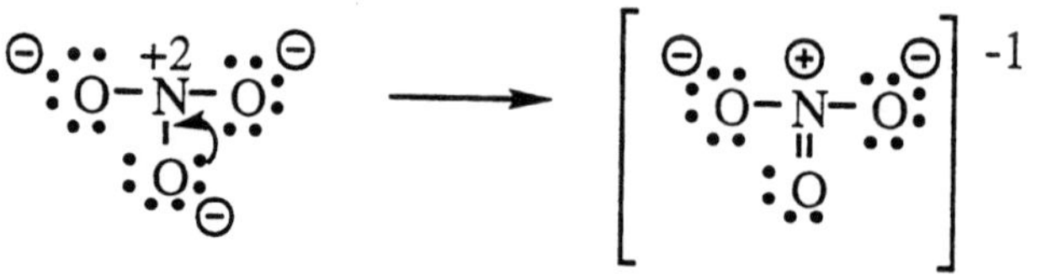

[1] The fact that we said either one of the three oxygens, implies that there are three different structures that may satisfy the Lewis structure.

These three structures are exactly equivalent, which means that we must acknowledge the existence of all three. You might think that all three are interconvertible by rotation, meaning that there is no need to draw all three. If this were true, then one of the bonds to an oxygen would be double, while the other two would be single. That would make the one bond significantly shorter than the other two.[1] In reality all three double bonds are identical in length (somewhere in between the length of a single and a double bond). So, the Lewis sketch for this ion should be drawn as follows. This concept is called **resonance**, and we will revisit it later.

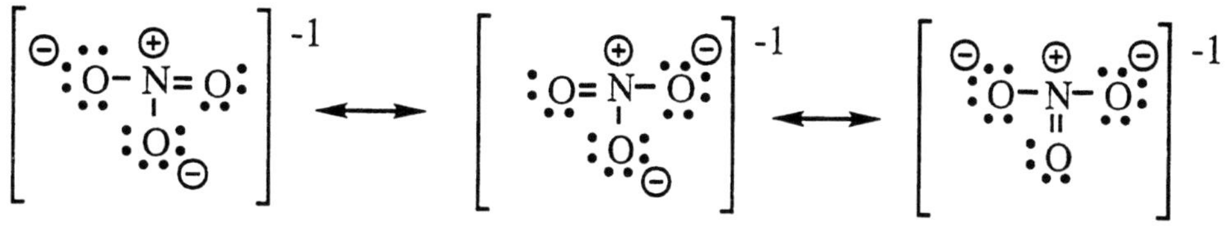

7. Sulfate ion, SO_4^{-2}.

As a last example, let's see an interesting case: that of the sulfate ion. We start in the same place, that is, counting the total number of valence shell electrons. We have 6 electrons from the sulfur, 24 electrons from the oxygens and two extra electrons from the charge of the ion, for a grand total of 32 valence shell electrons. Once more, we start by drawing the single bonds between all the oxygens and the central atom (sulfur).

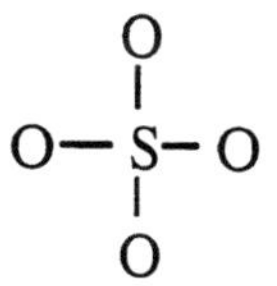

To this point, we have drawn 8 electrons, so we still need to draw 24 more. Next, we proceed to complete the octet for all the oxygens.

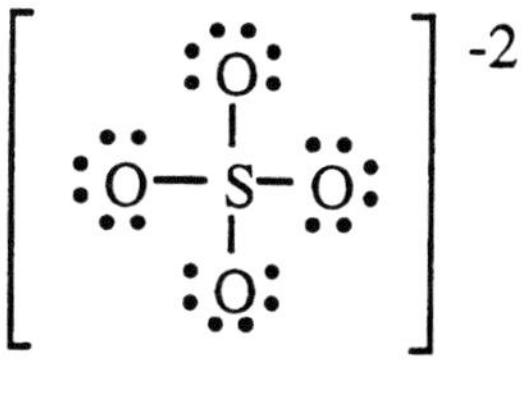

This actually completes all 32 electrons, so we are done. If we include the formal charges on each of the atoms, it would be:

---

[1] See section 8.3.

$$\left[\begin{array}{c} \overset{\ominus}{\ddot{\ddot{O}}} \\ | \\ \overset{\ominus}{\ddot{\ddot{O}}}-\overset{+2}{S}-\overset{\ominus}{\ddot{\ddot{O}}} \\ | \\ \underset{\ominus}{\ddot{\ddot{O}}} \end{array} \right]^{-2}$$

This is all we should draw this year. However, we should also mention in passing that since sulfur is an atom from the third period, it has a chance of expanding its octet, therefore, you will see many times that in Organic Chemistry, they draw this ion in a different way. This way, we minimize the formal charges on the atoms (notice that the formal charges disappeared from two oxygens and from the sulfur).

$$\left[\begin{array}{c} \ddot{\ddot{O}} \\ \| \\ \overset{\ominus}{\ddot{\ddot{O}}}-S-\overset{\ominus}{\ddot{\ddot{O}}} \\ \| \\ \ddot{\ddot{O}} \end{array} \right]^{-2}$$

8.10 Three Dimensionality of Lewis Structures.

Generally speaking, as we went through all these structures, we should notice that we paid close attention to a few things. First, we have tried to give atoms formal charges that are as close to zero as possible. Second, the most electronegative atom is the one that normally carries negative formal charges, if there are any left.

Now, how do we draw these molecules or ions accurately in *three dimensions*? Let's go back to something we mentioned at the beginning of the previous section. There is an inherent repulsion between the electron pairs (whether they are lone pairs or bond pairs). Therefore, these will try to go as far away as possible from one another in order to minimize electronic repulsion. This means that when we draw all the bonds and lone pairs around a central atom, we want to place them as far away as possible from one another.

For example, let's draw the molecule $BeCl_2$. When we follow the rules we learned in the previous section, we will see that there are only two bonds in the molecule (a single bond between the beryllium atom and each of the chlorines. Also, when we draw all the electrons, we find that there are no lone pairs at all on the beryllium. In conclusion, the only thing we have around the beryllium is two bond pairs. So here is the new concept: we will draw these two

Chapter 8: Bonding I.

bond pairs as far away as possible from one another. How far is that? It turns out to be 180° apart. Therefore, the molecule actually would be linear, with a bond angle of 180°, as seen below.

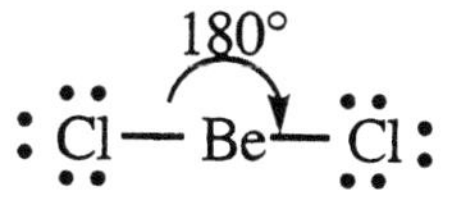

$$:\ddot{Cl}-Be-\ddot{Cl}: \quad 180°$$

If instead we had three bond pairs and nothing else around the central atom (case of BF_3), then we would have to place these three bond pairs as far away as possible from one another. That would convert the structure on the left (an example from the previous section) to the one on the right (**trigonal planar** with 120° bond angles).

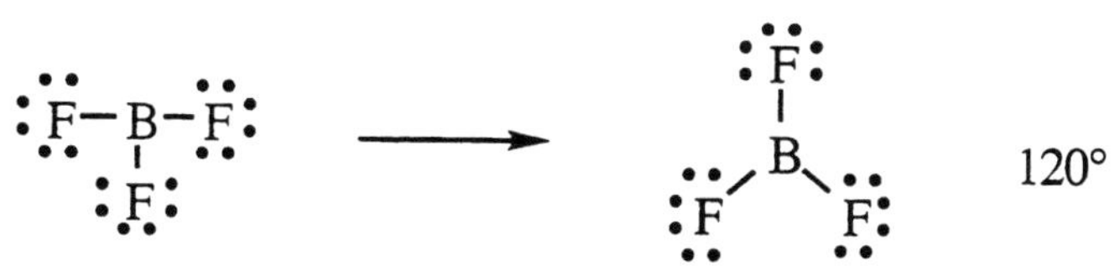

$$:\ddot{F}-\underset{\underset{\ddot{F}:}{|}}{B}-\ddot{F}: \quad \longrightarrow \quad \underset{:\ddot{F} \quad \ddot{F}:}{\overset{:\ddot{F}:}{\underset{|}{B}}} \quad 120°$$

How about if we had four electron pairs around the central atom? This is the case of the CH_4 molecule (called methane). In this case, we have four bond pairs around the C. The furthest away we can place four electron pairs is in a tetrahedral configuration with bond angles of 109.5°. Thus methane is a tetrahedral molecule with bond angles of 109.5°. How can we picture a tetrahedron? It can be visualized as a pyramid with the H's in each the four corners. That leaves the C in the very center of this pyramid, bonded to all corners, as seen below.

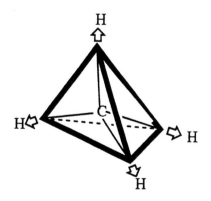

However, this is very difficult to draw and we need some kind of convention to sketch in three dimensions a molecule that has some atoms on the plane of the paper, some atoms coming at you from the plane of the paper, and some atoms going into the plane of the paper. For that purpose, we will use the following convention:

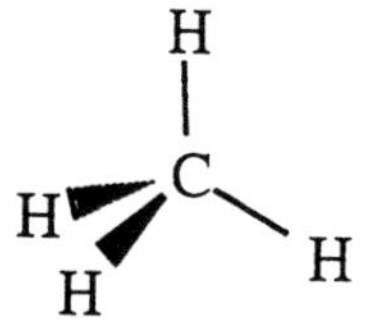

Chapter 8: Bonding I.

```
        H
        |
        C
    H◄──┼──H
    H  /
```

where: | and \ represent being on the plane of the paper

/ represents something coming at you (out of the plane)

▼ represents something going into the plane of the paper.

This is a very symmetrical geometry, so all four positions are identical and the bond angle is the same between any four of them. Convince yourself with your molecular models, or by seeing them in the network.[1]

Very quickly, let's go over one more example of a molecule that we drew previously with four electron pairs, but that had three bond pairs and one lone pair (NH_3). Since it has four electron pairs, the electron configuration around the N would be tetrahedral once again. However, in practice, when we determine experimentally the structure of any molecule, we find that we can only see the nuclei (the atoms themselves), as they diffract X-Rays and by the diffraction pattern, we can extrapolate and understand in what geometrical shape these atoms were bonded together. The key is that we can only see the nuclei, not the electrons. Therefore, lone pairs are not part of the geometry, per se. This means that the actual geometry of the ammonia molecule is not tetrahedral, but rather pyramidal.

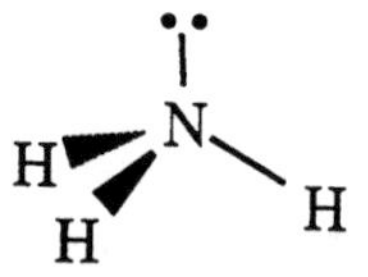

```
        ..
        N
    H◄──┼──H
        H
```

Does this mean that the lone pair does not play a role in the geometry? No! The lone pair has a stronger repulsion to the bond pairs than the bond pairs have amongst themselves. Therefore, the lone pair squeezes the bond pairs together and the bond angles observed for the ammonia molecule are 107° (instead of 109.5°).

---

[1] http://www.chem.memst.edu/hak/vespr.html

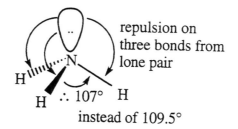

How about a molecule like PF$_5$? This molecule has an expanded octet around the P (five single bonds to the P). How do we draw this molecule in three dimensions? What is the furthest apart we can draw these five electron pairs? We do so as seen below and we refer to this shape as **trigonal bipyramidal**. The trigonal bipyramidal geometry consists of a trigonal planar base, one bond 90° up and another 90° down.

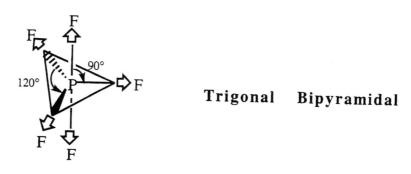

Trigonal Bipyramidal

Finally, how about the case where we have six electron pairs around the central atom? That is the case of the SF$_6$ molecule, in which we have six single bonds between the S and each of the F's. What is the furthest apart we can place these six electron pairs? In a geometrical shape that has the S bonded to the four corners of a square, with the fifth bond up from that square, and the sixth bond down from that square. We call this geometry octahedral, and all the bond angles are either 90° or 180°.

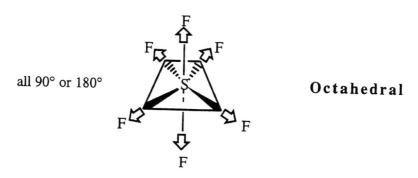

Octahedral

8.11 Expected Geometries With Lone Pairs.

Most of the geometries that we have discussed so far are extremely symmetrical so that if we have to place a lone pair it can be drawn anywhere (*i.e.* all positions are equivalent). Let's look at a few examples.

a) Three electron pairs, one of them a lone pair.

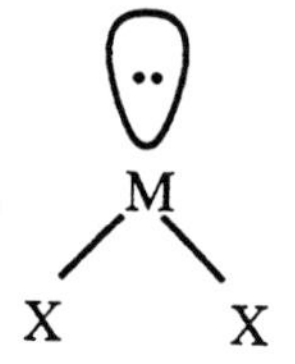

∴ **Angular**

b) Four electron pairs, one of them a lone pair.

∴ **Pyramidal**

c) Four electron pairs, two of them lone pairs.

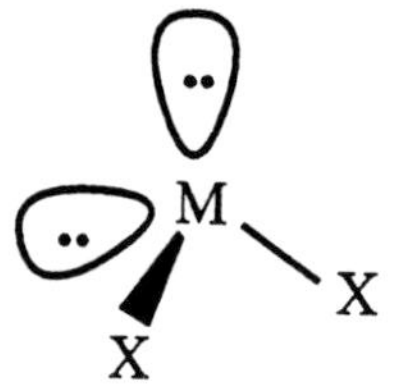

∴ **Angular**

d) Five electron pairs, one of them a lone pair.

The basic geometry for five electron pairs (trigonal bipyramidal) is not as symmetrical as the others, so, we will observe that for the first time placing the lone pair in a particular position **is** of utmost importance. We have two distinct possibilities: either the lone pair goes above (or below) the planar triangle OR it goes in any of the three positions directly on the triangle. Let's look at both cases separately.

Chapter 8: Bonding I.

1. LP up or down:

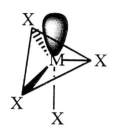

The LP is 90° from <u>three</u> bond pairs and 180° from the other one.

or

2. LP on the triangle:

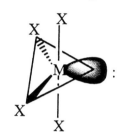

The LP is 90° from <u>two</u> bond pairs and 120° from the other two.

The second possibility has lower energy (minimized the repulsions) and is therefore what we will observe. In conclusion, any time we have this situation, the lone pair will be drawn anywhere on the triangle. This leads to the geometry seen below.

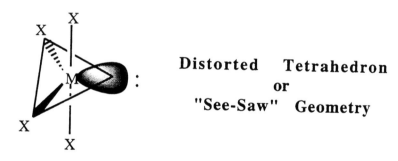

Distorted Tetrahedron or "See-Saw" Geometry

Rotating it, we can envision the "see-saw" geometry.

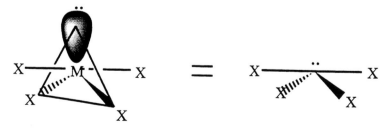

e) Five electron pairs, two of them being lone pairs.

Chapter 8: Bonding I.

Since the first lone pair went into the triangle, the second one must also go into the triangle so as to be as far away as possible from the first one. That would place the second lone pair 120° away from the first one, as opposed to 90° away from the first one. We do not place them across (at 180°) because the first one must go on the triangle.

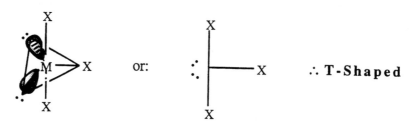 ∴ **T-Shaped**

f) Five electron pairs, three of them being lone pairs.

 ∴ **Linear**

g) Six electron pairs, one of them being a lone pair.

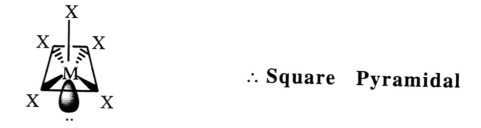 ∴ **Square Pyramidal**

h) Six electron pairs, two of them being lone pairs.

 ∴ **Square Planar**

Let's try to summarize what we have said so far and then go back to cover all the other examples we did before. The table below does just that, giving you some examples for each of

Chapter 8: Bonding I.

these geometries. We should double check and make sure that we understand why each example exhibits the particular geometry indicated in the table.

Summary of Lewis Structures

Total Number of Electron Pairs	Number of Lone Pairs	Predicted Geometry	Example
2	0	Linear	$BeCl_2$, CO_2
3	0	Trigonal Planar	BF_3, SO_3
	1	Angular	$SnCl_2$, NO_2^{-1}
4	0	Tetrahedral	CH_4, SO_4^{-2}
	1	Pyramidal	NH_3
	2	Angular	H_2O
	3	Linear	HF
5	0	Trigonal Bipyramidal	PCl_5
	1	Distorted Tetrahedron or "See-Saw"	$TeCl_4$
	2	T-Shaped	ICl_3
	3	Linear	XeF_2
6	0	Octahedral	SF_6
	1	Square Pyramidal	$BrCl_5$
	2	Square Planar	XeF_4

How do we explain the case of CO_2? We see from the table above that CO_2 is linear and it is classified as having two electron pairs. In reality we already saw that in the final version, there are two double bonds with the C. **For the purposes of determining geometries based on the number of electron pairs around the central atom, count any multiple bond as a single pair.** The carbon in the carbon dioxide has two pairs (one for each double bond). Therefore, the molecule is linear.

What will be the geometry for the other examples we did previously? These were: PO_3^{-3},

NO_3^{-1}, and SO_4^{-2}. Answer: Pyramidal, sp^3; Trigonal Planar, sp^2; Tetrahedral, sp^3.

8.12 Polarity vs. Geometry.

As we have seen before, for molecules containing more than two atoms, the resulting dipole moment µ (which determines if a molecule is polar or non-polar) depends on the magnitude and direction of the individual bond dipole moments, the angle between the bonds, and the lone pairs. If the bond dipoles cancel out and there is no effect of the lone pairs, then µ = 0 and the molecule is non-polar. Take the example of carbon dioxide which is linear and non-polar. This molecule consists of two polar bonds but the geometry is such that the two dipoles cancel out (magnitude of both is the same, but being in exactly opposite directions gives µ = 0).

$$\longleftarrow + \longrightarrow = \underline{0\,D} \quad \sqrt{}$$

If the bond dipoles do not cancel out completely, then the resulting molecular dipole will be larger than zero. Let's take as an example the molecule SCO, and compare it to carbon dioxide gas. The structure of SCO is the same as that of carbon dioxide. This means that we have two double bonds, one between the carbon and the sulfur and the other one between the carbon and the oxygen. In what way do these molecules differ? In the electronegativity of the atoms: C (2.50), O (3.50), S (2.55). The electronegativity difference (ΔEN) for the C-O bond is 1.00 whereas the ΔEN for the C-S bond is essentially zero. This means that when we add the vector dipoles for the carbon dioxide molecule, they will cancel out (this can be easily confirmed with elementary geometry). However, the addition of the two vectors for the SCO molecule, gives rise to a resulting vector that points in the direction of the Oxygen atom. This molecule will therefore be polar since its dipole moment will be larger than zero.

$$\ddot{\underset{..}{O}} = C = \ddot{\underset{..}{O}}$$
$\longleftarrow + \longrightarrow = \underline{0\,D}$ ∴ non-polar

$$\ddot{\underset{..}{S}} = C = \ddot{\underset{..}{O}}$$
$\longleftarrow + \longrightarrow > 0\,D$ ∴ polar

Don't be misled by the belief that a linear molecule is non-polar because it is very symmetrical. We must look at the molecule and decide.

Conclusion: Even an extremely symmetrical geometry (like linear) does not immediately

Chapter 8: Bonding I.

mean that it will yield a non-polar molecule. When all the substituents (atoms bonded to the central atom) are identical (or of same electronegativity) you will get a non-polar molecule. RULE: A molecule will be non-polar if it has one of the basic geometries (linear, trigonal planar, tetrahedral, trigonal bipyramidal, octahedral and square planar) **AND** if all the substituents are the same (same electronegativity). If all the substituents are the not the same, there is a possibility that it will be non-polar, if they are distributed in such a way so that the dipoles will cancel. Below you may find a small sample of some molecules in which you will find indicated their structure and polarity. Convince yourselves that these are correct.

a)	CO_2	linear	non-polar
b)	SCO	linear	polar
c)	BF_3	trigonal planar	non-polar
d)	CCl_4	tetrahedral	non-polar
e)	$HCCl_3$	tetrahedral	polar
f)	PCl_5	trigonal bipyramidal	non-polar
g)	SF_6	octahedral	non-polar
h)	XeF_4	square planar	non-polar
i)	SF_4	"see-saw"	polar
j)	XeF_2Br_2	square planar	may be either polar or nonpolar (why??)

Summary.

a) All the following geometries (linear, trigonal planar, tetrahedral, trigonal bipyramidal, octahedral, and square planar) are highly symmetrical and lead to non-polar molecules **if** all the bonds are identical.

b) The bonds to these geometries may be somewhat different, but still create such symmetry to give rise to a non-polar molecule. Ex. XeF_2Br_2.

c) All other geometries lead to polar molecules (like pyramidal, T-shaped, etc.).

Let's do some examples from scratch to make sure that we understand this method. It is important that you learn how to draw these geometries.

Chapter 8: Bonding I.

- HCN

    The first thing we must do is to actually draw the Lewis structure for the molecule, just like we learned in section 8.9. The total number of valence shell electrons in the molecule is ten (one from the H, four from the C, and five from the N). Our first attempt at drawing the molecule is seen below (with formal charges).

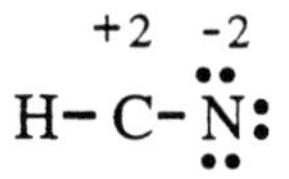

$$\overset{+2\ \ -2}{H-C-\ddot{\underset{\cdot\cdot}{N}}:}$$

    Since the carbon does not have a complete octet and has a positive (actually +2) formal charge, while the nitrogen has a negative (-2) charge, we can form two extra bonds between the carbon and the nitrogen and in the process "kill" the formal charges. This leaves us with the structure seen below.

    What about the geometry? Is it linear or angular? According to what we have said so far, we can treat the multiple bond as a single unit around the carbon, so we can say that there are two "electron pairs" around the carbon (one is the single bond and the other the triple bond). The furthest away we can place these is 180°, therefore, we can say that the geometry must be linear.

$$H-C\equiv N:$$

    Finally, is the molecule polar or non-polar? There are two atoms bonded to the carbon: a hydrogen and a nitrogen. Since the nitrogen is more electronegative than the hydrogen, then the molecule is polar with the dipole pointing towards the nitrogen (the most electronegative).

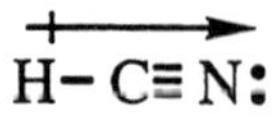

$$\overset{\longrightarrow}{H-C\equiv N:}$$

- SnCl$_2$

    Sn (Tin) is in the carbon family, therefore it has four valence shell electrons. There is a total of 18 valence shell electrons in the molecule. With this information, we can draw the Lewis structure like this.

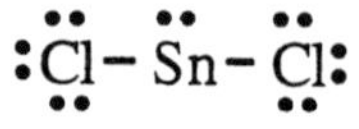

$$:\!\ddot{\underset{\cdot\cdot}{Cl}}-\ddot{\underset{\cdot\cdot}{Sn}}-\ddot{\underset{\cdot\cdot}{Cl}}\!:$$

Chapter 8: Bonding I.

There are no formal charges on any of the three atoms, therefore, there is no compelling reason to draw any more bonds between these atoms. Looking at the central atom (the tin), we can see that there are three electron pairs around this atom (one lone pair and two bond pairs – one to each chlorine). The furthest apart we can draw three electron pairs around an atom is in a trigonal planar geometry, with 120° bond angles. The structure is seen below.

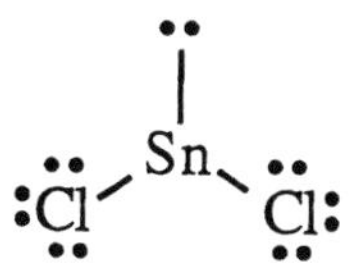

Is the molecule polar or non-polar? Answer: polar! The Sn-Cl bonds are polar, since the Cl is more electronegative than the Sn. When we add those two vectors, we get a resulting vector right in between the two chlorines, as seen below.

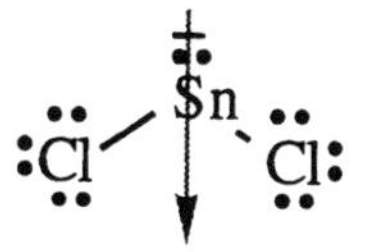

- PCl_3Br_2

There are 40 valence shell electrons in the molecule. This will eventually lead us to drawing five bonds to the phosphorus, thus we can immediately see that we have an expanded octet. Since there are no lone pairs on the P, then the geometry for the molecule must be trigonal bipyramidal, as seen below.

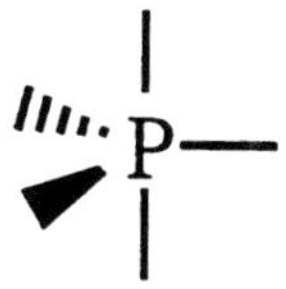

The only thing we have left to do is to decide where to draw the three chlorines and the two bromines. There are a few possibilities (different isomers). We will choose to draw[1] two in here: one that is polar and one that is non-polar.

---

[1] For the sake of simplicity, we will skip drawing the lone pairs on the halogens, but remember that each one of them has a complete octet (three lone pairs).

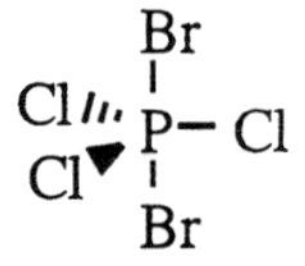

```
      Br                  Cl
      |                   |
Cl,, P—Cl           Br,, P—Cl
Cl▼ |               Br▼ |
      Br                  Cl

   Non-polar            Polar
```

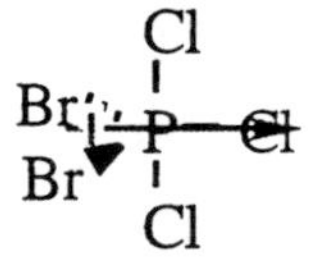

The one on the left is non-polar because the two bromines are 180° apart and cancel out, while the three chlorines are in a trigonal plane at 120°, therefore, they also cancel out. However, the one on the right is polar. Why? The two chlorines that oppose one another do cancel out (the one on top and the one on the bottom), however, the third chlorine is in the triangle, and it is more electronegative than the bromines. Therefore, the net dipole for this molecule is in between the two bromines, but pointing towards the chlorine.

- N_2O (laughing gas)

There are three atoms in this molecule: one of the nitrogens is the central atom, to which we have bonded another nitrogen and an oxygen. There are 16 valence shell electrons in the molecule. Our original Lewis sketch can be seen below.

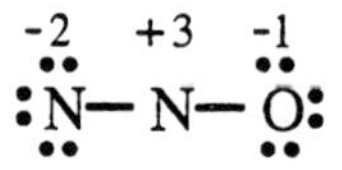

```
   -2   +3   -1
   ••        ••
  :N— N— O:
   ••        ••
```

The central nitrogen does not have a complete octet and has a positive formal charge, whereas both other atoms have negative formal charges. Therefore, it follows that formation of two extra bonds with the central atom will solve the problem of the octet. However, this leads to three distinct possibilities: 1) we can draw a double bond with the nitrogen on the left and one with the oxygen, or 2) we can draw a triple bond between the two nitrogens, or 3) we can draw a triple bond between the nitrogen and the oxygen. These three possibilities can be seen below.

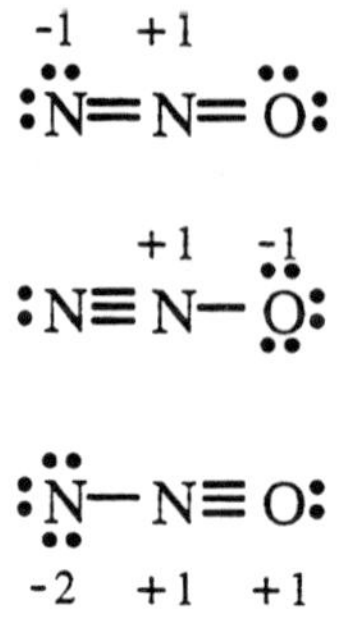

```
     -1    +1
     ••         ••
    :N= N= O:

           +1   -1
                ••
    :N≡ N— O:
                ••

     ••
    :N— N≡ O:
     ••
     -2   +1   +1
```

Once more, these are resonant forms of this molecule, but we will discuss this topic in Chapter 9. The molecule is linear, since we can see that there are only two "electron pairs" around the central nitrogen (there are only two bonding sites on that nitrogen and there are no lone pairs). Two pairs correspond to a linear molecule. The molecule is also polar, since the oxygen is more electronegative than the nitrogen.

Chapter 9: Bonding II. Advanced Concepts.

9.1 Hybrid Orbitals.

The Lewis method for drawing molecules is not a perfect one. We would have to make many refinements in order to be able to explain every single example that has been observed. However, it is a very simple and successful way of predicting the geometries of most molecules. We might ask ourselves the following question: How does the concept of orbitals (that we previously learned) apply to these geometries? No matter how simple a molecule you choose to draw, the orbitals of the atoms will never match the bond angles that we observe experimentally. For example, consider the molecule methane, CH_4. What did we just learn about methane? That it forms a tetrahedral molecule with four identical C-H single bonds and a characteristic bond angle of 109.5°. Is this consistent with the orbitals of the C and the H's? No!

Carbon has a configuration of $1s^2\ 2s^2\ 2p^4$. This means that it has four orbitals in the valence shell (the spherical 2s orbital, and the three orthogonal[1] 2p orbitals). On the other hand, all four H's are $1s^1$, so they must involve the 1s orbital in the bonding. Since we are forming covalent bonds between the C and the H's, we must have a good, solid overlap of the atomic orbitals. The arrangement of the orbitals in the C (the central atom) should be the ones that dictate the geometry of the resulting molecule. How are these four orbitals distributed in space? The 2s is a spherical orbital with spherical symmetry. This means that the 1s of the first H could bond anywhere around the C. In other words, a spherical orbital does not have any particular orientation. The other three H's would have to bond to the three[2] different 2p orbitals. What would be our immediate first impression? The C-H bond formed between the 2s of the C and the 1s of a H would be very different from all other three C-H bonds (formed between a particular 2p orbital for the C and the 1s for a given H). Furthermore, the three C-H bonds formed to the 2p's of the C would have to be orthogonal (exhibiting 90° angles). This is NOT the case! We know that all four bonds are IDENTICAL in every respect and that the bond angles are 109.5°. How can we explain this? We will try initially with a concept called hybridization of atomic orbitals.

In order to form identical orbitals that have the correct orientation (one that predicts the observed geometries) we must create new orbitals. Remember that orbitals are, after all, mathematical equations that describe 95% of the probability of finding an electron with a given energy. These new orbitals will allow for a vastly improved overlap of all the orbitals involved.

---

[1] Orthogonal means mutually perpendicular.
[2] One along the **x** axis, one along the **y** axis, and the other one along the **z** axis.

The greater the overlap between two orbitals in a bond, the more stable the bond will be. How do we obtain these new hybrid orbitals? By simply combining the mathematical equations for all the orbitals that we know are involved in the bonding situation for the central atom. For example, in the case of CH_4, we would combine the 2s orbital with the three 2p orbitals. Why these? Because those are the orbitals in the valence shell of the C atom. When you combine a certain number of atomic orbitals, you always obtain the same number of hybrid orbitals (the new ones), and these hybrid orbitals always turn out to be as far away as possible from one another. This method would then allow us to form four identical hybrid orbitals for the C, which would be arranged at 109.5° from one another. This process of combining the atomic orbitals of a central atom to form a set of hybrid orbitals is referred to as **hybridization**.

Let's look at some examples to learn how to tell the hybridization of any central atom in most molecules.

1. $BeCl_2$

We can easily draw the Lewis structure to be: Cl - Be - Cl, where the Cl's have a complete octet but the Be does not have one. We can also predict that the molecule is linear. However, how can we explain it from a hybridization point of view?

The ground state electronic configuration of Be is:

$$Be: 2s^2 \implies \underline{\updownarrow}_{2s}$$

Each chlorine has seven electrons, so each is missing just one. The problem is that both electrons in the Be are paired up, so for all practical reasons, they are unavailable. We should expect to see that Be forms no bonds (because all its electrons are paired up). However, in the gas phase, Be does form the molecule $BeCl_2$. How can we account for the formation of two covalent bonds? By unpairing the two valence shell electrons. In order to do that, one of the electrons must be promoted to the next available subshell *within the same valence shell!*

Energy must be supplied to do this ($E_{2p} > E_{2s}$) but it will be more than compensated upon the formation of the two covalent bonds. This electronic promotion explains the formation of two bonds **but** they are very different bonds (one with an **s** orbital and the other with a **p** orbital). In reality, both bonds are identical, as proven experimentally. This introduces us to a problem very similar to the one of C in CH_4, as seen above. We must remember, once more, that atomic orbitals really do not exist "per se" as they are only mathematical equations that gives us a high percentage probability of finding the electron with a given energy. We can therefore take these atomic orbital equations and combine them linearly (adding or subtracting them) to form a new set of equations (orbitals) which we will call hybrid orbitals. Since we will be combining one **s** orbital with one **p** orbital, we will get two hybrid orbitals, which we will call **sp hybrid orbitals** and they will be located as far away from one another as possible.

These new hybrid orbitals (**sp**) have 50% **s** character and 50% **p** character. This implies that the **sp** orbitals are much more spherical than a **p** orbital (which has two lobes). Actually the **sp** hybrid orbital looks something like this:

○ Be ◗

All hybrid orbitals look alike: one large lobe and a really tiny one which is a remnant - essentially, any hybrid orbital consists of a single lobe. In the case of the **sp** orbitals, they have a large spherical component. Others will have less of that spherical component and thus will look more elongated as we will see later.

Where is the other **sp** orbital? As far away as possible from the first one: 180° apart! Therefore **sp** hybridization will lead to **linear geometry**. Conclusion: The beryllium chloride molecule is linear and the Be[1] hybridizes sp in order to achieve that geometry.

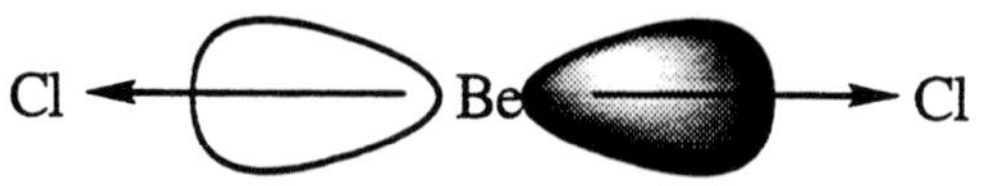

Cl ←◖)Be◗ → Cl

2. BF_3

This is a molecule we have drawn before, so we already know that it is trigonal planar

---

[1] Please, notice that the Be atom *does not* have a complete octet (configuration of a noble gas)! Therefore, this is an example of a molecule that is electron deficient (with an incomplete octet).

and non-polar, with all the bond angles being 120°. The ground state electronic configuration of the central atom (the B) is:

$$B: \ 2s^2 \ 2p^1 \implies \underset{2s}{\uparrow\downarrow} \ \underset{2p}{\uparrow \ \_ \ \_}$$

According to that configuration, B would form only one bond.[1] However, from the formula of the compound we see that B is forming 3 bonds (as it normally does). How can we account for the formation of three bonds? By promoting one of the electrons in the 2s orbital to one of the empty ones in the 2p subshell. This will then generate 3 unpaired electrons which will in turn explain the formation of 3 bonds.

$$\underset{2s}{\uparrow\downarrow} \ \underset{2p}{\uparrow \ \_ \ \_} \ \xrightarrow{+E} \ \left[\underset{2s}{\uparrow} \ \underset{2p}{\uparrow \ \uparrow} \right] \ \xrightarrow{hybridization} \ \underset{sp^2}{\uparrow \ \uparrow \ \uparrow} \ \overline{2p}$$

Experimentally, we find that all three bonds are identical, and that is why we must hybridize the three AO's to obtain the three identical hybrid orbitals: **sp²**, leaving behind an unhybridized, empty p orbital. These orbitals (sp²) have 33% **s** character and 66% **p** character. How are these three orbitals oriented in space? In such a way that will place them as far away as possible from one another: 120° apart in a plane, leading rise to a **trigonal planar geometry** for the molecule.

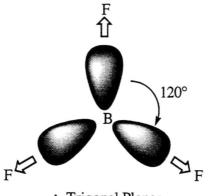

∴ Trigonal Planar

Notice that boron does not complete its octet either. Compounds that do not complete their octet and thus not attain the electronic configuration of a noble gas are referred to as **electron deficient compounds**. It should be fairly straightforward to ascertain that AlCl₃ is

---

[1] Because it only has one unpaired electron.

also electron deficient.

3. CH_4

This is another molecule that we have already discussed. We saw that it is tetrahedral with four identical bonds that are 109.5° apart. Let's try to corroborate this with an orbital approach. The ground state electronic configuration of the central atom is:

$$C: 2s^2\ 2p^2$$

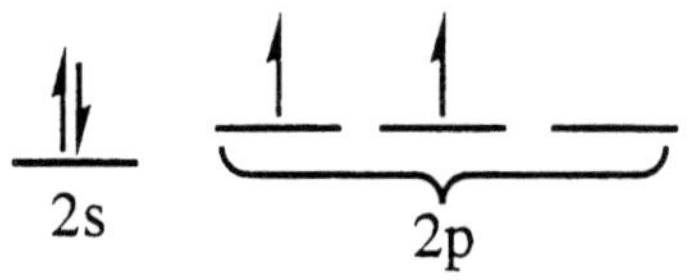

This configuration predicts the formation of two bonds, but we already know that C forms four bonds, CH_4. To explain the formation of four bonds, we must promote an electron from the 2s orbital into the 2p subshell. Just like the previous examples, the four bonds are determined experimentally to be identical so we must hybridize all four AO's and thus obtain four identical hybrid orbitals: **sp³**.

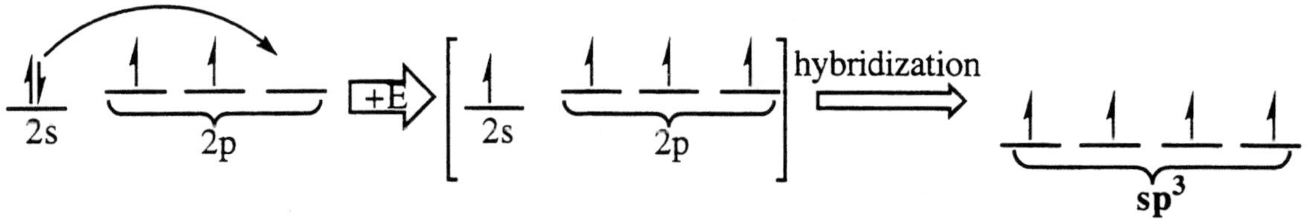

These new orbitals have 25% **s** character and 75% **p** character and are located 109.5° apart from one another in space forming a **tetrahedral geometry**. Thus, we can see that we can use hybrid orbitals to predict the shapes of molecules.

4. NH_3

The ground state electronic configuration for the N is:

$$N: 2s^2\ 2p^3$$

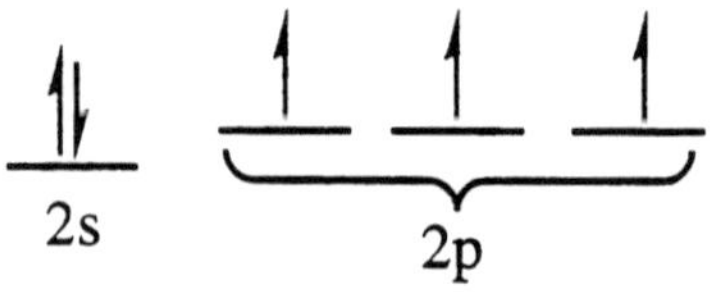

This time, we have three unpaired electrons and that corresponds with the three bonds we want to form. Therefore we do not need to promote electrons. The fact still remains that all three bonds are proven to be identical experimentally, therefore hybridization must still occur. All four orbitals will be hybridized to obtain four **sp³** orbitals which we already know are 109.5° apart in a tetrahedron configuration. There is a *big* difference this time, however. Not all four positions will be occupied by an H, hence we have a **lone pair**.

$$\underset{2s}{\uparrow\downarrow} \quad \underbrace{\underline{\uparrow} \; \underline{\uparrow} \; \underline{\uparrow}}_{2p} \quad \xRightarrow{\text{hybridization}} \quad \underset{}{\underbrace{\underline{\uparrow\downarrow} \; \underline{\uparrow} \; \underline{\uparrow} \; \underline{\uparrow}}_{sp^3}}$$

Since the hybridization is sp^3, the orbital geometry around the N must be tetrahedral with bond angles of 109.5°. However, one of these orbitals is not bonded to a H, and instead it has the lone pair. This means that the geometry of the molecule must be **pyramidal**. As previously seen, the lone pair repels the bond pairs and makes the bond angles a little smaller than 109.5° (technically it is about 107°, in this example). I want to point out that not all molecules that are sp^3 with one lone pair will have a bond angle of 107°. Only NH_3 does. As the orbitals get bigger and bigger, so will the repulsion between the lone pair and the bond pairs. Therefore, even though the molecule PH_3 is completely analogous to the NH_3 (both are sp^3 with one lone pair), the bond angle for the PH_3 is about 100°.

5. H_2O

The case of water is very similar to that of ammonia, so it will suffice to show in a schematic way how to get the geometry of the molecule.

$$O: 2s^2 \, 2p^4 \quad \Longrightarrow \quad \underset{2s}{\uparrow\downarrow} \quad \underbrace{\underline{\uparrow\downarrow} \; \underline{\uparrow} \; \underline{\uparrow}}_{2p}$$

and therefore:

$$\underset{2s}{\uparrow\downarrow} \quad \underbrace{\underline{\uparrow\downarrow} \; \underline{\uparrow} \; \underline{\uparrow}}_{2p} \quad \xRightarrow{\text{hybridization}} \quad \underbrace{\underline{\uparrow\downarrow} \; \underline{\uparrow\downarrow} \; \underline{\uparrow} \; \underline{\uparrow}}_{\mathbf{sp^3}}$$

The $\mathbf{sp^3}$ hybridization means that a tetrahedral distribution for the orbitals will be obtained, however this time we have two lone pairs. As expected, the two lone pairs have an augmented effect of repelling the bond pairs, therefore the actual bond angles in water are about 105° instead of 109.5°. The observed geometry for water is **angular**. Remember that the bond angle shown is unique for the water molecule. A very similar molecule (H_2S) has a much smaller bond angle (less than 100°).

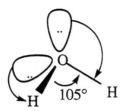

6. PF$_5$

Since there are only three unpaired electrons in the P, there is a need for promotion to a higher energy level in order to explain the formation of the three bonds.

P: $3s^2\ 3p^3$ ⟹ $\underset{3s}{\underline{\uparrow\downarrow}}\ \underset{3p}{\underline{\uparrow}\ \underline{\uparrow}\ \underline{\uparrow}}$

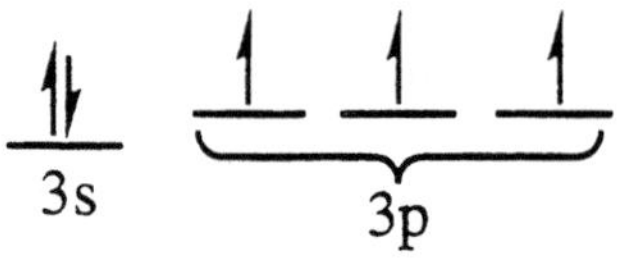

An interesting observation would be to say that the electron should be promoted to the 4s subshell, which energetically is the next one. However, the electron is promoted to the next available empty subshell **WITHIN THE SAME VALENCE SHELL**. This leaves us no choice but to promote the electron to the *3d* subshell as can be seen below.

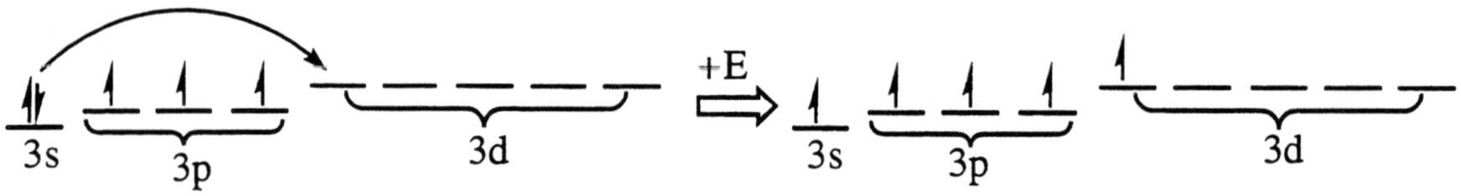

$\underset{3s}{\underline{\uparrow\downarrow}}\ \underset{3p}{\underline{\uparrow}\ \underline{\uparrow}\ \underline{\uparrow}}\ \underset{3d}{\underline{\ }\ \underline{\ }\ \underline{\ }\ \underline{\ }\ \underline{\ }}$ $\xrightarrow{+E}$ $\underset{3s}{\underline{\uparrow}}\ \underset{3p}{\underline{\uparrow}\ \underline{\uparrow}\ \underline{\uparrow}}\ \underset{3d}{\underline{\uparrow}\ \underline{\ }\ \underline{\ }\ \underline{\ }\ \underline{\ }}$

As we have seen quite a few times already, all orbitals containing electrons are hybridized. This will therefore lead to a new hybridization: **sp³d**.

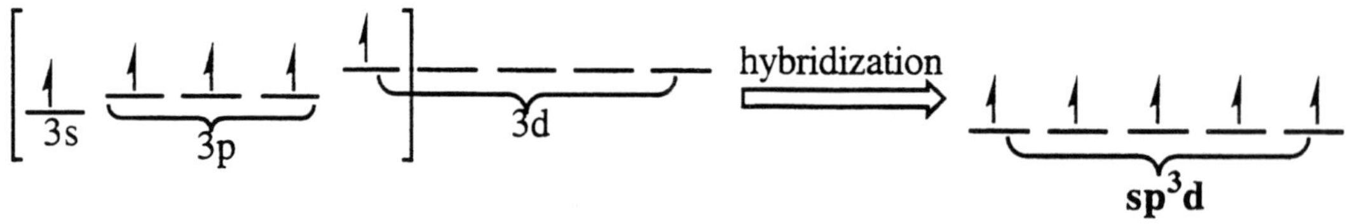

$\left[\ \underset{3s}{\underline{\uparrow}}\ \underset{3p}{\underline{\uparrow}\ \underline{\uparrow}\ \underline{\uparrow}}\ \underset{3d}{\underline{\uparrow}\ \underline{\ }\ \underline{\ }\ \underline{\ }\ \underline{\ }}\ \right]$ $\xrightarrow{\text{hybridization}}$ $\underset{sp^3d}{\underline{\uparrow}\ \underline{\uparrow}\ \underline{\uparrow}\ \underline{\uparrow}\ \underline{\uparrow}}$

The **sp³d** hybridization has 5 orbitals which are oriented in space as follows:

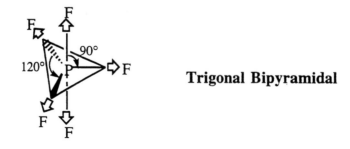

Trigonal Bipyramidal

The trigonal bipyramidal geometry consists of a trigonal planar base, one bond 90° up and another 90° down. This is exactly what we could have expected from the Lewis structure.

7. SF$_6$

Just like the case of P, we expect S to behave similarly since it has available an empty *3d* subshell to expand its octet.

S: $3s^2\ 3p^4$ ⟹ ↑↓   ↑↓ ↑ ↑
                         3s   3p

We must consider promotion of two electrons into the *3d* subshell because the S only has two unpaired electrons (and we need 6 unpaired electrons to explain the formation of 6 bonds). This will leave six orbitals with one electron in each which in turn are hybridized to form six identical hybrid orbitals: **sp^{3}d^2**. These six orbitals are arranged in space as far away as possible from one another giving a planar square base with the S in the center bonded to the four corners, one bond 90° up and the other bond 90° down. The resulting geometry is what we previously referred to as an **octahedron** and once more is a very symmetrical geometry. In the following scheme, we can see pictorially what we have just discussed in the previous paragraph. Note that this time we had to promote 2 electrons from the *3s* and the *3p* into the empty *3d* subshell.

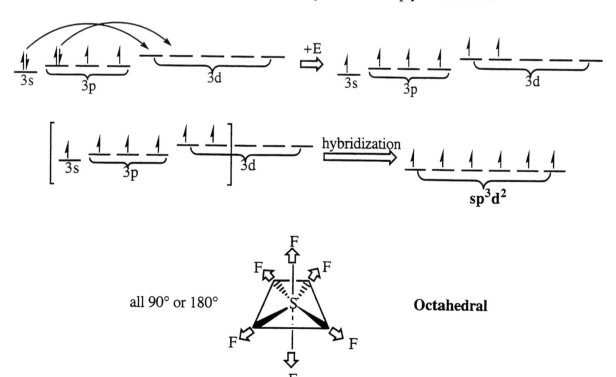

all 90° or 180°

Octahedral

We can now rewrite the Lewis summary, adding the hybridization of the central atom, as we just learned.

No. of Electron Pairs	Hybridization	No. of Lone Pairs	Geometry
2	sp	0	Linear
3	sp^2	0	Trigonal Planar
		1	Angular
4	sp^3	0	Tetrahedral
		1	Pyramidal
		2	Angular
		3	Linear
5	sp^{3}d	0	Trigonal Bipyramidal
		1	See Saw
		2	T-Shaped
		3	Linear
6	sp^{3}d^2	0	Octahedral
		1	Square Pyramidal
		2	Square Planar

Let's see a couple of examples of molecules that contain multiple bonds and see how we can use the hybrid orbitals to explain their Lewis structures.

a) CO_2. We already know that the molecule looks like this.

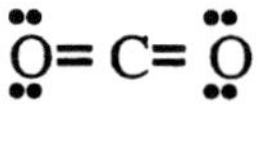

$$\ddot{\underset{\cdot\cdot}{O}} = C = \ddot{\underset{\cdot\cdot}{O}}$$

Let's try to predict its linearity. The electronic configuration for the C atom ready to form four bonds is:

$$C: \ 2s^1 \ 2p^3 \ \Longrightarrow \ \underset{2s}{\underline{\uparrow}} \ \underbrace{\underline{\uparrow} \ \underline{\uparrow} \ \underline{\uparrow}}_{2p}$$

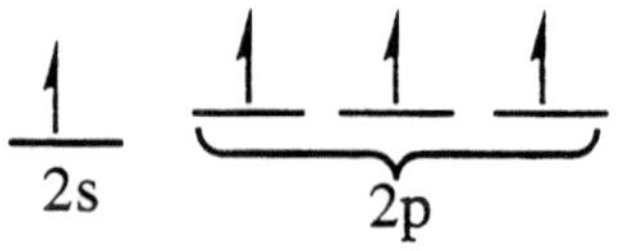

We also know that in any multiple bond[1], one of the bonds is sigma, while the rest are pi. This means that the C is involved in two pi bonds: one to the oxygen on the left and the other

---
[1] We saw this in Section 8.8.

one to the oxygen on the right. The formation of each pi bond requires an unhybridized p orbital from the carbon[1], as well as from the oxygen. This means that the carbon CANNOT hybridize two of its p orbitals in order to form the two pi bonds.

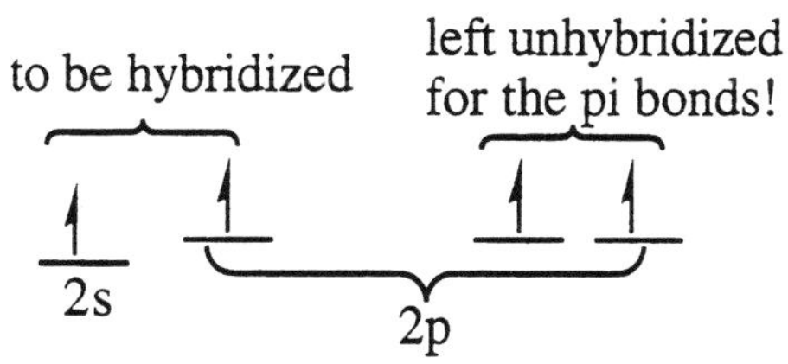

Therefore, the carbon will only be able to hybridize sp in order to leave behind two unhybridized p orbitals for the formation of the pi bonds. The oxygens, on the other hand, are only involved in one pi bond each. Therefore, the hybridization of each oxygen is sp^2, since they only need one unhybridized p orbital in order to form this pi bond. We can state that each double bond in the molecule is formed as follows: The sigma bond comes from the overlap of the sp^2 orbital of an oxygen with an sp orbital from the carbon. The pi bond is formed from the overlap of the unhybridized p orbital from an oxygen with one of the unhybridized p orbitals from the carbon. It is also true that the two pi bonds formed must be perpendicular to one another, because the two unhybridized p orbitals in the carbon are perpendicular to one another. In the diagram below, we can see the formation of one of the pi orbitals (from the unhybridized p orbitals of each atom).

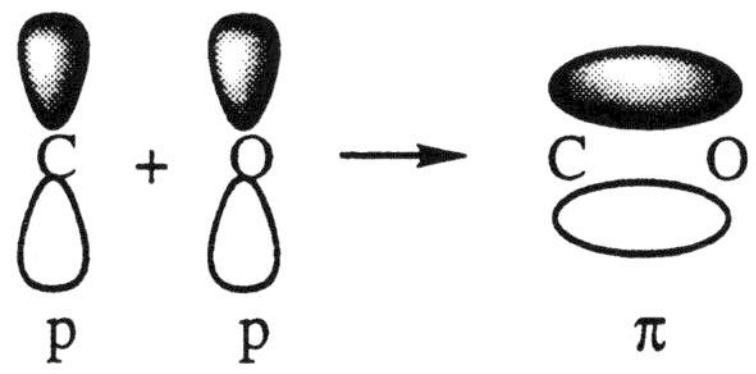

b) SO_2. When we draw the Lewis structure for this molecule we find the following.

These two structures have exactly the same energy and we call them <u>resonance structures</u>. At first glance it seems like both structures are one and the same (if you think in terms of grabbing one by one oxygen and flipping it upside down). However, if that were the structure of sulfur dioxide, then we would have a short bond (the double one) and a long bond (the single one). It has been determined experimentally that both S-O bonds are identical, their bond length

---

[1] We saw this in the same section, mainly:

corresponding to about 1.5 times the length of a single bond.

One way of explaining this (very popular although not true at all) would be as if these two structures were interconverting very rapidly. However, this is not the case. What we really have is **delocalization** of the double bond electrons (the π electrons) in between the two atoms. The real structure is what we call the **resonance hybrid**. Electron delocalization occurs because anytime we are able to spread a charge (whether it is positive, negative, or electrons in a bond), we will lower the energy of the system. The amount of energy lowered by the stabilization of the form below compared to any of the two resonance forms above is referred to as the resonance stabilization energy.

We also learned that to decide its geometry we count the double bond as one pair, so having three pairs around the sulfur makes it sp² and trigonal planar. Therefore, one of the resonance forms looks more like this:

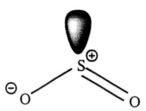

This resonance form should have a bond angle of approximately 120° (a little smaller due to the repulsion of the lone pair). We should point out that if you were asked to draw the Lewis structure for the SO$_2$ molecule, we would have to draw both resonance forms, not the resonance hybrid. We will simply point out that these resonance forms are very useful in explaining the reactivity of these molecules, but that they are not real structures.

Conclusion: In reality,

1. The molecule is said to be a *resonance hybrid* of the two resonance forms.

2. It is **not** considered as an interconversion from one resonance form to the other.

3. Neither structure alone represents the molecule.

However, we must draw the Lewis structures of the different resonance structures keeping in mind that the real structure is a weighted average of all the possibilities giving us a hybrid with a lower energy.

c) N₂O

Previously we saw that upon drawing this molecule, we encountered the three resonance structures seen below.

1   :N≡N⁺—Ö:⁻

2   ⁻:N̈=N⁺=Ö:

3   ⁽⁻²⁾:N̈—N⁺≡O:⁺

Let's analyze these three structures. Since the oxygen is more electronegative than either of the nitrogens, we expect structure **1** to be more probable than **2**. Similarly structure **3** must be very improbable since not only is the oxygen positive but one of the nitrogens has a charge of negative two! This goes against electronegativity. Furthermore, we should avoid drawing resonance structures in which two atoms that are bonded together carry the same formal charge. Therefore we pretty much neglect structure **3** by saying that its probability is minuscule compared with the other two.

The molecule N₂O is then said to be a resonance hybrid (weighted average) of structures **1** and **2**, although the hybrid itself should look a lot more like **1** than **2**. That is why we say that it is a weighted average of structures **1** and **2**. It also follows that since the central atom is involved in two π bonds, its hybridization will be **sp**, which will make the molecule's geometry linear. Below we can see a drawing of the weighted average, where we see that the pi electrons are more likely to be between the two nitrogens than between the nitrogen and the oxygen. This gives the oxygen more of a negative character.

We can use the concepts learned here to predict the geometry of molecules that are a lot more complex than the ones that have a central atom. Let's look at a couple of examples where we extend what we have learned so far into molecules that really do not have a central atom, per se.

d) Ethane, C_2H_6

We could very easily draw the Lewis structure for this molecule. This would look something like this:

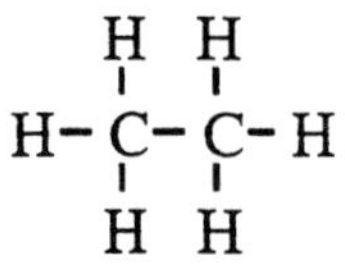

$$\begin{array}{c} \text{H H} \\ | \ | \\ \text{H-C-C-H} \\ | \ | \\ \text{H H} \end{array}$$

Since each C has a complete octet and is involved in single bonds only, then they must be **sp³** which means that every C moiety (sector surrounding the C) is tetrahedral (all bond angles approximately 109.5°).

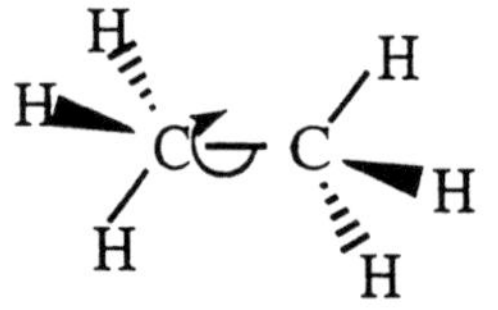

"free" rotation along C-C bond

This "free rotation" is what causes the three substituted positions in either C to be equivalent which explains why there is only one isomer for Chloroethane (C_2H_5Cl).

Don't forget (see section 8.8) that *ALL SINGLE BONDS ARE SIGMA (σ) BONDS!* In the example above, the C-C bond is a σ bond formed between two **sp³** orbitals, and each C-H bond is a s bond formed between an **sp³** orbital (C) and an **s** orbital (H).

e) Ethene, C_2H_4

Once more we start by drawing the Lewis structure of the molecule. We instantly recognize that we have a double bond between the two carbons and we apply the rule that we learned back in section 8.8 stating that *IN ANY MULTIPLE BOND, THE FIRST ONE IS SIGMA (σ) AND THE REST ARE PI (π)*. This gives us an **sp²** hybridization for each carbon with a trigonal planar geometry around every carbon moiety.

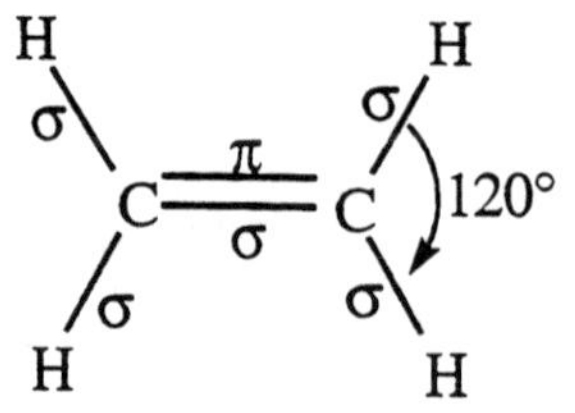

We observe that the molecule itself is planar and all bond angles are approximately 120°

with a π electronic cloud above and below the molecular plane. If we look at the molecule sideways it would look like this:

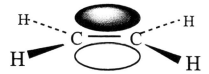

Due to the fact that in a π bond we have electron density above and below the plane, rotation along the C-C bond is impossible without breaking the π bond. This makes the two substituted positions on either C unequivalent. We will see exactly what that means next semester when we talk about geometric isomers (cis/trans).

f) Ethyne, C_2H_2, also called acetylene.

When we draw the molecule, we determine that there is a triple bond between the two carbons. According to what we have been discussing, the triple bond consists of one sigma bond and two π bonds. To form the two π bonds we need two <u>unhybridized</u> p orbitals in each C atom. That would only leave us with an s orbital and one of the p orbitals to hybridize. We should therefore expect the hybridization of each carbon atom to be **sp**. If this holds true then the geometry of the molecule would be linear. Experimentally it has been determined that indeed the acetylene molecule is linear.

$$H \overset{\sigma}{-} C \overset{\pi}{\underset{\sigma}{\equiv}} C \overset{\sigma}{-} H$$

The two π bonds in acetylene are perpendicular to one another and form a cylindrical electron density around the linear molecular axis.

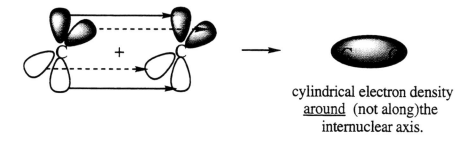

cylindrical electron density <u>around</u> (not along) the internuclear axis.

g) Benzene - C_6H_6. This is a very useful solvent and one of the top 20 chemicals[1] produced in the United States.

---

[1] About 12 billion pounds were produced in 1992.

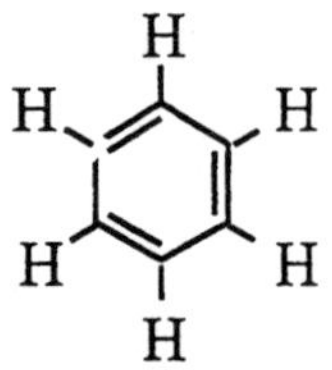

Chapter 9: Bonding II.

[benzene structure with alternating single and double bonds]

This structure shows a ring of 6 carbons with alternate single and double bonds. Since each Carbon atom is involved in a π bond, it should follow that its hybridization is sp², so that the geometry around any carbon moiety is trigonal planar with bond angles of 120°.

The structure shown above would imply alternate long and short bonds. However, this is not the case. All C-C bonds are equally long (somewhere between the length of a single and a double bond). This then suggests resonance or electron delocalization in the ring:

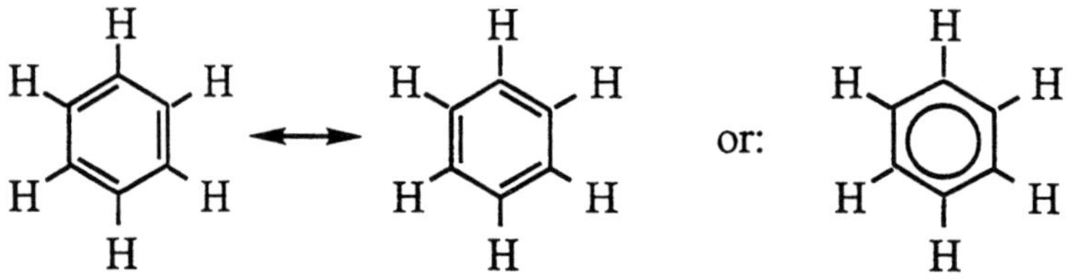

[two Kekulé resonance structures of benzene with double-headed arrow between them, followed by "or:" and a third structure with a circle inside the hexagon]

You can see that the electrons are delocalized all around the ring. This is a very large area for the electrons to delocalize into so it should be a very stable and fairly unreactive molecule. Sometimes benzene is drawn as you can see in the third representation above to show just that: *electron delocalization.* As you would expect, this molecule (which exists in the resonance hybrid shown as the third structure above) is more stable (*lower in energy*) than either of the two resonance structures would suggest. The resonance stabilization energy is determined experimentally to be 154 kJ/mol. This means that the resonance hybrid is 154 kJ/mol more stable than either of the two resonance forms. This helps to explain the unusual "unreactivity" of benzene, something that we will talk about more extensively in Chapter 13.

Finally, we should be able to draw much more complicated molecules in three dimensions, trying to be particularly careful with the bond angles and the geometries as we go along. For example, let's draw the molecule below in 3-D.

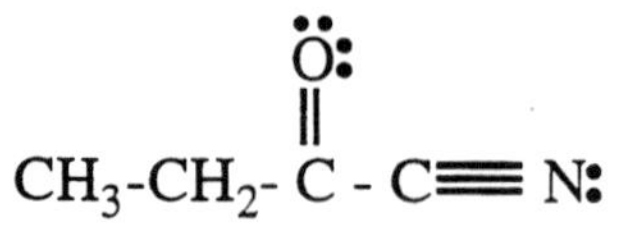

$$CH_3-CH_2-\overset{\overset{\ddot{\text{O}}:}{\|}}{C}-C\equiv N:$$

The first thing we do is identify the hybridization of each of the atoms in the molecules (except for the H's). We also notice that all the atoms in the molecule complete an octet. The first two C atoms are sp³ because they are not involved in any π bonds. This makes all the bonds coming out of each of these carbons at 109.5° angles. The third C and the O are sp² because they are both involved in a π bond. This makes the bond angles coming out of the C at 120° angles. It also makes the angle between the two lone pairs in the O to be 120°. Finally, the last C and the N are both sp, since there are two π bonds in the triple bond. This makes the bond angles coming out of the C to be 180°. It also makes the bond angle between the C, the N and the lone pair in the nitrogen 180°. The final version would look something like this.

9.2 Molecular Orbital (MO) Theory.

We have seen, from our previous discussion on the shapes of molecules, that whenever we can, we will draw as many resonance forms as possible because that generates delocalized electrons. Anytime you are able to delocalize electrons, the energy of the system goes down (it is more stable). As an example look at the example (Section 8.9) of the nitrate ion. As we saw then, there were three resonance forms for the nitrate ion, but the true form is a weighted average of these three forms, or a hybrid that show total electron delocalization between the three oxygens because this way the system has a much lower energy.

A similar concept can be applied to any molecule. Instead of thinking of individual atomic orbitals (or hybrid orbitals for that matter) bonding together to form bonds, we will think of orbitals that belong to the molecule as a whole. Such an orbital is said to be spread out (delocalized) over the molecule, certainly over more than one single atom, and we will call these molecular orbitals.

Postulates.

1. The orbitals are associated with molecules as a whole and not with individual atoms. When we think of MO's we have to think in terms of the molecule as a whole and not of the constituent atoms.

2. Ordering and filling of MO's is an Aufbau process similar to the filling of AO's. We will assign Greek letters to the MO's (σ - sigma, π - pi, δ - delta, etc.). Once we know the MO's and place them energetically in increasing order, we will fill them up just like we did for the AO's ($1s^2 2s^2 2p^6 3s^2$ ...). That is, start with the lowest one and place two electrons in that orbital with opposite spins, then go on to the next one, and so on...

3. MO's arise from the Linear Combination of Atomic Orbitals (LCAO). The combination of two AO's (which are mathematical equations[1]) will yield two MO's: one bonding (obtained by adding the AO's) and one antibonding (obtained by subtracting the AO's).

Interaction of Two **1s** Atomic Orbitals.

Let's start with the simplest case: the combination of a 1s orbital from one atom with the 1s orbital of another atom. Remember that both 1s orbitals are spherical in symmetry. What do we mean by combining these two atomic orbitals? Since these orbitals are nothing but a mathematical equation that describes 95% of the probability of finding the electron, all we must do is to add or subtract these two equations (addition and subtraction are linear combinations) in order to obtain two new equations. These new equations will be equations (or orbitals) for the molecule, and not for the individual atoms. The molecular orbital resulting from the addition (combination of the wave functions when they are in-phase) of the atomic orbitals will be lower in energy (stabilizing) than the individual atomic orbitals. These are referred to as **bonding** molecular orbitals. However, the molecular orbital resulting from the subtraction (combination of the wave functions when they are out-of-phase) of the atomic orbitals will be higher in energy than the individual atomic orbitals. These are referred to as anti-bonding molecular orbitals and are symbolized with a star (*). Below, we may see the shapes of the molecular orbitals obtained from the combination of the two 1s orbitals plotted against energy.

---

[1] Remember that AO's are pure mathematical equations (Ψ^2 with a given **n**, ℓ, and **m**) which tells us the region in space where there is a 95% probability of finding the electron. When you combine linearly two such equations (one for one atom and one for the other) you get two new equations which apply to the whole molecule and will give you the region in space where there is a 95% probability of finding the electron for the molecule!

Chapter 9: Bonding II.

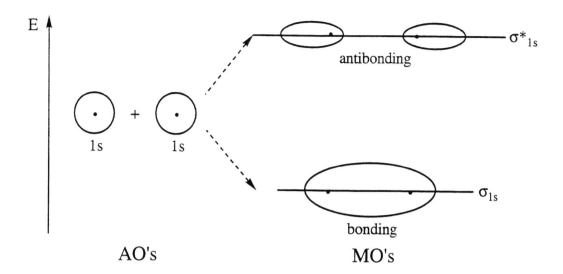

Notice that we called these molecular orbitals sigma (σ). Why is that so? We have already seen (see Section 8.8) that a sigma bond results when the electron density is a maximum between the two nuclei. Therefore, the electron density is symmetrical about the internuclear axis. This is exactly what we observe in the bonding orbitals: the electron density is maximum in between the two nuclei.

It would be very tedious to have to draw the diagrams above every time we would be looking at molecular orbital diagrams. So, we will do exactly the same thing we did when we were dealing with atomic orbitals: just draw a horizontal line to represent an orbital and arrows to represent the electrons with their spin. The diagram we have just seen can be simplified as follows (we will call it the molecular orbital energy level diagram):

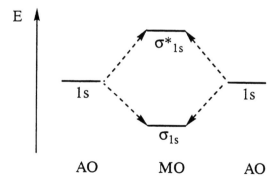

We will fill the molecular orbitals in exactly the same way we did with the atomic orbitals.

1. The maximum number of electrons in a molecular orbital is <u>two</u>, and they must have opposite spins.

2. First fill up the molecular orbital with the lowest energy (σ_{1s}) and once it is full, then start filling up the next one (σ^*_{1s}).

3. If you ever find degenerate molecular orbitals (with the same energy; there are none on the diagram above), use Hund's Rule - fill them up with one electron each first and then go back and add another electron to each degenerate orbital until they are all full.

4. We will define bond order as the number of bonds found in the molecule. We can calculate it as:

B.O. = 1/2 [# of bonding electrons - # of antibonding electrons].

Let's do some examples with this simplest possible diagram, the one above, just to see what we mean by those rules.

- H_2 molecule.

We already know that each hydrogen has a configuration of $1s^1$, therefore the total number of electrons in the molecule must be two. Where would we place these electrons in the molecular orbital diagram? In the σ_{1s} orbital for a configuration of: $(\sigma_{1s})^2$ as seen below.

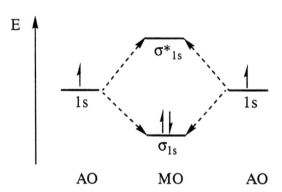

Let's determine the bond order for this molecule.

B.O. = 1/2 [# of bonding electrons - # of antibonding electrons]

= 1/2 (2 - 0) = <u>1</u>.

This confirms that there is only a single bond between the two hydrogens in this molecule.

- He_2 molecule.

The configuration of each helium atom is $1s^2$. Therefore, we must have four electrons in the molecule itself. Where would we place these four electrons? Two of them would go in the (σ_{1s}) molecular orbital to fill it up, but the remaining two would have to go in the other orbital, the (σ^*_{1s}). Therefore the configuration would be: $(\sigma_{1s})^2(\sigma^*_{1s})^2$. The molecular orbital energy level diagram may be seen below.

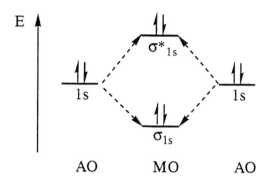

The bond order turns out to be zero (= $1/2 (2 - 2)$ = $\underline{0}$). This means that this molecule should not exist according to quantum mechanics. This is exactly what we knew already, since He is a noble gas.

Now that we have seen this case of the helium molecule, it should be apparent that the molecular orbital method can be used to predict if some molecules exist or not. Let's use this concept to predict whether the Li_2 molecule exists.

- Li_2 molecule.

The configuration of each lithium atom is $1s^2 2s^1$. This presents a small problem. In order for us to combine these two atoms, we are going to be forced to combine the 1s orbitals from each lithium (just like we did above), but we will also have to combine the 2s orbitals from each lithium atom. Since the 2s orbitals are also spherical in symmetry, it should follow that the interaction of these two orbitals will yield identical results to the ones seen for the interaction of the 1s orbitals, but at a higher energy.[1] Therefore, the molecular energy diagram looks like two of the ones we saw for the hydrogen molecule, with the energy order for the molecular orbitals being: $(\sigma_{1s}) < (\sigma^*_{1s}) < (\sigma_{2s}) < (\sigma^*_{2s})$. Where would we place the six electrons found in the lithium molecule? The first two will go in the (σ_{1s}) orbital, the

---

[1] The energy of the molecular orbitals resulting from the 2s atomic orbitals will be higher in energy than the ones resulting from the 1s because the 2s is higher in energy than the 1s.

next two will go in the (σ^*_{1s}) and the last two will go in the (σ_{2s}). The diagram itself can be seen below. Configuration: $(\sigma_{1s})^2 (\sigma^*_{1s})^2 (\sigma_{2s})^2$.

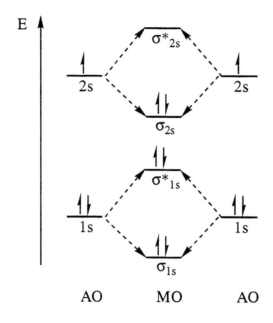

Calculation of the bond order shows that we should have a single bond between the two lithiums (BO = 1/2 (4 - 2) = 1). Therefore, this molecule should exist according to quantum mechanics. Indeed, the molecule has been observed in the gas phase.

Notice that the determination of the bond order could have been done using exclusively the part of the diagram that deals with the interaction of the 2s orbitals, ignoring completely the interaction of the 1s orbitals (BO = 1/2 (2 - 0) = 1, see?). *Conclusion: It is NOT necessary to include in the molecular orbital energy diagram atomic subshells that are completely full for both atoms, as they will have no bearing in the bonding situation.*

- Be_2 molecule. Does it exist?

We have seen that the only thing we must do in order to answer a question like this, is to determine the molecular orbital energy level diagram for the molecule in question and then determine the bond order. The configuration of each beryllium atom is: $1s^2 2s^2$. Therefore, we have a total of eight electrons in the molecule. Of course, the configuration would be: $(\sigma_{1s})^2 (\sigma^*_{1s})^2 (\sigma_{2s})^2 (\sigma^*_{2s})^2$, as seen below.

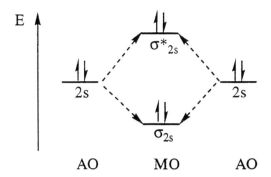

Since the bond order is zero, we do not expect this molecule to exist. It is not energetically feasible.

Interaction of Two 2p Atomic Orbitals.

Sometimes molecules will combine and other orbitals will have to interact (other than the s orbitals). For example: in order to study the molecular orbitals for a molecule like nitrogen gas would have to consider the interaction between 2p orbitals. Remember that each atom has three different 2p orbitals, which are located in the three axis: the x, the y, and the z axis. These three 2p orbitals are orthogonal (mutually perpendicular). Therefore, upon approaching the three orbitals from one atom and the three orbitals of the other atom, we can see that there are going to be two types of interactions. One of the orbitals will encounter its counterpart "head to head". We will call this orbital the $2p_z$ orbital, since we will be approaching the two atoms along the z axis. This interaction would yield two molecular orbitals, as seen below.

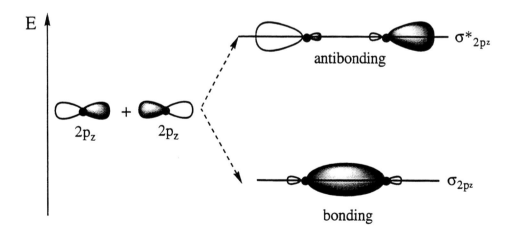

Due to the shape of the molecular orbitals (where the electron density is a maximum between the two nuclei, we will also call these molecular orbitals sigma molecular orbitals (therefore, we just generated the σ_{2pz} and the σ^*_{2pz} molecular orbitals).

Chapter 9: Bonding II.

The other two orbitals (the $2p_x$ and the $2p_y$) will interact with both lobes and will give rise to a very different kind of molecular orbital: one where the electron density is maximum above and below the internuclear axis. We call this a pi (π)[1] molecular orbital.

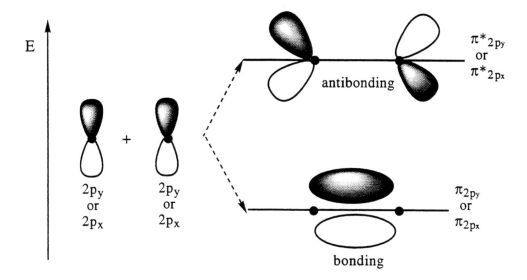

Since there is absolutely no difference (in energy) between the molecular orbital formed from the interaction of the $2p_x$'s with the interaction of the $2p_y$'s, then we should expect to have two degenerate molecular orbitals for both the pi (π) and the pi stars (π^*). Their only difference is that they will be in two different planes, 90° apart.

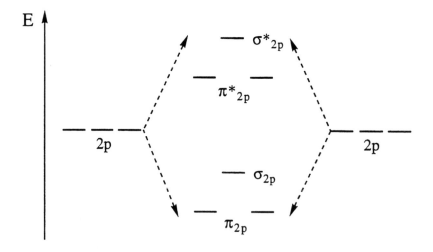

This diagram represents the molecular orbital energy level diagram for the 2p portion only. Please, notice that in this diagram the degenerate pi molecular orbitals are lower in energy than the sigma molecular orbital. This is actually true for molecules like B_2, C_2, and N_2. For

---

[1] See Section 8.8.

molecules like O_2, F_2, and Ne_2, it is the opposite, that is, the sigma molecular orbital is actually a little lower in energy than the degenerate pi molecular orbitals. However, since for our purposes it absolutely makes no difference whatsoever, we will use ALWAYS the molecular energy level diagram shown above.

- B_2 molecule.

The configuration for boron is $1s^2\ 2s^2\ 2p^1$. We should immediately notice that the first two subshells are completely filled for both atoms, therefore, we will concentrate only on the 2p portion of the diagram. In this portion, the molecule has two electrons (each atom is $2p^1$). Where would we place these two electrons? Since the molecular orbital energy level diagram shows that the lowest energy orbitals are the two degenerate pi MO's, then we should place one electron in each, as shown below.

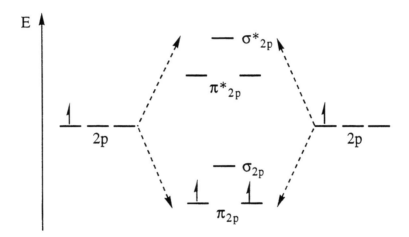

Let's look at this diagram. We should notice that the bond order is one (BO = 1/2 (2 - 0) = 1). Therefore, the molecule should exist with a single bond between the two borons. It has been observed in the gas phase and the information we get from the MO diagram is indeed correct. Not only that, but the diagram shows something unusual: two unpaired electrons in that only bond. This should make the molecule paramagnetic. Upon experimental inspection, it has been determined that the molecule is indeed paramagnetic and its magnetic moment corresponds to having two unpaired electrons. One more thing that seems unusual is that there is only a single bond and it is NOT a sigma bond (the electrons are in π MO's). This is indeed unusual, since we have previously stated that all single bonds are sigma in character. This is an exception and there are few of these cases (where single bonds are pi instead of sigma).

This last example clearly shows to us that this method is a lot more exact than the previous methods we were using to identify the geometry of molecules. The only problem is that it is too complicated (mathematically) to be practical at this level. That is why we limit our discussion to diatomic molecules. The purpose of the method (for us) is therefore to observe and predict details other than geometry, since all diatomic molecules are linear by definition. However, we will be able to tell about the magnetic properties of the molecules. The next example shows that the previous methods are actually incorrect many times when it comes to predicting the magnetic properties of molecules.

- O_2 molecule.

    Each O is $1s^2 2s^2 2p^4$. Once more, the 1s and the 2s subshells for both atoms are completely full, therefore, we will only draw the 2p portion of the diagram. Any information we require at this level is found in that section. Since each atom is $2p^4$, then we have eight electrons in the MO diagram, as seen below.

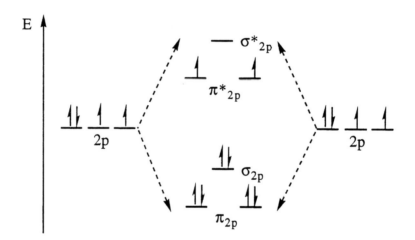

A quick calculation of the bond order reveals that there is a double bond between the two oxygens (BO = 1/2 (6 - 2) = 2). This is consistent with the Lewis diagram for the oxygen molecule that we have previously seen.

$$:\ddot{O}=\ddot{O}:$$

One thing that the Lewis structure does not predict correctly, however, is the fact that the molecule is paramagnetic (two unpaired electrons in the pi star antibonding degenerate orbitals). Once more, this serves to illustrate the fact that the MO method is a lot more revealing about the magnetic properties of the molecules.

- O_2^{-1}, the Superoxide ion.

    From the formula of the ion, it should be clear that the superoxide ion is really an oxygen molecule that has an extra electron in it. Therefore, you can visualize it as being formed from the combination of an oxygen atom and an O^{-1} ion, as seen below.

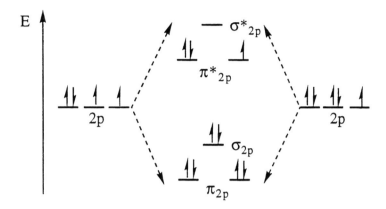

    What is this MO telling us? Among other things, it tells us that this ion is also paramagnetic, with one unpaired electron. It also allows us to determine that there are one and a half bonds between the two oxygen atoms (BO = 1.5). This would be very difficult to draw using the Lewis method (you would have to think of a creative way to show the two atoms sharing three electrons instead of two).

    We have stated that MO's will tell us a lot more about the molecules than any other method. We also mentioned that due to its complexity, we would limit its discussion to diatomic molecules. How is this applied to polyatomic molecules? In Organic chemistry, we will indeed apply it to polyatomic molecules and we will see that in order to use this method, we will have to make some major approximations. For example, if we were to use it for molecules that have π electrons, we will start by recognizing that these are the most reactive electrons in the molecule (the easiest to excite, by far). Therefore, the first approximation that we would make would be to look only at the π electrons in the molecule. This will certainly simplify the MO diagram considerably and make it very accurate to predict not only whether the molecule is diamagnetic or paramagnetic (which is what we are doing at this point), but also to predict whether some reaction are allowed quantum mechanically or not.

9.3 Heteronuclear Diatomic Molecules.

    Do we have to always work with homonuclear diatomic molecules? Of course not. As long as the two atoms forming the molecules have the same type of orbitals available for bonding, then the energy level diagram is similar to that of homonuclear diatomic molecules, therefore, we can use the exact same diagram. Let's look at one specific example.

- NO molecule.

    The configuration of the nitrogen is 1s² 2s² 2p³ whereas the configuration for the oxygen is 1s² 2s² 2p⁴. We immediately notice that both atoms have full 1s and 2s subshells. So, we should only look at the contribution from the 2p subshells. Since they are both in the same level (2p), we can use the same diagram as before in order to determine the bond order and the magnetic properties of this molecule. The diagram can be seen below.

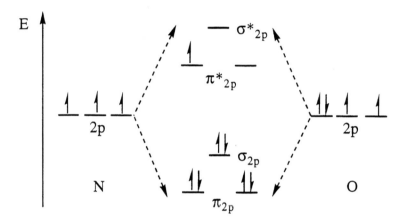

    Once more, we can see that this is a paramagnetic molecule (one unpaired electron in the π^*_{2p} MO). We can easily determine also that five electrons (two and a half bonds) are shared between these two atoms. It would be difficult to try to draw this molecule using the Lewis method, but this more accurate method is telling us what is occurring at the between the atoms: they are sharing five electrons!

- Which of the following species would you expect to have the strongest bond, O_2, O_2^+, or O_2^-?

    Just like in every other example, we should start by drawing the molecular orbital energy level diagrams for all three species. The difference between the oxygen molecule and the positive ion, is that the second one is missing an electron, whereas for the superoxide ion, we have an extra electron (as compared to the oxygen molecule). All three diagrams may be seen below. The diagrams include the bond order for each. We will base our conclusions on exactly that: the bond order. We have already learned that a double bond is stronger and shorter than a single bond.[1]

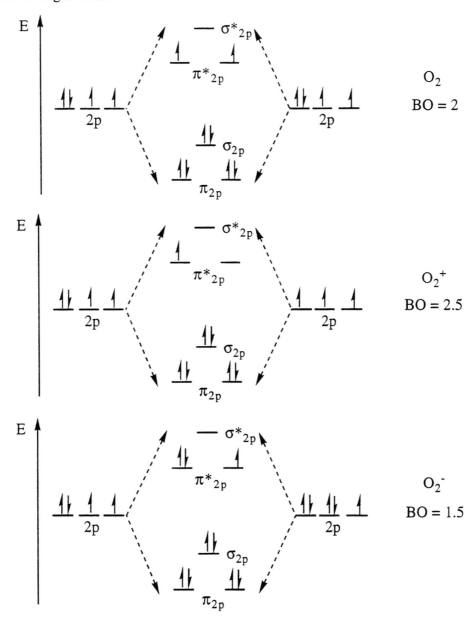

Since O_2^+ has the highest bond order (2.5), then its bond is stronger than the others (double and one and a half bonds).

---

[1] See Section 8.3.

Chapter 10: Condensed Phases.

10.1 Intermolecular Forces (IMF).

Let's first be very clear about what we mean by intermolecular forces. When we talk about intermolecular forces, we are referring to the strength of the attractions in between different molecules. A typical mistake is to think in terms of the actual bonds within the molecule. We are referring to attractions **between** one molecule and another. This should depend on one of two things: either an electrical attraction (+ ······ -) or a magnetic attraction (N ······ S). The only important one for molecules will be electrical attractions. The stronger the electrical charges in a molecule, the stronger the attraction between one molecule and another.

Is there a relationship between the different states of matter and the strength of the intermolecular forces of attraction? We have already seen that gases consist of tiny molecules that are very relatively far from one another, moving around at very high speeds and in a random fashion. Molecules in the gas phase have a lot of energy. We said back in Chapter 4 that there are essentially no intermolecular forces of attraction between gas molecules. Actually, we mentioned that ideal gases have absolutely no attractions nor repulsions between their molecules, and that all real gases do behave ideally under conditions of high temperature and low pressure. *Conclusion: There are very few intermolecular forces of attraction in the gas phase.*

Liquids, on the other hand, consist of molecules that are very close to one another but still have some translational energy. What keeps these liquid molecules close to one another? Intermolecular forces of attraction! So intermolecular forces for liquids must be stronger than the intermolecular forces for gases. If we define (rather crudely for the time being) the boiling point (BP) as the temperature at which a liquid is converted into a gas, then we can visualize the boiling point as the temperature at which we finally break "all" the intermolecular forces of attraction that kept the molecules close together in the liquid phase.

How about the solid phase? It is the most ordered of all the three phases. There is no translational energy and molecules (or ions) occupy a specific place in the crystal lattice. The intermolecular forces are so strong that molecules are no longer able to move out of position. For the time being we can define the melting point (MP) as the temperature at which the solid is converted into a liquid. What happens when we melt a solid? Basically two things: we break the crystal lattice and we also break some of the intermolecular forces of attraction (but not all since we know that there are still many intermolecular forces of attraction in the liquid phase).

Let's compare the boiling point to the melting point. It should be clear from the preceding discussion that the boiling point is going to be a much better "indicator" of the strength of the intermolecular forces in a substance since going into the gas phase requires breaking essentially all of the intermolecular forces. *Conclusion: The melting point is a good indicator but the boiling point is more indicative of the strength of the intermolecular forces of attraction.*

Solids sometimes go directly from the solid phase into the gas phase at a given temperature. We call this process <u>sublimation</u>. The sublimation point is also an excellent indicator of the strength of these intermolecular forces of attraction. The reason is the same as it was for the boiling point.

We should discuss at this point the different types of intermolecular forces that are present in different molecules. It is important that we understand which forces are present in which molecules, and when there are more than one type of forces present in a system, which are the predominant forces. Why is this important? Because these forces that will have a deep impact on the state of matter in which the substance exists and some of the properties of this material.

1. <u>Ion-Ion Forces</u>.

We will start with these simply because they are the easiest to understand. We will observe ion-ion forces only in ionic substances. If we have an ionic compound (for example: NaCl) we have pure ions (some are cations and the others are anions – in this example we have Na^+ and Cl^-). How are these ions attracted to one another? By purely electrostatic forces of attraction, that is, the positive charge is attracted to the negative charge, as seen below.

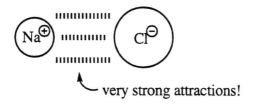

— very strong attractions!

These forces are gigantic compared to the other forces we will be referring to shortly, which helps to explain why <u>almost all salts are solids are room temperature</u>! Not only are they solids, but they have very high melting points, which are indicative of strong intermolecular forces of attraction.

Let's go back to the NaCl example. In the solid phase, each sodium ion is not limited to one chloride ion. There are six chloride ions in the immediate vicinity[1] of the sodium ion. The same thing is true for the chloride ion. Therefore, it would almost be like having each ion "bonded" to six opposite ions. This should give us an idea of the incredible order and the strength of the intermolecular forces present in ionic solids. Below, we can see a table with melting points of some common ionic compounds.

Salt	Melting Point (°C)
NaF	993
NaCl	801
NaBr	1390
$MgCl_2$	714
MgO	2852
Na_2O	1275 (subl)
CaO	2614
Fe_2O_3	1594
Al_2O_3	2072

2. <u>London Forces or Van der Waals Forces.</u>

We have seen the types of attractions present in ionic compounds in the previous section. These are rather simple to visualize since we are talking about full ions. However, what kinds of forces are present in non-polar molecules? Remember that these molecules do not have a dipole present and $\mu = 0$. There are many examples of non-polar molecules, but it would be best to try to understand the existing forces using a monoatomic substance, which by definition has to be non-polar. Therefore, let's use Helium gas as an example.

---

[1] One in the front, one in the back, one on either side (left and right), one above and one below.

Although this substance is non-polar, there has to be some kind of interaction between the helium atoms. How do we know that? Because if we cool this substance enough, it will eventually become a liquid (at 4 K!). This means that at this temperature, the helium atoms no longer possess enough thermal energy to break all the attractions between the molecules and escape into the gas phase. What possible attractions exists between two helium atoms if they do not have a permanent dipole (no charge separation)? We will call these very weak forces **London forces** (sometimes also called Van der Waals dispersion forces).

We will explain the generation of the London forces as instantaneous dipoles that form for a fraction of a second and immediately disappear as the electrons vibrate (or sway back and forth) relative to the nucleus. Remember that the electrons move much faster than the nucleus, which is part of the reason that we do not even attempt to describe the path of these moving electrons. Instead, we chose to talk about a 95% probability region where these electrons are most likely to be found. The electronic cloud for this molecule happens to be what we call a completely full 1s orbital. These electrons are very close to the nucleus (in relation to other atomic substances or molecules) and very strongly attracted to it (since there are no inner shell electrons, which means that they are staring directly at the two protons). However, the electronic cloud can and will vibrate relative to that nucleus, particularly when it comes in close proximity to another atom, as seen below.

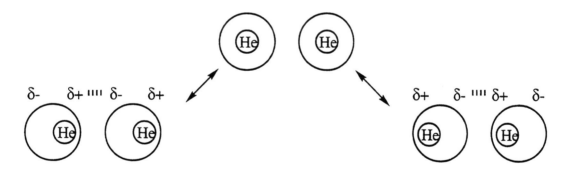

This way, a very weak dipole is generated in each atom and a weak attraction is formed. In this particular case, when we write δ+, we are referring to a very small fraction of a positive charge (an almost infinitesimal fraction of the kind of charge you have in an ion). Since the atom itself is neutral, if you form a very small positive charge on one end of the molecule, then you must have an equally tiny negative charge on the other end. This charge separation (dipole) will obviously not be permanent, so it will immediately return to the normal state (no dipole).

The smaller the atom, the weaker the δ+/δ- separation. This should make sense since for a small atom, the electrons are very close to the nucleus thus they are held much tighter and therefore are harder to vibrate or sway relative to the nucleus. If we are able to follow that, then we should expect that that boiling point of the noble gases should increase as we go down the family, that is, we should expect for helium gas to have the lowest boiling point and for Radon gas to have the highest boiling point. Remember that for the larger one (Rn), the electrons are much farther away from the nucleus and therefore, the vibration relative to the nucleus should be a lot easier. We say that the larger atom is more polarizable[1]. Let's look at the boiling points of all six noble gases[2].

	He	Ne	Ar	Kr	Xe	Rn
boiling point:	4 K	27 K	87 K	121 K	165 K	211 K

In general, the strength of the London forces is dependent on the number of electrons on the molecule (or the MW) and the total surface area available for the formation of forces. This means that if we are to compare two different molecules, the one that has a higher molecular weight, or the one that offers more surface area for the formation of these instantaneous dipoles, should have the higher boiling point. Let's look at a few examples to see if this pattern holds.

- Compare the boiling points of the hydrides of the fourth family (IVA), carbon's family.

A hydride is the name given to the compound formed when an atom of a given element bonds to as many hydrogens as possible. Therefore, since all the members of this family form four bonds, we would be referring to CH_4, SiH_4, GeH_4, and SnH_4. It is also true that all these molecules are non-polar, because the central atom is sp^3, and therefore the molecules are tetrahedral and do not have a dipole. We have previously learned that the atoms get larger when we go down a family, therefore, we can predict that the boiling point of these compounds should increase as we go down the family. This is because we would have a higher molecular weight, plus an increased surface area for the formation of these London forces. This is exactly what we see in the data shown below.

	CH_4	SiH_4	GeH_4	SnH_4
boiling point:	90 K	161 K	183 K	221 K

- Determine the relative boiling points of the halogens.

The halogens are family VIIA (the family of fluorine). All of them have a configuration ending in p^5, which means that they all exist in a diatomic state: F_2, Cl_2, Br_2,

---

[1] It has a better ability to form these instantaneous dipoles.
[2] They are all gases at room temperature (298 K) since they all boil below 298 K.

and I_2. We will not deal about astatine (At) because it is radioactive, however, if does follow the same pattern. Once more, we should expect for the boiling point to increase as we go down the family. Let's look at the data.

	F_2	Cl_2	Br_2	I_2
boiling point:	85 K	248 K	331 K	456 K
melting point:				386 K

The data indeed confirms what we suspected. The boiling point does go up as we do down the family. However, there is something that might surprise us initially. We should notice that the last two molecules have a boiling point above 298 K, which means they will not be gases. Actually, bromine is a liquid and iodine is a solid! This is almost incredible at first because we have termed these forces as extremely weak, however, neither bromine nor iodine have enough thermal energy to break all of these forces and escape into the gas phase even at 298 K. Conclusion: The fact that a substance is non-polar does not necessarily mean that it is going to be a gas. It will depend on the molecular weight and the size of the area that is available for the formation of the London forces. Actually, iodine is so large that it has a lot of surface area available to form these weak forces, and once you generate a lot of these, they can add up to a sizable overall force. Remember that the iodine atoms are very large and we have already stated that large atoms are very polarizable, that is, easy to form stronger, instantaneous dipoles (London forces).

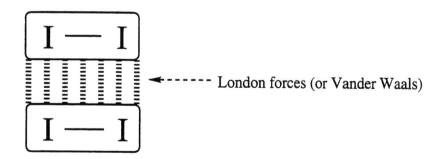

- Which of the following isomers has a higher boiling point, n-pentane (**1**) or 2,2-Dimethylpropane (**2**)?

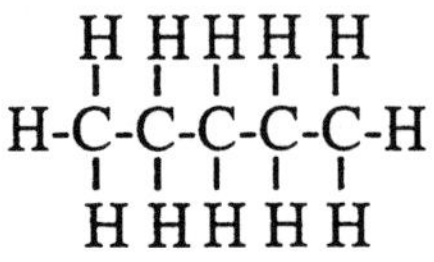

```
    H H H H H                    H-C-H
    | | | | |                 H   |   H
  H-C-C-C-C-C-H              H-C - C - C-H
    | | | | |                 H   |   H
    H H H H H                    H-C-H
                                    H
        1                            2
```

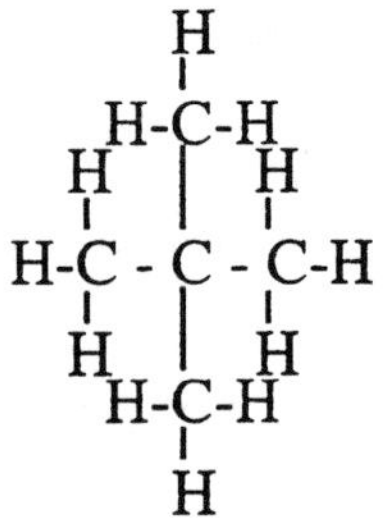

These are isomers because they both have the same molecular formula: C_5H_{12}. We already know that these molecules are not flat as shown, but rather tetrahedral around every Carbon moiety[1] since all carbons are **sp**^{3} (no π bonds). Looking at the three dimensional structures shown below, we should expect that isomer **1** should have a larger boiling point than isomer **2**, since the first one has more surface area available for the formation of these London forces.

In the following structures, only the C's are shown (at every corner or end). They also give a fairly good approximation of the bond angles between the C's.

The attractions between two molecules would therefore be as follows:

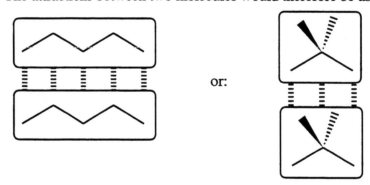

This explains the data below.

Isomer	Boiling Point	
1	36.1°C	∴ a liquid
2	9.4°C	∴ a gas

---

[1] The word moiety means half, but it is used in the sense of "part", that is, around each carbon atom, we see a tetrahedral configuration.

3. Dipole-Dipole Forces.

If a system is going to exhibit these forces, the molecules must have a permanent dipole. This is the case for polar molecules. Therefore, all polar molecules exhibit dipole-dipole forces. These are rather simple to visualize. The negative end of a dipole will attract to the positive end of another dipole.

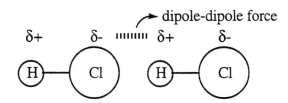

These charge separations in polar molecules ($\delta+/\delta-$) can be fairly large. In the case of HCl there is a 17% charge separation (in other words, the $\delta+$ is 17% of a positive charge). In the case of HF the charge separation is much larger (41%)[1]. This makes sense since F is more electronegative than Cl. Therefore, the dipole-dipole forces in HF are much stronger than those in HCl. The larger the charge separation, the stronger the dipole (remember that the dipole moment is directly proportional to this charge separation). The stronger the dipole, the stronger the attraction between two different molecules, and the higher the boiling point should be.

There is an important concept that is usually misunderstood. <u>ALL</u> molecules may form London forces, polar or non-polar (and even ionic). However, in the case of a non-polar molecule those are the *only* kinds of forces existing. In the case of a polar molecule they are both present but usually the dipole-dipole forces are much stronger and thus predominating. However, if the molecule is really large and not terribly polar then the London forces will predominate. The important point to make is: *IF WE COMPARE MOLECULES OF <u>SIMILAR</u> SIZE AND WEIGHT (ONE POLAR AND ONE NON-POLAR), THE POLAR MOLECULE SHOULD HAVE STRONGER FORCES OF ATTRACTION AND THEREFORE A HIGHER BOILING POINT.*

Let's compare the following three molecules: HCl, F_2, and HF. Which one do we think should have the highest boiling point?

In order to answer a question like this, we should start by determining whether each molecule is polar or non-polar. Of these, the only non-polar one is the fluorine molecule. Since their molecular weights are not dramatically different, we would expect for the fluorine molecule

---

[1] See Section 8.6.

to have the lowest boiling point, and it does. Now, the other two molecules are polar, therefore, both should exhibit dipole-dipole forces. Given that, we should expect that the difference between them would simply be their London forces, therefore, we would expect for the larger molecule (HCl) to have the larger boiling point. Let's check these results.

Molecule	Polarity	M.W.	B.P.
HCl	polar	36.5	188 K
F_2	non-polar	38.0	85 K
HF	polar	20.0	292 K

This is not what we observe. For some reason, the HF molecule has a significantly higher boiling point than the HCl molecule. This is due to a very strong kind of dipole-dipole intermolecular force of attraction which we will give a special name: Hydrogen Bonding. Let's define Hydrogen-bonding as an attractive force between a small and highly electronegative atom in a molecule and a polarized hydrogen atom in another molecule. The only atoms that are small and electronegative enough in order to form hydrogen bonding are: fluorine (F), oxygen (O), and nitrogen (N).

To observe hydrogen bonding in a system, we need two things: 1. a polarized hydrogen (one that is partially positive) and 2. either a fluorine, and oxygen, or a nitrogen in a molecule. Whenever we have this combination, then the polarized hydrogen of one molecule will attract very strongly to the partially negative fluorine (or oxygen, or nitrogen) of another molecule. How strong is this attraction? It is in the order of 13-40 kJ/mol and this is smaller but comparable to the strength of a real covalent bond, thus the name hydrogen **bonding**. Let it be clear, however, that we are talking about an intermolecular force of attraction, and not a real bond! Because of this unusually high forces, systems that exhibit them will have boiling points that are also unusually high. In conclusion, the HF molecule has a higher boiling point than the HCl molecule due to hydrogen bonding.

The water molecule is the perfect example of a system that is hydrogen bonded. This molecule has two hydrogens bonded to an oxygen, therefore, both of these hydrogens are polarized. We should stress once again that these two oxygen-hydrogen bonds are not the hydrogen bonds we are referring to in the previous discussion.

Chapter 10: Condensed Phases.

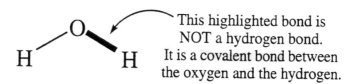

So, where are the hydrogen bonds? Between one of the hydrogens in a given molecule and the oxygen of another molecule, as seen below.

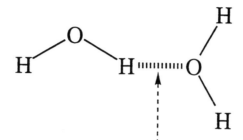

This IS a hydrogen bond: an attraction between the hydrogen of one molecule with the oxygen of another molecule.

We should be able to predict when a system is going to exhibit hydrogen bonding. Let's look at a few examples to see how we should reason whether these forces are present or not.

- Methanol: CH_3OH.

    The boiling point of methanol is 65°C. Let's analyze the molecule itself.

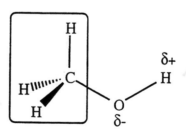

The three hydrogens bonded to the carbon (the boxed ones in the diagram above) are not polarized enough to participate in hydrogen bonding. In general, we should learn that hydrogen and carbon have very similar electronegativities, and they form non-polar bonds always. However, the fourth hydrogen is bonded to the oxygen. This hydrogen is very polarized because of the large difference in electronegativity between the oxygen and the hydrogen. Therefore, this molecule has all the requisites to exhibit hydrogen bonding.

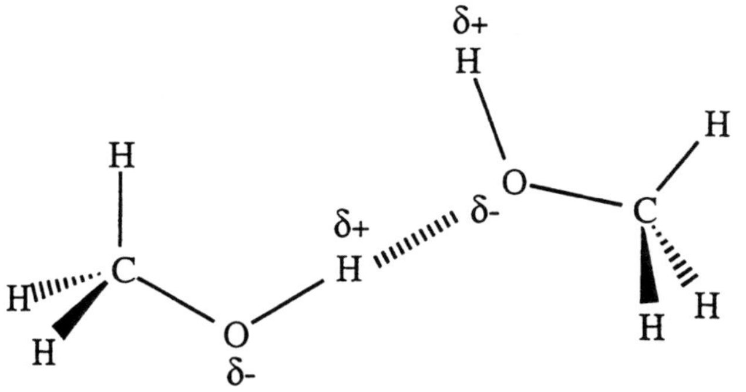

[Diagram: Two methanol molecules hydrogen bonding, with δ+ on H and δ− on O labels]

- Dimethyl ether, CH_3-O-CH_3.

    The boiling point of this molecule is -24°C. Since it is even heavier than methanol and it boils at a considerably lower temperature, this must mean that there is no hydrogen bonding for dimethyl ether.

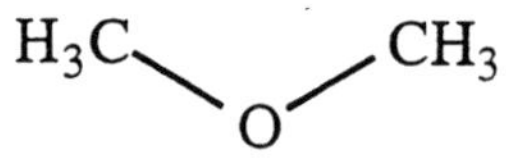

    [Diagram: H_3C—O—CH_3 structure]

    We can see that the molecule is not linear (the oxygen is sp^3 in hybridization). This makes the molecule angular and polar. Therefore, the forces present are dipole-dipole forces. Since there are no polarized hydrogens in the molecule (all six hydrogens are bonded to carbon), then we will not observe hydrogen bonding and the boiling point will be rather low.

- Ethanol: CH_3CH_2-O-H.

    This molecule is an isomer of the dimethyl ether, since they have the same molecular formula. Its boiling point is 78°C. This is a very similar case to the methanol example. In here we have a polar "head" and a non-polar "tail". The tails will form weak attractions with other tails, whereas the head will form hydrogen bonds with other heads. Due to the hydrogen bonds, this molecule should have a fairly large boiling point. The reason the boiling point is higher for ethanol than it was for methanol is because of London forces. This molecule is larger and heavier than the other, therefore, it forms stronger instantaneous dipoles in the tails.

- What is the relative boiling points of ammonia, water and hydrogen fluoride?

    These three molecules (NH_3, H_2O, and HF) have the ability to hydrogen bond. Therefore, the trend should be in the exact order as they were given because the electronegativity of fluorine is higher than that of oxygen, which in turn is higher than that of nitrogen. This means that the hydrogen bonds formed by the fluorine should be the strongest hydrogen bonds, followed by those formed by oxygen, and those by nitrogen should be the weakest. The actual boiling points can be seen below.

Molecule	Boiling Point
NH_3	- 33.3°C
HF	19.7°C
H_2O	100.0°C

This is not exactly what we expected. Ammonia does have the lowest boiling point. However, the HF does not have the highest boiling point! The arguments we made before are very valid, that is, the hydrogen bonds formed by the fluorine are indeed stronger than those formed by the oxygen. However, it turns out that the water molecule has the capability to form four hydrogen bonds per molecule[1], whereas the HF molecule only has the capability to form only two per molecule[2]. Four slightly weaker hydrogen bonds add up to a stronger overall interaction than the two hydrogen bonds in HF. We should point out that water does not always form four hydrogen bonds. Actually, only in the solid form (ice) does it form four hydrogen bonds. In the liquid phase, water forms on an average three hydrogen bonds per molecule.

We should also point out that the boiling point for ammonia is a lot higher than it would have been had it not exhibited hydrogen bonding. The boiling points of the hydrides for the nitrogen family should increase as we go down the family. Let's look at the data.

Molecule	Boiling Point
NH_3	- 33.3°C
PH_3	-133°C
AsH_3	-116.3°C
SbH_3	-88.0°C

We can immediately see that the ammonia does not fit the pattern. It should have the lowest boiling point from the family, and instead, it turns out to be the highest. This is due to hydrogen bonding.

---

[1] Two with the oxygen (through the two lone pairs) and one with each polarized hydrogen.
[2] One with the fluorine and one with the polarized hydrogen.

- An equimolar mixture of chloroform, $CHCl_3$, BP = 51.2°C and acetone, $(CH_3)_2C=O$, BP = 56°C, will boil at a much higher temperature than the boiling point of either component. Why does that happen?

    Both of these molecules are polar and they exhibit only dipole-dipole forces of attraction.

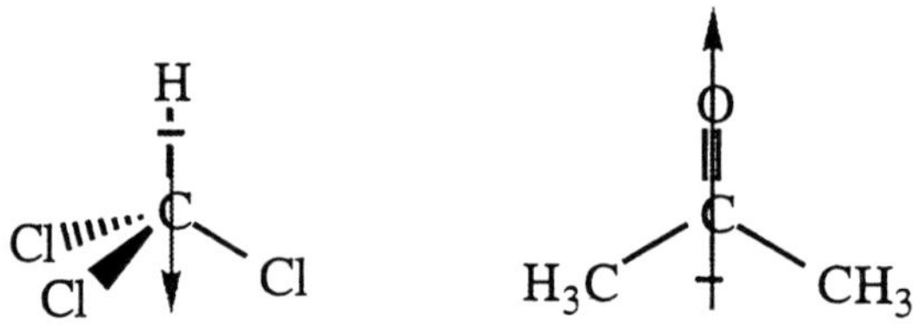

    However, the chloroform has a very polarized hydrogen. At first glance, this seems improbable because the hydrogen is bonded to a carbon. Yet, that carbon is also bonded to three chlorines, which are highly electronegative atoms. Notice from the picture above, that the positive end of the dipole goes right through the hydrogen. What happens when we mix these two polar molecules? Of course, the partially negative oxygen in the acetone is going to form a hydrogen bond with the polarized hydrogen in the chloroform. Therefore, the forces are stronger in the mixture than in the individual liquids, which is why the solution has a higher boiling point.

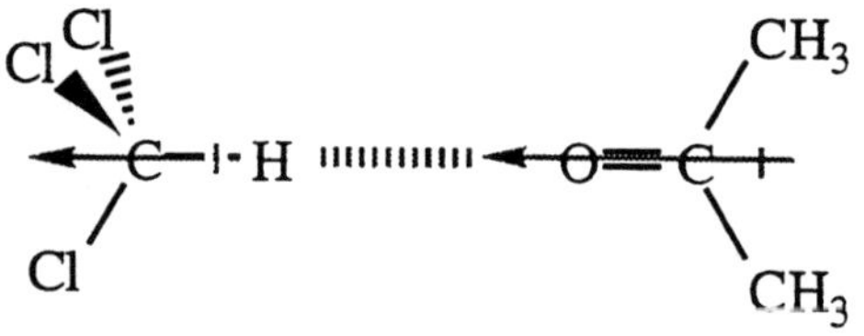

    One final comment. Hydrogen bonding is extremely important in the structure of proteins and nucleic acids. The four bases found in the DNA[1] molecule are Guanine, Cytosine, Thymine and Adenosine. These bases interact with one another through hydrogen bonding. Even though the proportion of bases vary from one DNA to another, it is found that guanine always pairs up with cytosine, whereas thymine always pairs up with adenosine. In order to maximize the hydrogen bonds between the bases, Watson and Crick[2] reached the conclusion that two DNA strands will pair up through these hydrogen bonds (the strongest of which were shown by Pauling to be N---H---N or N---H---O) they proposed the famous double helix. Below you will see the hydrogen bonds created by the interaction of these pair of bases. Try to convince yourself that the combination of guanine with adenosine, or for that matter, any other

---

[1] Deoxyribonucleic acid.
[2] For this work, they received the Nobel Prize in 1962.

combination but the ones shown, do not yield the appropriate H-bonding (the sites do not match).

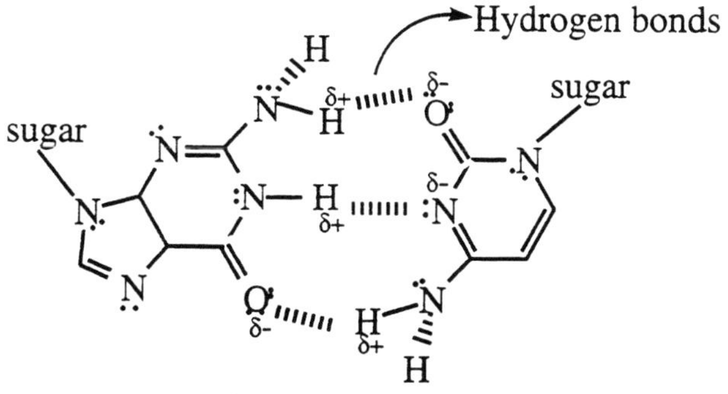

GUANINE   CYTOSINE

and

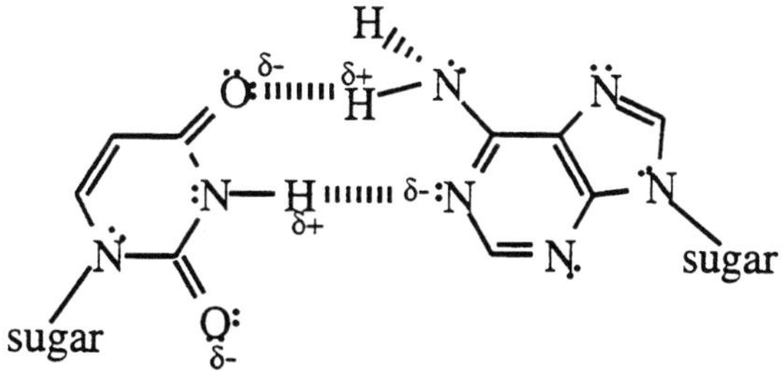

THYMINE   ADENOSINE

10.2 Solids.

So far we have only discussed the gaseous phase, which is a fluid state, just like the liquid one. This fluidity, or ability to flow, suggests that the individual particles that make up these phases are not locked into definite places. In the solid state, however, we do not need a container to hold up the shape of the solid form.

There are two types of solids: crystalline solids and amorphous solids (glasses). Many centuries ago, people discovered that if we heat a mixture of soda (sodium oxide), lime and sand to the melting point, a rigid substance that looks smooth and "glassy" is produced. This substance certainly looks like a regular crystalline solid. However, earlier in the 20th century, when we started to get a better idea about the structure of the atom, people noticed that there was a remarkable difference between these two types of solid structures. The glassy state is not as distinctly defined, and we do not know as much about it as we do about the crystalline state. Some people even visualize a glass as a liquid that has an unusually high viscosity and therefore, will hold its shape for a very long time. This is a little dangerous because it certainly IS a solid:

it has a rigid shape, *but* a great deal of disorder exists in its structure.

How can we distinguish, then, between crystals and glasses? A very efficient way of doing this is determining its melting point. Crystalline solids have a definite, sharp MP. A pure crystalline solid will melt at a constant temperature. For example, ice will melt at 0.0°C if it is pure (a melting point range usually indicates impurities). Glass, on the other hand, does not go through a clearly defined transition between its solid and liquid phases. Its viscosity simply decreases with increasing temperature, until the glass softens and eventually becomes a "puddle".

Why do liquids solidify into a crystalline solid? In order to answer this question, let's use water as an example. As the water molecules cool down, they rearrange themselves in order to maintain a point of minimum energy (by maximizing attractions and minimizing repulsions). What is then the problem with the glasses? Well, glasses fail to reach this minimum point. This is because the liquid cools rapidly and the viscosity increases dramatically. When this happens, the molecules do not have enough time to relax into the positions that are customary in the crystalline solid state. This has taught us something important: . Molecules always do have the tendency to join in an orderly fashion that minimizes its overall energy, but you must give them time to take their places in the crystal lattice. . As they are moving to reach their places, the viscosity has increased so much that, before they have taken their places in the regular crystal lattice, the whole thing is "frozen solid". These "glassy" solids are referred to as **amorphous** solids. As opposed to crystalline solids, these amorphous solids have no definite crystal lattice and their structure is pretty disorganized. There are "clusters" of organized molecules in a "sea" of unorganized molecules. This is why there is no definite melting point.

We will spend most of the time talking about the other type of solids, the crystalline solids. There are three types of crystalline solids: 1. ionic solids, made up of ionic substances (example: NaCl), 2. molecular solids, made up of covalent molecules (example: sucrose), and 3. atomic solids, made up pure elements and therefore, its basic components are atoms (example: Na metal, Carbon graphite, etc.). The difference in the properties between all these types and even within the different substances in a given type is due to the intermolecular forces of attraction and bonding. We will try to sort these out later.

Some of the distinguishing aspects of crystalline solids are the following:

1. For a given pure substance, all crystals (large or small) have the same shape, as long as they were prepared under identical experimental conditions.

2. Crystals have a high degree of symmetry.

3. An important way to distinguish between different types of crystals is by their faces and edges in their solids regular arrangement.

4. If a crystal is shattered, it will normally break along a plane that is parallel to the faces of the original crystal. This will tend to produce many smaller pieces that have the same shape as the original crystal. These planes along which the crystal breaks are called **cleavage planes**.

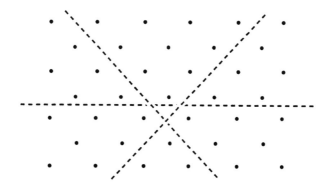

Crystalline solids are incredibly ordered in the solid phase. Therefore, we should expect to have a *basic arrangement* in the crystal that is then repeated over and over in a regular way as the size of the crystal increases. Above we can see a two-dimensional representation of this regular arrangement (which we will call the **crystal lattice** in 3-D). The lines drawn suggest possible directions of cleavage for this hypothetical 2-D lattice.

The fact that we have a basic arrangement repeats in all directions suggests that there should be a basic repeating unit, which we will call a **unit cell**. This unit cell is a very small portion of the entire structure, which is formed by connecting what we call lattice points. A lattice point is a point in the crystal lattice where we find an atom, an ion, or a molecule[1]. How do we repeat this unit cell in order to form the overall lattice? We do so by a process called translation – a simple shift of position which involved no other movements (such as rotation). Of course, since the lattice points of the initial unit cell are shared with the adjacent unit cells, we have to be particularly attentive as to how many cells are sharing a given lattice point. Why is this so? Because if an atom in a given point is shared by eight cubes, then there is only an eighth of an atom contributed to the unit cell by that lattice point. We will see this in more detail below.

The positions of the atoms in the crystal lattice can be determined experimentally with a

---

[1] This is not always true, because sometimes there is a cluster of atoms arranged around a particular lattice point, but for the sake of simplicity, we will assume that there is either an atom, an ion, or a molecule at every lattice point.

technique called X-Ray Crystallography. This is not something we will cover in this course, but it suffices to say that with this technique, we can predict the relative position of all atoms in a solid.

There is one more thing that we should point out about the unit cell. The unit cell itself must contain all of the components of the crystal as a whole and in their proper stoichiometric ratio. For example, the unit cell of magnesium fluoride (1 Mg to 2 F's) must have twice as many fluoride ions as it has magnesium ions.

To define a unit cell we need to specify six crystal parameters which are, length of each side: a, b, and c, and three angles: $\underline{\alpha}$, $\underline{\beta}$, and $\underline{\gamma}$.

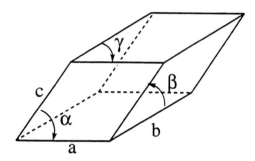

There are seven possible geometries for unit cells and their relative parameters are shown below.

Unit Cell	Edges	Angles
Triclinic	$a \neq b \neq c$	$\alpha \neq \beta \neq \gamma \neq 90°$
Rhombohedral	$a = b = c$	$\alpha = \beta = \gamma \neq 90°$
Monoclinic	$a \neq b \neq c$	$\alpha \neq \gamma \neq 90°, \beta = 90°$
Orthorhombic	$a \neq b \neq c$	$\alpha = \beta = \gamma = 90°$
Hexagonal	$a = b \neq c$	$\alpha = \beta = 90°, \gamma = 120°$
Tetragonal	$a \neq b = c$	$\alpha = \beta = \gamma = 90°$
Cubic	$a = b = c$	$\alpha = \beta = \gamma = 90°$

Of all these seven, we will go only over the cubic ones, simply because they are easier to see and they will illustrate the concepts. Let's look at a cube first.

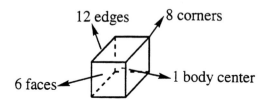

Chapter 10: Condensed Phases.

There are three kinds of cubic unit cells:

1. Simple Cubic (sometimes called primitive).
2. Body Centered Cubic.
3. Face Centered Cubic.

Before we attempt to describe these cubic lattices, let's first determine how many cubes share each part of one cube.

a) <u>Corner Atoms</u> - Each corner atom is shared by eight cubes. Another way of looking at it is that an atom only contributes an eighth of an atom to the unit cell. The following figure illustrates this point.

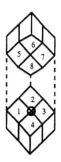

b) <u>Edge Atoms</u> - Each edge atom is shared by four cubes, so it will contribute only one fourth of an atom to the unit cell.

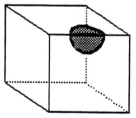

c) <u>Face Atoms</u> - Each face is shared by two cubes and an atom in the center of a face will contribute only one half of an atom to the unit cell.

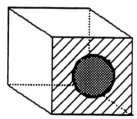

d) <u>Center Atom</u> - An atom in the very center of the cube will not be shared with any other cube, therefore it contributes the whole atom to the unit cell.

Chapter 10: Condensed Phases.

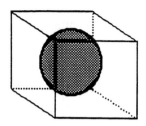

Summary

Location of Atom	Portion of Cubic Unit Cell
Corner	1/8
Edge	1/4
Face	1/2
Center	1

Let's now calculate the effective number of atoms in the three cubic unit cells.

a) <u>Simple Cubic</u> - In this structure we have one atom occupying each one of the eight corners of the unit cell.

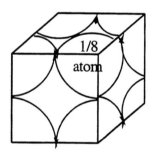

∴ 8 corners x 1/8 = <u>1 atom</u> in unit cell.

b) <u>Body-Centered Cubic</u> - This is similar to simple cubic but it also has one atom in the very center of the cube.

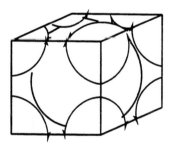

∴ 8 corners x 1/8 =   1 atom
+ <u>1 center x 1</u> =   <u>1 atom</u>
                                    <u>2 atoms in unit cell</u>.   ✓

c) <u>Face-Centered Cubic</u> - In this case we have one atom in every corner and one in the center of every face. Below you can see one of the six faces.

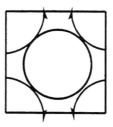

$$\therefore \text{ 8 corners } \times \text{ 1/8 } = \quad \text{1 atom}$$
$$+ \underline{\text{ 6 faces } \times \text{ 1/2 }} = \quad \underline{\text{3 atoms}}$$
$$\underline{\text{4 atoms in unit cell}}. \quad \checkmark$$

The information obtained from lattice parameters could enable us to calculate the density of a solid. Let's look at one example to understand how this is possible.

- Neon crystallizes in a face-centered cubic lattice, and the edge of the unit cell is 452 pm (or 4.52 Å). The atomic weight of Ne is 20.2. What is the density of crystalline Ne (in g/mL)?

In order to find the density, we need the mass and the volume of the unit cell. The entire key to this problem is to try to figure out how to get this information. Well, the volume is easy to find because we know the length of an edge, and since it is a cubic unit cell, all we have to do is cube that length.

$$V = (\text{edge})^3 = (4.52 \times 10^{-8} \text{ cm})^3$$

$$V = 9.23 \times 10^{-23} \text{ cm}^3.$$

Now, to find the mass of the unit cell, we need to know how many atoms you have in that small space. Since it is a face-centered unit cell, we already know that we have <u>four atoms</u>.

$$\text{mass} = 4 \text{ atoms} \times \frac{1 \text{ mol}}{6.022 \times 10^{23} \text{ atoms}} \times \frac{20.2 \text{ g Ne}}{1 \text{ mol}}$$

$$= 1.34 \times 10^{-22} \text{ g of Ne}.$$

Finally, Density = mass / volume = $(1.34 \times 10^{-22} \text{ g}) / (9.23 \times 10^{-23} \text{ cm}^3)$

$$= \textit{1.45 g/cm}^3 \quad \checkmark$$

10.3 Metals and Closest Packing.

It is possible that the first thing that we may think about when we say metals are those elements to the left of the dark staircase in the periodic table. These elements have a few things in common. First of all, most of them are solids, all of them conduct electricity and they have rather small ionization energies and large electron affinities. We tied all these with the idea that metals tend to react by giving away electrons, that is, they tend to form positive ions. Metals are also **malleable** (capable of being reshaped by hammering, etc.) and **ductile** (can be drawn into thin wires). We will attempt to explain why metals behave the way they do based on the type of bonding in these crystals.

Let us imagine that particles that make up the solid metal at the microscopic level are solid spheres. Now let's try to imagine the way in which we would pack these spheres, just as we would pack tennis balls in order to form a uniform three dimensional structure. There are various ways in which this task can be accomplished. Let us start with the simplest scenario.

Imagine that the spheres are just touching and forming uniform layers. There are two distinct ways in which you can pack these spheres into one layer, as shown below. Please note that all the spheres are touching. This is the basis of our discussion ahead. We will always find an arrangement in which the spheres are in direct contact. The big question is, what happens as we try to change it from the two dimensions shown into three dimensions?

 or

Well, as a first example, let us take the arrangement on the left and assume that all the other layers are identical and placed in such a way that the spheres in one layer are directly above and below the two surrounding layers. Try to visualize in your mind the meaning behind this. If you consider any one sphere in a given layer (let's say the one marked with a * above), how many other spheres are in direct contact with it? It is in contact with four in its own layer, one to its right, one to its left, one on top and on the bottom. Is it in contact with any other spheres? Yes! It is also touching the spheres directly above and below it for a grand total of **six**. This is normally referred to as the *coordination number*. Thus, we can say that for this example we are discussing, the coordination number is 6.

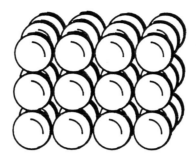

The simplest repeating unit in this lattice is a simple cubic unit cell (below). We have already seen this before but now we can see how we can "extract" these from a lattice.

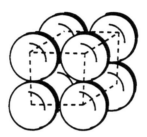

Is this the most efficient way of packing the spheres? In other words, is there another way in which we could pack these spheres in which they would occupy less space? In order to analyze this, let us learn how to calculate the packing efficiency for a given crystal. In order to do this we will call **a** the length of a side in the unit cell, and **r** the radius of the spheres.

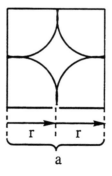

This figure shows that two times the radius of the atoms is equal to the length of an edge (a).

$$\therefore a = 2r$$

or: $\quad r = a/2 \quad \checkmark$

How effectively is this system packed? Is there a way to actually determine the efficiency of the packing process for the simple cubic system? In order to do this, we need two things. One of these is the volume of the unit cell itself, and the other is the volume that each of the atoms in the unit cell would occupy.

The volume of the unit cell is very simple to calculate since this is a cubic system. This should simply be equal to the length of an edge cubed. How about the volume of each individual atom? When we started this section, we mentioned that we would consider the atoms to be spheres. Since we know that the formula for the volume of a sphere, we can easily determine the volume of an atom. This will be in terms of the radius of an atom. However, since we just learned that two times the radius of the atom is the length of an edge, then we can write this in terms of **a**, as seen below.

$$\text{Volume of the unit cell} = a^3$$

$$\text{Volume of one sphere} = \frac{4}{3}\pi r^3 = \frac{4}{3}\pi \left(\frac{a}{2}\right)^3$$

With this information available, how can we determine the packing efficiency? The **packing efficiency** (in percentage) is define as:

$$\text{Packing Efficiency} = \frac{\text{Volume of all spheres in the cell}}{\text{Volume of the unit cell}} \times 100 \quad (\text{in \%})$$

How many spheres are there in a simple cubic unit cell? Previously, we learned that there is only ONE sphere in a simple cubic unit cell (because of the fact that each corner is shared by eight cubes). Therefore, the packing efficiency for a simple cubic unit cell is:

$$\text{Packing Efficiency} = \frac{\frac{4}{3}\pi \left(\frac{a}{2}\right)^3}{a^3} \times 100 = \frac{\pi}{6} \times 100\%$$

$$\text{Packing Efficiency} = 52.4\%$$

In contrast, what is the packing efficiency of a face centered cubic structure? Remember that in a face center structure there are FOUR spheres per unit cell. Which are the spheres that are in direct contact in a face centered cubic cell? It is not the corner spheres, as in the case of the simple cubic. Instead, the spheres that are touching are the ones along the diagonal of the face.

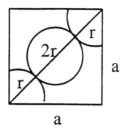

Using the Pythagorean Theorem,

$$(\text{hypotenuse})^2 = (\text{side})^2 + (\text{side})^2$$

Therefore,

$$(\text{face diagonal})^2 = a^2 + a^2 = 2a^2$$

Face diagonal $= a(2)^{1/2}$

But since the face diagonal is equal to four times the radius of the atom, then:

$$4r = a(2)^{1/2}$$

$$r = \frac{a\sqrt{2}}{4}$$

Let us determine the packing efficiency for the face centered cubic cell.

$$\text{Packing Efficiency} = \frac{4\left[\frac{4}{3}\pi\left(\frac{a\sqrt{2}}{4}\right)^3\right]}{a^3} \times 100\% = 74.0\%$$

The numbers tell us right away that this face centered cubic conformation is a lot more efficient when it comes to packing the spheres, and thus should form denser solids. Remember that the more efficient the packing is, the more mass you will have per unit volume. The coordination number for the FCC is actually 12, but we shall return to address this a little later.

Finally, let us look at the example of a body centered cubic unit cell and calculate its packing efficiency. How many spheres would be touching in this case? Well, the sphere in the very center of the cube is in direct contact with all eight corners (coordination number is 8). Let us look at the center sphere as it touches the two from opposite corners along the diagonal of the cube.

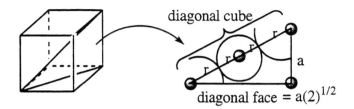

$$(\text{diagonal of the cube})^2 = a^2 + 2a^2 = 3a^2.$$

$$\therefore \text{ diagonal of the cube} = a(3)^{1/2}.$$

However, from the diagram above we can see that we can fit four atomic radii along the diagonal of the cube. Therefore,

$$\text{Diagonal of the cube} = 4r$$

$$4r = a(3)^{1/2}.$$

Solving for r, we get:

$$r = \frac{a\sqrt{3}}{4}$$

We can calculate the packing efficiency in the same manner, keeping in mind that there are only TWO spheres in a body centered cubic cell.

$$\text{Packing Efficiency} = \frac{2\left[\frac{4}{3}\pi\left(\frac{a\sqrt{3}}{4}\right)^3\right]}{a^3} \times 100\% = 68.0\%$$

From our discussion it is evident that the most efficient way to pack these spheres is in the face centered cubic lattice (74%) and the least efficient is the simple cubic unit cell (52.4%). How do we relate this to our original discussion on packing? We already saw how the spheres are packed in the simple cubic structure. The body centered unit cell is built from a similar basis as the simple cubic, except the atoms are not touching in the original layer. You might be thinking that we mentioned before that atoms would have to touch. This is true, but in this case the atoms touch through the second layer, since it will be placed directly in the depressions of the first layer.

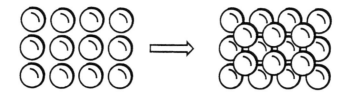

The very next layer is exactly the same as the original layer and it will be placed in such a way that it eclipses the first one. From the unit cell drawn below, it should be pretty clear that the coordination number for the body center cubic lattice is 8.

Before we go on with the generation of the face centered cubic unit cell, we should look at the following page in the internet: http://chemwww.cwru.edu/CWRU/Chime/solids.html. Once you are there, click on the link called "Cubic Unit Cells". There is a very similar discussion that will allow you to see the structures we have been displaying here, with the added benefit of being able to rotate them at will.

Now, how about the one that is most efficient, that is, the face center cubic lattice? How can we explain the generation of this unit cell? It turns out that the original layer for the face centered cubic system is different from the other two. The original layer is show below.

The immediate question is, where should we place the second layer? The second layer goes on the depressions of the first layer as shown below. Due to the size of the atoms as compared to the holes themselves, we can see that we were not able to cover all the holes completely.

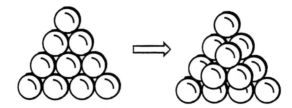

At this point we are ready to place a third layer of atoms. There are two choices and both of them will give an equal maximum efficiency for the packing (74%). One of the ways is to repeat the first layer so that it is exactly eclipsing the first layer. This way it yields what we call a hexagonal lattice (hexagonal unit cell which we will not discuss in this course since we are limiting our study to cubic unit cells). The other option places the third layer in the depressions of the second layer and we do not get a repetition until the fourth layer, in which we place layer 1 eclipsed once more.

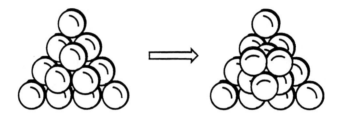

A side view is shown below, where the layers have been labeled A, B, C, and A.

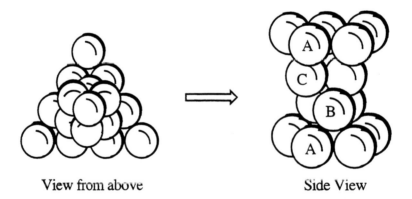

View from above   Side View

How can we explain a coordination number of 12? Refer back to the original layer. Please, note that any one atom in this layer is in direct contact with six other atoms in this same layer. When you place the next layer right above it, that one atom will be in contact with three others of the layer above. It should be evident that the same thing is true with the layer below it. That way we can easily see how we obtain a coordination number of 12.

10.4 Bonding in Metals.

One of the first few models developed for metallic bonding is referred to as the **electron sea model.** In this model, we envision all the metal atoms as having being ionized already, therefore, it would consist of positive ions. The question is, where are the electrons? You cannot have a solid made up of positive ions exclusively. The electrons are "free" to move about the crystal and serve as a "sea" of electrons or as a glue that keeps all the positive ions together. Therefore, the solid is composed of a regular, three-dimensional arrangement of positive ions in a "sea" of "free" valence electrons. These "free" valence electrons would explain why metals conduct heat and electricity. The mobility of these electrons also help to explain their malleability and ductility. As the positive ions are shifted around, the "free" electrons can easily hold them together, since they are free to move around the crystal. A non-metal (such as sulfur) would easily crumble under these conditions.

Another similar model is the **band model**. This is a little more detailed since it is based in molecular orbitals (MO's). We have said many times that metals have low ionization energies. They also have <u>few valence shell electrons</u>. Low ionization energies indicate a weak hold on the valence shell electrons, as well as low tendency to attract or add electrons to the valence shell. Because of that, not only do these atoms fail to form negative ions, but formation of covalent bonds (shared electrons pairs) provides little stabilization. Therefore, it is not surprising that we do not observe diatomics such as Na_2 (as seen below) or Pb_2. Even under favorable conditions, these molecules are not very stable. If we were to look at the MO diagram for the Na_2 molecule, we would see that both electrons from the 3s orbitals occupy a bonding σ MO, so we should expect this to be favorable to bonding. However, although this is true, it turns out that the bonding MO is only slightly lower in energy than that of the individual atomic orbitals (weak stabilization), therefore, the covalent bond formed is very weak.

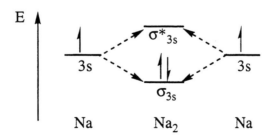

In reality, metals do not form stable diatomic molecules and instead, they exist as crystalline solids. We should point out that the existence of this molecule has been observed in the gas phase, but not in a condensed phase. Once we allow the diatomics to interact, they will aggregate and form clusters of atoms until they eventually form the crystal lattice. In this solid form, each sodium atom is surrounded by many other sodium atoms (in general the number is 8 or 12). Of course, there is no metal that possesses 8 or 12 valence shell electrons that would allow us to explain the formation of 12 bonds to adjacent neighbors in the solid. We would then be forced to conclude that the available electrons must be delocalized in such a way that they would keep the structure together. Keep in mind that this concept is rather different from our initial view of molecular orbitals, where the electrons were in a very distinct orbital.

Let's look at a very specific example. Let's go back to the case of Na_2. Its molecular orbital energy level diagram can be seen in the figure below and it is marked as (1). The two individual 3s orbitals of both sodium atoms form the two molecular orbitals, one of them is σ and the one with the higher energy is the σ*.

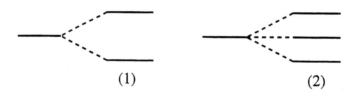

(1)   (2)

If instead we were to add another atom to form Na_3 {see figure above marked (2)}, now we have three molecular orbitals, the middle one being a non-bonding[1] molecular orbital. Two of the three electrons here occupy the lowest molecular orbital (the bonding one), while the third one occupies the non-bonding molecular orbital. As the number of atoms increases, the number of molecular orbitals produced increases accordingly. For a very large number of atoms (see the figure below), we have a very large number of MO's and they are in very close proximity.

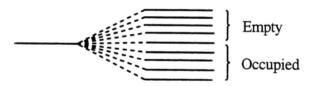

} Empty

} Occupied

The valence electrons occupy the lower portion of this band of orbitals. Movement of electrons through the metal for conduction of electricity or heat is easy, because the electrons near the top of the band can jump to a slightly higher level that is empty and move about the structure.

We can use this band theory to explain another observation concerning metals. All metals are **opaque**. This means that they absorb light and not let it through. The electron (or

---

[1] A non-bonding MO is one that is still in its atomic state (in this case, 3s).

electrons) in the highest occupied molecular orbital above is so close to the empty molecular orbitals that they are able to absorb essentially any frequency of light to get excited into one of these empty orbitals. This is why the metal is able to absorb all frequencies of light.

There is one more thing we should mention. Metallic bonding is a very strong type of bonding. This is clearly seen from the boiling points and melting points of metals. Actually, all metals have high boiling points, whereas their melting points vary immensely. For example, mercury is a liquid at room temperature, whereas tungsten melts at 3410°C.[1] A table with melting and boiling points of some selected metals appears below.

Metal	Melting Point	Boiling Point
Li	181°C	1346°C
Mg	649°C	1105°C
Al	660°C	2468°C
Fe	1535°C	2750°C
Ag	962°C	2212°C
Au	1064°C	2808°C
Hg	-39°C	357°C
Pb	328°C	1740°C
W	3410°C	5660°C

10.5 Ionic Crystals.

The basic components of an ionic crystals are the individual ions. This means that the crystal lattice, and consequently the unit cell, will be composed of two types of ions: the anions (typically the larger of the two) and the cations. The unit cell, as we mentioned previously, must have all the components that make up the salt with the *correct stoichiometric ratio*. The question

---

[1] This is the metal with the highest melting point.

then becomes, how do these ions arrange themselves in order to maximize the electrostatic attraction between the oppositely charged ions while minimizing the repulsions between ions with like charges? As we will see, <u>typically</u> the larger ions will configure themselves into one of the closest cubic pack structures (either hcp or ccp) and the smaller ions will fit into the holes of the closest pack structure. A lot of times we will be able to predict which unit cell a particular salt is going to crystallize in based on the relative sizes of the ions and the sizes of the holes in the close pack structures. Therefore, the main goal of this section will be to understand where these holes are located and how we can predict which will be filled and under which circumstances.

Which kinds of holes are talking about? There are three types of holes in the closest packed structures.

1. **Trigonal holes** are formed by three spheres in the same layer.

2. **Tetrahedral holes** are formed when a sphere sits in the dimple of three spheres in an adjacent layer.

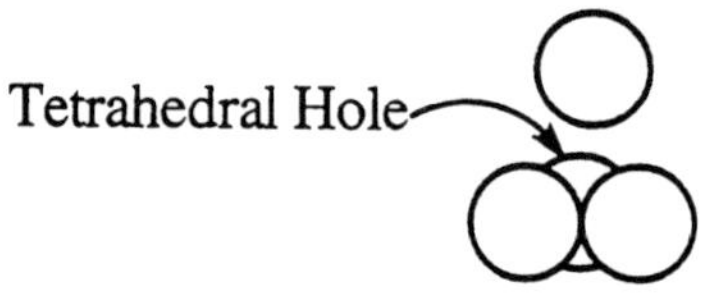

Tetrahedral Hole

3. **Octahedral holes** are formed between two sets of three spheres in adjoining layers of the closet packed structures.

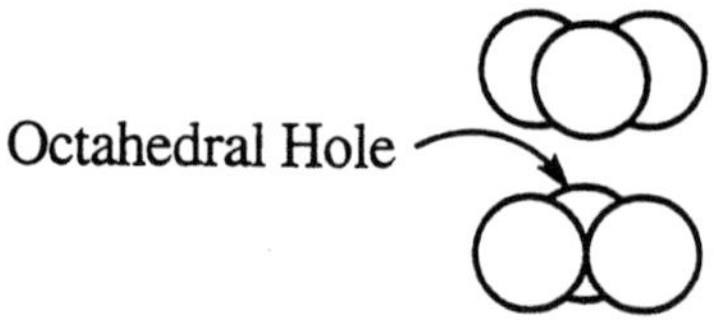

Octahedral Hole

For spheres of a given diameter, the holes increase in size as follows:

trigonal < tetrahedral < octahedral

Actually, trigonal holes are so tiny that they are never occupied in binary ionic compounds.

Whether the tetrahedral of octahedral holes in a given binary ionic compound are occupied depends mainly on the relative sizes of the anion and cation.

Octahedral holes.

An octahedral hole is at the center of six equidistant spheres whose centers define an octahedron. Three of the spheres lie in one of the planes of the closest pack structure, whereas the other three are in the next layer.

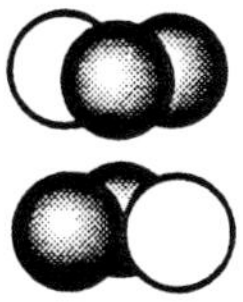

We could rotate it in such as way as to place the four dark spheres in a plane with one of the clear ones up and the other one down. That is what we call an octahedron. Anyway, the hole is such that we could theoretically fit a small ion in the center of that square. How big of an ion can we fit into that small hole? Let's look at the following diagram to figure that out. In this diagram, r is the radius of the small ion, whereas R is the radius of the large ion (the one that occupies the main sites of the face centered cubic system.

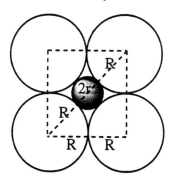

We can see that the ions are "touching" through the diagonal of the square.

$$\therefore d^2 = (2R)^2 + (2R)^2 = 8R^2$$

but: $$d = 2R + 2r$$

Substituting and solving for r we get:

$$r = 0.414\, R \quad \checkmark$$

This means that an octahedral hole has a radius that is about 41.4% the size of the radius for the other ion (R).

Tetrahedral Holes.

A tetrahedral hole is the space in between the four spheres whose center define a tetrahedron. Three of these spheres are in a given layer of the closest pack[1] whereas the fourth one is from the next layer, resting in the dimple formed by the original three. The center of the tetrahedral hole can be visualized as being in the center of a cube that has four spheres as shown below.

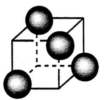

Now, the ion that we would fit into this hole will touch all four of these spheres. We can see that they will be touching along the diagonal of the cube as shown below.

This means that HALF of the diagonal is equal to the radius of the larger ion (R) plus the radius for the smaller ion (r).

$$\therefore 0.500 \, d = R + r$$

Using simple geometry, we can prove that:

$$\mathbf{r = 0.225 \, R} \quad \checkmark$$

This proves that indeed the tetrahedral holes are much smaller than the octahedral holes.

$$r_{tetra} = 0.225 \, R \qquad r_{octa} = 0.414 \, R$$

We need to figure out a way to decide if some salt will crystallize by placing the smaller ions into the tetrahedral holes or the octahedral holes. Let's assume that we have a 1:1 salt where the positive ion is smaller than the negative ions. We want to maximize the +/- interactions and minimize the -/- and the +/+ interactions. We can do this by placing the

---

[1] Basically meaning face centered cubic.

positive ions into holes that are slightly smaller than the size of the positive ions. This causes the negative ions to be pushed apart a little and under these conditions, the positive ions will be touching the negative ions, but the negative ions will NOT be touching the negative ions.

These are the guidelines that we will be using. If the radius of the positive ion falls in between the following range: $0.225\,R^- < r^+ < 0.414\,R^-$, then we will place the positive ions in the tetrahedral holes. This way the positive ions are bigger than the tetrahedral holes themselves and forces the negative ions not to touch (exactly what we want).

As the size of M^+ increases, eventually they will be able to fill and exceed the octahedral holes. In other words, if the size of the cation is larger than $0.414\,R^-$, then we place the cations in the octahedral holes. Does that mean that ANY cation larger than $0.414\,R^-$ will occupy the octahedral holes? No. There is an upper limit. After this limit is surpassed, something else will happen. For very large cations, the X^- ions rearrange themselves into a simple cubic arrangement, with the M^+ ions occupying the center of every cube. We could describe it, if we wished, as an M^+ occupying a **cubic hole**. It turns out that:

$$r_{cubic} = 0.732\,R^-.$$

This would imply that for a solid MX, the M^+ ions would occupy the octahedral holes is the range:

$$0.414\,R^- < r^+ < 0.732\,R^-.$$

Summary

Size of M^+	Type of Hole Filled
$0.225\,R^- < r^+ < 0.414\,R^-$	Tetrahedral
$0.414\,R^- < r^+ < 0.732\,R^-$	Octahedral
$r^+ > 0.732\,R^-$	Cubic

We should point out that these guidelines are not always obeyed. All along we made one assumption: that the ions are solid spheres that have only ionic forces. We know that we always have London forces also, therefore, we may not always obey these guidelines, but they give us a good estimate of what particular unit cell we are going to observe for some binary salts.

At this point it would be important to have a clear idea of where the different holes are in the face-centered cubic unit cell (close packed). The location of the tetrahedral holes may be seen in the home page[1] for CHEM 105/6. You should spend some time viewing all the structures depicted there, so that you can visualize where these holes are in the different lattices. Nevertheless, a tetrahedron can be formed by the center of three adjacent faces in a close-pack cubic structure, and the corner in between these three faces. The hole that lies in the center of this tetrahedron is what we call a tetrahedral hole. If we think about this for a while, we can see that there are eight (8) tetrahedral holes inside of the face-centered unit cubic cell, because we can form eight different tetrahedrons with each corner and the corresponding three face centers that it touches. If we were to connect these holes together, it would form a smaller cube inside of the unit cell. The meaning of all this is that we have **twice** as many tetrahedral holes as we have spheres packed into the unit cell. If we were to fill all of these holes, we would be placing **eight** oppositely charged ions into the unit cell.

The location of the octahedral holes is easier to see. These holes turn out to be at the center of the face-centered cubic unit cell and along all the twelve (12) edges. You can prove that the number of octahedral holes is equal to the number of spheres in the unit cell (one in the center and three from the edges). I urge you once more to look at the home page of the CHEM 105 course and check the unit on solids to see exactly the location of all of these holes.

Let's look at a few examples of some common salts.

1. <u>Sodium Chloride</u>.

The relative size of the ions is: $r_{Na^+} = 0.66\ R_{Cl^-}$. According to our guidelines, we should observe a face-centered cubic unit cell of chloride ions with the sodium ions occupying some, if not all of the octahedral holes. This is exactly what we observe experimentally. Remember that since the Na^+ ions are larger than the octahedral holes, this forces the Cl^- ions apart, thus minimizing the repulsion between them. That is why, if you look carefully, you will notice that the chloride ions are actually not touching. Since the number of octahedral holes is the same as the number of Cl^- ions in the unit cell, then it should follow that ALL the octahedral holes must be filled with Na^+ in order to obey stoichiometry (we need to have as many Na^+ as we have Cl^- because it is a 1:1 salt, NaCl).

---

[1] http://chemwww.cwru.edu/CWRU/Chime/solids/xtal1.html.

Chapter 10: Condensed Phases.

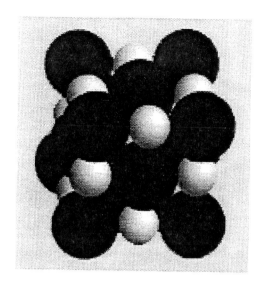

Further analysis:

∴ 8 corners × 1/8 Cl⁻ = 1 Cl⁻ ion
+ 6 faces × 1/2 Cl⁻ = 3 Cl⁻ ion
4 Cl⁻ ions in unit cell. ✓

and 12 edges × 1/4 Na⁺ = 3 Na⁺ ion
+ 1 center × 1 Na⁺ = 1 Na⁺ ion
4 Na⁺ ions in unit cell. ✓

Most of the alkali metal halide salts have the same structure (examples: LiCl and KCl). However, CsCl has a different structure.

2. Cesium Fluoride.

Cesium fluoride has the same structure as NaCl but the ions are reversed, because this time the Cs^+ ion is the large one and the F^- ion is the smaller one. Therefore, in this case, the Cs^+ form the face-centered cubic unit cell and the F^- ions fill out all the octahedral holes.

3. Cesium Chloride.

Now, this is a totally different case because these two ions have essentially the same size. What does that mean? According to what we have stated previously, the Cl^- ions should form a simple cubic unit cell with the Cs^+ ion occupying the cubic hole (in the center of the cube). The salts CsBr and CsI also have this structure.

$$\therefore 8 \text{ corners} \times 1/8 \text{ Cl}^- = 1 \text{ Cl}^- \text{ ion}$$
$$+ \underline{1 \text{ center} \times 1 \text{ Cs}^+ = \underline{1 \text{ Cs}^+ \text{ ion}}}$$
$$\underline{1 \text{ CsCl unit in unit cell}}. \quad \checkmark$$

4. <u>Zinc Sulfide</u>.

When we look at the sizes of the ions, we see that $r_{Zn^{+2}} = 0.35 \, R_{S^{-2}}$. According to our guidelines, then the sulfide ions should define a close-packed unit cell with the zinc ions occupying the tetrahedral holes. This is exactly what we observe in the ZnS case, however, don't forget that we mentioned that there are twice as many tetrahedral holes are there are sulfides in the unit cell. This means that if we were to fill them all out, then we would have twice as many zinc ions as we have sulfides. This would not agree with the stoichiometry of the salt, therefore, we observe that only one-half of the tetrahedral holes are filled with the zinc ions.

Remember that there are two close packed unit cells, one that is face-centered cubic and one that is hexagonal. This would then explain easily why we observe two different forms of ZnS in nature, one that is called Zinc Blende (ccp) and one called Wurtzite (hcp). The salts ZnO and CdS also crystallize in the same way showing both structures (zinc blende and wurtzite).

10.6 Liquids.

Most chemical reactions involving both living and non-living systems occur in liquid solutions. Therefore it is important to try to understand at least some of the properties of liquids. This is the hardest phase to describe because it is much harder to establish a model to describe molecules that are in constant random motions, but in which we have an incredible number of interactions between them. Liquids have low compressibility, lack of rigidity, and high density compared to gases. Let's mention some of the properties of liquids.

Properties of Liquids.

a) Similar to those of gases:

1. They are fluids - the individual molecules have freedom to move and diffuse.

2. They can be randomly dispersed.

b) Differing from those of gases:

1. They occupy a definitive volume. Take the case of one mole of water (MW = 18.0) at 25°C. Since the density of water is 1.00 g/mL, then it will occupy 18.0 mL, no matter what the size of the container is. The same would not be true for a gas since we know that a gas will occupy the entire volume.

2. They are not easily compressed.[1]

There are other properties of liquids that we should also mention at this point.

A. Viscosity.

It is one of the transport properties for molecules. It is due to intermolecular forces of attraction as well as long range forces. Molecules tend to push past other molecules with varying degrees of ease. Things like maple syrup flow slowly, meaning that there is a high degree of

---

[1] For all practical purposes, we will say that both liquids and solids are incompressible.

friction between adjacent layers of molecules. Water, however, has low level of friction between the layers of molecules and flows quite freely. We will define **viscosity (η)** as the measure of the internal liquid friction opposing any change in the liquid's movement. This phenomena is temperature dependent, as we can see below.

Temperature Dependence of the Viscosity of Water

T(°C)	0.0	20	25	55	100
η (centipoise)	1.8	1.0	0.9	0.5	0.3

We can see in this table that at higher temperatures, there is a decrease in the viscosity of water. This makes sense because at a higher temperature, the water molecules have more thermal energy to break some of the intermolecular forces of attraction, and thus flow more freely. Let's point out that the unit of viscosity is the **poise**, after Jean Louis Poiseuille (1 poise = 0.10 kg/m·s). Below you can see viscosity values for some common substances.

Viscosities for Some Substances @ 25°C (in centipoise)

Substance	Viscosity
$N_{2(g)}$	0.019
$O_{2(g)}$	0.020
$C_6H_{6(\ell)}$	0.60
Water	0.90
H_2SO_4	19.0
Olive Oil	80.0
Glycerin	954.0

How can we determine the viscosity of a substance? We do it with a viscometer (shown below).

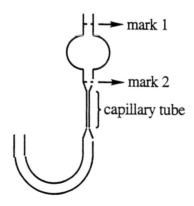

The liquid is pulled above mark **1** and then it is allowed to fall. Since the liquid has to go through a capillary tube, it will take a considerable amount of time to flow. The larger the viscosity, the longer it will take. A liquid with high viscosity is held together by strong intermolecular forces of attraction so it will show a considerable resistance to flow through the capillary tube. Generally, what we saw for water is true: as you increase the temperature, the viscosity will decrease. This is due to the fact that at higher temperatures the molecules have more energy, so <u>some</u> of the intermolecular forces are broken and therefore the liquid can flow easier.

B. <u>Surface Tension</u>.

If we want to get any kind of understanding of this property, we have to think about the thin layer that separates the liquid from the gas above it, the surface. Think about this for a second. There are dramatic differences between liquids and solids, yet, it is in this layer where the transition between them takes place. Therefore, we should expect to see molecules with properties that are intermediate between them. Keep in mind that we have mentioned that it is a thin layer, and is only a few molecules thick. As we go from the liquid into the gas, we will see a dramatic change in the density and an obvious change in molecular energies.

A molecule inside the bulk of the liquid is attracted to other molecules in a very symmetrical fashion (in all directions). However, a molecule on the surface of the liquid has resulting forces that are unequal, and their average is directed toward the body of the liquid. Because of this *inward* pull from the other molecules, there is a **surface tension** that results in: (1) a membrane-like character to the surface, which has an unusually high resistance to penetration, and (2) a tendency for drops of a liquid to become spherical in shape. Surface tension is a temperature-dependent property of the liquid surface, decreasing with increasing temperature. The surface tension is measured in units of energy per unit area.[1]

<u>Surface Tension of Water/Air Interface at Various Temperatures</u>

T(°C)	20	40	60	80	100
Surface Tension (dynes/cm)	72.8	69.6	66.2	62.6	58.9

---

[1] Do this experiment at home. Take a glass of water and *very carefully* place a <u>dry</u> paper clip on the surface of the water. Why does it float? If you think about it is pretty amazing since the steel paper clip is much denser than water, therefore it should sink. This is due to the surface tension. If you now add a drop of soap (dish-washing detergent), the paper clip immediately sinks. This is because some of the H-bonds are broken to attract to the soap molecules which in turn breaks much of the surface tension thus allowing the paper clip to sink.

- Which liquid should have a lower surface tension, water or hexane (C_6H_{12})?

Surface tension generally goes hand in hand with the strength of the intermolecular forces of attraction. Water is very polar, whereas hexane would be totally non-polar[1]. Therefore, we should expect for hexane to have a lower surface tension than water. This is exactly the case, as we can see from the table below.

Surface Tension (dynes/cm) of Some Fluids at Liquid/Air Interface @ 25°C

$(CH_3CH_2)_2O$	17.1
Hexane	18.4
Ethylene Glycol	47.7
Glycerol	63.4
Water	72.0

The molecules of a liquid exhibit **cohesive** and **adhesive** forces. Attractive forces between the molecules of a liquid are called **cohesive** forces. Forces existing between the molecules of the liquid and those of the container are called **adhesive** forces. The spherical shape that we see in drops is due to the cohesive forces of attraction between the molecules of the liquid. On the other hand, we observe that if we get a notebook wet, the pages are hard to separate and this is because of the adhesive forces between the liquid molecules and the material of the paper.

Capillary action is exhibited by polar liquids in a capillary (glass) tube. We will define this as the spontaneous rising of a liquid in a narrow tube. This is due to both adhesive and cohesive forces. We observe that when water comes in contact with the glass, it forms strong interactions with the oxygen atoms in the glass, thus creating a strong attraction to them. This makes the liquid creep up the walls of the tube where the water surface touches the glass. This, in turn, tends to increase the surface area of the liquid, which is opposed by the cohesive forces (H-bonds) that are trying to minimize the surface area. Thus, because water has both strong cohesive and adhesive forces, it "pulls itself" up a glass capillary tube to a height (h) where the weight of the column of water just balances the water's tendency to be attracted to the surface of the glass. The **concave** shape of the meniscus demonstrates that the adhesive forces are stronger than the cohesive forces (see figure below). A nonpolar liquid, such as mercury (Hg), shows a lower level in a glass capillary tube and a **convex** meniscus. This is what happens when you have a liquid that has stronger cohesive forces than adhesive forces.

---

[1] Remember that the hydrogens bonded to carbon are always not polarized. This is true for all hydrocarbons.

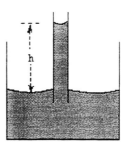

Below we can see both the concave and the convex shapes for a meniscus. Remember that the concave shape of the meniscus is attained when the molecules of the liquid are strongly attracted to the walls of the container, whereas the second case (convex) is exhibited in liquids whose molecules show little attraction to the walls of the container.

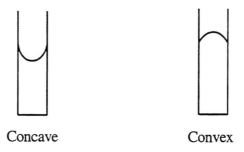

        Concave   Convex

10.7 <u>Vapor Pressure</u>.

This is one of the most important properties of a condensed phase (it applies to liquids and solids). What do we mean by vapor pressure? Let us start by proving the existence of vapor pressure with the following experiment. We will first introduce a certain amount of liquid water into an evacuated container having a fixed volume and at a constant temperature of 25°C. Stopcock A is closed so that a good vacuum remains in the right bulb. After a while, the pressure starts to rise until it equals 23.8 torr. This must be due to water molecules in the gas phase. But how can we explain this? Some molecules on the surface of the liquid have enough energy at 25°C to break all the intermolecular forces of attraction (H-bonds) and thus escape into the gas phase.

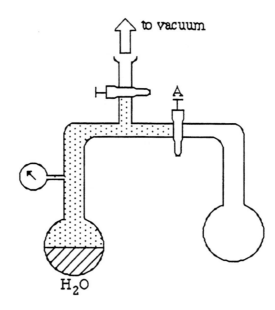

It might be clearer if we see the Maxwell-Boltzmann molecular energy distribution.

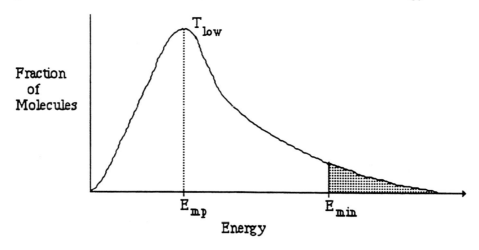

The shaded area represents the fraction of liquid water molecules that have that minimum energy (or more) necessary to break all the intermolecular forces of attraction and thus escape into the gas phase. Therefore, we can see that although the vast majority of molecules do not have that much energy, some of them indeed do, and that is why we are able to measure an actual vapor pressure.

If we now open stopcock A, the pressure will drop instantly (increasing the volume decreases the pressure) but will rise again slowly until it reaches 23.8 torr. Therefore, the vapor pressure (VP) is *independent* of the volume above the liquid.

In a closed container, the molecules that escape into the gas phase will ALSO distribute into a Maxwell-Boltzmann distribution and so, some of them will be moving slowly and will be trapped by the intermolecular forces of attraction in the liquid when they collide with the surface of the liquid. We obtain the previously mentioned vapor pressure (23.8 torr at 25°C) when the rate of vaporization is equal to the rate of condensation! At this point we will not observe a net change in the pressure above the liquid; it is a constant (the vapor pressure) at a given temperature. This condition is called *EQUILIBRIUM*. For the time being, just remember that an equilibrium state is a dynamic state.

We could easily prove experimentally that increasing the temperature increases the vapor pressure. For example, if we repeat the same experiment at other temperatures we obtain the following results:

Temperature	Vapor Pressure
25°C	23.8 torr
50°C	92.5 torr
75°C	289.0 torr
100°C	760.0 torr

This is due to the fact that as you increase the temperature you are increasing the fraction of molecules that have enough energy to escape into the gas phase. The following plot proves that point. Notice that at a higher temperature, the shape of the curve changes enough so that the tail end of the curve is elongated and the shaded area increases.

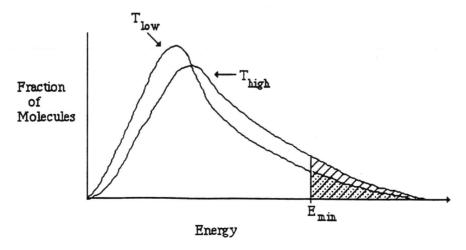

10.8 Temperature Dependence of the Vapor Pressure.

The temperature dependence of the vapor pressure is a characteristic property of a liquid (i.e., the vapor pressure of two different liquids will have different temperature dependence). The graph below shows exactly what we stated previously: *as the temperature increases, so does the vapor pressure.*

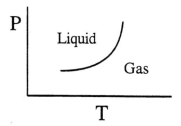

We can predict exactly what the value of the VP will be at a given temperature for a given liquid. Quantitatively, the vapor pressure depends on temperature according to the following equation:

$$\ln\left(\frac{P_2}{P_1}\right) = -\frac{\Delta H_{vap}}{R}\left(\frac{1}{T_2} - \frac{1}{T_1}\right)$$

where: ΔH_{vap} = heat of vaporization, J/mol
R = gas constant (8.314 J/molK)
P_2 is the VP at the second temperature
P_1 is the VP at the first temperature
Temperatures are always in Kelvin.

- Determine the vapor pressure of water at 69°C, given that the heat of vaporization is 40.7 kJ/mol and that the vapor pressure at 25°C is 23.8 torr.

We will use the equation above in order to determine the vapor pressure of water at 69°C. In this example, P_1 = 23.8 torr, P_2 is what we are looking for, R = 8.314 J/molK, T_1 = 298 K, and T_2 = 342 K. Keep in mind that since R is in units of J/molK, we are forced to change the heat of vaporization into similar units (so that they will cancel) of J/mol. So,

$$\ln\left(\frac{P_2}{23.8 \text{ torr}}\right) = \frac{-40{,}700 \text{ J/mol}}{8.314 \text{ J/molK}}\left[\frac{1}{342 \text{ K}} - \frac{1}{298 \text{ K}}\right]$$

$$\therefore P_2 = 197 \text{ torr} \quad \checkmark$$

The vapor pressure will indeed give you a good idea of the strength of the intermolecular forces of attraction in a liquid. The stronger the intermolecular forces, the smaller the fraction of molecules that will be able to "break" these forces and escape into the gas phase (i.e., the lower the vapor pressure). As we will see next, the boiling point (BP) depends on the vapor pressure. Let's see a couple of definitions, one for boiling and another one for boiling point.

a) <u>Boiling</u> - when bubbles of vapor appear all throughout the liquid. This is a very simplistic definition and it simply gives an obvious interpretation of what we visualize as boiling.

b) <u>Boiling Point</u> - the temperature at which the vapor pressure of a liquid is equal to the pressure of the surrounding atmosphere. This is an important prerequisite for boiling point: *In order for a liquid to be boiling, its vapor pressure must be identical to the external pressure.*

<u>Boiling Point of Water at Some Locations</u>

Location	Feet above Sea Level	P_{atm}	BP
Top of Mt. Everest, Tibet	29,028	240 torr	70°C
Top of Mt. McKinley, Alaska	20,320	340 torr	79°C
Top of Mt. Whitney, Calif.	14,494	430 torr	85°C
Leadville, Colorado	10,150	510 torr	89°C
Boulder, Colorado	5,430	610 torr	94°C
Cleveland, Ohio	850	760 torr	100°C
Death Valley, California	-282	770 torr	100.3°C

The implication of these definitions is that a liquid can boil at almost any temperature as long as the atmospheric pressure is equal to the vapor pressure! If you recall, we said that at 25°C the vapor pressure of water is 23.8 torr so if we generate a vacuum in the apparatus shown earlier, water will BOIL (!!) at room temperature. This also explains why at the top of a very high mountain, where the atmospheric pressure is a lot lower than at sea level, water will boil at a temperature lower than 100°C. Therefore, to compare boiling points of different liquids we need a standard pressure.

c) <u>Normal Boiling Point</u> - the temperature at which the VP of a liquid is exactly equal to 1 atmosphere (760 torr). Therefore, the **normal** boiling point of water is 100°C!

As we have already seen, at a given temperature there is always a certain fraction of molecules with enough energy to escape into the gas phase. Why is it that we do not observe intense bubbling (a sign of boiling) until we reach the boiling point? If the temperature is lower than the boiling point, then the atmospheric pressure is larger than the vapor pressure, therefore the atmospheric pressure will *crush* any attempt of bubble formation.

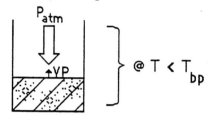

Let's now look at one particular example where we have drawn the Liquid/Gas equilibrium curves for three different liquids and see how we can identify the three liquids based on this data only. The three liquids chosen are ethanol (CH_3CH_2-O-H), water (H-O-H), and diethyl ether (CH_3CH_2-O-CH_2CH_3). The question becomes, which curve corresponds to which liquid?

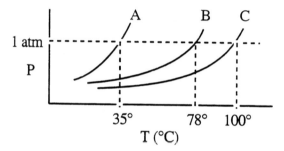

Before we attempt to answer this question we must first analyze the liquids in two respects: molecular weight and intermolecular forces of attraction. In general, the larger the molecular weight the higher the boiling point (for similar intermolecular forces) and ALSO, the stronger the intermolecular forces, the higher the boiling point. First of all, notice that there are no π-bonds on the O's, therefore they are all sp^3 with two lone pairs, thus angular and polar. Conclusion: the three molecules are POLAR. The ethanol and the water can H-bond whereas the ether has no polarized H's, $\therefore$ it cannot H-bond.

	M.W.	I.M.F.'s	
CH_3CH_2 O H	46	H-bonds	$\therefore$ B
H O H	18	H-bonds	$\therefore$ C
CH_3CH_2 O CH_2CH_3	74	Dipole-Dipole	$\therefore$ A

It is interesting to note that the heaviest molecule has the lowest boiling point because it has the weakest forces! How come then if both the ethanol and the water H-bond, the lighter one (water) has the higher BP? Because the water has <u>two</u> polarized H's that are available to form H-bonds whereas ethanol only has one! This example illustrates what your thinking process should be whenever you are trying to decide the order of increasing boiling point for different liquids. In general: intermolecular forces predominate *unless* there is a large difference in molecular weight. *Ex.:* I_2 is a *solid, non-polar* and the MW = 254.

And how about melting points? As the temperature of a liquid decreases two things decrease also:

    1. the vapor pressure.
    2. the motion of the molecules (kinetic energy).

Therefore, at a certain point the attractive forces between the molecules become very dominant and the translational motion stops. The molecules become fixed in specific geometric arrangements. The temperature at which this occurs is called the normal freezing point (or the *normal melting point*) if the external pressure is 1 atm (760 torr).

What is the relationship between the melting points and the intermolecular forces of attraction? We have stated previously that the melting point is not the best indicator since when a solid is molten you only break the crystal lattice and a few of the intermolecular forces. Even so, in general it is true that the stronger the intermolecular forces of attraction, the higher the melting point. The table below illustrates that point. Non polar substances (like He and octane) melt at lower temperatures than similar polar substances. Of course, the larger they are, the stronger the London forces and the higher the melting point. This explains why the octane molecule has a higher melting point than helium. On the other hand, it is surprising that water's melting point is higher than that of octane. The reason is hydrogen bonding, but it is still surprising considering that water weighs so much less than octane. Finally, we see that the ionic one has a much higher melting point.

Substance	MP	Comments
He	1 K	extremely weak IMF's, ∴ very low MP.
n-C_8H_{18} (straight chain)	216 K	non-polar, ∴ London forces; low MP for a heavy molecule.
H_2O	273 K	H-bonds; high MP for a light molecule.
NaCl	1074 K	ionic compound: pure electrostatic forces of attraction (ion-ion), ∴ very high MP.

How about the vapor pressure for a solid? Is there such a thing? Certainly! Energy is transmitted in a solid through vibrations. Therefore, some surface molecules have enough energy to escape directly into the gas phase, thus creating a vapor pressure. This is seen in the following equilibrium:

$$\text{Solid} \rightleftharpoons \text{Gas} \qquad \textit{Sublimation.}$$

However, the vapor pressure of a solid is inversely proportional to the strength of the intermolecular forces of attraction. Therefore, ionic crystals have a very low vapor pressure (negligible). In contrast, I_2 has an appreciable vapor pressure since the intermolecular forces of Iodine are London (thus weak). Therefore, surface I_2 should not have much trouble breaking away from the crystal and thus generating a vapor pressure.

It should be clear that most solids melt as we add heat, and upon further heating, the liquid will warm up, then it will boil and vaporize. Let us try to follow this process with a particular example, that of water, and see how much heat is required in order to carry out this process (and how it can be determined).

10.9 Heating Curves.

Let us take 10.0 g of ice all the way from -20°C (where it will be a chunk of ice) to +120°C (where it will be steam). How is the heat absorbed? What is the relationship between the heat absorbed and the obvious changes that will occur? We can divide this process into five steps (see graph below).

1. Warming up the solid from -20°C to 0°C. This is a physical change so:

$$q_1 = m\, c_{solid}\, \Delta T$$

2. Melting the ice at 0°C. We already know that this occurs at a constant temperature since all the heat (**q**) added is used up to break the crystal lattice and some of the intermolecular forces of attraction.

$$q_2 = \text{mass} \times \Delta H_{fus}$$

3. Warming up the liquid from 0°C to 100°C. Again, this is a simple physical process of warming up a substance, so:

$$q_3 = m\, c_{liquid}\, \Delta T$$

4. Vaporization of the liquid. Once more, this process occurs at constant temperature, 100°C.

$$q_4 = \text{mass} \times \Delta H_{vap}$$

5. Warming up the steam from 100°C to 120°C, the final temperature. This will depend on the specific heat of the gas:

$$q_5 = m\, c_{gas}\, \Delta T$$

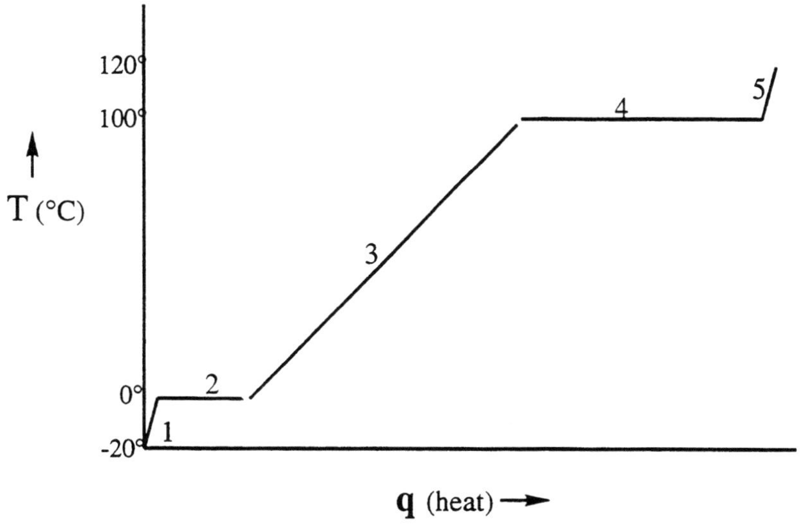

Let's analyze the plot above. Initially, we have pure ice which will slowly warm up as heat is absorbed. This is due to the fact that molecular vibration increases with increasing temperature. The part of the plot that indicates the absorption of heat for the ice is labeled **1**. Once you reach 0°C, an equilibrium between the solid ice and the liquid water is established. At this point, the energy added is going into breaking some of the hydrogen-bonds and the crystal lattice. This equilibrium will continue until all of the solid has melted. This is why we observe a plateau for this portion of the heating curve (labeled **2**). Once all the solid has liquefied, then the pure liquid absorbs heat in order to increase its temperature (section labeled **3**). Notice that the slope is smaller than it was in the first part of the curve. This is because the liquid has a much higher specific heat, thus it absorbs more heat in order for its temperature to increase a degree Celsius. How much longer will the liquid absorb heat? Until it reaches its boiling point (100°C). At this point, all the energy supplied will go into breaking essentially all of the hydrogen bonds in the liquid, which are significant, and change the water into the vapor phase. Thus, at the part labeled **4**, we have both the liquid and the vapor phases present at equilibrium. Once again, this equilibrium will be maintained until all of the liquid has been converted into steam. Finally, once all the liquid has been vaporized, the heat added will go into increasing the motion of the molecules in the gas phase, and thus, its temperature.

Let's calculate the total amount of heat (q_{tot}) absorbed by the 10.0 g of ice as we bring it from -20°C to +120°C. We need the following information in order to perform this calculation:

$$c_{solid} = 2.10 \text{ J/g deg}$$

$$c_{liquid} = 4.184 \text{ J/g deg}$$

$$c_{gas} = 2.00 \text{ J/g deg}$$

$$\Delta H_{fus} = 333 \text{ J/g}$$

$$\Delta H_{vap} = 2258 \text{ J/g}$$

◊ Notice that we can work with **q** or **ΔH**. As we mentioned before, under conditions of constant pressure $\Delta H = q$.

We should do all the calculations independently, that is, we should calculate the amount of heat required to go through each of the steps in the conversion. These are shown below.

1. $$\Delta H_1 = m \, c_{ice} \, \Delta T$$
$$= 10.0 \text{ g } (2.10 \text{ J/g deg}) [0 - (-20)\text{deg}]$$
$$= 420 \text{ J}$$

2. $$\Delta H_2 = \text{mass} \times \Delta H_{fus}$$
$$= 10.0 \text{ g } (333 \text{ J/g})$$
$$= 3{,}330 \text{ J}$$

3. $$\Delta H_3 = m \, c_{liquid} \, \Delta T$$
$$= 10.0 \text{ g } (4.18 \text{ J/g deg}) [100° - 0°] \text{ deg}$$
$$= 4{,}180 \text{ J}$$

4. $$\Delta H_4 = \text{mass} \times \Delta H_{vap}$$
$$= 10.0 \text{ g } (2258 \text{ J/g})$$
$$= 22{,}600 \text{ J}$$

5. $$\Delta H_5 = m \, c_{gas} \, \Delta T$$
$$= 10.0 \text{ g } (2.00 \text{ J/g deg}) [120° - 100°]$$
$$= 400 \text{ J}$$

In order to get the total heat absorbed by the chunk of ice to be converted into steam, we would have to simply add all five components (Hess' Law).

$$q_{tot} = \Delta H_{tot} = \Delta H_1 + \Delta H_2 + \Delta H_3 + \Delta H_4 + \Delta H_5$$

$$= 420 \text{ J} + 3{,}330 \text{ J} + 4{,}180 \text{ J} + 22{,}600 \text{ J} + 400 \text{ J}$$

$$= +30{,}930 \text{ J} = \mathit{30.93 \text{ kJ}} \checkmark$$

- Why is the heat of vaporization (ΔH_{vap}) so much larger than the heat of fusion (ΔH_{fus})?

    This turns out to be true for every liquid! The reason is that when you determine the heat of fusion (ΔH_{fus}) you are measuring the heat required to break the crystal lattice and some of the IMF's whereas when you determine the heat of vaporization (ΔH_{vap}) you are breaking essentially <u>all</u> the intermolecular forces of attraction to escape into the gas phase. It should take a lot more energy to break essentially every intermolecular force of attraction in order to bring the molecules into the gas phase, than to break just a few and the crystal lattice.

10.10 Phase Diagrams.

It would be advantageous if we could have a plot that compiles all the information we actually need for the different phases of a given pure substance. This is exactly what we obtain with a phase diagram. **Phase diagrams** are temperature/pressure diagrams for a given substance which illustrate the conditions under which that substance exists as a solid, a liquid or a vapor as well as conditions that bring about changes of state. Let's look as a first example at the phase diagram for water.[1]

---

[1] This phase diagram is not drawn to scale.

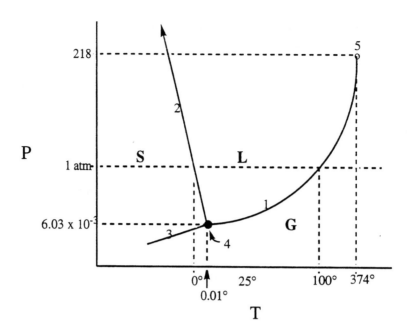

Interpretation.

a) We can see that under normal conditions (1 atm and 25°C) water is a liquid. The normal boiling point of water can be seen directly in the phase diagram (100°C when P = 1 atm). The normal melting point can also be obtained from the phase diagram (0°C when P = 1 atm).

b) Curve **1** represents the boiling curve (vapor pressure as a function of temperature), that is, the conditions of temperature and pressure upon which we observe an equilibrium between the liquid and the gas phase. It is also called the vapor pressure curve, since it tells you the vapor pressure at any temperature. This is exactly the same curve we saw back in Section 10.8.

c) Curve **2** represents the melting curve, that is, the conditions of temperature and pressure upon which we observe an equilibrium between the solid and liquid phase.

d) Curve **3** represents the sublimation curve, that is, the conditions of temperature and pressure upon which we observe an equilibrium between the solid and the gas phase. If you lower the pressure enough you can convert ice directly into vapor without melting it.

e) These three curves meet in point **4** and this is a very particular set of conditions under which the three phases (solid, liquid and gas) coexist in equilibrium. It is called the triple point. It must be pointed out that in general a triple point is one where a total of three phases coexist in equilibrium but they do not need to be solid, liquid and gas. Let's say that you have a solid which crystallizes in three different ways: simple cubic, face-centered cubic and body-centered cubic. You could therefore have a triple point where these three phases coexisted in equilibrium! Another point that we should make clear is that there is a very particular (just one) set of conditions to get a triple point (in the case of water it is 6.03×10^{-3} atm and

0.01°C). If the temperature and the pressure are not exactly equal to that, you will NOT have the three phases at equilibrium. The same thing is NOT true for melting, boiling or sublimation. There are lots of points (anywhere along the curve) in which you can have any two phases in equilibrium (*ex.* liquid and gas coexist in equilibrium at 25°C if the pressure is 23.8 torr; also at 100°C if the pressure is 760 torr). In more advanced courses, the number of possibilities under which we can have a certain number of phases at equilibrium is referred to as *degrees of freedom*.

f) Degrees of Freedom (F) are calculated with the Phase Rule, first proposed by J. Willard Gibbs[1] in 1876. It states that the number of degrees of freedom that we have in a system is given by the following expression:

$$F = C - P + 2$$

where C is the number of components and P is the number of phases. For the case of water, we have only one component. Our phase diagram shows various regions where only one phase exists, mainly the solid, the liquid and the gas phases. On these regions, we have two degrees of freedom (1 - 1 + 2). What is the meaning of that? This means that in order for us to unequivocally locate a point in the solid region (for example), we need to specify two variables: the pressure and the temperature. Specifying only one variable (let's say a pressure of 1 atm) will give us an infinite number of points where the solid is stable at 1 atm. In order for us to know which point we are referring to, we need to also specify the temperature. Along the equilibrium lines, let's say the liquid/vapor equilibrium line as an example, we have ONE degree of freedom (1 - 2 + 2). This means that when we have two phases at equilibrium, we only need one variable in order to know unequivocally which point we are referring to in the phase diagram. If we specify the pressure to be 1.00 atm, then the only point at that pressure where we have liquid and vapor at equilibrium occurs at 100°C. Finally, at the triple point, where all three phases are at equilibrium, there are ZERO degrees of freedom (1 - 3 + 2) and there is no need to specify anything because there is only one such point in the graph.

g) The boiling curve ends in point **5** which is called the critical point. For water it occurs at 374°C and 218 atm. Above this critical temperature (374°C), no matter how hard you try to compress a gas it will not liquefy. At the critical point, the densities of the liquid and the gas are identical so you cannot distinguish between the two phases.

h) The slope of the melting line is negative. This implies that as you increase the pressure at

---

[1] He was a professor of Mathematical Physics at Yale from 1871 to 1903. We credit him with establishing the foundations of many areas of thermodynamics, particularly as they apply to Chemistry.

any given constant temperature, the ice will melt. Therefore, the liquid must be more dense (more compact) than the solid since increasing the pressure is equivalent to getting the molecules closer together. As you get the molecules closer and closer (denser), the ice is converted into a liquid.

The reason that solid water is less dense than liquid water can be found in its structure. In the solid, each water molecule is H-bonded to four other molecules. Since the oxygen in water is hybridized **sp^3**, then there is a slight positive charge in each hydrogen and a slight negative charge in every lone pair in the oxygen and they are all at about 109° from one another. The four H-bonds formed should therefore be in a tetrahedral configuration. The common form of ice shown below has open, cage-like channels running through the structure, and the water molecules can get closer together if they break the framework in the solid and pack together in a more random manner in the liquid phase. This is the reason why ice is less dense than liquid water and floats on the surface.

This concept has been used in the past *erroneously* to explain ice-skating. People thought that when you ice skate you are applying a lot of pressure on the tip of the blade, so ice melts and you slide in the water. As soon as you skate past over a given point, the water freezes right back (because the pressure is released). This is not true because a normal person is not heavy enough to increase the pressure on the solid enough to make it cross the equilibrium line. Furthermore, researchers[1] at the University of California at Berkeley have observed that bombarding a thin layer of ice with a stream of electrons did not scatter in the way they would have had it been a regular solid surface. What they saw was a scattering pattern that was consistent with a constantly changing surface. The most obvious explanation for this phenomenon is that the surface layer of ice is not really frozen, but it is rather a very thin film of liquid water. This film would therefore serve as a lubricant. They also discovered that at approximately –22°C, the phenomenon ceases to exist and ice then behaves like any other solid. That is, at that low temperature the thin film of water finally freezes, thus it then behaves like any other solid.

---

[1] Gabor Somorjai (chemist) and Michel Van Hove (physicist).

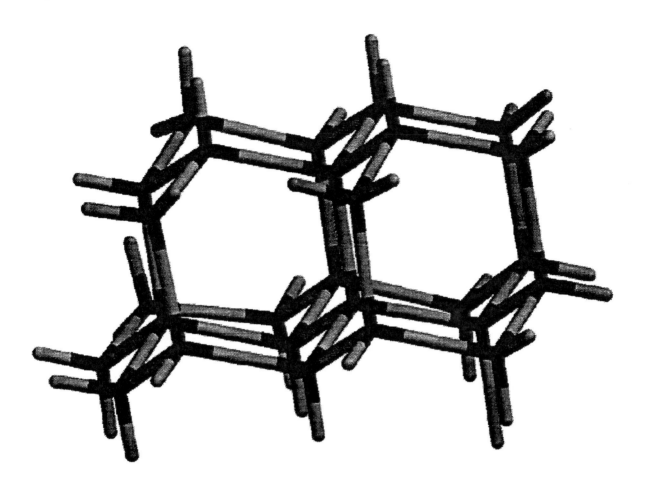

We should point out that most phase diagrams do not look like the one we just described (the one for water). Essentially all phase diagrams have a positive slope for the solid-liquid equilibrium curve. This would immediately imply that the solid is more dense than the liquid phase. This is actually what you normally would expect for any substance; water happens to be the exception due to its unusual structure. Remember that water has a phase diagram where the solid-liquid equilibrium line has a negative slope and almost all other substances will have this solid-liquid equilibrium line with a positive slope.[1]

Below you can see the phase diagram for carbon dioxide gas and we will use this one as an example of a typical one. There are more differences (than just the slope of the melting curve) between the phase diagram of water and that of carbon dioxide. Another obvious one is the fact that at ambient conditions, carbon dioxide is a gas and it cannot be liquefied unless its pressure is increased dramatically. An analysis of the phase diagram will elucidate these points, as seen in the observations following the figure.

---

[1] Bismuth and antimony also have solids that are less dense than their liquids.

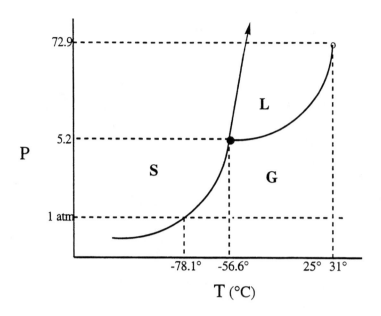

Interpretation of the phase diagram of CO_2

a) Under normal conditions (25°C and 1 atm) carbon dioxide is a <u>gas</u>.

b) At the triple point (5.2 atm and -56.6°C), the three phases (in this case solid, liquid and gas) coexist at equilibrium.

c) Lowering the temperature of carbon dioxide gas at a constant pressure of 1 atm will convert it into a solid directly (@ -78.1°C), therefore at -78.1°C the solid carbon dioxide is in equilibrium with the gas with no liquid present. That is why they call this DRY ICE.

d) You cannot obtain liquid carbon dioxide at pressures lower than 5.2 atm. This gives you an idea of how weak the intermolecular forces for carbon dioxide are. It takes quite a bit to finally have these forces take control and thus become a liquid. Even at 1 atm pressure you must lower the temperature all the way down to -78.1°C to finally obtain a condensed phase (the solid).

e) The critical point for carbon dioxide occurs at 72.9 atm and 31°C. This means that you can never liquefy carbon dioxide above 31°C, no matter how much you increase the pressure.

<u>The Sulfur System.</u>

Sulfur has two possible and stable solid forms, the rhombic form is stable at normal temperatures, whereas the monoclinic form is stable at higher temperatures. The phase diagram for Sulfur looks different because we have to include both solid forms, the liquid and the vapor form of the substance. What are the possible equilibria established by these? Well, we can have the following scenarios: Two phases at equilibrium: rhombic solid/vapor, monoclinic solid/vapor, rhombic/monoclinic, rhombic/liquid, monoclinic/liquid, and liquid/vapor; Three phases at equilibrium: Rhombic/monoclinic/liquid, rhombic/liquid/vapor, monoclinic/liquid/vapor, and rhombic/monoclinic/vapor; Four phases at equilibrium: Rhombic/monoclinic/liquid/vapor.

According to the phase rule, we cannot have less than zero degrees of freedom, and since we have only one component, all four phases CANNOT be at equilibrium (1 - 4 + 2 = -1), therefore, we never will have a quadruple point in a phase diagram for a one-component system. A schematic form for the phase diagram for sulfur is shown below. Answer the following questions: How many triple points do we see in this phase diagram? Can you locate the critical point? Will the Rhombic solid sink or float in liquid Sulfur?

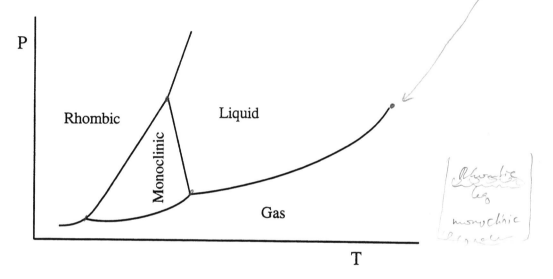

Answers: Three; At the end of the liquid/gas equilibrium line; Sink.

Chapter 11: Solutions and Their Properties.

11.1 Classification of Solutions.

We defined solutions the first day of classes as homogeneous mixtures, that is, two or more components mixed in a single phase. Why is it so important to understand solutions? Because 90% of all chemistry occurs in liquid solutions (including most of the body chemistry).

As we shall see shortly, certain properties of solutions depend on the relative amount present of the individual components. For example, the melting point of a solvent is affected by the amount of solute dissolved in it. It is observed that as the concentration of solute increases, the melting point of the solution decreases (later on we will discuss the reason as to why this happens).

Concentration of NaCl (g/100 mL of water)	Melting Point (°C)
0 (pure water)	0.0°
0.2 (blood)	-0.2°
Sea Water	-1.5°
Dead Sea	-3.7°
Great Salt Lake	-15.8°
Natural Salt Brine (ª25%)	-23.9°

How does the solution process work at the molecular level? Basically, the solvent molecules come to the surface of the crystal and pull out the molecules, atoms, or ions (depending on the type of crystal we are referring to at that point), and surround them in an orderly fashion. At this point, they are taken away from the vicinity of the crystal and becomes an independent entity (a solvated unit[1]). Of course, the energetics must be such so that the new attractions formed (between the solvent molecules and the solute) will be at least close to equivalent with those between the solute particles themselves. In other words, for a solute to be dissolved three things must take place: 1) we must break some solvent-solvent attractions, 2) we must break some solute-solute attractions, and 3) we must form some solvent-solute attractions. The overall energy process can be either negative or positive. When the newly formed attractions are stronger than the older ones it is negative and it is positive when the opposite[2] is true. At first, it seems incredible that a solution in which the ΔH of the solution process is positive even takes place, since it seems to be destabilizing. However, as we will learn in

---

[1] These may include various solvent layers around the molecule, atom or ion.
[2] The newly formed forces are weaker than the older ones.

Chapter 15, there are basically two factors that control whether a process occurs or not. One of these is the one we discussed above (the energetics of the process). The other one deals with randomness. The actual process of dissolving the solute particles into the solution is normally accompanied by an increase of the randomness of the system. It is the combination of these two factors that will explain why a solute actually dissolves in a given solvent.

Normally there is a limit as to how much of a solute you can dissolve in a given quantity of solvent at a given temperature. This limit is referred to as the *solubility*. We can actually define solubility as the maximum concentration of a solute that can be mixed with a given quantity of a solvent and still maintain a homogeneous mixture.[1]

Before we start giving examples let's discuss the <u>first law of solubility</u>: LIKE DISSOLVES LIKE! What do we mean by *like*? We are referring to substances of similar intermolecular forces of attraction. Basically, we could rephrase this and state that polar solvents will dissolve polar solutes and that non-polar solvents will dissolve non-polar solutes. The following table illustrates this point.

Solvent	Soluble Solutes	"Insoluble" Solutes
H_2O (polar, H-bonded)	CH_3CH_2OH NaCl $CH_3CH_2\text{-}O\text{-}CH_2CH_3$[2]	CCl_4 oil
C_6H_6 (non-polar)	I_2 CCl_4 oil	NaCl H_2O

Notice that one of the substances that is substantially soluble in water is salt (NaCl). Although the forces for NaCl are much stronger than the hydrogen bonds in water, it is soluble because the surface ions of the NaCl are less strongly held than the ones further inside the crystal. If we ever wanted to dissolve a salt (not just NaCl) in a given solvent, we would choose a highly polar solvent. Keep in mind that for all practical purposes, salts are insoluble in non-polar solvents.

Which solvent is most suited for dissolving NaCl, water or CH_3OH? We could look at the dipole moments of these two liquids to determine this, but the dipole moments of most molecules have not been measured. Instead, we measure a quantity called the **dielectric constant** for a given solvent. We can roughly define the dielectric constant as the ability of the

---

[1] Still maintaining a single phase.

solvent to isolate opposing charges from one another. In other words, the dielectric constant of a solvent basically measures its polarity. Therefore, a solvent with a high dielectric constant should be a better solvent for ionic species than a solvent with a lower dielectric constant. Keeping that in mind and looking at the table below, it is clear that water is a better solvent to separate charges in solution than methanol.

Dielectric Constants for a Few Common Solvents.

Solvent	Formula	Dielectric Constant
Water	H_2O	80
Methanol	CH_3OH	33
Ethanol	CH_3CH_2OH	24

- Why is the dielectric constant of Ethanol lower than that of methanol?

When we look at these two molecules, we notice that these have basically two parts: a polar head (the –OH) and a non-polar tail (the hydrocarbon part). The longer the hydrocarbon tail, the less polar the molecule will be and the lower the dielectric constant. In conclusion, the ethanol would be a less suitable solvent to dissolve salts than methanol, and both of them are significantly less suitable than water.

When we say that a substance is soluble or insoluble, what are we referring to exactly? Nothing is 100% **in**soluble. There is always a certain degree of solubility which can be determined experimentally.

However, some liquids are homogeneous in all proportions (infinite solubility). We say that they are totally **miscible**. As an example take the case of water and ethanol (CH_3CH_2-OH).

	% Ethanol
Beer	≈ 3%
Wine	≈ 10%
Geritol	≈ 20%
Gin	≈ 30%
Bourbon	≈ 45%
White Lightning	≈ 75-80%
Grain Alcohol	95%

The reason for their total miscibility is due to their similar intermolecular forces of attraction. Water is H-bonded and so is ethanol. The strength of an attraction between one water molecule and another water molecule is pretty much the same as that between a water molecule and an ethanol molecule, therefore energetically this is feasible. In contrast, oil and water do not mix to any appreciable extent. Why is that so? Oil is non-polar and thus its intermolecular forces are weak London forces. Water's intermolecular forces are much stronger so you need to add a lot of energy to break them apart. The formation of the intermolecular forces between water and oil molecules would release very little energy (weak attraction) and therefore energetically this is not very probable. Thus, water and oil form two *immiscible* layers.

As we just said, not all substances are homogeneous in all proportions, i.e. many substances exhibit only limited solubility. It is important to specify the temperature at which the solubility is measured since the actual amount of solute dissolved depends on it.

Substance	Solubility (g/100 mL of water @ 20°C)	
Sucrose ($C_{12}H_{22}O_{11}$)	204.00	∴ Soluble
Li_2CO_3	1.33	∴ Slightly Soluble
CCl_4	0.08	∴ Very Slightly Soluble
Octadecane ($C_{18}H_{38}$)	0.00*	∴ "Insoluble"
component of motor oil	* not exactly zero, but zero to two decimal places.	

When the concentration of a substance equals its solubility, we say that the solution is *saturated*. In general, increasing the temperature increases the solubility of a substance, however *this is not always the case*. This will depend on whether the ΔH of solution, is endothermic or exothermic. We will learn about how to predict exactly this in Chapter 16, when we talk about Le Chatelier's Principle. Let's look at the example of the solubility of common table sugar (sucrose) in water.

Sucrose in Water.

Temperature	Solubility (g/100 mL of water)
20°C	204.0 g
80°C	362.0 g

This table tells us that 100 mL of water have the ability to dissolve 204.0 g of sucrose at 20°C, and if we warm up the solution, we will be able to dissolve an extra 158 g of sucrose. Conversely, if we have 100 mL of a saturated[1] solution of sucrose at 80°C and allow it cool

---

[1] A solution with the maximum concentration at that temperature.

down to 20°C, we would observe the precipitation of 158 g of sugar from the solution.

11.2 Henry's Law.

How about if we were to dissolve gases in polar liquids? In general, we do not expect gases to be very soluble in a polar liquid, especially a liquid with a high dielectric constant like water. We have already discussed that solubility is dependent on the kinds of intermolecular forces of attraction. Gases have weak intermolecular forces whereas water's intermolecular forces are strong hydrogen-bonds. Therefore a gas like nitrogen (N_2) is very "insoluble"; only about 1.8×10^{-4} g of N_2 dissolve in 100 g of water at 20°C and under one atmosphere of pure nitrogen gas! This is typically the case *unless* the gas reacts with water. If there is a reaction taking place, then the solubility would be much higher. As an example, look at the case of carbon dioxide.

$$CO_2 + H_2O_{(\ell)} \rightleftharpoons H_2CO_{3(aq)} \leftarrow \text{carbonic acid!}$$

We would therefore expect the solubility of CO_2 gas in water to be much higher than that of N_2 gas, and it indeed is!

Solubility of CO_2 = 0.15 g CO_2 / 100 g of water @ 20°C and 1 atm of CO_2.

The solubility of a gas in a liquid depends on two factors: the temperature and the pressure of the gas over the solution. This is why these numbers above specify both of these factors.

The solubility of gases in liquids usually decreases with increasing temperature. This is because at a higher temperature, the gas molecules have more energy and are able to escape from the solution. As an example think about a pan of water close to the boiling point (≈90°C). You observe bubbles forming and adhering to the sides or the walls of the container. These can not be water vapor since you have not reached the boiling point yet. These bubbles consist of air[1] that was dissolved in the water at the lower temperatures and as the solution gets warmer it is less soluble, so it comes out of solution.

William Henry[2] studied the dependence of the solubility of a gas in a solvent as a function of its partial pressure. He found that at a given temperature, the solubility of gases which do not react extensively with the solvent is directly proportional to the partial pressures of

---

[1] Mostly nitrogen and oxygen gases.
[2] An English chemist who lived from 1775 until 1835.

the gases above the solution. Calling the gas **A**,

$$[A]_{\text{in solution}} = k\, P_A \qquad \leftarrow \textit{Henry's Law}$$

Therefore, if the solubility of a gas is known (or experimentally determined) at one pressure, it can be calculated at any other pressure.

The best example is probably a can of soda or "pop". Before we open it, we have CO_2 at a pressure of about 2 atm above the liquid. Since CO_2 forms carbonic acid in water, a lot of the CO_2 is dissolved in the soft drink. H_2CO_3 is acidic and this gives the characteristic taste to carbonated beverages. When you open the can and release the pressure, the partial pressure of CO_2 will go down from 2 atm to about 0.03 atm (the normal partial pressure of CO_2 in air). Therefore, the solubility of the CO_2 decreases dramatically in the drink and, if you wait a while, it will go "flat". This means that most of the carbon dioxide will be gone, which implicitly also means that the carbonic acid is also gone.

- The solubility of N_2 in water is 2.2×10^{-4} g per 100 of water at 20°C when the pressure of N_2 above the solution is 1.2 atm. What is the solubility of N_2 at 20°C when the pressure is 10 atm?

We can use Henry's Law, which is:

$$[N_2]_{\text{soln}} = k\, P_{N_2}$$

To avoid using the constant (k), we can derive the following relationship:

$$\frac{[N_2]_1}{P_1} = \frac{[N_2]_2}{P_2}$$

Substituting the values, we get:

$$\frac{2.2 \times 10^{-4} \text{ g}/100 \text{ g water}}{1.2 \text{ atm}} = \frac{x}{10 \text{ atm}}$$

$$\therefore [N_2]_2 = 0.0018 \text{ g per 100 g of water.} \quad \checkmark$$

11.3 Concentration Units.

It is important to understand different concentration units if we want to discuss some quantitative aspects concerning solubility or other properties. We have already discussed and used three common concentration units[1] and now we will learn two more. From these five, only two are temperature dependent since they are dependent on the volume of the solution (the volume changes with the temperature, although just slightly).

1. Temperature Dependent.

    a) *Molarity (M)*

    $$M = n_{solute} / V_{solution \text{ (in L)}}$$

    b) *Normality (N)*

    $$N = \text{no. of equivalents} / V_{solution \text{ (in L)}}$$

    An equivalent is defined in terms of either acid/bases or redox. We will only mention the acid/base equivalents at this point. An acid equivalent is that fraction of a mole that donates one mole of H^+ ions. Similarly, a base equivalent is that fraction of a mole that donates one mole of OH^- ions. For example, HCl can donate one mole of H^+, therefore one mole has one acid equivalent. Similarly, one mole of H_2SO_4 has two equivalents of H^+. Let us take a simple example so that we can learn the relationship between Molarity and Normality.

- Determine the normality of a 2.50 M solution of phosphoric acid (H_3PO_4).

    Just looking at the molecular formula, it should be evident that this acid has three acidic Hydrogens, that is, it is able to donate three protons (H^+). Therefore, one mole of phosphoric acid has three equivalents of acid. We may use this fact as a conversion factor.

    $$\frac{2.50 \text{ moles } H_3PO_4}{1 \text{ L of solution}} \times \frac{3 \text{ equivalents of acid}}{1 \text{ mol } H_3PO_4} = 7.50 \text{ equivalents/L} = 7.50 \text{ N}$$

2. Temperature Independent.

    a) *Mass Percent*

    $$\% A = [\text{grams}_A / g_{total \text{ of solution}}] \times 100$$

    where A is the solute.

    b) *Mole Fraction (X)*

    $$X_A = n_A / n_{total \text{ of solution}}$$

    c) *Molality (m)*

---

[1] Molarity, mole fraction and mass percent.

Chapter 11: Solutions and Their Properties.

$$m = n_{solute} / kg_{solvent}$$

This last concentration unit (m) is the only one of the five that depends directly on the amount of SOLVENT used to prepare the solution! Molality will therefore be very important when we talk about some of the properties of solvents that are affected by the presence of a solute. Before we go on any further, we should practice on interconverting these concentration units. Remember that the only way to do this easily is by memorizing the definitions for all the concentration units, as presented on the previous page.

- A solution of 20.0% ethanol, C_2H_5OH (MW = 46.1) and 80.0% water has a density of 0.966 g/mL at 25°C. Find the mole fraction (X), the molality (m) and the molarity (M) of ethanol in the solution.

What is the meaning of a 20.0% ethanol solution? It means that for every hundred grams of solution, we have 20.0 g of ethanol and 80.0 g of water. Since for most these units we need to find the moles of each component, let's start by doing just that.

$$20.0 \text{ g } C_2H_5OH \quad \text{and} \quad 80.0 \text{ g water}$$
$$\Downarrow \quad\quad\quad\quad\quad\quad\quad\quad \Downarrow$$
$$0.434 \text{ mol } C_2H_5OH \quad\quad 4.44 \text{ mol water}$$

Now we can proceed to determine each of the concentration units based on their definitions.

$$X_{C_2H_5OH} = \frac{n_{C_2H_5OH}}{n_{C_2H_5OH} + n_{H_2O}} = \frac{0.434}{0.434 + 4.44} = 0.0890$$

$$m = \frac{n_{C_2H_5OH}}{kg_{water}} = \frac{0.434 mol}{0.0800 kg} = 5.42m$$

$$M = \frac{n_{C_2H_5OH}}{V_{solution}} = \frac{0.434 mol}{??}$$

Now, in order to calculate the molarity of this solution, we need to figure out first the volume of the solution. That information was not provided directly but we were given more than enough information in order to do this. The volume of the solution can be calculated from the density of the solution (0.966 g/mL). Since we have assumed all along that we have 100.0 g of solution, then:

$$100.0 g_{solution} \times \frac{1 mL_{solution}}{0.966 g_{solution}} \times \frac{10^{-3} L}{1 mL} = 0.104 L$$

Finally,

$$[C_2H_5OH] = \frac{0.434 mol}{0.104 L}$$

$$\therefore [Ethanol] = 4.17 M \quad \checkmark$$

- Determine the Molarity for an aqueous NaOH solution that is 5.50 molal given that the solution's density is 1.25 g/mL.

    In order to solve this problem, we start by asking ourselves, what is the meaning of 5.50 molal? This means that we have 5.50 moles (or 220 g) of NaOH in every kilogram of water (1,000 g). In order to determine the volume of the solution (NaOH PLUS water) we need to know the total mass of the solution. This should not be a problem, since we know that we have 220 g of NaOH and 1,000 g of water for a grand total of 1,220 g of solution.

$$1,200 g_{solution} \times \frac{1 mL_{solution}}{1.25 g_{solution}} \times \frac{10^{-3} L}{1 mL}$$

$$V = 0.960 L \quad \checkmark$$

Finally, $\quad$ M = moles of NaOH/Volume of solution

$$M = 5.50 \text{ mol} / 0.960 \text{ L}$$

$$\therefore M = 5.73 M \quad \checkmark$$

- Determine the Molarity of a 2.34 N solution of $Ca(OH)_2$.

    Calcium hydroxide is a base, and from the formula we can see that 1 mol $Ca(OH)_2$ = 2 base equivalents!

$$\frac{2.34 equivalents}{1 L} \times \frac{1 mol}{2 equivalents} = 1.17 mol/L$$

$$\therefore [Ca(OH)_2] = 1.17 M \quad \checkmark$$

Chapter 11: Solutions and Their Properties.

- How many mL of 3.33 N Mg(OH)$_2$ contain 23.4 g of Mg(OH)$_2$? MW = 58.3.

    Once more, one mole of Mg(OH)$_2$ has two equivalents of base. So,

$$\frac{3.33 \text{ equivalents}}{1L} \times \frac{1 \text{ mol}}{2 \text{ equivalents}} = 1.67 \text{ mol}/L$$

We can use this information as a conversion factor that tells us that in every liter of solution, we have 1.67 moles of magnesium hydroxide.

$$23.4 g_{Mg(OH)_2} \times \frac{1 \text{ mol}_{Mg(OH)_2}}{58.3 g_{Mg(OH)_2}} \times \frac{1000 mL}{1.67 \text{ mol}_{Mg(OH)_2}}$$

$$\therefore \text{ Volume } = 240 \text{ mL} \quad \checkmark$$

- How many mL of 1.250 N H$_2$SO$_4$ are required to react completely with 25.00 mL of 1.765 N NaOH?

$$2 \text{ NaOH} + H_2SO_4 \Rightarrow 2 H_2O + Na_2SO_4$$

    Since we are working with equivalents, it is clear that at any point in this reaction, the equivalents of acid are equal to the equivalents of base. How do we get equivalents? By multiplying the normality by the volume. So,

$$N_{acid} \times V_{acid} = N_{base} \times V_{base}$$

$$\therefore V_{acid} = \frac{N_{base} \times V_{base}}{N_{acid}}$$

$$= \frac{1.765 N (25.00 mL)}{1.250 N}$$

$$V_{acid} = 35.30 \text{ mL} \quad \checkmark$$

11.4 Net Ionic Equations.

When we write a balanced chemical equation, often there is a lot more information that we really want. Many times we are only interested in knowing only what is exactly going on, that is, what is happening as we move from the reactants into the products. In the case of ionic substances, these are theoretically dissociated 100% in water.[1] In here the word dissociation simply tells us that the ions are completely separated from one another by the solvent. In other

---

[1] Due to the large dielectric constant of the solvent.

words, if we were talking about a salt like NaCl (which is very soluble), then we assume that once it goes into solution, all of the ions are separated well by the water. On the other hand, if we were talking about a slightly soluble salt, most of this salt is not going to go into solution and will remain as a solid in the bottom of the flask, but whatever microscopic amount is dissolved, that will be totally dissociated into the ions and separated by the water. Because of this, we give salts the name of strong electrolytes.

How can we distinguish the soluble salts from the slightly soluble? Typically by the subscript written next to the salt in the chemical equation. If it says aqueous, then it is already dissolved in water (it is soluble). If, on the other hand, it says solid, that means that this is a slightly soluble salt and it is remaining as a solid in the bottom of the flask.

Ex. $NaCl_{(aq)}$ forms $Na^+_{(aq)} + Cl^-_{(aq)}$ ions

$K_3PO_{4(aq)}$ forms $3 K^+_{(aq)} + PO_4^{-3}{}_{(aq)}$ ions

$AgCl_{(s)}$ does not dissociate.

$BaSO_{4(s)}$ does not dissociate.

It is also advantageous to know two simple rules that will allow you to tell some of the salts that are always soluble (more than 1 mg of the salt per liter of solvent). These rules are:

1. All <u>alkali metal salts</u> (like **NaCl** or **K₂CO₃**) are soluble in water.

2. All <u>nitrates</u> {like **AgNO₃** or **Al(NO₃)₃**} are soluble in water.

What happens when we mix a solution of $AgNO_{3(aq)}$ with one of $KCl_{(aq)}$? We obtain a white precipitate, which is $AgCl_{(s)}$. The overall reaction can be seen below.

$$AgNO_{3(aq)} + KCl_{(aq)} \Rightarrow AgCl_{(s)} + KNO_{3(aq)}$$

Let us proceed to split all strong electrolytes into the individual ions. Remember that the key is to look at the subscripts. If it says aqueous, then it is in solution, and if on top it is a salt, then it will be split into the ions. For the example above, we would get:

$$Ag^+ + NO_3^- + K^+ + Cl^- \Rightarrow AgCl_{(s)} + K^+ + NO_3^-$$

Notice that the AgCl was not split into ions because it had as a subscript an **s**, which means that it is an insoluble solid. Now, let's cancel any ions that are common on either side of the reaction. That takes care of two ions: the potassium and the nitrate ions, leaving the following net ionic equation.

$$Ag^+ + Cl^- \Rightarrow AgCl_{(s)} \quad : \textit{Net Ionic Equation}$$

The reason we can delete the potassium and the nitrate ions is that they are actually not involved at all in the "reaction". They are just there, in solution, "observing" what is going on between the ions that are actually reacting. We refer to these as *spectator ions*.

11.5 Raoult's Law (1886).

As we mentioned before[1], the properties of a solvent (such as the melting point or the vapor pressure) are modified by the presence of a solute. We mentioned that as we add more and more salt to water, the melting point of the solution keeps getting lower and lower. Specifically there are four such properties (like the melting point) that are affected by the presence of a solute. As we shall see, these properties will be affected by the *number of particles of solute* dissolved and NOT on their nature. In other words, the most important thing is to keep in mind is how much solute is dissolved in the solvent and not what the solute actually is. Such properties are referred to as **colligative** properties.

We will start our discussion with the *lowering of the vapor pressure*. The addition of a non-volatile[2] solute to a liquid decreases the vapor pressure of the solvent at any temperature. This is because the surface no longer has only solvent molecules and therefore fewer molecules of solvent escape into the gas phase at a given temperature. The vapor pressure of the solvent must therefore be a function of how much solvent we have relative to the whole solution ($X_{solvent}$).

That was the line of reasoning that Raoult[3] followed. Let's identify the solvent with the letter A. He proposed the following relationship, which holds true at any given temperature.

$$P_A = X_A \times P_A^\circ$$

This is known as Raoult's Law. In this equation, P_A is the vapor pressure of A over the solution at a given temperature, X_A is the mole fraction of A in solution, and P°_A is the vapor pressure of pure[4] A at that same temperature.

---

[1] See Section 11.1.
[2] A non-volatile substance is one that has a negligible vapor pressure.
[3] François M. Raoult was a French chemist who lived from 1830 until 1901.
[4] The implication of pure A is that there is nothing else mixed with it.

Let's take the case of water as the solvent. At 25°C, we have seen that the pure vapor pressure ($P°$) of water[1] is 23.8 torr. What happens when we add just a little bit of a given solute to water? Well, the mole fraction of the water will be just a little less than one[2], therefore the vapor pressure of water over that new solution would be just a little less than 23.8 torr. To get exact numbers, we would need to know exactly how much solute we added, so that we could then calculate the actual mole fraction.

This is a very straight forward law, very easy to understand and work with. However, when we go into the laboratory and measure the vapor pressure of the solvent over the solution, in most cases we observe that it is not followed exactly. We therefore say that Raoult's Law works only for **ideal** solutions. And what do we mean by an ideal solution?

An ideal solution can be defined as one that obeys Raoult's Law. It is one which can be formed from its components with no evolution or absorption of heat and whose total volume is exactly equal to the sum of the volumes of the components. At first, that may sound a little bit strange. As we mentioned before, when we dissolve a solute in a solvent some intermolecular forces of attraction of the type solvent-solvent and solute-solute will be broken (heat absorbed) to form solute-solvent intermolecular forces (heat released). If the ΔH of this process is zero (no evolution or absorption of heat), that implies that the intermolecular forces of the solvent and the solute are IDENTICAL. This is the case for ideal solutions. Pictorially (A = solvent, B = solute):

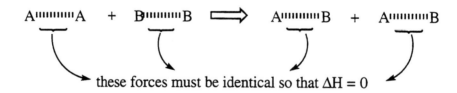

these forces must be identical so that $\Delta H = 0$

In real solutions the total volume is not equal (close, but not equal) to the sum of the volumes of the components. In other words,

10.0 mL of water + 10.0 mL of ethanol ≠ 20.0 mL of solution

Actually, it is a little less than 20.0 mL of solution. Again, this is due to the difference in their intermolecular forces of attraction (they are very close but not identical).

It should be pointed out, however, that all real solutions approach ideal behavior when

---

[1] See Section 10.7.
[2] Remember that: $X_{water} = n_{water} / \{n_{water} + n_{solute}\}$

they are *very dilute* (low concentration of solute). So, independently of what kind of forces we have in a real solution, as the concentration of the solute decreases, the solution will behave more ideally.

In general, we can say that real solutions exhibit either positive deviations or negative deviations from Raoult's Law.

a) Positive Deviations – These are observed when the pressure over the solution is actually higher than the one calculated with Raoult's Law. This would be the case when we mix two molecules with very different intermolecular forces of attraction, for example, the case of a polar molecule {like acetone, which is $(CH_3)_2$-C=O} with a non-polar molecule like CS_2. Since the new intermolecular forces (those between the carbon disulfide and the acetone) are going to be weaker than some of the initial forces, then more molecules will have the ability to escape into the gas phase than we thought initially (from Raoult's Law).

$$:\ddot{S}=C=\ddot{S}: \quad + \quad \underset{CH_3 \quad CH_3}{\overset{\overset{\ddot{O}:}{\|}}{C}}$$

non-polar polar

b) Negative Deviations – These are observed when the opposite is true, that is, when the pressure over the solution is actually less than that calculated from Raoult's Law. This would be the case when you mix two molecules that form a stronger attraction between each other than between themselves. In the example seen below, we can see the combination of chloroform ($CHCl_3$) with acetone. Both of these molecules are polar, but both also fail to form hydrogen bonds with molecules like themselves, because each one is lacking what the other has. In the case of the chloroform, it has a polarized hydrogen, but no small, highly electronegative atom (F, O, or N). On the other hand, the acetone has the oxygen atom, but does not have polarized hydrogens. The combination of these two molecules will form a hydrogen bond. Since this attraction is stronger than the original forces, fewer molecules will escape into the gas phase.

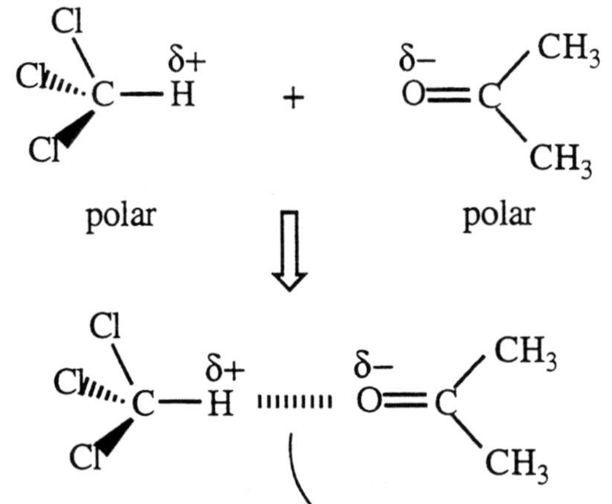

[Chemical structures showing CHCl₃ (polar) + (CH₃)₂C=O (polar) forming a hydrogen-bonded complex]

Strong H-"bond" (IMF) in solution!

11.6 Raoult's Law for Two Volatile Components.

The last two examples we saw consisted of two volatile components. How does Raoult's Law apply in such a case? Components A and B are both volatile so both have a measurable vapor pressure at a given temperature: $P_A°$ and $P_B°$. The total pressure over the solution will be given by the sum of the partial pressures (Dalton's Law):

$$P_{total} = P_A + P_B$$

but according to Raoult:

$$P_A = X_A P_A°$$

and: $$P_B = X_B P_B°$$

$$\therefore \boxed{P_{total} = X_A P_A° + X_B P_B°} \checkmark$$

This equation describes the total pressure over a liquid solution of liquids A and B, which in the liquid have the composition given by their respective mole fractions.

- Consider the two liquids C_6H_6 (benzene) and CCl_4 (carbon tetrachloride) at 25°C. These are both non-polar liquids and have very similar intermolecular forces of attraction, so they will form essentially an ideal solution. At this temperature: $P°_{bz}$ = 270 torr and $P°_{ct}$ = 320 torr. Let's prepare a liquid solution consisting of 1.00 mol of benzene and 3.00 mol of CCl_4. Determine the total pressure over this solution.

The first thing we may notice is that both of these liquids are non-polar and they actually have very similar intermolecular forces of attraction. This makes for a solution that is essentially ideal, therefore, it should obey Raoult's law. Another thing that we may notice is that since the pure vapor pressure of benzene is lower than that of carbon tetrachloride, then the intermolecular forces of attraction for the benzene must be just a little stronger than those of the carbon tetrachloride.

Since we want the total pressure over the liquid solution, we should determine both mole fractions first.

$$X_{bz} = n_{bz} / (n_{bz} + n_{ct})$$

$$= 1.00 / 4.00 = 0.250 \checkmark$$

and: $X_{ct} = 1 - X_{bz} = 0.750 \checkmark$

Since we have already established that both liquids are volatile, then the total pressure over the solution will be given by the expression we derived earlier.

$$P_{total} = X_{bz} P_{bz}° + X_{ct} P_{ct}°$$

Substituting the numbers we have calculated for the mole fractions, and the given pure vapor pressures, we get the following.

$$= 0.250 \, (270 \text{ torr}) + 0.750 \, (320 \text{ torr})$$

$$= 67.5 \text{ torr} + 240 \text{ torr}$$

$$\therefore P_{total} = 308 \text{ torr} \checkmark$$

We should point out that these two numbers (67.5 torr and 240 torr) represent the actual vapor pressures of the two substances, benzene and carbon tetrachloride, respectively, over the solution. This is important because with this information we could theoretically determine the composition of the vapor phase.

How can we do that? Can we determine the composition of the vapor phase? Yes, we can. Let's derive an equation that will enable us to do so. For that particular purpose, we will continue with the example we started above.

- Determine the composition of the vapor phase over the liquid solution in the previous example.

Is the composition of the vapor phase identical with that of the liquid solution? No!

It cannot be, since these liquids have different pure vapor pressures. Therefore, one of them has a higher tendency to escape into the gas phase than the other. How could we determine the mole fraction of benzene in the gas phase? After all, we have no idea as to how many moles of benzene will have escaped from the liquid into the vapor. All we know is the vapor pressure of benzene over the solution. By definition the mole fraction of benzene in the gas phase is given by the expression below.

$$X_{bz}^g = \frac{n_{bz}^g}{n_{bz}^g + n_{ct}^g}$$

In this equation, the superscript **g** reminds us that we are determining the mole fraction in the **g**as phase. If we assume that the gases behave ideally[1] (which is an excellent assumption), then:

$$X_{bz}^g = \frac{P_{bz} \times \frac{V}{RT}}{P_{bz} \times \frac{V}{RT} + P_{ct} \times \frac{V}{RT}}$$

The factor of (V/RT) cancels out, so we are left with the expression below. Notice that this is completely equivalent to Dalton's Law of partial pressures.

$$X_{bz}^g = \frac{P_{bz}}{P_{bz} + P_{ct}} = \frac{P_{bz}}{P_{total}}$$

Substituting into this equation, we can find the mole fraction of benzene in the gas phase, which is:

$$\therefore X_{bz}^g = \frac{67.5 torr}{308 torr} = 0.219$$

Let's compare the numbers we have for the mole fraction of benzene in the liquid phase and in the vapor phase. The mole fraction in the gas phase (0.219) is actually smaller than that in the liquid phase (0.250). This means that there is more carbon tetrachloride (relative to the benzene) in the gas phase that in the liquid solution. Does this make sense? Definitively *yes!* *The vapor phase is always richer (relative to the liquid solution) in the more volatile component.* The more volatile component (carbon tetrachloride) is able to escape easier into the gas phase than the benzene, therefore, you will have more carbon tetrachloride, relative to the solution, in the gas phase.

In order to get a better picture of what is going on in this solution, let's calculate the

---

[1] Remember that moles (n) are equal to: PV/RT.

Chapter 11: Solutions and Their Properties.

partial pressures of benzene and carbon tetrachloride for different solutions (with benzene mole fractions of 0.000, 0.200, 0.400, 0.600, 0.800, and 1.000). Let us also determine the composition of benzene in the gas phase for each of this solutions. These calculations can be done in exactly the same way as we did for the example above. The results are tabulated below.

X_{bz}	P_{bz}	P_{ct}	P_{total}	X^g_{bz} (pressures in torr)
0.000	0	320	320	0.000
0.200	54	256	310	0.174
0.400	108	192	300	0.360
0.600	162	128	290	0.559
0.800	216	64	280	0.771
1.000	270	0	270	1.000

As always, it is always easier to understand the patterns and even to get more information when the results are expressed in a graphic format. Below we can find a graph of total pressure over a liquid solution of benzene and carbon tetrachloride that was plotted with the information above. We should point out that this plot is not to scale and it is shown only to see the kind of information we can potentially extract out of the graph.

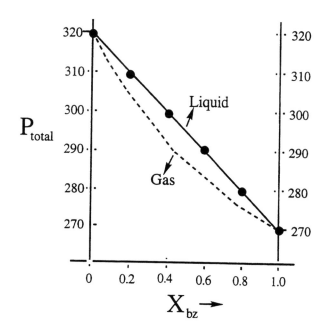

What is this previous graph telling us? The composition (mole fraction) of both the liquid layer and the gas phase at any total pressure. In order to read the graph just choose the total pressure in question and read directly the mole fraction of the liquid and the gas phases.

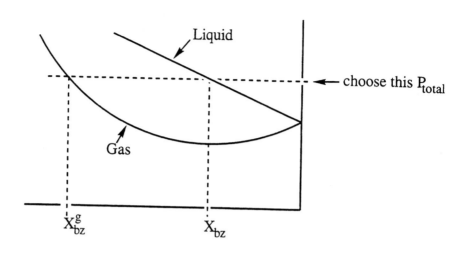

We can see that at the chosen total pressure (the same is true at any total pressure) there is less benzene (relative to CCl_4) in the gas phase than in the liquid solution, just like we expected.

This discussion has a direct bearing on **fractional distillation**, a procedure for separating liquid components of a solution that is based on their different boiling points. The concept is simple: If you take a solution of benzene and toluene (boiling points are 80.1°C and 110.6°C, respectively) and heat it up, we know that the vapor will be richer in the more volatile component, benzene. If we were to condense the vapor, the new liquid will be more concentrated in benzene than the original liquid. If we now heat up this new liquid, its vapor will be even richer in benzene, thus, upon condensation of this new vapor, we will get an even more concentrated solution of benzene. If we keep doing this over and over, we will eventually be able to get pure benzene separated from the original solution.

Experimentally, how do we accomplish this? We need to use at least a simple fractionating column[1]. This column would roughly look like the one shown below. In the round bottom flask, we would place the mixture of the two liquids (benzene and toluene) and proceed to heat it up. At a given temperature, the vapor in equilibrium with the liquid will be richer in benzene (the more volatile component). This vapor would then rise into the column, which has many colder glass beads, which will in turn lead to condensation of the vapor. This new liquid formed in the column is already richer in benzene. As time goes by, some of the fresh vapor coming into the column will create more evaporation of the previous droplet, which now will be even richer in benzene. This new vapor will travel even further up the column and recondense. This will go on and on, and if the column is long enough, by the time we get to the thermometer, we will see that the liquid that is making it up there has a boiling point of 80.1°C, that is, it is

---

[1] Like the one seen below, which has been taken from Pavia, Lampman, Kriz, and Engel, "Introduction to Organic Laboratory Techniques", First Edition, Saunders (1990), page 680.

pure benzene. We then allow the vapor to go into the condenser. This condenser is bathed on the outside with cold water, thus it will make the vapor condense, and then we proceed to collect the pure liquid.

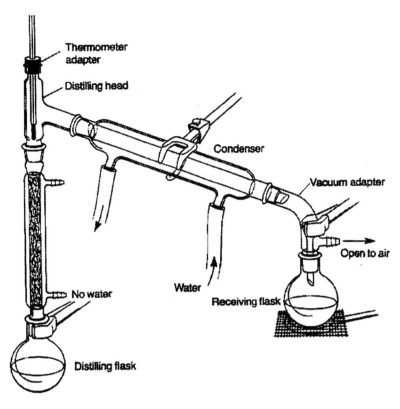

11.6A Colligative Properties.

There are four colligative properties[1]; we have already discussed the theory behind the first one. The colligative properties are:

1. Lowering of the Vapor Pressure
2. Elevation of the Boiling Point.
3. Depression of the Melting Point.
4. Osmotic Pressure.

When we discussed Raoult's law, we explained why the vapor pressure of a liquid is lowered in the presence of a solute. This is true at any temperature. Therefore, at any temperature, the vapor pressure of the solvent in a solution is lower than if it were pure. How much lower? It depends on the amount of solute dissolved in it.[2] The more solute we have, the

---

[1] Properties of solutions that are affected by the *number of particles of solute* dissolved and NOT on their nature.
[2] We can actually calculate the vapor pressure using Raoult's Law, which depends on the mole fraction of the solvent.

larger the lowering of the vapor pressure. This can be expressed mathematically as follows:

$$\Delta P_{solvent} = X_{solute} \, P°_{solvent}$$

This is the equation for the first colligative property.[1] At any rate, let's concentrate our discussion using water as a solvent and any non-volatile solute. If we take a solution of molality **m**, then the vapor pressure of water over the solution will be less than the pure vapor pressure of water. This will be true at any temperature, therefore the whole liquid gas equilibrium curve is shifted in the phase diagram (as shown below):

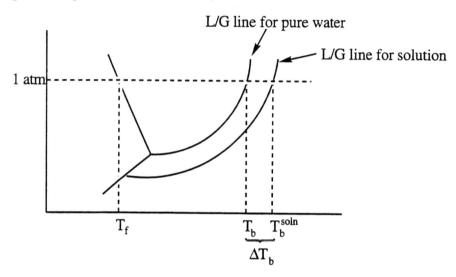

The lower liquid/gas curve is the one for a solution of concentration **m**. If the molality were any larger, then the curve would be further down. We can see how the normal boiling point of the solution (at 1 atm) is higher for the solution that for pure water. The elevation of the boiling point depends on the number of particles of solute dissolved in the solvent.

$$\therefore \Delta T_b \, \alpha \, m$$

or $\quad \Delta T_b = K_b \, m \quad \checkmark$

where K_b = ebullioscopic constant

From the phase diagram above you can also notice that since the whole Liquid/Gas curve was shifted, this also had an effect on the position of the triple point. Consequently, the entire Solid/Liquid equilibrium curve is also shifted. The following phase diagram for water shows not only the normal phase diagram for water but also the phase diagram for an aqueous solution of

---

[1] This equation looks different from Raoult's Law, but it is exactly the same; it is just expressing the **lowering** of the vapor pressure, rather than the actual vapor pressure. As a curious mathematical exercise, try to derive one from the other.

molality **m**. The larger the molality of the solution, the larger the difference between the melting point of the solution as compared to that for the solvent. From this graph note that the solute affects the melting point much more than it does the boiling point, *i.e..*, the effect is much more pronounced in the melting point than in the boiling point.

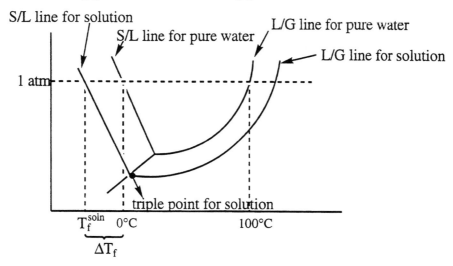

We can see that the freezing point is therefore lowered and that the **depression of the freezing point** (ΔT_f) is also dependent on the molality (**m**) of the solution.

$$\therefore \Delta T_f \propto m$$

or $\quad \Delta T_f = K_f m \quad \checkmark$

where K_f = cryoscopic constant

The table below, gives us the values of these constants for some common solvents used in laboratories.

Boiling Point Elevation and Freezing Point Depression Constants.

Solvent	Formula	T_b (°C)	K_b (Kkg/mol)	T_f (°C)	K_f (Kkg/mol)
Benzene	C_6H_6	80.1	2.53	5.50	4.90
Ethanol	C_2H_5OH	78.4	1.22	-114.7	1.92
Carbon Tet.	CCl_4	76.7	5.03	-22.9	32.1
Naphthalene	$C_{10}H_8$	-	-	80.5	6.8
Camphor	$C_{10}H_{16}O$	-	-	179.0	39.7
Water	H_2O	100.0	0.512	0.00	1.86

- Determine the freezing point and the boiling point of a solution containing 6.50 grams of ethylene glycol ($C_2H_6O_2$, MW = 62.1) in 200 grams of water.

Water is the solvent, so we need its cryoscopic and ebullioscopic constants. These can be seen in the table above. Since for either one of them we need the molality, we can calculate that first.

$$6.50 \text{ g}_{eg} = 0.105 \text{ mol ethylene glycol}; \therefore m = (0.105 \text{ mol})/(0.200 \text{ kg}) = 0.525 \text{ m}$$

Now, the freezing point depression is given by:

$$\Delta T_f = K_f \, m = (1.86°C/m) \, 0.525 \text{ m} = 0.977°C.$$

and the actual freezing point would be:

$$\therefore T_f = -0.977°C \quad \checkmark$$

The boiling point can be calculated in a similar fashion, but using the ebullioscopic constant, therefore,

$$\Delta T_b = K_b \, m = (0.512°C/m) \, 0.525 \text{ m} = 0.269°C.$$

$$\therefore T_f = +100.269°C \quad \checkmark$$

- Determine the freezing point of the resulting mixture when you dissolve 30.0 g of solid $C_5H_8O_7$ (MW = 180.12) with 200 g of solid camphor, liquefy it, and then slowly allow it to freeze.

Once more, we need to start by finding the molality of the solute, but this time the solvent is camphor, which has a much higher cryoscopic constant (39.7) than water.

$$30.0 \text{ g solute} = 0.167 \text{ moles } C_5H_8O_7$$

$$\therefore m = 0.167 \text{ moles} / 0.200 \text{ kg camphor} = 0.835 \text{ m} \quad \checkmark$$

The decrease in the freezing point is calculated with the equation for the colligative property, that is:

$$\Delta T_f = K_f \, m = (39.7°C/m) \, 0.835 \text{ m} = 33.1°C.$$

Finally, to determine the freezing point of the solution itself, we need to know the normal melting point of camphor. From the table above, we can see that it is 179.0°C. Since the freezing point was lowered by 33.1°C, then the freezing point of the solution must be *145.9°C*.

11.7 Osmotic Pressure.

This last colligative property is a little different from the rest, and it requires that we explain what it is and how it is measured. Consider the following system consisting of a solution of a non-volatile solute of molarity M separated from another liquid consisting of just pure solvent (water) by a semi-permeable membrane. This membrane will allow the passage of solvent molecules but not solute molecules.

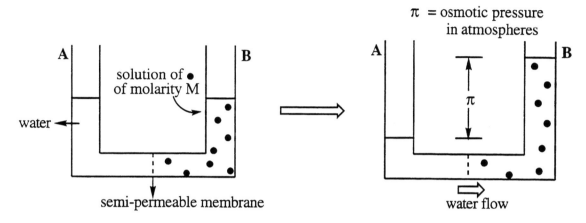

There is a natural tendency for these solutions, which happen to be in contact, to attain equal concentrations. The easiest way to achieve this would be to get half of the solute molecules across the membrane, but they cannot penetrate the membrane. Therefore, water molecules will flow from left to right to try to dilute the solution. As the solvent moves across, the meniscus on the left compartment drops while the one on the right side goes up. This generates an imbalance since then column B would weigh more than column A. The flow of water from left to right will continue until the weight imbalance is exactly equal to this tendency to make both solutions equimolar. This tendency of the solvent to flow through the membrane is what we call the osmotic pressure and it is measured in terms of the weight or height difference between the two columns, much like a barometer, and expressed in atmospheres.

Mathematically, it turns out that the equation for osmotic pressure is very similar to that for ideal gases.

$$\Pi V = n R T$$

In this equation, Π represents the osmotic pressure of the solution (in atmospheres), V is the volume of the solution, n is the number of moles of solute, R is the universal gas constant (in this case it is 0.0821 Latm/Kmol), and T is the absolute temperature.

rearranging: $\Pi = (n/V) R T$

Chapter 11: Solutions and Their Properties.

but: (n/V) = Molarity

Therefore, the equation for this last colligative property would be:

$$\therefore \Pi = MRT \quad \checkmark$$

- Calculate the osmotic pressure at 20°C of a solution prepared by dissolving 64.3 grams of sucrose, $C_{12}H_{22}O_{11}$ (MW = 342), in 200.0 grams of water. The density of the solution is 1.10 g/mL. $\Pi = MRT$. Clearly, 64.3 g sucrose $\Rightarrow$ 0.188 mol sucrose.

This amount of sucrose (64.3 g) represent 0.188 moles of sucrose, which is our solute. In order to calculate the molarity of sucrose, we need to know also the volume of the solution. All we know about the solution is that it weighs 264.3 grams (the mass of the sucrose plus the mass of the water). Using the density, we get:

$$264.3 \text{ g soln.} \times \frac{1 \text{ mL}}{1.10 \text{ g}} \times \frac{1 \text{ L}}{1000 \text{ mL}} = 0.240 \text{ L}$$

Therefore, the concentration of sucrose is calculated as seen below.

$$[\text{Sucrose}] = 0.188 \text{ mol} / 0.240 \text{ L} = 0.783 \text{ M}.$$

Finally, the osmotic pressure is determined with the equation for this colligative property. Remember that R, the gas equation, must be in units of Liters atmospheres per Kelvin mole. Therefore,

$$\Pi = (0.783 \text{ mol/L}) (0.0821 \text{ Latm/Kmol}) (293 \text{ K})$$

$$\therefore \Pi = 18.8 \text{ atm} \quad \checkmark$$

The colligative properties provide us with an excellent experimental method to determine the molecular weight of an unknown substance. Let's look at a couple of problems to see how we can accomplish this.

- A solution containing 6.79 g of an unknown, non-volatile, non-dissociating solid (A) in 50.0 g of camphor is prepared and its melting point is determined to be 135.0°C. Determine the MW of A.

Let's first find the drop in the melting point for the camphor. This is done by looking at the difference between the normal melting point of the solvent and the melting point of the solution.

$$\Delta T_f = |179.0 - 135.0| = 44.0°C$$

Using the equation for the depression of the melting point, we can solve for the molality of the solute.

$$\Delta T_f = K_f \times m$$

$$\Delta T_f = K_f \times \frac{n_{solute}}{kg_{camphor}}$$

$$\Delta T_f = K_f \times \frac{g_{unknown}/MW_{unknown}}{kg_{camphor}}$$

$$\therefore MW_{unknown} = \frac{K_f \times g_{unknown}}{kg_{camphor} \times \Delta T_f}$$

And substituting the values, we get:

$$\therefore MW_{unknown} = \frac{39.7 K kg/mol \times 6.79 g}{0.0500 kg \times 44.0 K}$$

$$= 123 \text{ g/mol} \quad \checkmark$$

- A non-volatile, non-dissociating solid is determined to be 54.6%C, 9.09%H and 36.4%O. An aqueous solution is prepared by dissolving 2.000 g of this compound in 25.000 g of water and its melting point is determined to be -1.13°C. Determine the MF of this compound.

We should start by determining the empirical formula. We use the elemental composition given in order to do this. We discussed this in Section 2.4.

	grams	moles	SR
C	54.6 g	4.55 mol	2
H	9.09 g	9.09 mol	4
O	36.4 g	2.28 mol	1

$\therefore$ EF is C_2H_4O which weighs 44 amu.

Our next step should be the determination of the molecular weight of the solute. From the information given, we can determine it by the depression of the melting point. This time the solvent is water.

$$MW_{unknown} = \frac{K_f \times g_{unknown}}{kg_{water} \times \Delta T_f}$$

Substituting the values, we get a molecular weight of 132 g/mol.

$$MW_{unknown} = \frac{1.86 K kg/mol \times 2.00 g}{0.0250 kg \times 1.13 K}$$

$$MW = 132 \text{ g/mol} \checkmark$$

Finally, now that we have both the molecular weight and the empirical weight, we can figure out how many times we can fit the empirical weight into the molecular weight.

$$132/44 = 3$$

$$\therefore \text{Molecular Formula is } C_6H_{12}O_3. \checkmark$$

We keep mentioning the term *non-dissociating* substance in some of these problems. Why? We must remember that colligative properties depend on the number of particles dissolved and not on their nature. So if a solute were to dissociate, it would generate more particles than we would expect and the net result would be a larger effect on the colligative properties themselves. That is exactly what we observe in the case of salts (strong electrolytes[1]). This means when we dissolve one mole of NaCl in water, we will generate two moles of particles: one mole of sodium ions and one mole of chloride ions. Similarly, if we were to dissolve one mole of $AlCl_3$ in water, we would not get one mole of particles, but rather four moles of particles: one mole of aluminum ions and three moles of chloride ions. Therefore, all the colligative properties must be corrected by this factor (**i = van't Hoff factor**) to account for the increased number of particles. This factor represents the total number of moles of particles that a certain salt will dissociate into. As an example, for NaCl, and $AlCl_3$, the factor (**i**) is equal to 2 and 4, respectively.

How do we account for this factor in the different colligative properties? The part of the equation that is affected is the number of moles of solute. Technically, it should be the number of moles of particles, therefore:

$$\text{moles of particles} = i \cdot \text{moles of solute.}$$

---

[1] See Section 11.4.

Now, where do the moles of solute appear in the molality? It does in the numerator.[1] Therefore, the **i** will also be in the numerator.

$$\Delta T_f = i\, K_f\, m$$

$$\Delta T_b = i\, K_b\, m$$

Where are the moles of solute in the molarity? Once more, it appears in the numerator. Therefore,

$$\Pi = M R T\, i$$

Finally, where are the moles of solute in the mole fraction? Remember that the colligative property for the decrease of the vapor pressure depends on the mole fraction of the solute.

$$\Delta P_{solvent} = X_{solute}\, P°_{solvent}$$

But, what is the mole fraction of the solute? It is defined as follows:

$$X_{solute} = \frac{n_{solute}}{n_{solute} + n_{solvent}}$$

Everywhere we see the moles of solute, we should multiply by **i**. Therefore, the equation becomes

$$X_{solute} = \frac{i \times n_{solute}}{i \times n_{solute} + n_{solvent}}$$

Let's do a problem where we will have to calculate all the colligative properties, but we will have to be careful because it is a salt (**i** ≠ 1).

- Calculate the theoretical melting point, boiling point, the osmotic pressure, and the vapor pressure of water over the solution at 75°C for a 20.0% aqueous $Al_2(SO_4)_3$, MW = 342, solution if its density is 1.05 g/mL. The vapor pressure of water at that temperature is 289 torr.

    Aluminum sulfate is a salt, therefore it should dissociate 100% into its ions to form individually solvated ions.

---

[1] Remember that molality is the moles of solute divided by the kilograms of solvent.

$$Al_2(SO_4)_{3(aq)} \Rightarrow 2\, Al^{+3} + 3\, SO_4^{-2}$$

$$\therefore i = 2 + 3 = 5 \checkmark$$

Let's start by figuring out the molality of the solution. We are told that it is 20.0% in aluminum sulfate. This means that for every hundred grams of solution, 20.0 g will be aluminum sulfate and the remaining 80.0 g will be water. Molality is defined as the moles of aluminum sulfate per kilogram of water, therefore:

$$m = \frac{20.0g / 342 g/mol}{0.0800 kg} = \frac{5.82 \times 10^{-2} mol}{0.0800 kg}$$

$$\therefore m = 0.731 \text{ molal} \checkmark$$

With that information in hand, we can immediately determine the freezing point and the boiling point of this solution.

$$\Delta T_f = 5\, K_f\, m$$

$$= 5\, (1.86°C/m)\, (0.731\, m) = 6.80°C$$

$$\therefore T_f = -6.80°C \checkmark$$

and

$$\Delta T_b = 5\, K_b\, m$$

$$= 5\, (0.512°C/m)\, (0.731\, m) = 1.87°C$$

$$\therefore T_b = 101.87°C \checkmark$$

To figure out the osmotic pressure we need the molarity of the solution.

$$M = \text{moles solute / Volume solution}$$

But we have 100.0 g of solution. Remember that in order to convert the mass into the volume, we need the density of the solution (which is given as 1.05 g/mL). Therefore,

$$V = 100.0 \text{ g soln } (1 \text{ mL}/1.05 \text{ g soln}) (10^{-3} \text{ L}/1 \text{ mL})$$

$$\therefore V = 0.0952 \text{ L}$$

$$\therefore M = 5.85 \times 10^{-2} \text{ mol}/0.0952 \text{ L}$$

$$= 0.614\, M \checkmark$$

The osmotic pressure can then be calculated with the following equation.

$$\Pi = 5MRT$$

$$= 5\,(0.614\text{ mol/L})\,(0.0821\text{ L atm/K mol})\,(348\text{ K})$$

$$= 87.7\ atm\ \checkmark$$

Finally, the colligative property that allows us to determine the decrease in the vapor pressure requires that we determine the mole fraction of the solute first.

$$X_{solute} = \frac{5 \times 0.0582\,mol}{5 \times 0.0582\,mol + \dfrac{80.0\,g}{18.02\,g/mol}}$$

$$X_{solute} = 0.0615$$

Finally,

$$\Delta P_{water} = X_{solute}\,P°_{water}$$

$$= 0.0615\,(289\text{ torr}) = 17.8\text{ torr.}$$

Therefore, $\quad P_{water} = 289.0\text{ torr} - 17.8\text{ torr}$

$$P_{water} = 271.2\ torr\ \checkmark$$

- What concentration of sodium chloride in water is needed to produce an aqueous solution that is isotonic[1] with blood ($\pi = 7.70$ atm at 25°C)?

Since NaCl is a salt, the osmotic pressure equation is affected by the number of ions this salt will dissociate into. Therefore,

$$\Pi = 2MRT$$

Substituting into this equation, we get:

$$7.70\text{ atm} = 2\,M\,(0.0821\text{ Latm/Kmol})\,(298\text{ K})$$

Therefore,

$$Molarity = 0.157\,M\ \checkmark$$

Let's go back to the previous problem (the one with aluminum sulfate) just for a moment and mention an important concept. We stated that the dissociation factor (**i**) is equal to five because the salt would dissociate into five ions. In reality, **i** will be a little smaller than 5 because some ions stay partially associated, somewhat separated by the solvent, but traveling as

---

[1] Isotonic solutions are solutions that have identical osmotic pressures at a given temperature.

a unit. We could avoid this problem with a very dilute solution of the salt, because then the solvent will have a much better chance of actually separating the ions until they essentially feel no attraction for one another and just interact with the solvent molecules. Ions of charges than one tend to form this kind of pairing, which we call **ion pairing**, much easier than ions of charge of one.

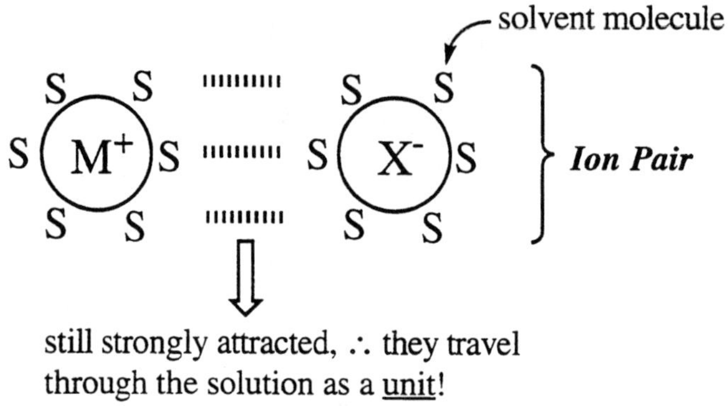

$$\left. \begin{array}{c} \text{S} \quad \text{S} \\ \text{S} \, (M^+) \, \text{S} \; \text{||||||||} \; \text{S} \, (X^-) \, \text{S} \\ \text{S} \quad \text{S} \quad \quad \quad \text{S} \quad \text{S} \end{array} \right\} \textit{Ion Pair}$$

— solvent molecule

⇓

still strongly attracted, ∴ they travel through the solution as a <u>unit</u>!

In conclusion, a salt is a strong electrolyte and thus we can easily predict the numerical value of **i**. However, it must be kept in mind that when we experimentally measure the colligative properties, they typically differ slightly from the calculated ones (in which we assumed total ion separation). This is due to ion-pairing.

Chapter 12: Kinetics.

12.1 Introduction.

Chemical kinetics is the branch of physical chemistry that studies the speeds (or the rates) of chemical reactions. From elementary physics, we know that velocity is the change of distance with respect to time, or: dx/dt[1], where x is the position along that axis. In terms of chemical reactions, we do not measure displacement, but we need to figure out a way of determining how far the reaction has proceeded in a certain time. We can accomplish this by measuring the concentration of any species that is involved in the reaction as a function of time. Of course, if the species is a reactant, its concentration will decrease as a function of time, whereas if the species is a product, its concentration will increase as a function of time. Therefore, the actual data we collect in the laboratory is relatively simple: just the concentration at different intervals. As we will see in this chapter, there is an incredible wealth of information we can obtain from this apparently trivial data.

In Chapter 5 we stated that most thermodynamic parameters depend on the difference between the initial and the final state, *ie.,* Hess' Law. Thermodynamics is, therefore, a science that does not involve the actual pathway that the reaction follows.[2] Actually, thermodynamics is an "old science" in that it was developed before we believed in the existence of atoms. Therefore, it is exclusively a macroscopic science.

On the other hand, kinetic data will be used to propose the reaction pathway, which we will call the **reaction mechanism**. A reaction mechanism describes a series of steps that describe the molecules themselves as they collide with one another. These collisions may lead to bonds breaking, or bonds forming, or transfer of electrons, etc. We call each of these steps **elementary steps** because they describe molecules, not moles of molecules. For example, if a given step reads: $A + B \Rightarrow AB$, this is not telling us that one mole of A reacts with one mole of B to form one mole of AB, but rather that one molecule of A collides with one molecule of B to form a bond between the two atoms. In other words, kinetics is concerned with what is happening at the molecular level, and will give us an insight into the events that take place at this microscopic level.

There are several factors that affect the rate[3] of a chemical reaction. Four of these can be

---

[1] This would be the slope of a graph of displacement with respect to time, that is, the first derivative at any point (x).
[2] The pathway described how we go from the initial to the final state.
[3] That is, affect how fast the reaction will proceed as we go from reactants to products.

seen below. There are a few others, but these will suffice for the level we are aiming at in this course.

1. Nature of the Reactants.
2. Concentration or Pressures of *Reacting* Species.
3. Temperature.
4. Presence of a Catalyst.

Let's look at each one of these four factors individually.

12.2 Nature of the Reactants.

The rate of a chemical reaction will depend heavily on the state of the reactants. In order for a reaction to take place, the reacting molecules must come in contact with one another. Thus, if these are mobile (like gases or liquids) the reaction should take place faster than if they are not (as is the case when solids are involved). We should also point out that ions in solution are very mobile and normally yield faster reactions.

Probably some of the fastest reactions are those that involve simply an electron transfer (redox[1] reactions). If we were to compare such a reaction with one where covalent bonds are both broken and formed, it is easier to understand why electron transfer reactions occur much more rapidly. The first case below involves the transfer to two electrons from the zinc metal into the cupric ions, whereas in the second reaction, we need to break two H-I bonds to form an H-H and the I-I bonds.

Fast: $Zn_{(s)} + Cu^{+2}_{(aq)} \Rightarrow Cu_{(s)} + Zn^{+2}_{(aq)}$

Slow: $2\,HI_{(g)} \Rightarrow H_{2(g)} + I_{2(g)}$

It should be pointed out that the terms fast and slow are relative to one another at this point. We have not specified (nor will we at this level) actual cutoffs to decide what is slow or fast. The point we were trying to make here is typically, slow reactions involve the breaking and forming of covalent bonds whereas fast reactions are much less complicated (such as those that involve electron transfer).

12.3 Concentration of Reacting Species.

This second factor involves a lot more than the nature of the reactants. We should always

---

[1] We will discuss reduction reactions in Chapter 19.

remember that in order for molecules to react, molecules must come in contact with one another. The rate of the reaction will partially depend on how often these molecules encounter one another. Actually, we could say that the higher the concentration, the faster the rate of the reaction. This is generally true, but it is a lot more involved than it sounds. For example, it would be completely wrong to say that the rate of a reaction is directly proportional to the concentration of a reagent, that is, if we were to double the concentration of a given reagent, then the reaction would go twice as fast. This may or may not be true.

In order to understand how the concentration affects the rate of a reaction, let's look at the following hypothetical, one-step[1] reaction.

$$A_{2(g)} + B_{2(g)} \Rightarrow 2\,AB_{(g)}$$

If we were to initially place equal concentrations of both reacting gases in a given flask, but no initial amount of gas AB, it should be clear that as time goes by, the concentrations of gases A_2 and B_2 will decrease equally fast (since they react in a one to one mole ratio). It is also true that the concentration of the gas AB should increase as a function of time. However, since for every mole of A_2 that disappears we get 2 moles of AB, then the rate of appearance of AB should be twice as fast as the rate of disappearance of either A_2 or B_2.

As we mentioned in the introduction, rate is a measure of how fast is either A_2 or B_2 disappearing or how fast AB is being formed (the slope of either curve shown below).

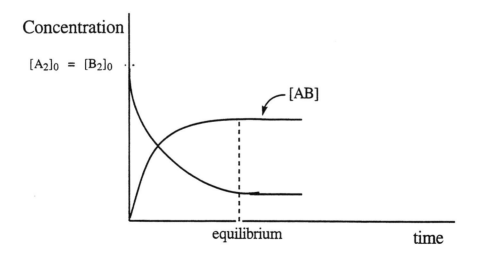

Notice that the first curve (the ones for the reactants) has a negative slope, whereas the second

---

[1] The term one-step refers to the mechanism of the reaction, that is, the reaction occurs in a single collision, in this case a bimolecular collision between each reacting molecule.

one (the one for the product) has a positive slope. We would like for the rate to be a positive quantity, therefore, our definition of rate will depend on whether the concentration we are actually monitoring as a function of time is a reactant or a product. To make the quantity positive, we will multiply the slope of the curve by (-1) when we measure the rate in terms of one of the reactants. In our example, this would give us:

$$\text{rate}^1 = -d[A_2]/dt$$

or $\quad\text{rate} = -d[B_2]/dt.$

How do we decide in the laboratory which substance's concentration to monitor as a function of time? There may be many reasons, but as an example, let's say that one of the reagents is blue, whereas the other one and the product are colorless. We could easily monitor the disappearance of the color as a function of time. Of course, maybe the one with the color is the product (AB), and not one of the reactants. In this case, we would monitor the appearance of the blue color as a function of time. However, since the rate of appearance of AB is twice the rate of disappearance of either A_2 or B_2, then we would have to multiply this quantity by one half[2] in order to make them equal.

$$\therefore \text{rate}^3 = +1/2\, d[AB]/dt$$

We mentioned earlier that the rate depends on the concentration of each reagent. What is this dependence? In the mechanism that we proposed (one-step), there is one molecule of A_2 involved in the collision. Therefore,

$$\text{rate} \propto [A_2]^1$$

The exponent, which we will call the **order** with respect to that reagent, depends on the number of molecules of that particular reagent involved in that one-step mechanism. Consequently, since there is also one molecule of B_2 involved in this particular step, then the rate should also be proportional to the concentration of B_2 to the first power.

$$\text{rate} \propto [B_2]^1$$

Mathematically, if X is proportional to Y and X is also proportional to Z, then X is proportional

---

[1] In this equation, the negative sign indicates disappearance of the reactant. Remember that the slope of the curve at any point in time is given by: $d[A_2]/dt$.

[2] The reason we are trying to make them equal is to avoid having to specify every time which substance we monitored.

[3] In this case, there is no need to multiply by –1 since the slope is already positive.

to the product of Y times Z. Therefore,

$$\text{rate} \propto [A_2][B_2]$$

This leads to the following equation.

$$\textbf{rate} = \textbf{k} [\textbf{A}_2][\textbf{B}_2] \Rightarrow \text{Rate Equation}$$

In this equation, k is referred to as the rate constant, and it must be determined experimentally. As a matter of fact, the orders of each reagent cannot be deduced from the balanced equation, therefore, they also have to be determined experimentally.

For this hypothetical reaction, we say that the reaction is <u>first order</u> with respect to A_2 (by just looking at the exponent of A_2 in the rate equation), <u>first order</u> with respect to B_2, and <u>second order</u> overall. The *overall order* is the sum of all the individual orders.

The <u>only reason</u> we know these orders is because we know the mechanism of the reaction, *ie.*, we know it occurred in one-step. Otherwise we would not be able to predict the orders of A_2 and B_2. If we would not have specified that the reaction occurred in one step, all we would have been able to say is:

$$\text{rate} = k [A_2]^x [B_2]^y$$

where: **x** and **y** are unknown orders.

Let's look at a couple of examples, to see how accurately we can predict the orders based on a balanced chemical equation.

1. $$2 N_2O_{5(g)} \Rightarrow 4 NO_{2(g)} + O_{2(g)}$$

If we look at the reaction and assume it occurs in one step, like the hypothetical example above, then we would predict that the reaction is second order with respect to the nitrogen pentoxide. Experimentally, we determine the rate law to be:

$$\text{rate}_{experimental} = k [N_2O_5]$$

This tells us that the reaction actually is <u>first order</u>.

2. $$2 NO_{(g)} + 2 H_{2(g)} \Rightarrow N_{2(g)} + 2 H_2O_{(g)}$$

From the balanced reaction, we could guess that the reaction is fourth order, second with respect to nitrogen oxide and second with respect to hydrogen gas. Experimentally, we determine the rate law, which is:

$$\text{rate}_{\text{experimental}} = k\,[NO]^2\,[H_2]$$

This tells us immediately that the reaction is third order.

Conclusion: It will be wrong most of the time to assume that the reaction orders are the same as the coefficients in the balanced reaction. *The order must be determined experimentally!*

How do we determine these orders experimentally? There are various methods and we will learn a couple of them in the next couple of sections: the initial rate method and the graphical method.

12.4 Initial Rate Method.

This is one of the most common methods used to determine the order and the numerical value of the rate constant, k. Before we go into the method itself, how would we measure the initial rate? By definition, it is the slope of a graph of concentration vs. time close to the initial concentration (as $t \to 0$).

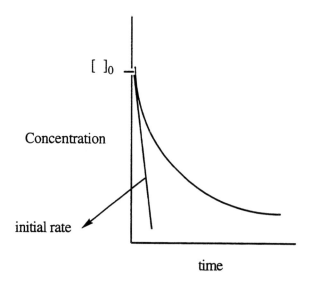

The method consists of measuring the initial rate under various experimental conditions. If the experiment is repeated while we vary the concentration of one of the reactants and leave the rest unchanged, any observed change in the initial rate will be due to the reactant we varied. Let's look at one example.

Chapter 12: Chemical Kinetics.

- Consider the following reaction:

$$A_{(g)} + B_{(g)} \Rightarrow \text{Products}$$

for which rate = $k[A]^x[B]^y$.

Experimental Data

Run	$[A]_o$, M	$[B]_o$, M	Initial Rate (mol/L s)
1	0.010	0.020	7.0×10^{-6}
2	0.010	0.060	63.0×10^{-6}
3	0.020	0.020	14.0×10^{-6}

Determine the rate equation and the numerical value for the rate constant, **k**.

In order to do this problem we need to find out first what the rate equation is, in other words, find the orders with respect to A and B. Using the rate equation we can write the following:

Run		
1	7×10^{-6}	$= k (0.010)^x (0.020)^y$
2	63×10^{-6}	$= k (0.010)^x (0.060)^y$
3	14×10^{-6}	$= k (0.020)^x (0.020)^y$

We notice that there was a change in the initial rate for this reaction between runs one and two. What is the difference between those two runs? The only thing we changed was the concentration of B (which we tripled). The experimental observation is that upon tripling the concentration of B, the rate increases by a factor of nine. The relationship between these two numbers (3 and 9) is: $3^2 = 9$. The implication is that the order with respect to B must be two. Does this actually make sense mathematically? Let's divide run two by run one.

$$\frac{63 \times 10^{-6}}{7 \times 10^{-6}} = \frac{\cancel{k}\,\cancel{(0.010)^x}\,(0.060)^y}{\cancel{k}\,\cancel{(0.010)^x}\,(0.020)^y}$$

$$9 = 3^y$$

or $y = 2.$ ✓

Now that we have found the order with respect to B, we are met with the challenge of finding the order with respect to A. Which two experiments change the concentration of A while keeping that of B constant? Those would be runs 1 and 3. In here, the concentration of A is doubled and we see that the rate is also doubled. This tells us immediately that it should be first order with respect to A: $2^1 = 2$. The mathematical proof can be seen below.

$$\frac{14 \times 10^{-6}}{7 \times 10^{-6}} = \frac{\cancel{k}(0.020)^x \cancel{(0.020)^2}}{\cancel{k}(0.010)^x \cancel{(0.020)^2}}$$

$$2 = 2^x$$

or $x = 1.$ ✓

With this information, we can now write the rate equation.

$$\text{Rate} = k[A][B]^2 \quad \checkmark$$

This is a third order equation, first with respect to A and second with respect to B. However, we still do not know the rate constant, k. We can calculate k by substituting the values of any one of the three runs into the rate equation and solving for k. Let's use run #1.

$$7 \times 10^{-6} \text{ mol/L s} = k(0.010 \text{ M})(0.020 \text{ M})^2$$

$$\therefore k = 1.8 \; L^2/mol^2 s \quad \checkmark$$

The units for the rate constant can be determined easily. In the example above, they came from a simple algebraic deduction (where M = mol/L). It would be convenient to have an equation that would allow us to know the units of k directly.

$$k(units) = \left(\frac{L}{mol}\right)^{n-1} \left(\frac{1}{time}\right)$$

where: **n** = total order for the reaction.

For the example above, since it was a third order reaction, the exponent would have been two.

- If we were to mention that the rate constant for a given reaction is 0.0234 min^{-1}, could we tell the order for the reaction?

    Of course. The only thing left is the inverse time, which means that the first factor is gone. This only happens when the exponent is zero[1]. In order for the exponent to be zero, **n** must be one. Therefore, it is a first order reaction.

If we are dealing with pressures instead of molar concentration, then the units of the rate constant will be a little different and will be given by:

$$k(units) = \left(\frac{1}{atm}\right)^{n-1}\left(\frac{1}{time}\right)$$

12.5 Solutions to Differential Rate Equations.

In order to talk about the graphical method, we need to make a short mathematical interlude and talk about the solutions to the rate equations. What are rate equations? They are simple differential equations which have a simple mathematical solution. However, in order to keep them simple, we will only deal with reactions of the type seen below. These reactions have only one reagent, which decomposes to form the products.

$$A_{(g)} \Rightarrow \text{Products.}$$

For this reaction, the best we can do for the rate equation would be: rate = k $[A]^x$. Let's find the solution to this equation for various possible values of **x** (the order).

Zero order.

These are highly unusual reactions, and seldom see a reaction in a laboratory that has an overall order of zero. However, it will illustrate the mathematical method itself. If the reaction were actually zero order, then the exponent for the concentration of A would be zero, which means that the reaction is independent of the concentration of A. This would yield the following rate equation.

$$rate = -\frac{d[A]}{dt} = k[A]^0$$

$$\therefore -\frac{d[A]}{dt} = k$$

Separating variables, we get:

$$- d[A] = k \, dt$$

---

[1] Any number raised to the zero power is equal to one.

$$-\int_{[A]_o}^{[A]_t} d[A] = k \int_0^t dt$$

$$[A]_t - [A]_o = -kt$$

In this equation, $[A]_t$ is the concentration of A at time t, whereas $[A]_0$ is the initial concentration of A (at time zero).

This is the equation of a straight line, where the variables are the concentration of A and time. A plot of [A] vs. time would yield a straight line if the reaction actually followed zero order because it would then obey that equation.

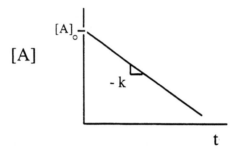

First Order.

If the reaction is first order, then the exponent (**x**) would be equal to one. Therefore, the rate equation would be:

$$rate = -\frac{d[A]}{dt} = k[A]^1$$

$$\therefore -\frac{d[A]}{dt} = k[A]$$

Separating variables, we would get:

$$\frac{d[A]}{[A]} = -k\,dt$$

$$\int_{[A]_o}^{[A]_t} \frac{d[A]}{[A]} = -k \int_0^t dt$$

Chapter 12: Chemical Kinetics.

$$\boxed{\ln \frac{[A]_t}{[A]_o} = -kt}$$

Once more, this is an equation for a straight line. At first glance it does not look that way, but we must remember that the natural logarithm of a ratio {for example: $\ln(a/b)$} is equal to the difference between the logarithm of each component {$\ln(a) - \ln(b)$}. Therefore,

$$\underbrace{\ln [A]_t}_{y} = \underbrace{-k}_{m} \underbrace{t}_{x} + \underbrace{\ln [A]_o}_{b}$$

Therefore, if our reaction were first order, it would have to obey that equation and then a plot of $\ln[A]$ vs. time would have to be linear.

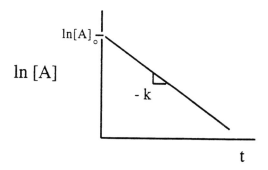

If that plot turns out to be linear, then we can find the rate constant from the slope of the line.

Second Order.

If the reaction is second order, then the rate equation would be:

$$-\frac{d[A]}{dt} = k[A]^2$$

Separating variables and integrating on both sides yields the following solution to this differential rate equation.

$$-\int_{[A]_o}^{[A]_t} \frac{d[A]}{[A]^2} = k \int_0^t dt$$

$$-\left(-\frac{1}{[A]}\right)\Bigg|_{[A]_o}^{[A]_t} = kt$$

$$\boxed{\frac{1}{[A]_t} - \frac{1}{[A]_o} = kt}\; \text{2}^{nd}$$

Once more, this is the equation of a straight line. Therefore, if our reaction were second order, it would obey that equation and a plot of: $1/[A]$ vs. time would be linear. If the plot turns out to be linear, then we could obtain the value for **k** from its slope[1].

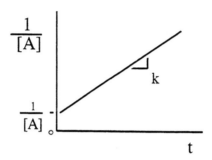

Third Order.

Using an identical mathematical method, we can solve for the third order differential rate equation and obtain the following result.

$$\boxed{\frac{1}{[A]_t^2} - \frac{1}{[A]_o^2} = 2kt}\; \text{3}^{rd}$$

If our reaction were third order, it would obey that equation and a plot of: $1/[A]^2$ vs. time would be linear. If we were to make such a plot and it turns out to be linear, then the reaction is third order and the numerical value of the rate constant can be determined from the slope.[2]

12.6 Graphical Method.

Now that we have all that information, let's use these plots for any given reaction in order to determine the reaction order and the numerical value for the rate constant, **k**. Experimentally, all we need to do is go into the laboratory and carry out the reaction while we collect kinetic data. Remember that kinetic data consists basically of two items: 1) the concentration of the substance we will monitor and 2) time. We will know that it will be one[3] of these possibilities: zero[4] order, first order, second order, or third order. This method consists of determining which

---

[1] In this case, the slope is equal to +k.
[2] In this case, the slope is equal to 2k.
[3] Actually, a reaction can have fractional orders (like 1/2), but that is outside of the scope of this course.
[4] Once more, highly unlikely.

of the four solutions we have already learned fits the experimental data. What do we mean by "fitting" the data? The one that when plotted yields a straight line. This means that we will have to do the four plots we discussed in the previous sections. Only one of those will turn out to be linear and from that linear plot, we will determine the correct order and the numerical value for the rate constant.

- The reaction: $A_{2(g)} \Rightarrow 2 A_{(g)}$ has been studied in the laboratory and the following data was collected. Determine the order for this reaction, together with the numerical value for the rate constant, using the graphical method.

Kinetic Data.

$[A_2]_o$, M	t, minutes
8.000	0.00
2.353	15.00
1.379	30.00
0.976	45.00

The data, as collected, is all we need to make the first plot: [A] versus time.

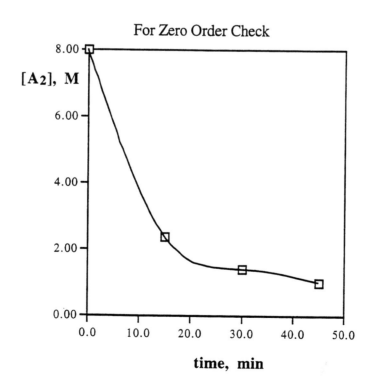

We can immediately tell that this is not a zero order reaction. As we can see, this plot is not linear. It must therefore be either first order, second order, or third order. In order for us to be able to do this, we need to perform a few calculations first. Remember that for the first order plot, we need the natural logarithm of the concentration of A, for the second order plot we need the inverse of the concentration of A, and for the third order plot, we need the square of the inverse of the concentration of A. The results of all these calculations are shown in the table below.

$[A_2]_o$, M	t, min	$\ln[A_2]$	$1/[A_2]$, M^{-1}	$\{1/[A_2]\}^2$, M^{-2}
8.000	0.00	2.0794	0.125	0.01563
2.353	15.00	0.8557	0.425	0.8557
1.379	30.00	0.3214	0.725	0.5256
0.976	45.00	-0.0243	1.025	1.0506

Let's check to see if it first order or not. Below, we can see the plot of: $\ln[A_2]$ vs. time.

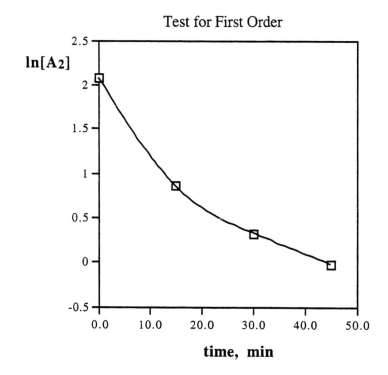

Once more, we can see that this is not a straight line, therefore, our reaction is not first order either. Therefore, it is either second or third order. Let's plot both of these to see which is the one that fits the data. Actually, we should always plot all four, since sometimes it is a close call. Drawing it approximately on poor graph paper will many times give an erroneous answer.

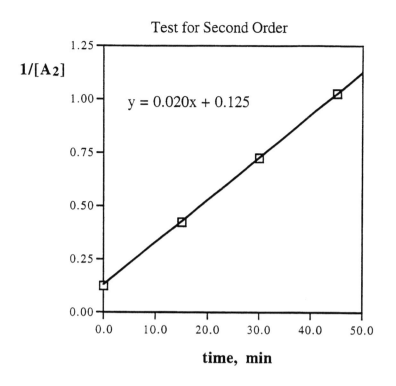

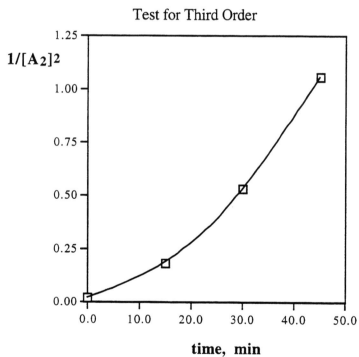

We should learn how to make plots using a computer. There are numerous programs that will allow us to do this. These graphs have been drawn using a program for the Macintosh called *Graph III*. A very similar program for the IBM and IBM compatible machines is one called *Origin*.

Let's analyze the plots first. We can see that the reaction is not third order, as this plot is not linear either, however, it should be clear that the second order plot is indeed a straight line. This means that this reaction follows second order kinetics. The slope of this line should give us the numerical value for **k**. These computer programs also tell us directly the slope of the line. We can see (in the second order plot) the equation for the line that fits our graph. This equation is given by:

$$y = 0.020\,x + 0.125$$

Therefore, the slope for this line is 0.020, and that must be the numerical value for the rate constant, **k**.

Summary of the graphical method.

Order	Plot	Slope
Zero	[A] vs. t	$-k$
First	ln[A] vs. t	$-k$
Second	1/[A] vs. t	$+k$
Third	$1/[A]^2$ vs. t	$+2k$

The solutions to the differential equations are useful not only for the graphical method but to solve any kind of problem in which we need to know the actual concentration of A at any time during the course of a given reaction.

- Consider the reaction $2\,A_{(g)} \Rightarrow$ Products, for which $k = 3.00 \times 10^{-3}\,s^{-}$. How long will it take for the reaction to go 90.0% to completion?

The first thing we notice is that the reaction follows first order kinetics, due to the units of k (inverse seconds). Therefore, we can use the following equation:

$$\ln \frac{[A]_t}{[A]_o} = -k\,t$$

We are told that the reaction has gone 90.0% to completion. This means that we have left 10.0% of the original amount. We do not know the original amount, but that is the beauty of a first order reaction: since its solution involves a ratio, we do not need to know its original concentration.

$$\ln\left(\frac{0.100\,[A]_o}{[A]_o}\right) = -3.00 \times 10^{-3}\,s^{-1}\,t$$

Solving for **t**, we get:

t = 768 s or *12.8 minutes*. ✓

- Consider the following reaction: $A_2B_{(g)} \Rightarrow A_{2(g)} + B_{(g)}$. This reaction has the following rate equation: rate = k $[A_2B]^2$, where k = 5.20 x 10^{-3} L/(mol min). If we start with the concentration of A_2B at 2.50 M, how long will it take for the concentration of A_2B to be exactly 1.00 M?

    This time we are told directly that the reaction is second order with respect to the reagent. Therefore, it obeys the following equation.

    $$\frac{1}{[A_2B]} - \frac{1}{[A_2B]_o} = k\,t$$

Substituting the values directly, we can solve for the time, **t**.

$$\frac{1}{[1.00]} - \frac{1}{[2.50]_o} = 5.20 \times 10^{-3}\,t$$

∴ **t = *115 minutes*.** ✓

Half Lives.

The term half-life refers to the time it takes for the concentration of a given reagent to drop to exactly one half of the initial value. This has been a very useful concept that has helped many scientists to investigate the kinetics of a wide variety of reactions. The only time that the half-life is independent of the actual numerical value for the initial concentration of the reactant is when the order with respect to that reactant is first order. We shall therefore limit our discussion to only first order reactions when we are talking about half lives.

We will use this concept to allow us to determine quickly the numerical value for a first order reaction. Many times we do not know the initial concentration of the reactant exactly, and as long as the reaction is first order, it turns out we do not really need to know it. Let's say that the reagent is colored, whereas the products are colorless. We could follow the reaction rate by simply measuring a property we call **absorbance** (A). As we have seen before[1], substances interact with light and the end result is that it may absorb a photon of light to get excited electronically. Well, substances that exhibit a particular color do so because it is absorbing a particular wavelength of light, and in the process it reflects the complimentary color. For example, when a substance absorbs a red photon of light, it will appear to be green in color. Green is the complimentary color for red. The color wheel below shows the main colors and the complimentary colors are directly across from them on the wheel.

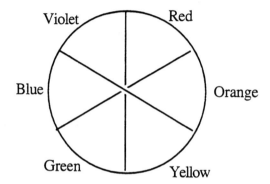

- What is the color of a solution that absorbs yellow light?

    From the color wheel, we can see that the complimentary color for yellow is violet, therefore, the solution will appear to be violet.

    The instrument used to measure the absorbance of a solution is called a **spectrophotometer**. In a very simplistic view, let us say that we will shine a beam of light through the colored solution whose absorbance we want to measure. The logarithm of the ratio of the intensity of the light as is comes into the solution ($I°$) to the intensity of the light once is comes out of the solution (I) is the way we define absorbance technically.

$$A = -\log (I/I°).$$

The quantity ($I/I°$) represents the fraction of light that went through the solution, or the Transmittance (T). So, another way of writing this equation would be:

$$\mathbf{A = -\log T}$$

The absorbance is directly proportional to the concentration of the colored substance. Therefore, the higher the absorbance, the larger the concentration of the substance. Because of

---

[1] See Section 6.7.

this, we can carry out a simple experiment where we would simply have to measure the absorbance of the solution as a function of time. Not having to know the initial concentration of the reagent[1], will allow us to do this experiment without having to calibrate the instrument. Therefore, it we were to make a plot of absorbance versus time, we could find the time required for the concentration (which is proportional to the absorbance) of the reagent to be halved. This is the time we call the half-life, symbolized by: $t_{1/2}$.

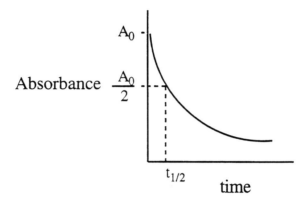

Since it is first order, then:

$$-\ln \frac{A}{A_o} = kt$$

We know that the absorbance at the half-life is exactly one half the initial absorbance (1/2 A°), therefore, the equation becomes:

$$-\ln \frac{(1/2)A_o}{A_o} = k t_{1/2}$$

$$k = \frac{0.693}{t_{1/2}}$$

- A given reaction is first order with respect to the only reagent, A. What is the numerical value of **k** if its half-life is 5.00 minutes? What fraction of the original concentration will be left after 50.0 minutes?

    We can determine the rate constant (k) with: $k = 0.693/t_{1/2}$.

    $\therefore k = 0.693 / 5.00 \text{ min} = 0.139 \text{ min}^{-1}$.

Since it is first order, then:

---

[1] Remember that we are dealing with first order reactions only when working with half-lives.

$$\ln \frac{[A]_t}{[A]_o} = -kt$$

After 50.0 minutes, we have:

$$\ln \frac{[A]_t}{[A]_o} = -0.139 \text{ min}^{-1} (50.0 \text{ min}) = -6.95$$

$$\therefore \frac{[A]_t}{[A]_o} = 9.59 \times 10^{-4}$$

This number represents the fraction of the initial concentration that is left over after that time (50.0 minutes).

- If a given decomposition reaction follows first order kinetics and it has a rate constant of 0.0100 days⁻, what is its half-life? How long will it take for the original concentration of a given sample to come down to 12.5 % of the original?

    With the rate constant, we can quickly determine half-life for this reaction.

    $$k = 0.693 / 0.0100 \text{ days}^{-1} = 69.3 \text{ days.}$$

    What is the meaning of this half-life? That after 69.3 days we will have exactly one half of the initial concentration of the reagent. It is also true that after an additional 69.3 days, the initial concentration will have dropped to one quarter of the initial value. Since 12.5% of the initial concentration happens to be one eighth of the initial concentration, we should recognize this as three half lives. Therefore, it should take approximately 208 days for the concentration to drop to one eighth of the initial value.

- Consider the following first order reaction which has a half life of 10.0 minutes; rate = k [A].

    $$A_{(g)} + 2 B_{(g)} \Rightarrow AB_{2(g)}.$$

If you place initially an equimolar mixture of gases A and B in a container at a total pressure of 5.00 atm, what will be the total pressure in the flask after 25.0 minutes?

    The first thing that we should do is to determine the numerical value of k, since we know the half-life.

$$k = \frac{0.693}{10.0 \text{ min}} = 0.0693 \text{ min}^-$$

From the rate law, we can see that the reaction depends only on A. This means that although B is also disappearing as a function of time, its presence does not affect the rate of the reaction. All we need to know in order to understand the behavior of this reaction as a function of time is the pressure of gas A. However, the <u>initial total pressure</u> was given instead, plus the question deals with the total pressure after a certain time. Therefore, we need to be able to *figure out the pressure of A as a function of the total pressure and the time*. In order to do this, let's look at the reaction and try to understand what is happening to the GASES as the reaction proceeds. We see that every time three moles of gases combine (one of A and two of B), we form one mol of gaseous product (AB). This means that the total pressure in the container is going to decrease as a function of time.

Since we do not know exactly by how much this pressure is decreasing nor how fast, we can say that every time **x** atm of A disappear, **2x** atm of B also disappear and **x** atm of AB appear. We need to keep track of this and the easiest way of accomplishing this is by making a table, as shown below (right under the reaction).

	A	+	2 B	$\Longrightarrow$	AB$_2$
init:	2.50 atm		2.50 atm		0 atm
@time t:	2.50 - x		2.50 - 2x		x

What is the total pressure at any time?

$$P_{tot} = P_A + P_B + P_{AB_2}$$

$$= (2.50 - x) + (2.50 - 2x) + x$$

$$\therefore \mathbf{P_{tot} = 5.00 - 2x}$$

and: $\quad \mathbf{P_A = 2.50 - x}$

Let's go back to the problem. We know how the pressure of A changes as a function of time (since we were told that it was first order), therefore:

$$\ln \frac{[A]_t}{[A]_o} = -kt$$

$$\ln\left(\frac{[A]}{2.50 \text{ atm}}\right) = -0.0693 \text{ min}^- (25.0 \text{ min})$$

$$\therefore [A]_{25 \text{ min}} = 0.442 \text{ atm} \quad \checkmark$$

Since we are asked for the total pressure, all we need at this point is to figure out the value for **x**.

$$x = 2.50 - P_A = 2.50 - 0.442 = 2.06 \text{ atm}$$

Substituting this value into the equation for the total pressure, we get:

$$P_{tot} = 5.00 - 2x = 5.00 - 2(2.06 \text{ atm})$$

$$\therefore P_{tot} = 0.88 \text{ atm}. \quad \checkmark$$

- Consider the reaction below, for which rate = 0.0234 atm^{-1}min^{-1} $(P_A)^2$.

$$2 A_{(g)} + 3 B_{(g)} \Rightarrow C_{(s)} + 3 D_{(g)} + 4 E_{(g)}$$

If we place gases A, B and D in a given flask at initial partial pressures of 1.00, 3.50, and 2.00 atm, respectively, what will be the total pressure in that flask after 3.00 hours?

The first thing we may notice is that there is some gas D present in the mixture initially. Is that actually possible? Of course it is. We can place inside of the flask anything we want, including some of the product materials. However, the initial amount of gas D is not going to actually react; for that matter, it will behave as an inert substance. As the reaction progresses, some more D will be formed, so at the end, we should have more D than the amount we started with.

To actually start the problem, we should notice that the rate of the reaction depends only on gas A. According to the information given, the reaction is second order with respect to gas A, therefore, it obeys the following equation.

$$\frac{1}{[A]} - \frac{1}{[A]_o} = k t$$

Although we know the pressure of gas A initially, we do not know its pressure after 3.00 hours. This is precisely the kind of information we can get from this equation. Therefore,

$$\frac{1}{[A]} - \frac{1}{1.00 \text{ atm}} = 0.0234 \text{ atm}^{-1}\text{min}^{-1} \, (180 \text{ min})$$

$$\therefore [A]_{3 \text{ hrs}} = 0.192 \text{ atm} \quad \checkmark$$

Now, we want to know the total pressure in the flask after three hours, and all we have been able to determine so far is the concentration of A at that time. What is the relationship between the concentration of A at any time and the total pressure? In order to answer that question, we need to keep track of the partial pressures of all gases as a function of time, as we did in the previous problem.

	2 A	+	3 B	$\Longrightarrow$	$C_{(solid)}$	+	3 D	+	4 E
init:	1.00 atm		3.50 atm		0		2.00 atm		0 atm
@time t:	1.00 − 2x		3.50 − 3x		−		2.00 + 3x		4x

Therefore, at any time t, the total pressure in the flask should be equal to the sum of all the partial pressures (that of gases A, B, D, and E). These partial pressures are expressed above in terms of x. Therefore,

$$P_{tot} = (1.00 - 2x) + (3.50 - 3x) + (2.00 + 3x) + 4x$$

$$\therefore P_{tot} = 6.50 + 2x$$

We can also see that the pressure of A at any time is given by:

$$P_A = 1.00 - 2x.$$

We know the pressure of A after three hours (0.192 atm). Substituting into the equation for the pressure of A at any time, we can figure out the numerical value for x.

$$0.192 \text{ atm} = 1.00 - 2x$$

$$\therefore x = 0.404 \text{ atm} \quad \checkmark$$

Finally,

$$P_{tot} = 6.50 \text{ atm} + 2 \, (0.404 \text{ atm})$$

$$\therefore \boldsymbol{P_{tot} = 7.31 \text{ atm}} \quad \checkmark$$

12.7 Effect of Temperature on Rate.

It has been observed experimentally that all chemical reactions go faster as the temperature increases. The purpose of this section is to try to understand why this is the case. Back in Chapter 4, we learned the Kinetic Molecular Theory of Gases, which is based in the following statements.

1. Gases consist of molecules in constant, random motion.
2. There are no forces of attraction or repulsion between the molecules.
3. Molecules are insignificantly small compared to the volume of the container.
4. Collisions are elastic, which means that there is no kinetic energy lost during the collision. Energy may be transferred from one molecule to another, but none will be lost.
5. The average kinetic energy of the molecules is proportional to the absolute temperature. Actually, it was stated that the average kinetic energy[1] of the molecules is given by:

$$\overline{KE} = \frac{3}{2}RT$$

Let's see if this theory is sufficient to explain the experimental observations. For a typical reaction, increasing the temperature by 10 degrees has the effect of doubling the reaction rate. At first glance, we can immediately say that this is true because it should be clear that at a higher temperature, the molecules will have a higher average kinetic energy, and since they are moving faster, they will collide more often and yield the product at a faster rate. Of course, in order for that reasoning to be correct, the increase in the average kinetic energy must be in the order of two (should be twice as large). Let's see by what factor the average kinetic energy increases.

$$\overline{KE}_{35} / \overline{KE}_{25} = \frac{\frac{3}{2}(8.314 \frac{J}{Kmol})(308K)}{\frac{3}{2}(8.314 \frac{J}{Kmol})(298K)} = 1.03$$

We can see that the average kinetic energy only increased by about 3%! Therefore, this argument is not sufficient to explain the experimental result. Let's look at a little further into the idea that molecules need to collide, or come in contact, in order for a reaction to occur. We describe this in something we call the collision theory.

---

[1] Remember that not all molecules have the same kinetic energy at a given temperature.

12.8 Collision Theory.

1. For molecules to react they must at least come in contact.
2. The rate of the reaction will depend on the number of collisions per second.
3. Not all collisions are effective, but if we increase the number of collisions, we can expect more effective collision also.

Let's see if this theory is enough to explain the experimental observations. In order to analyze this, let's go back to our hypothetical one-step reaction.

$$A_{2(g)} + B_{2(g)} \Rightarrow 2 AB_{(g)}$$

Since the reaction occurs in one single step, we can visualize this reaction at the molecular level, that is, one molecule of A_2 (the while circles) collides with one molecule of B_2 (the dark circles) and in that simple bimolecular collision, we observe directly the result: two molecules of AB are formed.

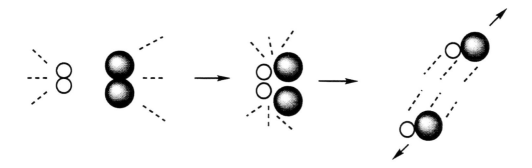

Common sense dictates that the rate of the reaction is proportional to the number of collisions per second, that is, the speed of the reaction will depend on how often the molecules of the reactants come in contact with one another. What will the effect be on the number of collisions per second, if we were to increase the temperature by about 10°C? We should expect to have more collisions per second at a higher temperature, simply because molecules are moving faster and therefore, will encounter one another more often. It turns out that an increase in temperature by 10°C does indeed increase the frequency of collision, but only by about 2%. Once more, this increase is certainly not enough to explain by itself the fact that the rate is doubled! We should point out that at room temperature and about 1 atm pressure, there are about 1×10^{31} collisions in every second. If each collision were effective, the reaction would be over in a fraction of a second. This is normally not the case. Why is it that not all collisions are effective? This is mainly due to two factors: an alignment factor and an energy factor, both of which are very important.

1. Alignment Factor.

In order for a collision to be effective, the molecules must approach one another at a particular angle, otherwise they will bounce off unreacted. We could see above that the collision between the two molecules had to be such that at the exact moment of contact, each atom of A was in contact with one atom of B. If that were not the case, then the reaction would have been taking place in steps, therefore, it would not occur. Below, we can see an example of the improper orientation when the collision takes place.

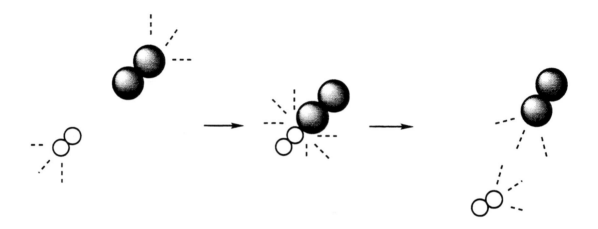

2. Energy Factor.

Slow moving molecules will rebound unreacted even if they are properly aligned (repulsion of electronic clouds). You need a **minimum** amount of energy for the electronic clouds to penetrate and thus yield the product if they have the correct alignment. We call this minimum energy the **Activation Energy**, Ea.

The Maxwell-Boltzmann[1] distribution showed (as can be seen in the plot below) that not all molecules in a given sample have the same energy. In this illustration, we have plotted the fraction of molecules with a given energy versus the kinetic energy of the molecules. We can see that the largest fraction of molecules have an energy equal to what we call the most probable energy (E_{mp}), but we also have a few slow moving molecules (on the left side) and some fast moving molecules (on the right side). If we call the minimum amount of energy necessary for the collision to be effective Ea and we arbitrarily say that Ea is much higher than the most probable energy, then we can see that only a small fraction of molecules (shaded area) have enough energy for the collision to have a chance of being effective. To make it worse, remember that from those collisions, only a small fraction will be effective after all, because most of them

---

[1] See Section 4.6.

will have the incorrect alignment.

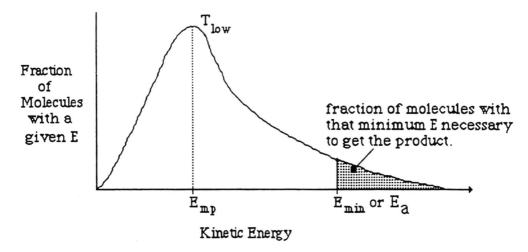

The plot we see above is at a given temperature, let's say 298 K. Now, let's increase the temperature by ten degrees, to 308 K. At a higher temperature, the shape of the distribution (the plot) changes in the following manner: the maximum of the curve shifts to the right by about 3 % only, and there are less molecules with that particular energy (the curve is flatter), plus we observe an increase of high energy molecules which actually turns out to be about twice as large as it was at 298 K.

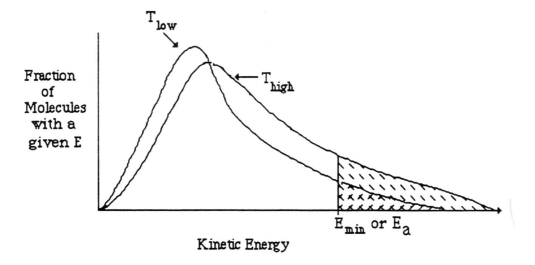

Therefore, the main reason as to why the rate is doubled for a typical chemical reaction with a ten degree increase in the temperature is that the *number of high energy molecules is approximately doubled* (the average kinetic energy increased only slightly and so did the number of collisions per second).

12.9 Temperature Dependence on **k**.

We have seen that the rate equation relates the concentration of the reacting species to the rate of the reaction. Can this same rate equation explain the influence that the temperature has on the rate of the reaction? It should! Let's continue our discussion with the same hypothetical one step reaction. For this reaction, we have already seen that the rate equation is given by:

$$\text{rate} = k\,[A_2]\,[B_2]$$

Where is the temperature dependence? It is in the constant itself, **k**. The rate constant is indeed at constant at a given temperature, but if we change the temperature, so will the value of **k**.

Arrhenius[1] tried to explain the temperature dependence of **k** by proposing the following equation:

$$k = \rho A e^{-(E_a/RT)}$$

In this equation, the ρ represents the steric factor which is related to the shape and orientation of the molecules, the A is a constant which is characteristic of the particular reaction, Ea is the activation energy in units of J/mol, R is the universal gas constant (8.314 J/molK), and T is the absolute temperature (K).

It is quite inconvenient to work with that equation as written, so let's modify it slightly. If we were to take natural logarithms on both sides of the equation, a new equation would look like:

$$\ln k = \ln(\rho A) - E_a/RT$$

Once more, this is the equation for a straight line in which *ln k* is y, *1/T* is x, *ln (ρA)* is the intercept, and – *Ea/R* would be the slope. Therefore, if we make a plot of *ln k* vs. *1/T* we will always get a straight line.

---

[1] Svante Arrhenius was a Swedish chemist who lived from 1859 until 1927. He won the Nobel prize for Chemistry in 1903.

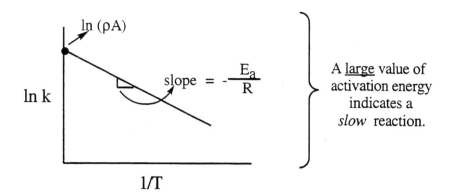

A large value of activation energy indicates a *slow* reaction.

We have just seen a very useful way in which to determine the activation energy for a chemical reaction. It involves going into the laboratory and measuring the rate constant at various different temperatures. Remember that to measure the rate constant at different temperatures, all we need to do is repeat the graphical method[1] at the desired temperatures. If instead we want to *calculate* the activation energy (rather than measure it experimentally), all we need to remember is the definition for the slope of that curve.

$$Slope = -\frac{Ea}{R} = \frac{\Delta y}{\Delta x} = \frac{\Delta \ln k}{\Delta(1/T)}$$

From there, we can derive the following equation, which will allow us to calculate the numerical value for the activation energy if we are given two points in the graph (that is, the rate constant at two different temperatures).

$$\ln(\frac{k_2}{k_1}) = -(\frac{Ea}{R})(\frac{1}{T_2} - \frac{1}{T_1})$$

- The rate constant for a given reaction is found to be 55.2 s⁻ at 27°C and 100.0 at 37°C. Calculate the activation energy for this reaction (in kJ/mol).

    This problem should be very simple, and all it requires is the direct substitution into the equation we just derived. However, we must keep in mind that the temperatures must be in Kelvin always!

$$\ln(\frac{100.0 s^{-1}}{55.2 s^{-1}}) = -(\frac{Ea}{8.314 J/molK})(\frac{1}{310K} - \frac{1}{300K})$$

This gives us an activation energy of,

---

[1] Or for that matter, do the initial rate method at various temperatures.

$$Ea = 45{,}900 \text{ J/mol or } 45.9 \text{ kJ/mol} \quad \checkmark$$

- The activation energy for a given reaction is 67.8 kJ/mol. How much faster is the reaction going at a temperature of 57°C than at 27°C?

Once more, we have to use the same equation in order to answer this question. However, it may not be clear as to what we are looking for. The question is asking how much faster the reaction is going. That is exactly the meaning of the ratios of the two constants (k_2/k_1). Since k depends on temperature, as the temperature increases, so does k. Therefore, an increase in the speed of the reaction due to the temperature will be reflected as an increase in the numerical value for k. At any rate, the ratio represents the fractional increase of the speed of the reaction.

$$\ln\left(\frac{k_2}{k_1}\right) = -\left(\frac{67{,}800 \, J/mol}{8.314 \, J/molK}\right)\left(\frac{1}{330K} - \frac{1}{300K}\right)$$

The ratio itself is equal to 11.8. This means that the reaction is going almost twelve times as fast at 57°C than it was at 27°C.

12.10 Transition State Theory.

We have seen what the energy requirements are in order to obtain a successful collision. We described that a certain minimum amount of energy was necessary for this purpose (the activation energy). Once more, let's use the example we established from the start, the hypothetical one step reaction. Since this reaction involves a single bimolecular collision between A_2 and B_2, we established that in that single collision we must achieve quite a few results. We must break the A-A bond, we must break the B-B bond, and we must form the two A-B bonds. These three seemingly separate events must take place simultaneously in order for this reaction to occur in one step. This means that right at the moment of the collision, we will form an unstable (high energy) species that shows all of these events in process. We will call this species the **activated complex**.

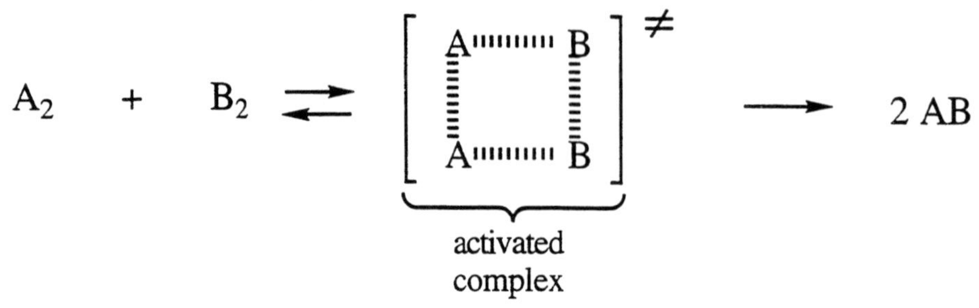

$$A_2 + B_2 \rightleftharpoons \left[\begin{array}{c} A \cdots B \\ \vdots \quad \vdots \\ A \cdots B \end{array}\right]^{\neq} \longrightarrow 2 \, AB$$

$$\underbrace{}_{\text{activated complex}}$$

This activated complex is a transition state of high energy. It looks somewhat like an intermediate state in between the reactants and the products (A-A and B-B bonds are half broken and the A-B bonds are half formed). Keep in mind at all times that in order for the reactants to form the products, they must form this activated complex shown above. Activated complexes cannot be isolated; they can either go back to the reactants or form the products but they are too unstable (high energy) to exist for a long time. The energy necessary to go from the reactants to the transition state is what we call the activation energy. This can all be seen graphically below. In this diagram, we plot the potential energy of the system as function of the advancement of the reaction. In other words, we can see how the potential energy changes as the **molecules** of the reactants interact to form the **molecules** of the products. The difference in energy between the reacting molecules and the transition state is what we call the activation energy of the forward reaction. The difference in energy between the reacting molecules and the product molecules (initial and final state) is the ΔE (or the ΔH) of the reaction. If the molecules of the product were to react with one another to form the reacting molecules (the reverse reaction), then the activation energy would be the difference in energy between the products and the transition state. Keep in mind that the transition state for the forward and the reverse reactions would be exactly the same. The activation energy for the reverse reaction is, therefore, equal to $E_a + \Delta E$.

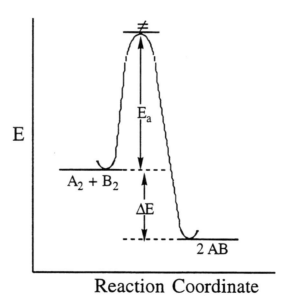

Reaction Coordinate

Many times reactions have very high activation energies, which make these reactions rather slow. It would be convenient to find a way to go from the reactants to the products with lower energy requirements, that is, with a lower activation energy. This way more molecules would have enough energy to react and the reaction would go faster. This is impossible to achieve unless we can figure out how to carry out the reaction with a totally different mechanism that will have an overall lower activation energy. This can be achieved with a suitable catalyst. A **catalyst** is a substance that increases the rate of a reaction without being consumed in the reaction itself. It provides an alternate mechanism[1] with a lower activation energy (the graph below shows the difference between a catalyzed reaction [dotted line] and one that is non-catalyzed [solid line]). We should point out that the purpose of this plot is to show that indeed, the activation energy does come down with a catalyzed reaction. However, since the mechanism is going to change, the actual reaction path will be a bit more complicated. In the example below, we will see what is meant by a more complicated path. Keep in mind that at the very least, the catalyst must get involved in a given step, and then it must be regenerated in another step. This would yield two clear activation energies (two bumps)[2] for the catalyzed reaction.

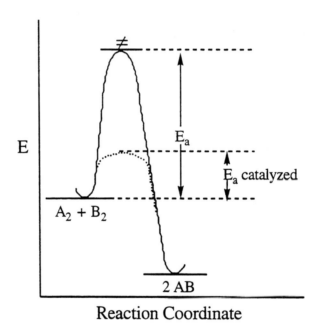

Reaction Coordinate

As an example of catalysis, let's take the case in which a metal surface serves as a catalytic surface for the reaction between A_2 and B_2. We would have to search for a metal which could adsorb both molecules. Once the molecules are attracted to the metal surface, that attraction will in turn weaken the A-A bond and the B-B bond. It will also place the two molecules in the proper alignment, therefore the reaction will readily occur. This metal must be such that the product (A-B) will not be strongly attracted to the metallic surface, thus it will escape once it is formed. The newly recovered surface will be ready to accept more A_2 and B_2

---

[1] As we have previously mentioned, mechanisms refer to a sequence of elementary steps (molecules, not moles) that show how the reaction occurs at the molecular level. We will discuss this further in the next section.

[2] And this is for the simplest scenario. More likely than not, we will have multiple steps, and multiple activation energies in the catalyzed reaction. The important thing to remember is that all of those activation energies will be a lot lower than the one for the uncatalyzed reaction.

molecules to continue the reaction.

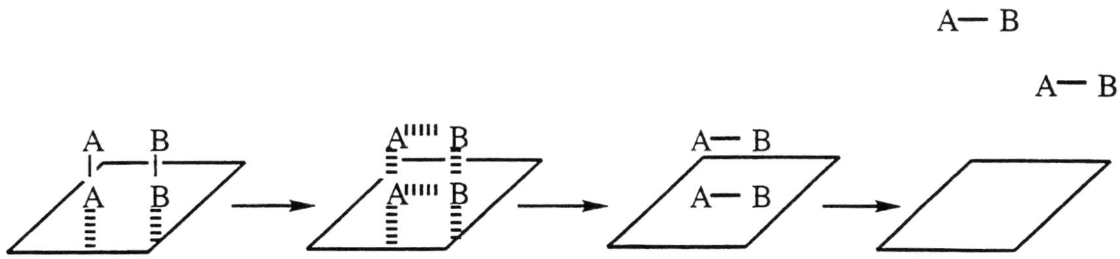

Let's talk about one more example. Consider the reaction of Ethene, commonly known as Ethylene, C_2H_4, with hydrogen gas. This molecule has a double bond between the two carbons, thus giving it one degree of unsaturation. If we wanted to saturate it with hydrogens, we should be able to simply add hydrogen gas (H_2). The reaction itself has a ΔH of about - 125 kJ/mol. However, the activation energy for this reaction is extremely large, which means that we could wait forever and it still would not occur. The reason for the reaction being so slow is that we would have to break an H-H bond, a very strong sigma bond, and that imposes a very large energy barrier (Ea). Can we get a suitable catalyst to speed up this reaction? Yes! One such catalyst could be finely divided[1] Nickel (Ni) metal.

Mechanism for the reaction of Ethene with Hydrogen gas with Nickel metal.[2]

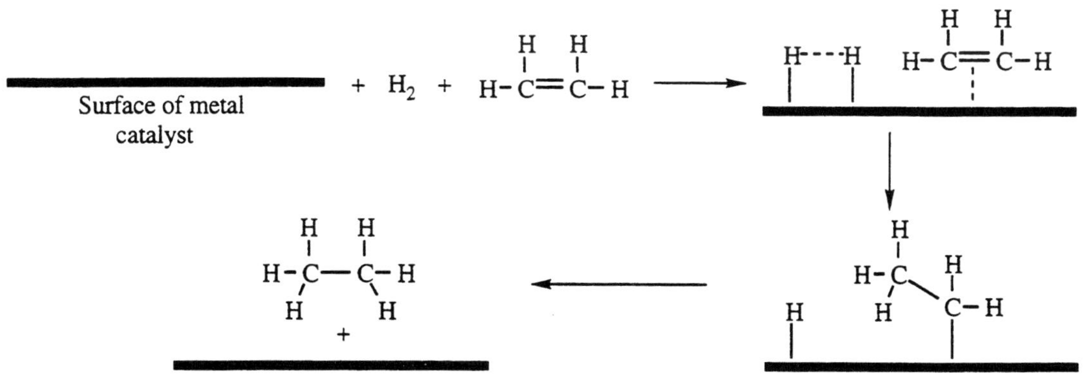

Other metals work as well, for example: Pd, Pt, etc. The hydrogen and ethylene molecules are adsorbed onto the surface of the metal. This will, for all practical purposes, break the bond between the two hydrogen atoms, because the metal-hydrogen bond is fairly strong. The pi bond in the ethylene is weakened simultaneously. As this molecules/atoms move along the surface of

---

[1] This would serve the purpose of increasing the surface area exposed to the gases.
[2] Concept originally seen in Solomons, T. W. Graham, Fundamentals of Organic Chemistry, 5th Edition, New York: John Wiley and Sons, Inc., 1997.

the metal, a stepwise transfer of hydrogen atoms takes place, and this produces the product (C_2H_6), which then leaves the surface of the metal. The diagram above shows these three steps in our new mechanism, that is: Step #1: Adsorption of the reacting molecules, Step #2: Addition of first hydrogen, and Step #3: Addition of the second hydrogen.

The energy graph below shows a comparison between the catalyzed and uncatalyzed reactions. Please, notice that the catalyzed reaction has three "bumps", which correspond to the three steps shown in the mechanism of the catalyzed reaction. Each of the three maxima represent the transition states of the three steps, and the minima, the intermediates. We can see that the three activation energies are a lot lower than the activation energy for the uncatalyzed reaction. Therefore, this path would lead to a much faster reaction.

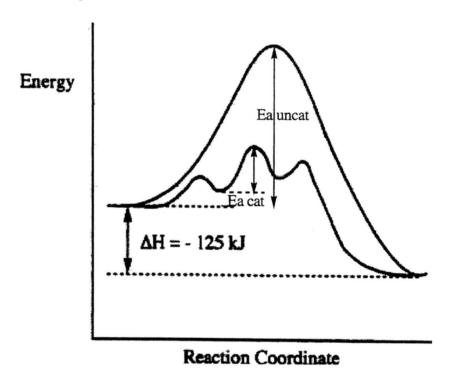

12.11 Reaction Mechanisms.

At this point we already have an idea as to what reaction mechanisms are, but we should concentrate now on two main goals: 1) being able to predict the rate equation from the mechanism of the reaction, and 2) being able to predict a plausible mechanism based on the experimentally determined rate equation.

We mentioned that catalysts affect the rate of a chemical reaction by generating a different reaction mechanism with a lower activation energy. In the example of the hydrogenation of ethylene[1], we saw that indeed a totally different mechanism is followed and that the overall activation energy of this "new process" is lower, therefore, the reaction is faster. Let's look at one more example.

Very high in the atmosphere, there is a layer of ozone (O_3) which protects us from the ultraviolet radiation (high ν) by absorbing ultraviolet (UV) light. Upon the absorption of a photon of light, a given ozone molecule will break and form a normal oxygen molecule and an oxygen radical[2].

$$O_3 \xrightleftharpoons{h\nu} O_2 + O\cdot$$

Oxygen radical = unpaired electron

This oxygen radical is a very reactive species, so we may be concerned with the following reaction, which could have the devastating effect of depleting the ozone layer.

$$O_3 + O\cdot \Rightarrow 2\,O_2$$

However, it turns out that this is a very slow reaction and the formation of ozone from the electrical storms at that altitude outruns this reaction. In conclusion, there is a very delicate balance between the destruction and the formation of ozone at that altitude. The introduction of foreign species at that layer could have an adverse effect on this delicate balance. For example, the presence of chlorofluorocarbons (for example: CF_3Cl) from aerosol cans in the upper atmosphere could lead to the following reaction.

$$F_3C\text{-}Cl + light \Rightarrow F_3C\cdot + \cdot Cl$$

This chlorine radical is extremely reactive and it will combine rather quickly with an ozone molecule to form a regular oxygen molecule and a reactive intermediate (ClO). This reactive intermediate will very quickly react with an oxygen radical to form another oxygen molecule and thus regenerating the chlorine radical. We can immediately see the problem we have generated. We have created a species (the chlorine radical) which will create havoc at this altitude because if will act as a catalyst in the destruction of the ozone layer.

$$Cl\cdot + O_3 \Rightarrow ClO + O_2$$

$$ClO + O\cdot \Rightarrow Cl\cdot + O_2$$

---

[1] See Section 12.10.
[2] An oxygen radical is simply an oxygen atom (with an incomplete octet, therefore it has unpaired electrons).

The overall reaction (the sum of the two steps) is the reaction that normally occurs at that altitude, but which has a low activation energy. In the presence of the chlorine radical, we have duplicated this reaction through a totally different mechanism with a much lower activation energy, and therefore, the reaction occurs much more rapidly. To make it worse, we do not need a whole lot of radicals in order to carry out this reaction, since only a minute amount can destroy many ozone molecules since it is constantly regenerated. This is another property of catalysts: we do not need a very large quantity of these because they are regenerated. In future courses, we will hear phrases like "the metal was added in catalytic amounts." This means that only a minute amount was needed in order to successfully achieve a reaction.

Let us refer now to the following reaction.

$$3 \text{ ClO}^- \Rightarrow \text{ClO}_3^- + 2 \text{ Cl}^-$$

By now we know that there is no way we can safely predict the rate equation for this reaction. The best we can do is to state that the rate equation would be:

$$\text{rate} = k\,[\text{ClO}^{-1}]^x$$

In order to determine the order with respect to the hypochlorite ion, we would need to do either the graphical method or the initial rate method. Of course, this has been done before, and the following is the actual experimentally determined rate equation.

$$\text{Rate} = k\,[\text{ClO}^{-1}]^2$$

This should mean something to us at this point. We should be certain that this reaction does not take place in a single termolecular collision between all three hypochlorite ions. If this were the case, then the reaction would be third order with respect to the hypochlorite ions. Instead, it is second order. What does that mean? That the mechanism is a lot more involved. In this more complicated mechanism there are going to be various steps (we cannot even start to imagine how many) that will take longer or shorter to occur, depending on what actually is happening in that particular step.[1] The slowest of these steps will determine how fast the reaction is going to go and it will be the actual step whose rate equation we measured. This means that the slow step of the mechanism (which we will call the **rate determining step – RDS**) involves the bimolecular collision of two hypochlorite ions. The job of the chemist is then to propose a mechanism that is consistent with the observed experimental rate equation. In order to propose a plausible

---

[1] If the step is "complicated", that is, chemically difficult, then it will take longer than others where very little is involved.

mechanism, we must make certain of two things.

1) The sum of all the steps shall give us the overall reaction.

2) The slow step (the rate determining step) should correspond to the observed experimental rate equation.

Let's say that a given chemist proposes the following two step mechanism for the reaction we have been discussing in this section.

$$2\ ClO^- \Rightarrow ClO_2^- + Cl^- \quad \text{SLOW}$$

$$ClO_2^- + ClO^- \Rightarrow ClO_3^- + Cl^- \quad \text{FAST}$$

We should immediately notice that both criteria are met. The sum of these two elementary steps do add indeed to the overall reaction, and the step we have called the slow step does involve the bimolecular collision of two hypochlorite ions. Therefore, this is a plausible[1] mechanism for this reaction. As a chemist, we just have to think logically as to what could possibly happen when these two ions collide. Which bonds are most likely to be broken and why? This is not the easiest thing to do, but we will practice with some very simple examples and will always leave it at that level.

Before we do such examples, we should stop for a second and ponder as to what is the significance of the orders, in light of the information we just learned. **The order of a reagent normally tells us the number of molecules of that reagent that are involved (or that collide) during the slow step of the mechanism.** That means that once we see the rate equation (experimentally determined), we should look at the order for all the reagents in that equation and try to combine that same number of molecules of each in the slow step of the mechanism. Let's do an example to see exactly what we mean by all of this.

- Propose a simple two step mechanism for the following reaction, knowing that the rate law is rate = k [AB] [D_2].

$$2\ AB_{(g)} + D_{2(g)} \Rightarrow 2\ ABD_{(g)}.$$

Where should we start? We should notice immediately that the order for the reagents

---

[1] We should comment that there is no such thing as a correct mechanism. There are simply mechanisms that are possibly correct. This is the implication of the word "plausible". In the American Heritage Dictionary of the English Language (1978) it is defined as "acceptable, seemingly correct".

is not the same as the molecularity (the coefficients in the balanced reaction). This means that the mechanism is a little more complex than a single termolecular collision. Of course, we knew that since we were asked to propose a two step mechanism. We know, for sure, that the slow step of the mechanism must involve a bimolecular collision between one molecule of AB and one molecule of D_2. This is the information we just obtained from the rate law. However, what could possibly happen to these two molecules upon this collision? In other words, which bonds will be broken and which bonds will be formed? We should get a clue from the eventual products of the reaction. Since we are going to end up with ABD eventually, then we could easily propose the following step as the rate determining step.

$$AB_{(g)} + D_{2(g)} \Rightarrow ABD_{(g)} + D_{(g)}$$

Since there is only one more step to go, and since we know exactly what the overall reaction must be, then all we have left to do is to figure out what is missing in order for the sum of the two steps to give the overall reaction. Therefore, the mechanism below is a plausible mechanism[1].

$$AB_{(g)} + D_{2(g)} \Rightarrow ABD_{(g)} + D_{(g)} \qquad \text{SLOW}$$

$$AB_{(g)} + D_{(g)} \Rightarrow ABD_{(g)} \qquad \text{FAST}$$

12.12 Equilibrium Elementary Steps.

Many times, a particular step in one of these mechanisms is a reversible step. A reversible step is one that could go both in the forward direction (towards the products) and in the reverse directions (towards the reactants). We call such a step an equilibrium step since the reactants will combine to form the product of that step, but as the concentration of the product starts to build up, some of it will react to form the reactants again. The apparent effect is to slow down the reaction, since we are forming back the reactants. Once we reach equilibrium, the concentrations of the reactants and the products will stay constant as a function of time. Always remember that the equilibrium state is a dynamic state. The reason the concentrations are not changing any more as function of time is because at equilibrium, *the rate of the forward reaction is equal to the rate of the reverse reaction.*

Let's consider the following reversible elementary step. Here one molecule of A breaks up to form two molecules of B. The forward reaction has a rate constant symbolized by k_1,

---

[1] Please, notice that the sum of the two steps will give us the overall reaction.

whereas the reverse reaction has a rate constant symbolized by k_{-1}.

$$A \underset{k_{-1}}{\overset{k_1}{\rightleftharpoons}} 2B$$

However, the rates of the forward and reverse reactions can be written immediately since these are elementary steps (these species represent molecules, not moles). In the forward reaction, we have one molecule of A, therefore it should be first order with respect to A. In the reverse reaction, we have two molecules of B reacting, therefore, the order with respect to B is two.

$$\text{rate forward} = k_1 [A]$$

$$\text{rate reverse} = k_{-1} [B]^2$$

We just stated that at equilibrium, the rate of the forward reaction is equal to the rate of the reverse reaction, therefore,

$$k_1 [A] = k_{-1} [B]^2$$

or, rearranging it, we get:

$$\frac{k_1}{k_{-1}} = \frac{[B]^2}{[A]}$$

As we shall see in the next example, this becomes very useful sometimes in trying to find a mathematical expression for a proposed mechanism in terms of stable species. It is actually important to always express a rate equation in terms of stable species because we should be able to measure their concentration. *Conclusion:* If we ever get a mechanism that has a radical in the slow step, we should use the other steps to find an expression for the concentration of that radical is in terms of more stable species.

- Consider the reaction below.

$$H_{2(g)} + I_{2(g)} \Rightarrow 2 HI_{(g)}$$

Determine the mechanism for this reaction, given that its rate equation is: rate = $k_{exp} [H_2][I_2]$.

This should require very little thought, as the molecularity corresponds with the orders. The immediate implication is that the reaction occurs in one single step: the bimolecular collision of a hydrogen molecule with an iodine molecule. This thought prevailed for many years until scientists discovered the presence of iodine radicals[1] every single time in the vessel where this reaction was carried out. This implies that somehow the radicals must be part of the mechanism and at the same time, the overall rate law must correspond to one molecule of hydrogen and one molecule of iodine. A new mechanism had to be proposed.

- Let's say that chemist #1 proposes the following mechanism for that reaction. Is this a plausible mechanism?

$$I_2 \underset{k_{-1}}{\overset{k_1}{\rightleftharpoons}} 2\, I\bullet \qquad \text{Fast}$$

$$H_2 + I\bullet \xrightarrow{k_2} HI + H\bullet \qquad \text{Slow}$$

$$H\bullet + I_2 \xrightarrow{k_3} HI + I\bullet \qquad \text{Fast}$$

In order to answer this question, we need to always start with the rate determining step, or the slow step. This is the second step and in here we can see two species colliding with one another: a hydrogen molecule and an iodine atom (a radical). This would lead to the following rate equation:

$$\text{Rate} = k_2\,[H_2]\,[I\bullet]$$

We are immediately tempted to say that it is not a good mechanism because it does not match the experimentally determined one (rate = $k_{exp}\,[H_2]\,[I_2]$). However, let's not jump to conclusions! Remember that we should always express the rate equation in terms of stable species. An iodine atom is not a stable species, therefore, we should look at an equilibrium step that involves the iodine radicals to see if we can make an algebraic substitution for the concentration of iodine radicals in terms of stable molecules (like the iodine molecule, the hydrogen molecule, or the hydrogen iodide molecule).

This is exactly what we observe in the first step: an equilibrium step that involves iodine radicals. From that step, we can conclude:

$$\frac{k_1}{k_{-1}} = \frac{[I\bullet]^2}{[I_2]}$$

---

[1] Radicals can be detected with an instrument called an electron spin resonance spectrometer.

Solving for the concentration of iodine radicals, we obtain the following expression.

$$[I\bullet] = \left(\frac{k_1}{k_{-1}}\right)^{1/2} [I_2]^{1/2}$$

Substituting this expression into the one we derived for the rate equation from the rate determining step we get the following rate law:

$$\text{rate} = k_2 [H_2] \left(\frac{k_1}{k_{-1}}\right)^{1/2} [I_2]^{1/2}$$

Grouping all the constants together, we can get it into a more familiar form.

$$\text{rate} = k_2 \left(\frac{k_1}{k_{-1}}\right)^{1/2} [H_2] [I_2]^{1/2} = k_{expt} [H_2] [I_2]^{1/2}$$

We can see that the order for the iodine molecule does NOT correspond to the experimentally determined order (which was one), therefore, this is NOT an acceptable mechanism.

- Is the following mechanism consistent with the experimental observations? Once more, we are referring to the same reaction.

$$I_2 \underset{k_{-1}}{\overset{k_1}{\rightleftharpoons}} 2 I\bullet \qquad \text{Fast}$$

$$2 I\bullet + H_2 \xrightarrow{k_2} 2 HI \qquad \text{Slow}$$

As always, we start with the rate determining step, which this time is the last step. From here, we can write the following rate law.

$$\text{rate} = k_2 [H_2] [I\bullet]^2$$

The first step, the equilibrium step, is the same as the one in the above example, therefore we already know the expression for the concentration of the Iodine radicals.

$$[I\bullet]^2 = \left(\frac{k_1}{k_{-1}}\right) [I_2]$$

When we substitute in the rate equation above, we get a rate law that is consistent with the experimentally observed one.

$$\text{rate} = k_2 \left(\frac{k_1}{k_{-1}}\right) [H_2][I_2] = k_{expt}[H_2][I_2]$$

- Consider the following reaction.

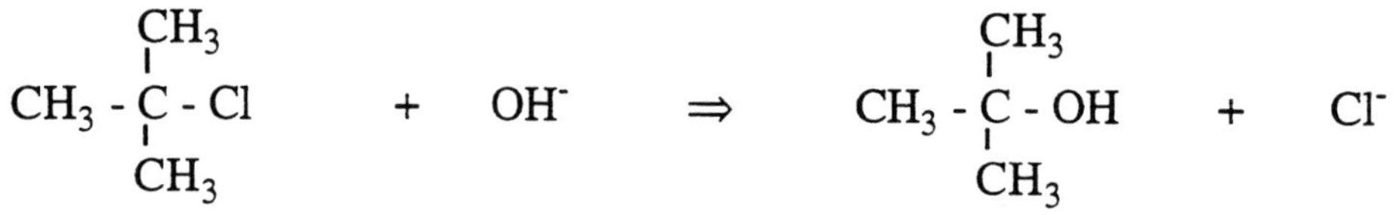

$$\underset{\underset{CH_3}{|}}{\overset{\overset{CH_3}{|}}{CH_3-C-Cl}} + OH^- \Rightarrow \underset{\underset{CH_3}{|}}{\overset{\overset{CH_3}{|}}{CH_3-C-OH}} + Cl^-$$

What would you say is the rate equation?

Certainly based on the fact that no more information is given, we are forced to say that the rate equation must be unknown, and the best we can do is to say that the rate equation is given by the following expression.

$$\text{rate} = k\,[(CH_3)_3CCl]^x\,[OH^-]^y$$

However, if you were told that experimentally, the rate equation is: rate = k [(CH$_3$)$_3$CCl], then we should be able to think about a plausible mechanism for this reaction. Suggest a two step mechanism that would work.

Based on the information provided, we know that only one molecule is involved in the slow step of the mechanism. Therefore, the rate determining step must be the following.

$$(CH_3)_3C\text{-}Cl \rightleftharpoons (CH_3)C^{+1} + Cl^{-1}.$$

How did we know that the bond broke in such a way so that the chlorine took both electrons to form a negative ion? Because we looked at the products formed and that is one of them.

This would immediately be followed by the fast step, which would have to be:

$$(CH_3)C^{+1} + OH^{-1} \Rightarrow (CH_3)_3C\text{-}OH$$

Chapter 13: Organic Chemistry.

13.1 Introduction.

We give the name of organic chemistry to that branch of chemistry that studies all carbon compounds. Why is there a special branch of chemistry dedicated to the study of just carbon compounds? Because of the versatility of carbon as an element. As we have previously seen, carbon can form single, double or triple bonds with other carbon atoms or with other atoms like O, S, N, etc. We will see that it also has the ability to form *long* chains with other carbon atoms (catenation). There are no other elements that can do that. It also has the ability to branch off these long chains and/or form cyclic compounds. Therefore, there are innumerable possibilities in which to combine carbon atoms to form these organic compounds.

Many things we see around us is organic in nature. A few examples are: proteins, sugars, starches, fats, hormones, vitamins, alkaloids (like nicotine and morphine), antibiotics (like tetracycline and penicillin), drugs (like aspirin, LSD, and cocaine), coal products, rubber, perfumes, insecticides, explosives (like TNT and nitroglycerine), plastics, synthetic fibers (like Nylon and Rayon), dyes, paints, and detergents.

There are many different places where we could start our overview of these organic compounds, but we will start by looking at some of the hydrocarbons first. The simplest family of all organic compounds are the alkanes.

13.2 Alkanes (C_nH_{2n+2}).

The general formula of these hydrocarbons is C_nH_{2n+2} and it consists of what we call *saturated hydrocarbons*. By saturated we mean a compound that consists exclusively of single bonds and having the maximum number of hydrogens[1] possible. Since all the bonds in the molecule are single and we know that carbon forms four bonds, then the hybridization of all carbons in an alkane must be **sp³**. Let's look at some of the members of this family:

1) Methane. - simplest member of the alkane family. Chemical formula: CH_4.

This simplest compound is a perfect tetrahedron because the hybridization of the carbon is sp³. We can notice that all bonds are completely equivalent and any H-C-H bond angle in the molecule is 109.5°.

---
[1] Or univalent atoms, which are those that form only one bond (like the halogens).

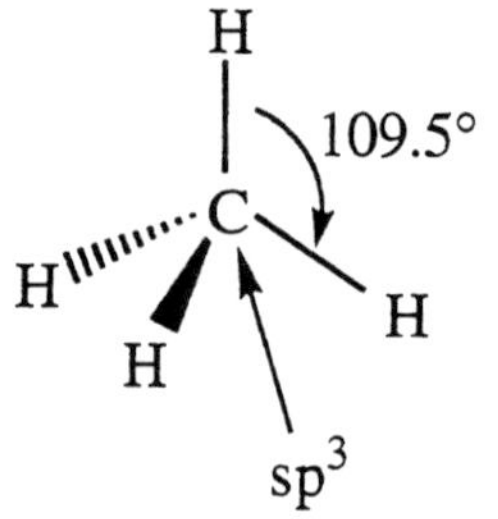

$$\text{H--C(109.5°)--H, H, H} \quad sp^3$$

2) <u>Ethane</u> - the second member of the alkane family. Chemical Formula: C_2H_6.

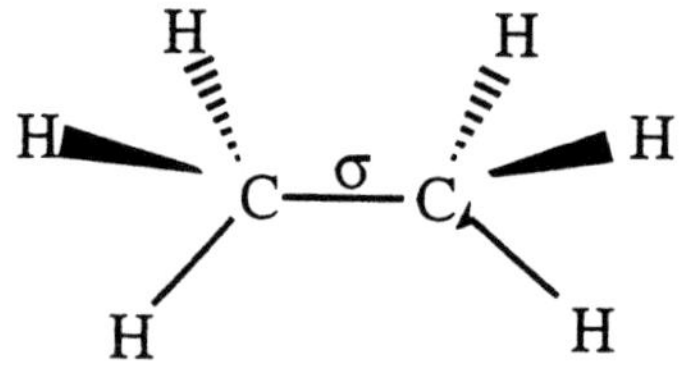

$$H_3C-\sigma-CH_3 \text{ (with all H's shown)}$$

Once more, all bond angles in the molecule are approximately 109.5°. We should remember that all bonds are also sigma (σ) bonds because they are single bonds. This allows for "free rotation" along the carbon-carbon single bond[1]. This is the main reason as to why all the positions are equivalent, so as we saw in Chapter 8, if we were to substitute any one of the hydrogens for another atom (let's say a bromine atom), we would only have one possible isomer. That is, rotation would interconvert all positions and make them equivalent.

3) <u>Propane</u>. Chemical Formula: C_3H_8. The main components of natural gas are methane, ethane, and propane, in decreasing order[2]. There are also very minor components of other small alkanes. Many times, for practical purposes, people assume that natural gas is methane gas.

$$CH_3CH_2CH_3$$

As we will see in organic chemistry, it is a lot easier (less involved) to write these structures with a line formula. In this notation, every end of a line and every corner represents a carbon atom and we indicate all the bonds to any atom except hydrogen. Using this notation, the line representation for propane is shown below. This notation exhibits approximate bond angles between the carbons. We can see that the C-C-C bond angle is about 109.5°. The carbon on the left is showing one bond to the middle carbon in that structure. Since all carbons form four bonds, then the three missing bonds must be three hydrogens. Therefore,

---

[1] We should build this molecule with the molecular models and convince ourselves that this is the case. The carbon atoms are the black ones, whereas the hydrogen atoms are the white ones (in your models).

[2] That is, methane is the largest component, followed by ethane, and so on.

that first carbon is a CH_3. The same thing is true about the last carbon. However, the structure represented shows that the middle carbon is involved in two bonds (to the two other carbons). Therefore, the remaining two bonds must be to two hydrogens. This makes the middle carbon a CH_2.

Let's look very quickly at the next few members of this hydrocarbon family. We will draw the compounds in both ways: the structural formula and the line formula.

4) <u>Butane</u>. Chemical Formula: C_4H_{10}.

$$CH_3CH_2CH_2CH_3$$

5) <u>Pentane</u>. Chemical Formula: C_5H_{12}.

$$CH_3CH_2CH_2CH_2CH_3$$

The rest of the family has names that should be obvious: Hexane, Heptane, Octane, Nonane, Decane, Undecane, Dodecane, etc.

In Chapter 8, we mentioned the existence of isomers. The first few members of this family have no isomers, but from butane on, we will see that there are various different ways in which we can combine the atoms that form that particular molecular formula. For example, in the case of butane, we have two possible isomers that have four carbons and ten hydrogens. These are:

    1        $CH_3CH_2CH_2CH_3$ butane

    2        $CH_3\text{-}\underset{\underset{CH_3}{|}}{CH}\text{-}CH_3$ isobutane

In order to differentiate them by name, one is called butane (many times referred to as "normal" butane and written as n-Butane), and the other one is called isobutane. The prefix "iso" simply tells us that it is the isomer of the normal butane. However, this kind of terminology (called the "common names") can get confusing pretty fast. As the hydrocarbons start to get more and more carbons, the number of possible isomers increase rapidly. Soon we will run out of prefixes in order to name all the possibilities. To prove this point, let's look at the case of C_5H_{12}. This is

Chapter 13: Organic Chemistry.

just one more carbon than the butanes, and we already have three isomers.

$$CH_3CH_2CH_2CH_2CH_3 \qquad CH_3\text{-}CH\text{-}CH_2CH_3 \qquad CH_3\text{-}\underset{\underset{CH_3}{|}}{\overset{\overset{CH_3}{|}}{C}}\text{-}CH_3$$
$$ \underset{CH_3}{|}$$

$$\text{n-Pentane} \qquad\qquad \text{Isopentane} \qquad\qquad \text{Neopentane}$$

Imagine how much more complicated the naming would be in order to name all the 75 isomers for $C_{10}H_{22}$ (Decane)! Therefore, we need a systematic approach to the nomenclature of organic compounds.

IUPAC[1] Nomenclature.

1) Select the <u>longest</u> continuous carbon chain in the molecule and use the parent hydrocarbon name of this chain as the base name.

2) Number the longest carbon chain so that the branching groups are on the LOWER (rather than the higher) numbered carbon atoms.

3) Locate the branching groups and name them (suffix: -yl).

4) Arrange the names of the branching groups in alphabetical order. Each one is prefixed with the number of the carbon atom to which it is attached.

Let's look at a couple of examples to see how to apply these rules.

- Below we see all the isomers for C_6H_{14}. Name these isomers with the IUPAC system.

$$CH_3CH_2CH_2CH_2CH_2CH_3 \qquad CH_3\text{-}\underset{\underset{CH_3}{|}}{CH}\text{-}CH_2CH_2CH_3 \qquad CH_3CH_2\text{-}\underset{\underset{CH_3}{|}}{CH}\text{-}CH_2CH_3$$

$$\qquad 1 \qquad\qquad\qquad\qquad 2 \qquad\qquad\qquad\qquad 3$$

$$CH_3\text{-}\underset{\underset{CH_3}{|}}{CH}\text{-}\overset{\overset{CH_3}{|}}{CH}\text{-}CH_3 \qquad\qquad CH_3CH_2\text{-}\underset{\underset{CH_3}{|}}{\overset{\overset{CH_3}{|}}{C}}\text{-}CH_3$$

$$\qquad 4 \qquad\qquad\qquad\qquad 5$$

We can see that there are five different isomers with that particular molecular formula. The easiest one to name is the first one: Hexane. It is just the six carbons in a row.

---

[1] International Union of Pure and Applied Chemistry.

Let's name the compound labeled **2**.

$$CH_3-\underset{\underset{CH_3}{|}}{CH}-CH_2CH_2CH_3$$

We notice that the longest continuous carbon chain is only five, therefore, this compound is a pentane. However, there is a CH_3 bonded to one of the carbon atoms in the chain. We must start numbering the carbon chain on the left side, so that the substituent will be on the lower number carbon atom. Therefore, the substituent is on carbon #2. How about the substituent? It consists of a single carbon with three hydrogens. If the carbon had four hydrogens, it would be called methane. When you remove a hydrogen from it, it becomes a substituent and we would then remove the ending of –ane and change it to –yl. Therefore, the substituent is called a methyl group. Therefore, the name of that isomer would be: 2-Methylpentane.

The isomer labeled #3 is very similar in structure, except the methyl group is on carbon 3, therefore, it is called: 3-Methylpentane.

Isomer #4 is a little different. The longest continuous carbon chain is only four carbons long, therefore, it is a butane.

$$\underset{\underset{CH_3}{|}}{CH_3-\overset{\overset{CH_3}{|}}{CH}-CH-CH_3}$$

Now we see that there are two substituents that need to be named. Both of them are CH_3's, therefore, both of them are methyl groups. Since in this case it is irrelevant as to where you start numbering the chain (it would give the exact numbers), then the compound would be called: 2,3-Dimethylbutane. The prefix di- is used to indicate that we had two different methyl groups in the molecule. Keep in mind that if we ever use that prefix, two numbers MUST precede it.[1]

Finally, isomer #5 also has four carbons as its longest continuous carbon chain and it would be called: 2,2-Dimethylbutane. Once more, notice that although both methyls are attached to the same carbon, we MUST use the number twice to indicate that this is where the methyls are located.

---

[1] The two numbers indicating the carbon atoms to which the groups are attached.

- Name the following molecule.

$$\begin{array}{c} CH_2CH_3 \\ | \\ CH_3CH_2CH_2 - C - CH - CH_3 \\ | \quad | \\ H_3C \quad CH_2CH_3 \end{array}$$

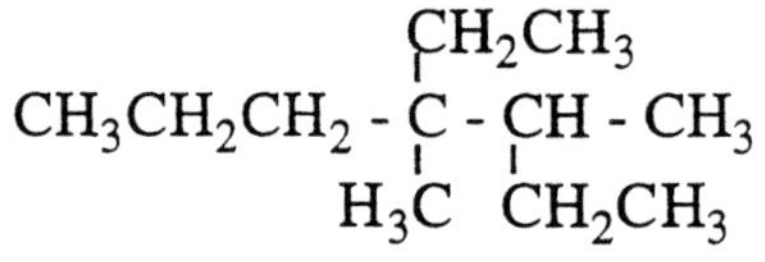

As always, we start by finding the longest continuous carbon chain, which in this case it is seven carbons long. Therefore, this is a heptane. Once we have located this chain, we proceed to number the chain. We must start on the right side, since that would find the first substituent on carbon 3, whereas on the other side, the first substituent appears on carbon 4. How about the substituents themselves? Two of them are methyl groups (we should recognize it easily by now), but the other one is different. It is a CH_3CH_2. We should see that this is almost the same as the ethane molecule, except it is missing a hydrogen. Therefore, we shall call it an **ethyl** group.

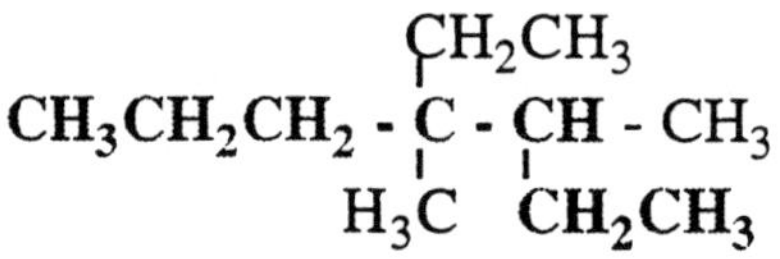

$$\begin{array}{c} CH_2CH_3 \\ | \\ CH_3CH_2CH_2 - C - CH - CH_3 \\ | \quad | \\ H_3C \quad CH_2CH_3 \end{array}$$

Therefore, the name of the molecule would be: 4-Ethyl-3,4-dimethylheptane[1].

- Name the following molecule.

$$\begin{array}{c} CH_3 \quad CH_3 \\ | \quad\quad | \\ CH_3 - CH - C - CH_3 \\ | \\ CH_3 \end{array}$$

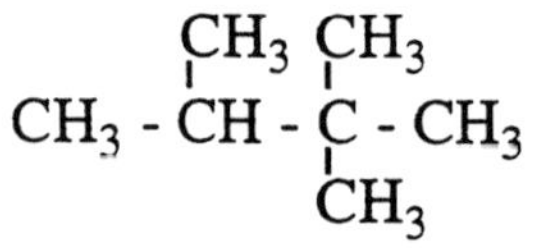

The easiest part is turning out to be finding the longest chain (this time it is four). Where should we start to number the chain? On the right side because that would give the smallest possible numbers for this compound.

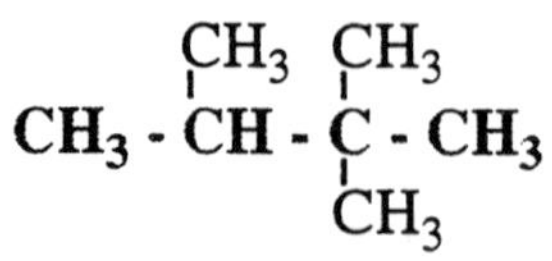

$$\begin{array}{c} CH_3 \quad CH_3 \\ | \quad\quad | \\ CH_3 - CH - C - CH_3 \\ | \\ CH_3 \end{array}$$

Therefore, the compound is called: 2,2,3-Trimethylbutane. This time the prefix is tri- which indicates the presence of three substituents. Just like before, the presence of that prefix necessitates the presence of three numbers to indicate the carbons to which these substituents

---

[1] Notice that the ethyl group was named first since it starts with an e (goes first alphabetically).

Chapter 13: Organic Chemistry.

are attached.

- Name this molecule.

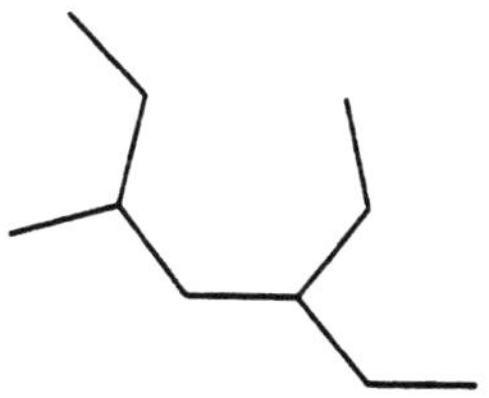

The first thing we notice is that the longest chain is seven carbons long, and that no matter where we start counting, we will come across a substituent on carbon 3. That is, no matter which side we start counting, the substituents are going to carry the numbers 3 and 5. This is another "rule" we should keep present at all times. Whenever this situation is encountered (only then!), the one that goes alphabetically first should get the smaller number. Make this very clear: this rule only applies to this very particular situation (having the same number carbon for the first substituent on either side). Since ethyl goes before methyl, then we should start numbering on the right side.

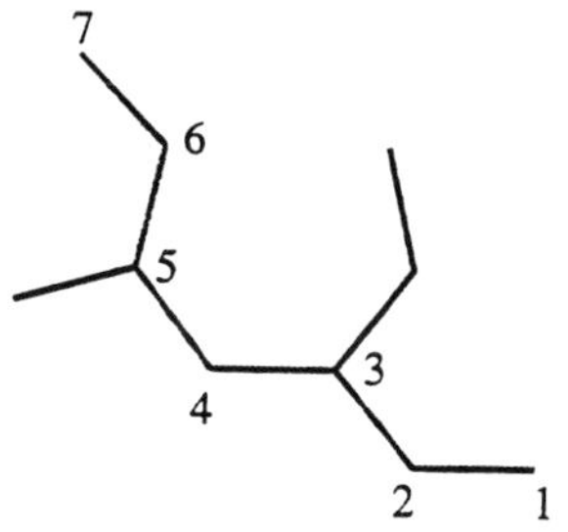

Therefore, the name of the molecule would be: 3-Ethyl-5-methylheptane.

- Name the following molecule.

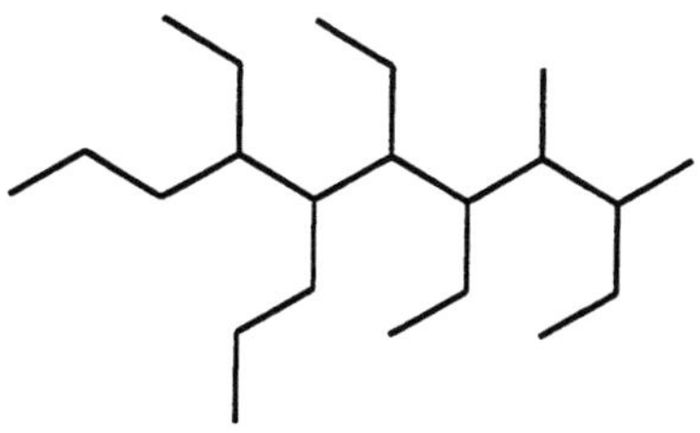

The longest continuous carbon chain is eleven. This is something that is very easy to miss, so we must be very careful. We will start numbering the chain on the right side, because that gives the first substituent the lowest possible number.

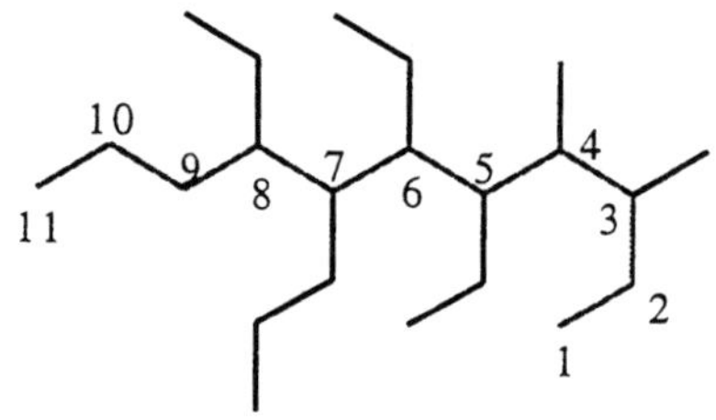

Therefore, the name of the molecule would be: 5,6,8-Triethyl-3,4-dimethyl-7-propylundecane.

In the last example, we saw a substituent we had not encountered before: the propyl group. This is the case of the $CH_3CH_2CH_2$ group. Notice that once more this is the propane molecule with a hydrogen atom missing on the first carbon. However, there is a different possibility for the propyl group. The hydrogen atom could have been missing from the second carbon (rather than from the first carbon). In that case, we would have called the substituent the isopropyl group. For example, the molecule below is called 4-Isopropylnonane.

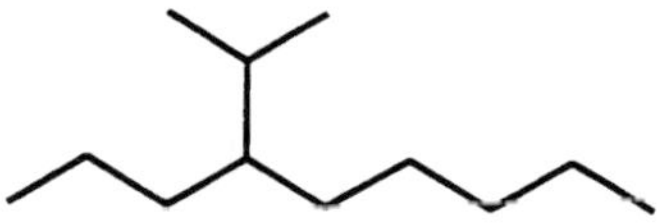

There are four more substituents that we should know how to name. We should memorize these so that we can quickly name organic molecules.

$CH_3CH_2CH_2CH_2$ --→ n-butyl

CH_3-CH-CH_2CH_3 sec-butyl

$$\begin{array}{c}CH_3\\|\\CH_3\text{-CH-}CH_2\end{array}$$ --→ isobutyl

$$\begin{array}{c}CH_3\\|\\CH_3\text{-C-}\\|\\CH_3\end{array}$$ --→ tert-butyl

In these structures, the arrows represent the missing bond. Remember that these substituents (called alkyl groups) are derived from a hydrocarbon in which one hydrogen has been removed. The position of the arrow indicates where the hydrogen is missing (and where we would have to bond them into the main carbon chain). For example, the name of the following molecule is 6-Isobutyl-5-sec-butyl-3-methyldodecane.

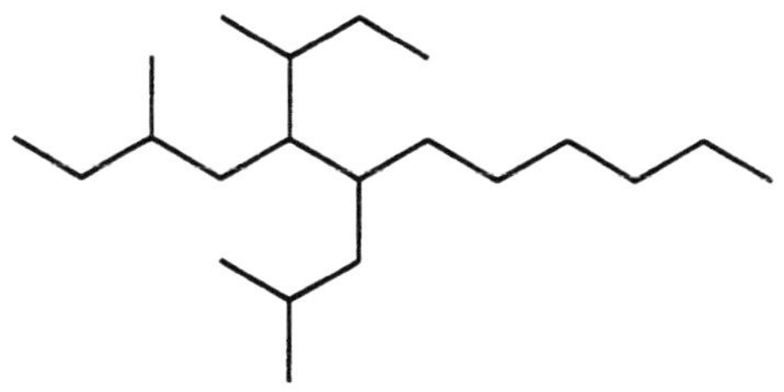

The properties of alkanes are consistent with what we have previously learned. An alkane molecule is held together by covalent bonds exclusively. Since the electronegativity between C and H is essentially the same, these compounds are for all practical purposes non-polar. As we discussed in Section 10.1, non-polar substances have London or Vander Waals forces as the only intermolecular forces of attraction. These are very weak forces and of very short range, acting only between the surfaces of molecules that may come in close contact. Within the alkanes, we would therefore expect that the larger the molecule (the larger the surface area), the stronger the intermolecular forces and thus, the higher the boiling point. This is exactly what we observe.

Alkane	BP (°C)
Methane	-162
Ethane	-88.5
Propane	-42
n-Butane	0
n-Pentane	36
2-Methylbutane	28
2,2-Dimethylpropane	9.5

Notice that the last three examples in the table are isomers. However, they have very different boiling points. This is because the straight chain isomer has more surface area available for the formation of the London forces, whereas the last one has the least surface area available.[1]

Alkanes are, in general, very unreactive hydrocarbons. This is mainly due to the fact that in order to react a molecule we would need a center in that molecule that has some kind of dipole so that other polar molecules can be attracted to that center and attack it. The problem is that the alkanes, being non-polar, do not have such center. However, alkanes can be made to react if you

---

[1] See Section 10.1.

Chapter 13: Organic Chemistry.

choose the right reagent. For example, we can substitute (not easily) a H from the alkane with a halogen. The reaction we are referring to is called **free radical substitution of alkanes** and a very specific example is shown below.

$$CH_4 + Br_2 \text{ (presence of light)} \Rightarrow CH_3\text{-}Br + H\text{-}Br$$

The question is, how does this happen? We just mentioned that methane is very unreactive, but under the right set of conditions, it can be made to react to form Bromomethane. In order to understand what it going on, it is imperative that we follow the mechanism of the reaction. If there is no light present (if the reaction is carried out in the dark), the molecules of methane and bromine will collide forever without leading to any reaction because they are both non-polar. However, one photon of light[1] will split up the Bromine molecule into two Bromine radicals[2]. What is a Bromine radical? It is a species that has seven electrons in its valence shell, therefore, it is missing one electron to have the configuration of a noble gas.

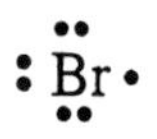

$$:\!\ddot{B}r\cdot \quad \text{or more commonly drawn as:} \quad Br\cdot$$

This Bromine radical is a very reactive species and it is looking around for an electron. When one of these radicals collides with a CH_4 molecule, it will abstract a Hydrogen from the CH_4 with ONE electron in it (since after all, that is what it is looking for: one electron).

Inititation Step: $\quad Br_2 \rightleftharpoons 2\, Br\cdot$

Propagation Step #1: $\quad CH_4 + Br\cdot \longrightarrow \cdot CH_3 + H\text{-}Br$

Propagation Step #2: $\quad \cdot CH_3 + Br\text{-}Br \longrightarrow CH_3Br + Br\cdot$

This forms the H-Br molecule, but it leaves behind a $\cdot CH_3$ radical, which is also a very reactive species. What will the CH_3 radical do? It will also seek an electron. The collision of a CH_3 radical with a Br-Br molecule will cause the CH_3 radical to abstract a Br from the molecule (with an electron), leaving behind another Br radical which starts the process all over again. This mechanism itself can be seen above. Notice that after the second step, we have regenerated the Br• which then goes around to find more CH_4 molecules to continue this radical propagation and in turn, the reaction.

---

[1] This photon of light must contain the exact amount of energy required to break the Br-Br single bond homolytically (each atom would keep an electron from the bond).
[2] The term radical relates to a species with one or more unpaired electrons.

Alkyl Halides.

The products of these reactions are the alkyl halides. One example of such a compound is shown below.

$$CH_3CH_2CH_2CH_2CH_2-Br \qquad \text{1-Bromopentane.}$$

Notice that non-carbon substituents are not suffixed with the –yl. Actually, all the halides end in –o. Therefore, the four commonly used halides would be: fluoro, chloro, bromo, and iodo.

If we were to change the position of the halogen along the carbon chain, then we would generate positional isomers. Let's look at some examples.

a) 2-Bromopentane.

$$CH_3-\underset{Br}{CH}-CH_2CH_2CH_3$$

b) 3-Bromopentane.

$$CH_3CH_2-\underset{Br}{CH}-CH_2CH_3$$

c) 2-Bromo-3-methylbutane.

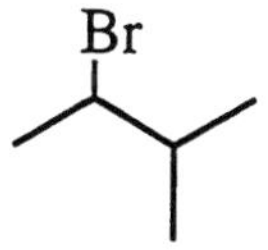

We should point out that the bromine atom does not take precedence in the numbering of the carbon chain. This last example simply illustrates the case of having the same numbers on the substituents independent on the side where you count, therefore, we start on the side that gives the one that goes alphabetically first the lowest number carbon. The proof is in the next example.

d) 5-Bromo-2-methylheptane.

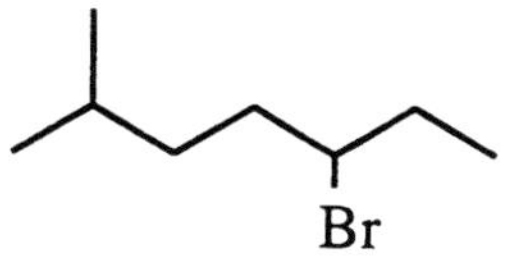

Chapter 13: Organic Chemistry.

This time, the bromine is simply on carbon # 5, because if we were to start on the right side, then the first substituent would be on carbon #3.

- What is wrong with the name 4-Chloropentane?

    The name of this compound should be 2-Chloropentane. We should have numbered the carbon chain starting at the other end in order to give the substituent the lowest possible number.

- What is wrong with the name 3,4-Difluoropentane?

    Once more, the problem is with the numbering of the carbon chain. We should start on the other end in order to place the fluorines in the smaller numbered carbons. The correct name would be: 2,3-Difluoropentane.

- Name this molecule.

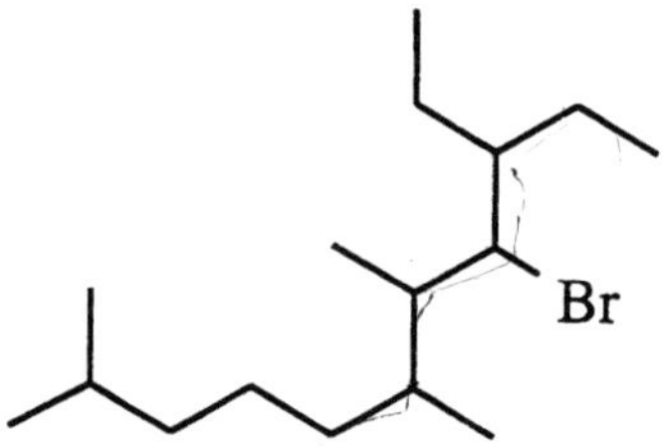

    How long is the longest carbon chain? It is 11 carbons long. We should start numbering the chain on the lower left side, because starting there, the first substituent (in this case a methyl group) appears on carbon 2. If we would have started on the other side, the first substituent (in this case an ethyl group) appears on carbon 3.

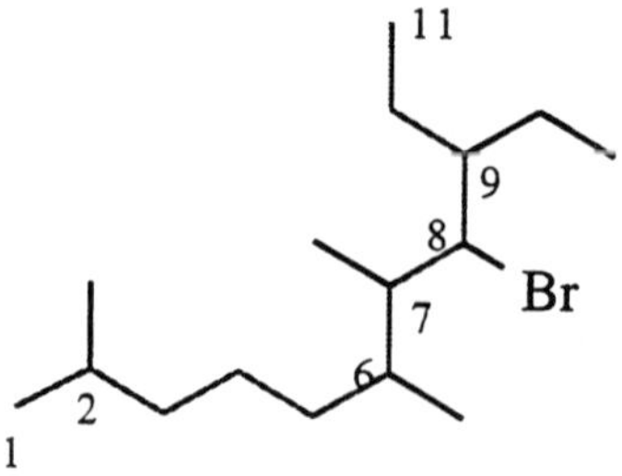

    The name would be: 8-Bromo-9-ethyl-2,6,7-trimethylundecane.

13.3 Alkenes (C_nH_{2n}).

This family of hydrocarbons contain less hydrogens, carbon for carbon, than the alkanes. Since these hydrocarbons contain less than the maximum quantity of hydrogens we refer to them as <u>unsaturated</u> hydrocarbons. Actually, alkenes are missing two hydrogens compared to the alkanes. Any compound that is missing two hydrogens compared to the alkanes is said to have *one degree of unsaturation*. This unsaturation makes them much more reactive than the alkanes.

The general formula is C_nH_{2n} and therefore the simplest member of the family is C_2H_4, <u>Ethene</u> (common name is ethylene). The distinguishing feature of all alkenes is a carbon-carbon double bond. The Ethene molecule would look like this:

$$\begin{array}{c} H \\ \diagdown \\ C = C \\ \diagup \quad \diagup \\ H \quad \quad H \end{array} \begin{array}{c} H \\ \diagup \\ \\ \diagdown \\ H \end{array}$$

sp²

Therefore, the molecule itself is planar. To get a better view of the molecule, let's rotate it 90°, so that we can see the π-bond.

As we can see from that diagram, the π-bond impedes the rotation along the carbon-carbon bond. Prove it with your models! This is an important concept which will introduce a new type of isomers, as we shall see in the next section, because it makes the two positions on each Carbon different. This is in direct contrast with the three positions in an sp³ Carbon, which due to "free rotation" makes them identical.

The nomenclature of alkenes is basically the same as alkanes but with two salient features: The longest continuous carbon chain **must** contain the carbon-carbon double bond. This double bond is between two carbons, but we only name the first carbon of the double bond. The name of the molecule ends in a different suffix than the alkanes. Since these are **alkenes**, we shall use the suffix of **–ene**. Finally, since the double bond changes the ending of the molecule, then this double bond should take precedence in the numbering of the chain and the double bonded carbons should have the lowest possible numbers. Let's look at some examples.

Chapter 13: Organic Chemistry.

- Name the following alkenes.

$$CH_3-CH=CH-CH_2CH_3$$
1

$$CH_3CH_2-CH=CH-CH_2CH_3$$
2

$$CH_3-\underset{\underset{CH_3}{|}}{C}=CH-CH_2CH_2CH_3$$
3

$$CH_3-CH=CH-CH_2-\underset{\underset{CH_3}{|}}{CH}-CH_3$$
4

The one labeled **1** is called 2-Pentene. The double bond is between carbons 2 and 3, but we only give the number of the first carbon involved in the double bond. The one labeled **2** is called 3-Hexene, for the same reason. In the one labeled **3**, we see that there is a substituent, a methyl, in carbon 2. Therefore, this compound is called 2-Methyl-2-hexene. Finally, the one labeled **4** proves the point we made before. The double bond takes precedence in the numbering of the chain, therefore, this isomer is called 5-Methyl-2-hexene.

13.4 Stereoisomers, Part I.

In alkene compounds, we find a new type of isomers that were non-existing in the alkane hydrocarbons. If we were to draw all the alkene isomers for C_4H_8, we would immediately find the following three isomers.

$$CH_2=CH-CH_2CH_3$$
1-Butene

$$CH_3-CH=CH-CH_3$$
2-Butene

$$CH_2=\underset{\underset{CH_3}{|}}{C}-CH_3$$
2-Methylpropene

However, in reality there are four isomers. Where is the fourth isomer? The problem is that there are two different isomers for 2-Butene due to the fact that there is no rotation along the carbon-carbon double bond.[1]

The isomer labeled **1** has both hydrogens on the same side of the double bond and we will call

---

[1] We should convince ourselves of this fact using the molecular models. Prove that the isomer labeled **1** cannot be converted into the one labeled **2** unless the π bond is broken.

this compound: cis-2-Butene. The one labeled **2** has the two hydrogens on opposite sides of the double bond and we shall call this one: trans-2-Butene. These two compounds have the same base name (2-Butene) but they are totally different compounds. The table below illustrates that point. Notice that these compounds have different densities, boiling points and melting points. There is not doubt that they are different compounds. However, the only difference between them is not the connectivity of the atoms themselves, but rather the spatial arrangement of these atoms. These kind of isomers are called **stereoisomers**. Stereoisomers have the same basic name (in this case 2-Butene) but they differ in their orientation in space, therefore, something must be added to the name in order to know which one in particular we are referring to. There are many different types of stereoisomers and in the course of this chapter we will see two of these types. These particular kind of stereoisomers (cis/trans isomers) are called **geometric isomers**. Geometric isomers are stereoisomers that cannot be converted into one another due to a hindered rotation. Due to the presence of the double bond, we cannot convert the cis into the trans.

Compound	Density	Boiling Point	Melting Point
Cis-2-Butene	0.621 g/mL	3.7°C	-138.9°C
Trans-2-Butene	0.604 g/mL	0.90°C	-105.5°C

- Name the following molecule, which has been carefully drawn around the carbon-carbon double bond.

$$H_3C-CH=CH-CH(CH_3)-CH_3$$

At first glance, we can see that this is a cis isomer. This is something we must be very careful with when trying to name alkenes. Sometimes there is a potential for a particular double bond to be either cis or trans. The longest chain that contains the double bond is 5, therefore, the name of this compound would be: cis-4-Methyl-2-pentene.

Not every single carbon-carbon double bond has the potential to have geometric isomers. In order to have cis/trans isomers, one very important requisite must be met. We must have one and only one hydrogen in each of the two carbons involved in the double bond. We then proceed to compare the orientation of these two hydrogens. If they are on the same side of the double bond, it will be a cis isomer whereas if they are on opposite sides of the double bond, then it will be the trans isomer.

- Name the following molecule:

$$\begin{array}{c} H \\ H \end{array} \!\!>\!\!=\!\!<\!\! \begin{array}{c} H \\ CH\text{-}CH_3 \\ | \\ CH_3 \end{array}$$

Of course, we see that there are two hydrogens on the first carbon of the double bond, therefore, this molecule does not have a chance of having cis or trans isomers. If we were to switch any of these two hydrogens, we still would have the exact same compound. Therefore, the name of this molecule would be: 3-Methyl-1-butene.

- How many of the following compounds could exhibit cis/trans isomerism?

$$CH_3CH_2\text{-}CH=CH\text{-}CH_3 \quad CH_3\text{-}CH=CH_2 \quad CH_3CH_2\text{-}CH=\underset{\underset{CH_3}{|}}{C}\text{-}CH_3 \quad CH_3CH_2\text{-}\underset{\underset{CH_3}{|}}{C}=\underset{\underset{CH_3}{|}}{C}\text{-}CH_3$$

Actually, only one of them: the first one. The other three have different problems. Propene has two hydrogens on one of the double bonded carbons. 2-Methyl-2-pentene has two methyls on the second carbon (the first of the double bonded carbons). The problem with 2,3-Dimethyl-2-pentene is exactly the same: two methyls on the second carbon.

13.5 Reactivity of Alkenes.

Alkenes are also hydrocarbons, but these are unsaturated hydrocarbons. The main feature of an alkene is the carbon-carbon double bond. What is so special about this bond? The bond and its immediate surrounding are flat, due to the sp² hybridization of the two carbons. This places the π electrons in an orbital that is sticking out of that plane (above and below the plane). This means that positive things will be attracted to those pi electrons. Therefore, we say that alkenes have the ability to attract electrophiles[1]. If a positive species (or an <u>electron deficient species</u>), which we can call an **electrophile**, collides with an alkene, it will be attracted to these π

---

[1] Greek for electron-lovers.

electrons (which can be visualized as reaching out to the positive species). This feature make these molecules fairly reactive, so the double bond turns out to be a *"functional"* part of the molecule. Because of this, we will see in a later course (organic chemistry) that alkenes undergo addition reactions. In these reactions, one atom is added to each carbon, thus breaking the pi bond and saturating the molecule. An example of such a reaction is the addition of bromine (Br_2) to an alkene. Let's look at the particular case of the addition of bromine to Ethene.

$$CH_2=CH_2 + Br_2 \Rightarrow Br-CH_2-CH_2-Br$$

How does this reaction occur? How can a non-polar Br_2 molecule get attracted to the Ethene molecule without even a catalyst? This can be understood if we try to follow the mechanism of the reaction.

We must remember that the fact that these two molecules react at all should be kind of surprising, to start with. These are non-polar molecules and therefore, the attraction between them should be minimal. However, we must keep in mind that the bromine molecule is composed of two very large atoms. The valence shell electrons of the bromine atoms are relatively far away from the nucleus and therefore, they are less strongly held. This means that these electrons will be able to sway back and forth from the nucleus with relative ease. This movement of the electronic cloud will generate relatively strong London forces, as evidenced from the fact that it is a liquid at room temperature. The terminology we use is to state that this molecule is **polarizable, or that it can create instantaneous dipoles** with ease. Now, as the alkene, with its pi electrons sticking out of the plane of the molecule, gets close to the bromine molecule, these negative electrons will repel the electronic cloud of the bromine and it will polarize the molecule making one bromine atom partially positive and the other one partially negative, as shown below.

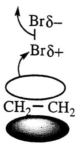

This leads to a heterolytic breakage of the Br-Br bond (one Br will take both electrons). The end result is the formation of a special carbocation and a bromide ion.

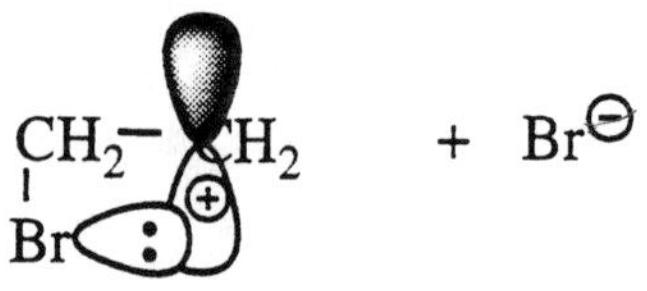

$$CH_2\text{—}CH_2 \quad + \quad Br^{\ominus}$$
$$|$$
$$Br$$

This special carbocation has a positive charge on one of the carbons and we see that the Br has attached to the other carbon. The two π electrons went to form the sigma single bond between the Carbon and the bromine, while the other carbon is now positive. However, the Br is so large that one of its three lone pairs interacts easily with the "empty" p-orbital in the sp^2 positive carbon. This interaction is so strong that we can visualize it as if the Br were forming a bridge between the two carbons. This is called a "**bromonium ion**" and many times is drawn as:

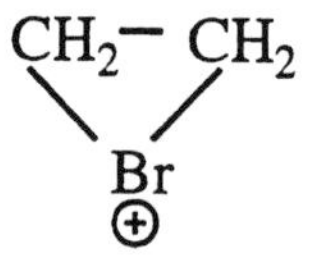

$$CH_2\text{—}CH_2$$
$$\diagdown\diagup$$
$$Br$$
$$\oplus$$

Bromonium Ion

Remember that having a positive Carbon in the molecule will make it susceptible to attack by something negative (referred to as a nucleophile). This means that the Bromide ion (Br^-) that we formed in the first step will come right back to attack that positive Carbon. From the diagram above, we should be able to see that the Br involved in the bromonium ion is effectively blocking one side of the molecule (in the diagram, it is blocking the bottom side), therefore, it forces the bromide ion to attack one of the carbons from the opposite side, as shown below. This is referred to as an ANTI addition, because both Br's were added from opposite sides of the double bond. Once more, the only reason we ended up with an anti addition was due to the bromonium ion, which effectively blocks one side of the molecule and forces the second Br to attack from the opposite side.

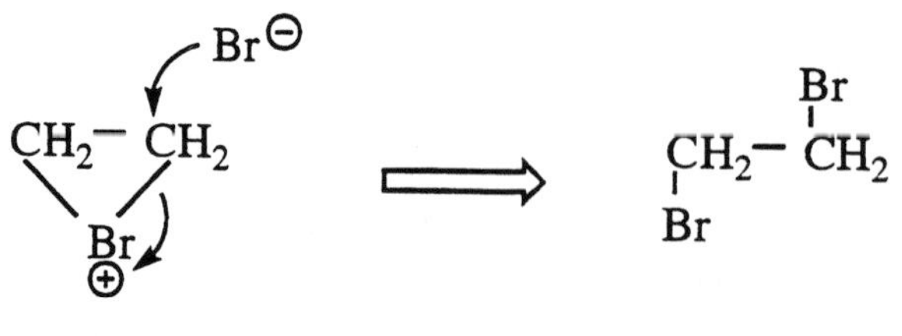

13.6 Alkynes (C_nH_{2n-2}).

The general formula (C_nH_{2n-2}) indicates that there are four hydrogens missing, as compared to the alkanes. This gives the alkynes *two degrees of unsaturation*. Once more, the unsaturation gives these compounds their chemical properties and makes them much more reactive than alkanes.

The simplest member of the family is <u>Ethyne</u> (common name: acetylene), of formula C_2H_2. Its distinguishing feature is a carbon-carbon triple bond, therefore in an alkyne, the carbons involved in the triple bond will hybridize **sp**, making the triple bond moiety linear.

$$H - C \equiv C - H$$

The nomenclature of these compounds is very similar to alkenes but the ending will be **-yne**. Since the triple bond changes the base name of the molecule, then the triple bond will take precedence in the numbering of the carbon chain.

- Name the following alkynes.

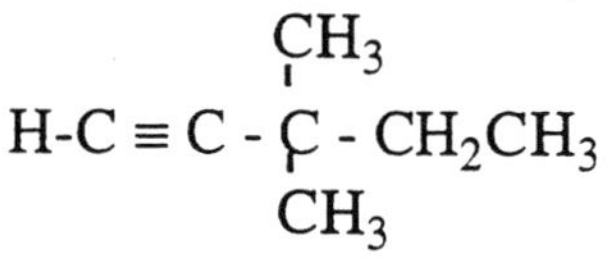

$$\underset{\textbf{1}}{H-C \equiv C - \underset{\underset{CH_3}{|}}{\overset{\overset{CH_3}{|}}{C}} - CH_2CH_3} \qquad \underset{\textbf{2}}{CH_3 - C \equiv C - \underset{\underset{Br}{|}}{CH} - CH_3}$$

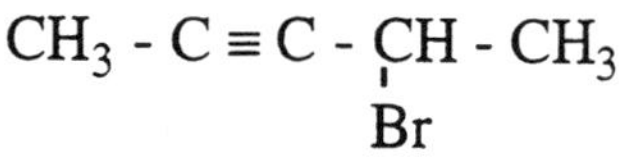

Seeing the carbon-carbon triple bond immediately identifies these as alkynes. Since the triple bond takes precedence in the numbering of the carbon chain, then the IUPAC names for these two molecules would be: **1** = 3,3-Dimethyl-1-pentyne, and **2** = 4-Bromo-2-pentyne.

We will not look at the reactions of the alkynes, but it will suffice to say that their reactions in many ways are similar to those of alkenes. For example, the addition of bromine in excess to a molecule like 1-Butyne, will form 1,1,2,2-Tetrabromobutane.

13.7 <u>Aromatics</u>.

Aromatic is the name given to compounds like benzene or any other compounds that resembles benzene in chemical behavior. Back in Section 9.1, we mentioned that benzene, C_6H_6, , exhibits resonance and that the best possible way to represent benzene is not by anyone of the two resonant forms but rather by the resonance hybrid.

However, in reality the π-electrons are delocalized over the entire ring.

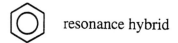 resonance hybrid

From the resonance forms, we can see that benzene has four degrees of unsaturation. Due to the stabilization generated by the delocalized π-electrons, benzene is much more unreactive than you would expect from an unsaturated compound.

The nomenclature of these compounds simply names the substituents on benzene. In the case of a monosubstituted benzene it is as simple as calling it Methylbenzene, or Ethylbenzene, etc. If we have a disubstituted benzene, then we must specify the carbons in which we find the substituents. There are only three possibilities for disubstituted compounds: 1,2 substitution (also called *ortho*), 1,3 substitution (also called *meta*), and 1.4 substitution (also called *para*). There are some common names that we should know, since they are very common and basically, every scientist uses them. Let's see some of these below.

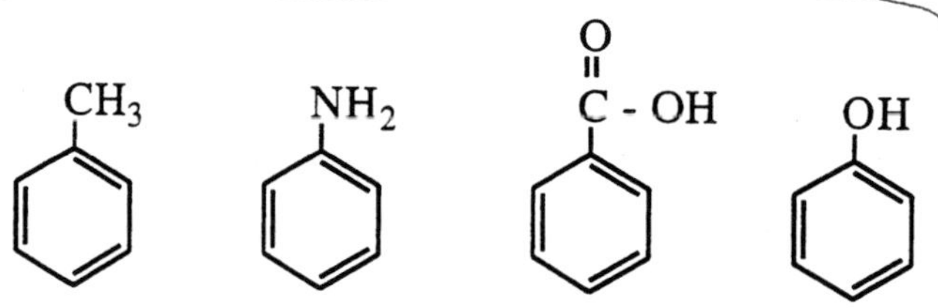

Toluene   Aniline   Benzoic Acid   Phenol

We often use these names as the base name for the molecule itself.

- Name the following molecule.

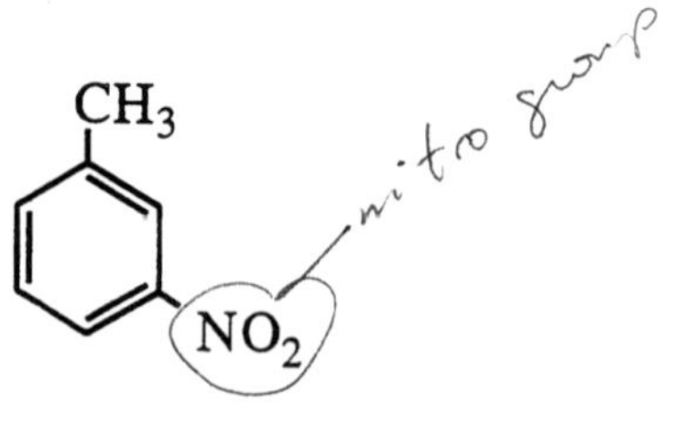

(nitro group — NO₂)

This molecule can be named in a variety of ways. For example, since we have two substituents in the benzene, then we can simply call this 1-Methyl-3-nitrobenzene, or *meta*-Methylnitrobenzene. However, we should recognize that a benzene with a methyl group in it is referred to as Toluene. Therefore, we could also (and preferably) name this molecule: 3-Nitrotoluene, or *meta*-Nitrotoluene.

We should point out that sometimes it is easier to name the benzene group as a substituent, rather than being the main part of the molecule. In these cases, the C_6H_5 group (a benzene with one hydrogen missing at one of the positions), is called a **phenyl** group. This is the case of a long and complicated molecule that has a phenyl group coming off one of its carbons. Imagine the problem if we had to name that long complicated molecule as a substituent in the molecule! The example below illustrates this concept.

- Name the following molecule.

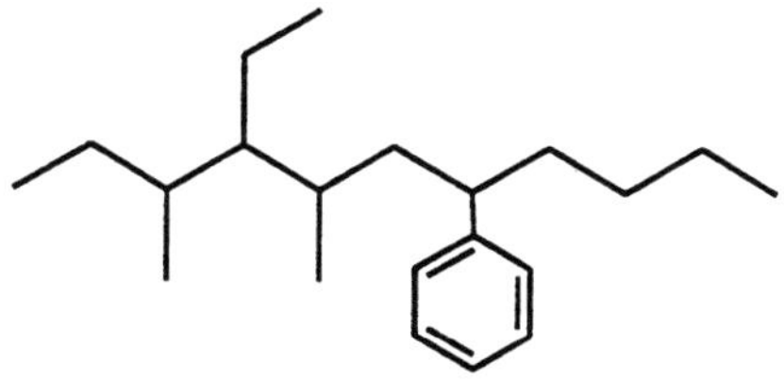

Once more, it is a lot easier to name the phenyl ring as a substituent, therefore, this molecule would be called: 4-Ethyl-3,5-dimethyl-7-phenylundecane.

Simply as another illustration, let's see the cases of the Xylenes. This is the common name given to the Dimethyl benzenes. They could be *ortho*-Xylene, *meta*-Xylene, or *para*-Xylene. These can be seen below with their other names.

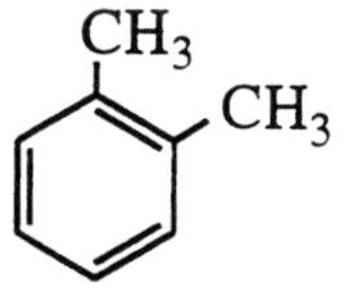

1,2-Dimethylbenzene
or: *ortho*-Dimethylbenzene
ortho Xylene

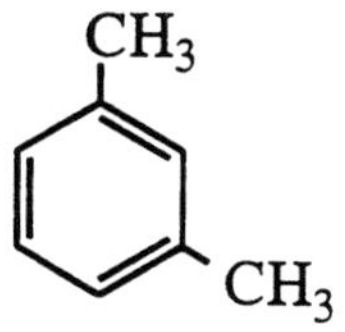

1,3-Dimethylbenzene
or: *meta*-Dimethylbenzene
meta Xylene

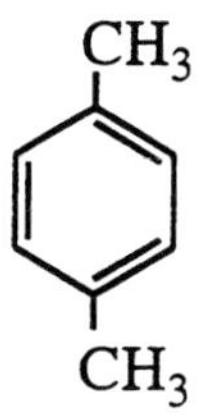

1,4-Dimethylbenzene
or: *para*-Dimethylbenzene
para Xylene

- Name this molecule.

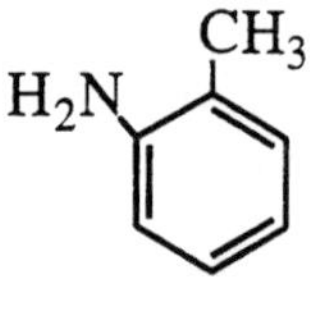

There are various possible names for this molecule. The most common would be: 2-Methylaniline or 2-Aminotoluene.

13.8 Cycloalkanes (C_nH_{2n}).

These are cyclic compounds, like benzene, but they do not contain any π bonds. Their name includes the word alkane because all the bonds are single, however, there *is one degree of unsaturation* since in order for the two terminal carbons on an alkane to bond to one another and form a ring, they must lose one hydrogen each.

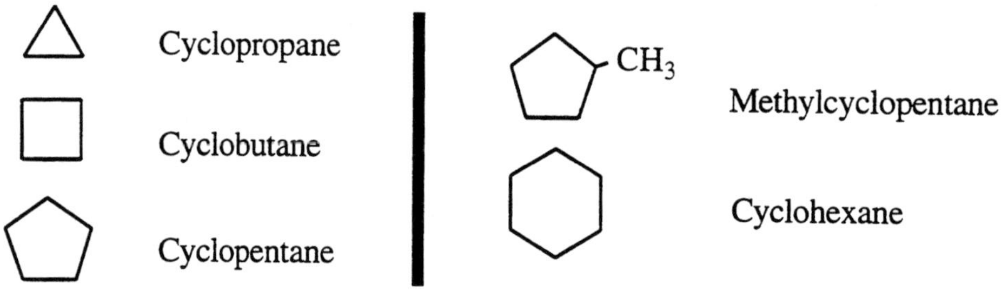

Cyclopropane

Cyclobutane

Cyclopentane

Methylcyclopentane

Cyclohexane

At first glance, this seems odd. All the bonds are single, which means that the hybridization of the carbons in the cyclic alkanes must be sp^3. How can that be possible? We know that the bond angle predicted by that hybridization is 109.5°. This is certainly not the case for the molecules shown above. Just as a simple example, the Cyclohexane appears to have 120° angles in the diagram above. However, we should keep in mind that these molecules are NOT flat, as shown above. For example, Cyclohexane exists in different conformations like the chair forms shown below. This way, the molecules remove some of the angle stress.

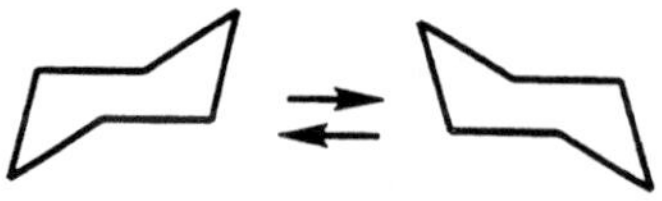

However, the smaller cyclic alkanes are planar because there is no way that the atoms can possibly get into some kind of distortion to relieve the stress. This creates a problem for the three, and four member rings, that is, Cyclopropane and Cyclobutane. Let's concentrate on

Cyclopropane as an example. Since the three carbons HAVE to be on the same plane (three atoms define a plane), then the apparent bond angle is 60°. This is very far away from the theoretical 109.5° that the hybridization predicts.[1] There is a way to explain this.

Any time we form a covalent bond, we will have orbitals of the two atoms that are bonding get close to one another and overlap. The greater the amount of overlap, the more stable the bond formed is going to be. In a regular bond (like on an open alkane), the two carbon atoms that will bond together will bring two of the sp³ hybridized orbitals and approach one another in such an angle to generate a great amount of overlap. Therefore, they form a strong σ bond between the two atoms.

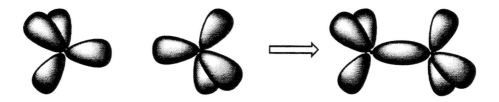

However, in the case of the cyclopropane, the orbitals are still 109.5° apart, therefore, what we observe is that the amount of overlap between the interacting orbitals is a lot less than in the previous example.

The apparent bond angle (remember that we can only see the nuclei) is 60°, but what we really have is poor overlap between the orbitals. This imparts a different reactivity to this molecule (it will be relatively easy to break the ring) when compared to Cyclohexane, which releases the bond stress by distorting out of the plane (see the boats shown above).

13.9 Functional Groups.

We have seen that different groups impart a certain level of reactivity on a molecule. The

---

[1] Let's try to build this molecule with the models and we will immediately see what we mean by the stress generated when we force this atoms to form this three member ring.

alkanes were the most unreactive of all the families we have seen so far. In contrast, a molecule that contains a pi bond (like alkenes or alkynes) will have these electrons in a π-orbital sticking out of the plane of the molecule. When an electrophilic substance comes along (looking for electrons), it will get to those π-electrons very easily. In the case of benzene it will be harder to react with an electrophilic species since the electron delocalization makes the molecule more stable and the π-electrons less accessible.

There are certain groups that have the ability to "activate" the molecule, or make it more reactive. We know that carbon has the ability to bond to a wide variety of different atoms, like F, Cl, Br, I, O, N, S, and P, for example. If carbon is bonded to a highly electronegative atom (like Cl), then the carbon itself will be partially positive:

$$C^{\delta+} - Cl^{\delta-}$$

This will make the carbon an easy target to a nucleophilic[1] species.

$$C^{\delta+} - Cl^{\delta-} + :Nu^{\ominus} \Longrightarrow C^{\delta+} - Nu^{\delta-} + Cl^{\ominus}$$

Therefore, the chlorine itself made the molecule reactive along the C-Cl bond.

The site of the substituted molecule which is reactive is called the functioning part of the molecule and the substituted group of atoms (or atom) is called the functional group.

Let's look at some of the most common functional groups. In these abbreviations, the letter R represents the hydrocarbon part of the molecule, sometimes called the alkyl group[2]. We should keep in mind that there are many more functional groups than the ones we will mention below, but we will learn the rest in another course, Organic Chemistry.

1) Alkyl halides.

We have seen this family before, when we substituted one of the hydrogens in an alkane

---

[1] The word literally means "nuclei lover", therefore, it refers to negative things, or at least, species that contain lone pairs, like ammonia.
[2] For example, Methyl, Ethyl, etc.

with a halogen. The general formula for the alkyl halides is: **R-X**. In this general formula, the X represents any of the halogens.[1]

As we saw in Section 13.2, the halogen itself does not take precedence in the numbering of the chain. This is because the basic name of the compound is just like an alkane (ends in **–ane**). We simply must comply with the rule that the substituents must have the lowest possible numbers.

- Name the following compounds. For the first two, also give the common name.

```
    Cl         Br                                         F
    |          |                                          |
   /\         /\/        \/\/\        \/\/\
                  |           |              |
                  Br         (Ph)            I
    1           2              3              4
```

Let's start with the first two, which are by far the simplest to name. Compound **1** is called 2-Chloropropane (IUPAC)[2], but the common name is Isopropyl chloride. The common name requires that we name the alkyl group itself with the common names (the only ones we know are: n-propyl, isopropyl, n-butyl, isobutyl, sec-butyl, and tert-butyl) and then name the halide (which will be either fluoride, chloride, bromide, or iodide). Compound **2** is called 2-Bromobutane (IUPAC) and its common name would be Sec-butyl bromide.

Compound **3** is called 2-Bromo-6-phenyloctane. The bromine did not take precedence in the numbering of the chain, but it has the lowest number simply because starting on the left side gave the first substituent (which happens to be the Br) the number two. Starting on the other end, the first substituent, which would have been the phenyl group, would have been on carbon number three.

Finally, compound **4** is called 4-Fluoro-5-iodo-3-methyloctane.

2) <u>Alcohols</u>.

This is one of the most important families in organic chemistry. In a subsequent course, we will see that many compounds can be synthesized from alcohols, and alcohols are relatively simple to obtain. The general formula for alcohols is: **R-OH**. This tells us that one of the

---

[1] F, Cl, Br, or I.
[2] Remember that the IUPAC names the halogen as a substituent, and its suffix is **–o** (like fluoro, chloro, bromo, and iodo).

hydrogens in an alkane has been replaced by an **OH**.

The rules for the nomenclature of these compounds are very similar to the others we have seen so far, with one major exception. In order to name these compounds, we will remove the ending –e from the alkane chain that contains the OH group and instead use the ending **–ol**. This is consistent with the overall name of the family: alco**hols**. Since the name of the basic chain is different now (for example: etha**nol** instead of etha**ne**), then the OH group will take precedence in the numbering of the carbon chain.

- Name the following alcohols. Give the common name for the first two.

Once more, let's start with the first two since we have to name these in two different ways. The IUPAC name for compound **1** would be 2-Propanol, whereas the common name would be Isopropyl alcohol. Compound **2** has an IUPAC name of 1-Butanol, whereas its common name is n-Butyl alcohol.

For compound **3**, we must start numbering the chain in the extreme right, because that will give the OH group the lowest possible number. Therefore, the name of this molecule is: 7,8-Dichloro-2-octanol.

Finally, compound **4** is called 5-Bromo-6-methyl-4-octanol. We were forced to start on the left because we HAVE to give the OH the lowest possible number.

3) <u>Ethers.</u>

This is another family of organic compounds that contains a singly bonded oxygen atom in it. In the case of an alcohol, the oxygen atom is bonded to a carbon on one side and to a hydrogen on the other side. The ethers have alkyl groups on both sides of the oxygen. This makes these compounds dramatically different. Alcohols exhibit hydrogen bonding, whereas ethers do not.[1]

The nomenclature for ethers end in the word **ether** and we simply must name the two

---

[1] Ethers do not have polarized hydrogens.

alkyl groups that are bonded to the oxygen atom. Since we are limited in our knowledge of the common names of alkyl groups, we will only be able to name very simple ethers. The general formula for ethers is: **R-O-R'**. The prime by the second R indicates that this second alkyl group may or may not be the same as the first one.

- Name the following ethers.

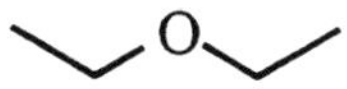

At this stage, we can only give the common names for ethers, therefore, these are called: **1** = Ethyl methyl ether. **2** = Diethyl ether. **3** = Sec-butyl tert-butyl ether. **4** = Diisopropyl ether.

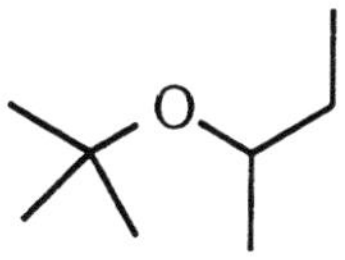

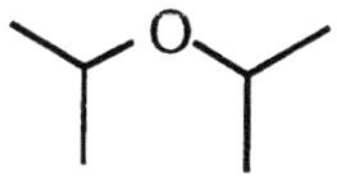

4) <u>Amines</u>.

The general formula for an amine is: **R-NH$_2$**. The **NH$_2$** group is called the **amino** group. It does not take precedence in the numbering of the chain because it will not change the basic name of the alkane. It is simply named as a substituent in whichever carbon it is found.

- Name the following amines.

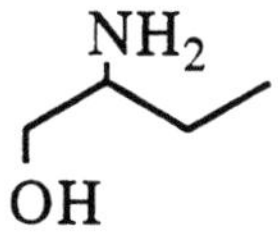

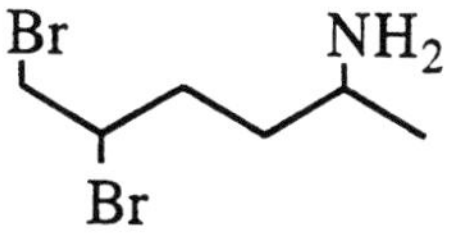

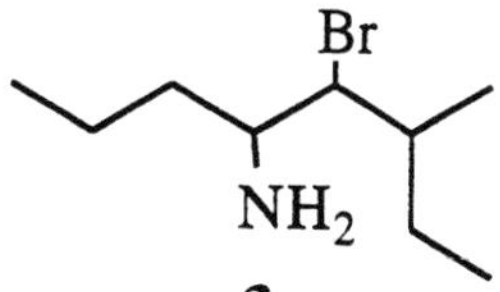

Compound **1** is also an alcohol. The name is: 2-Amino-1-butanol. Compound **2** is an example that illustrates that the amino group does not take precedence in the numbering of the chain. If we were to start to number the chain in the right side, the first substituent (the amino group) would appear in carbon 2. However, starting on the left side, places the first substituent (in that case a Br) in carbon 1. Of course, we must start numbering on the left side. The name of this compound is: 5-Amino-1,2-dibromohexane. Finally, compound **3** is called: 5-Amino-4-bromo-3-methyloctane. In this last compound, we can see that if we were to start numbering the chain on the left side, the first substituent (the amino) would appear on carbon 4. However, starting on the right side (lower side, actually), would place the first substituent (the methyl) on carbon 3. This is why we numbered the carbon chain starting on the lower right side.

Oxygen can form a double bond with a carbon atom, so there are many families of compounds that contain the **carbonyl group**. Let's look at the most important of these families of carbonyl containing compounds.

$$\overset{\underset{\parallel}{O}}{\sim\!\! C\!\!\sim}\quad \text{Carbonyl Group}$$

5) <u>Aldehydes</u>.

These families of carbonyl containing compounds have a general formula of **R-CHO**. This is an abbreviation for the real structure. The fact is that we have a carbonyl compound bonded on one side to the R group, and on the other side to a hydrogen atom. Therefore, the CHO part represents a carbonyl group bonded to a hydrogen atom.

The nomenclature of these compounds depends on the longest carbon chain that contains the aldehyde group (CHO). The actual name is determined by removing the ending –e from the alkane and adding the suffix **–al**. It is not necessary (and it would be actually incorrect to do so) to specify that the carbonyl group is carbon #1. This is because since it takes precedence in the numbering of the chain, if it is an aldehyde, by definition it will always be in carbon #1.

- Name the following aldehydes.

$$\underset{1}{CH_3CH_2-\overset{\underset{\parallel}{O}}{C}-H} \qquad \underset{2}{Br-CH_2CH_2CH_2-\overset{\underset{\parallel}{O}}{C}-H} \qquad \underset{3}{\underset{H_3C\ \ CH_3}{\overset{H_3C\ \ H}{CH_3-\overset{|}{C}-\overset{|}{C}-\overset{\underset{\parallel}{O}}{C}-H}}}$$

The first aldehyde is called: Propanal. This is because the longest chain is three carbons long. The second one is called: 4-Bromobutanal. Remember that we must number the chain in such way that the carbonyl compound is carbon #1 in an aldehyde. Finally, the third aldehyde would be called: 2,3,3-Trimethylbutanal.

6) <u>Ketones</u>.

This is the second family of carbonyl compounds we will see. It differs from the aldehydes in exactly the same way as the ethers differed from the alcohols. In other words, the carbonyl is bonded on both sides to an R group, rather than on one side to an R groups and on the other side to a hydrogen atom. The general formula for a ketone is:

$$R-\overset{\overset{O}{\|}}{C}-R'$$

The IUPAC name of a ketone depends on the longest carbon chain that contains the carbonyl group. The actual name is determined by changing the ending –e from the alkane chain and adding the suffix –one. It is necessary to specify the number of the carbon atom that corresponds to the carbonyl group.

- Name the following ketones.

  1. CH₃CH₂ - C(=O) - CH₂CH₃
  2. F-CH₂CH₂ - C(=O) - CH₃
  3. (phenyl-CH₂-C(=O)-CH₂CH₃)
  4. (2-ethylcyclopentanone)

Compound **1** is simply called 3-Pentanone.[1] Compound **2** is called 4-Fluoro-2-butanone. We must give the carbonyl the lowest possible number, which is why the fluorine is in carbon #4. Compound **3** is called: 1-Phenyl-2-pentanone. Finally, the one labeled **4** is called: 2-Ethylcyclopentanone. Since any carbon can be carbon #1 in a cyclic structure, by definition, the carbon with the carbonyl is carbon #1, therefore, there is no need to say 1-cyclo…

7) <u>Carboxylic Acids</u>.

This is yet another family of organic compounds that contain the carbonyl group. However, this family happens to have TWO oxygen atoms in the functional group. The general formula for carboxylic acids is **RCOOH**. The COOH group is just a carbonyl group bonded to an OH. Therefore, the carboxylic acid consists of a carbonyl group bonded on one side to the R group and on the other side to an OH group.

The acids are named by the longest chain that contains the **carboxyl** group (**COOH**). The actual name is given by taking away the ending -e from the alkane and changing the suffix

---

[1] A typical mistake is to call it 3-Pentone. Remember that we must remove the –e from Pentane, and then add –**one**.

to **-oic acid**. It is *not* necessary to specify (and it is actually incorrect to do so) that the carboxyl group is carbon #1.

- Name the following carboxylic acids.

$$CH_3CH_2-\overset{\overset{O}{\|}}{C}-OH \qquad Cl-CH_2CH_2CH_2-\overset{\overset{O}{\|}}{C}-OH \qquad CH_3-\overset{\overset{Br}{|}}{\underset{H_3C}{C}}-\overset{\overset{H}{|}}{\underset{CH_3}{C}}-\overset{\overset{O}{\|}}{C}-OH \qquad CH_3-\overset{\overset{H}{|}}{\underset{NH_2}{C}}-\overset{\overset{O}{\|}}{C}-OH$$

$$\quad\quad 1 \quad\quad\quad\quad\quad 2 \quad\quad\quad\quad\quad\quad 3 \quad\quad\quad\quad\quad\quad 4$$

Compound **1** is simply called Propanoic acid. Notice that we did indeed just remove the –e from propane and added **–oic acid** to it. Compound **2** is called: 4-Chlorobutanoic acid. Compound **3** is called: 3-Bromo-2,3-Dimethylbutanoic acid. Finally, the last one is a special kind of compound. It has an amino group and a carboxyl group. These are referred to as amino acids. This one in particular is called: 2-Aminopropanoic acid.

8) <u>Esters</u>.

This is the last family of organic compounds we will discuss. It is a variation on the carboxylic acids. We can get an ester if we replace the H in the OH for the acid for another R group, normally written as R'. This group may or may not be the same as the one on the other side of the carbonyl group. The general formula for an ester is: **R-COOR'**, where the whole idea is to have a carbonyl group with an R bonded to one side, and an OR' bonded on the other side.

Esters can be synthesized from the condensation of an acid with an alcohol, and the name is derived from these groups. What do we mean by the condensation of an acid with an alcohol? Let's look at the following example. This reaction depicts the formation of an ester from the condensation of Ethanoic acid with Ethanol.[1]

$$CH_3-\overset{\overset{O}{\|}}{C}-OH \;+\; HO-CH_2CH_3 \;\rightleftharpoons\; CH_3-\overset{\overset{O}{\|}}{C}-O-CH_2CH_3 \;+\; H_2O$$

The naming will be done as follows. The R' group is named ending in **–yl** and the rest is like the acid, removing the **–ic acid** and adding **–ate**. Therefore, the example we just synthesized would be called: **Ethyl** ethano**ate**.

---

[1] Notice that this reaction simply takes a water molecule out of the two initial molecules: the acid and the alcohol.

Chapter 13: Organic Chemistry.

When you react Ethanoic acid with ethanol, the molecules dehydrate (lose a water molecule in the process) and end up giving you the ester, Ethyl ethanoate.

- Name the following esters.

Compound **1** is called: Ethyl 2-chloropropanoate. The second one is called Sec-butyl propanoate. The one labeled **3** is called: Isopropyl pentanoate.

13.10 Stereoisomers, Part II: Chirality (Asymmetry).

We have already seen one type of stereoisomers: the geometric isomers or the cis/trans isomers[1]. These are found sometimes in alkenes and they exist because of the impossibility to rotate the carbon-carbon double bond. There are other types of stereoisomers and we will talk now about another one of these types. Once more, in order to fully understand the following discussion, it is imperative that we try to build these molecules with our molecular models.

Sometimes, the mirror image of a given object is a totally different object, and therefore, they are not superimposable. For example, the mirror image of your left hand is your right hand. Of course, you can easily identify a hand by just looking at it and state: "It is the left hand". If the hands were identical mirror images, then we would not be able to do so. If, however, we had a spoon, its mirror image would look exactly the same. Therefore, we could superimpose the mirror image of a spoon with the spoon itself. That is, we cannot look at a spoon and say that it is a left handed spoon.

Molecules that are not superimposable with their mirror images are said to be **chiral**[2] or asymmetric. Once more, what do we mean by superimposable? If we could bring the image from behind the mirror and make it coincide in all of its parts with the original molecule, then it is superimposable and therefore not chiral.

How can we tell when we have a chiral carbon (or a chiral center)? This will be the case every time that a given carbon is bonded to *four different groups*. This pre-requisite of having

---

[1] See Section 13.8.
[2] From the Greek word for hand (*cheir*).

four different groups bonded to a given carbon forces the carbon to be sp³. Of course, if the carbon were doubly bonded to an atom, then it would not have a chance of having four different bonds, since carbon cannot form more than four bonds (under any circumstances). The resulting non-superimposable isomers obtained when we have a chiral carbon are called **enantiomers**. Let's look at one particular example. The molecule shown below (with its mirror image) is called Bromochlorofluoromethane. These two compounds are not superimposable since the carbon has four different groups: a hydrogen, a chlorine, a fluorine, and a bromine.

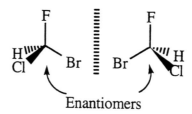

Enantiomers

These two enantiomers are identical in every respect (melting point, boiling point, index of refraction, density, color, taste, etc.) except for one thing. If we were to pass a beam of polarized light through a solution of one of them, it will to rotate the plane of polarization of the light a certain number (call it **x**) of degrees to the right. The other one will rotate the plane of polarization of the light **x** number of degrees to the left!

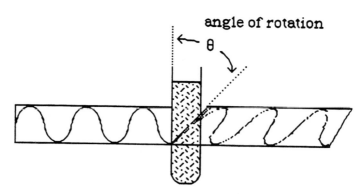

An obvious question that would follow this description would be, can we tell which isomer rotates the light to the right and which one rotates it to the left by just looking at the molecular drawings above? We cannot! However, we can do something else. We can assign an absolute configuration so that we know which isomer we are talking about. This absolute configuration will NOT tell us if the plane of light will be rotating to the left or to the right. It just helps us to identify one or the other on paper. We will use the letters **R**[1] or **S**[2] when we are using this absolute configuration that we just mentioned. Remember, this configuration is just

---

[1] From the Latin for right: Rectus.
[2] From the Latin for left: Sinister.

for the sake of drawing them in a piece of paper, but it is not going to assign the real rotation to the enantiomer in question.

In nature we find that living organisms have evolved in such a way that their enzymes (which are catalysts) can only "react" with the **R** isomers. Next we will learn how to assign the absolute configuration.

Absolute Configuration.

In order to name one of the enantiomers either R or S, we need to do a few things. First of all (and probably most important) is that the chiral center MUST be drawn in three dimensions. This is the only way that we will be able to tell if it is one or the other, since the only difference between them is the orientation of the four groups in space.[1]

Let's consider the following asymmetric molecule.

The following rules must be followed carefully.

1. Identify the four different substituents attached to the asymmetric carbon. To each group, assign a priority such that **a > b > c > d**.

2. Orient the molecule in space so that one may look down the bond from the asymmetric carbon to the substituent with the lowest priority: **d**.

3. Look at the asymmetric carbon with the three attached substituents **a, b, c** radiating from it. Trace a path from **a** to **b** to **c**. If the path describes a *clockwise* motion, then the asymmetric carbon is called **R**. If the path describes a *counterclockwise* motion, then it is called **S**.

Of course, there is a problem with these rules. What are the rules for the priorities, that is how can we tell which group has what priority?

Rules for Priorities. Remember that we are trying to pinpoint the differences between the four groups.

---

[1] That is true since they are stereoisomers.

Chapter 13: Organic Chemistry.

1) Look exclusively at the four atoms directly attached to the chiral carbon. Notice that we mentioned the atoms and not the group as a whole. If these single atoms are different, then the rule states the higher atomic number will have the higher priority.

- Let's look at one particular example to see how to apply this rule. Let's see if this enantiomer of Bromochlorofluoromethane is R or S.

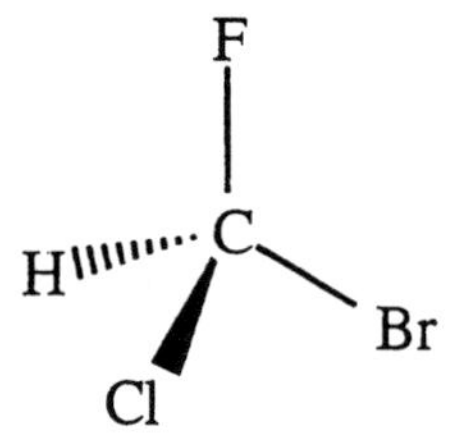

The first thing we notice is that there are four different atoms bonded to the chiral carbon. Since we are assigning the priorities based on the atomic number of these four atoms, then the priorities would be: **1** for Br, **2** for Cl, **3** for F, and **4** for H (the lowest priority). The next thing we notice is that the atom with the lowest priority (the H) is on the dotted line. This means that it is already away from us. This is exactly what we want. Now we trace a path from Br to Cl to F and we see that the path is clockwise. Therefore, the molecule has an absolute configuration of R.

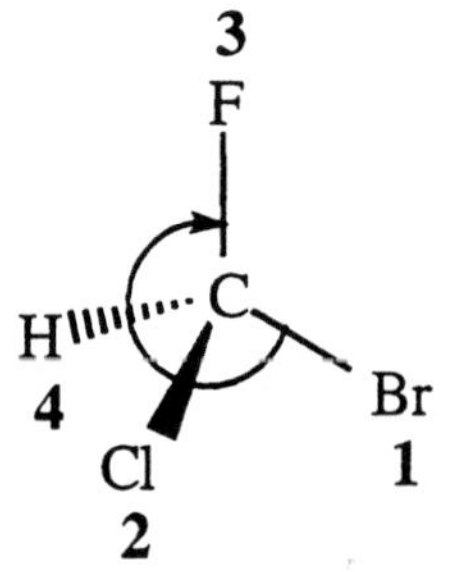

Name: (R)-Bromochlorofluoromethane. Of course, its mirror image would be (S)-Bromochlorofluoromethane.

- Name this particular enantiomer.

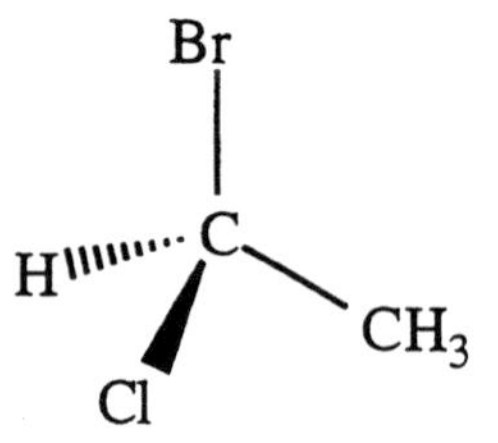

Once more, the first thing we notice is that the molecule is drawn in three dimensions along the chiral center. Also, we notice that all four groups bonded to the chiral carbon are different: a hydrogen, a chlorine, a bromine, and a methyl group. The last group we mentioned is not a single atom, but we will concentrate on the atom bonded to the chiral carbon: another carbon atom. Of course, the lowest priority is still that of the hydrogen, the highest is also that of bromine. Since chlorine's atomic number is higher than that of carbon, then the Cl has a higher priority than the methyl group.

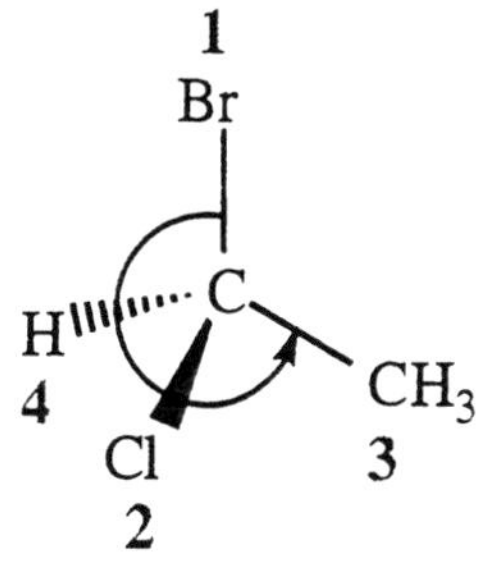

Since the hydrogen is on the dotted line (away from us), all that is left to do is to trace a path from Br to Cl to methyl and this path turns out to be counterclockwise. Therefore, the name of this particular enantiomer is: (S)-1-Bromo-1-chloroethane.

2) If two or more of the atoms bonded to the chiral center are the same, then we need to start moving out from those atoms until we find a difference. There HAS to be a difference at some point since all four groups are different. The priorities are then assigned by following the first rule *at the first point of difference*.

- Name the following molecule.

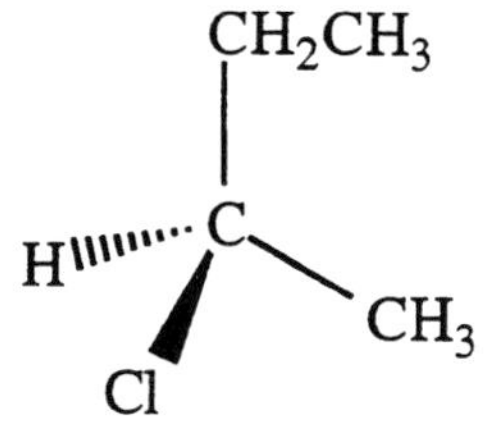

We need to apply this second rule because there are two groups whose atom directly bonded to the chiral carbon are also carbons. Which one has a higher priority, the methyl or the ethyl group? Let's compare the two groups side by side[1].

---

[1] The chiral carbon is indicated by a star.

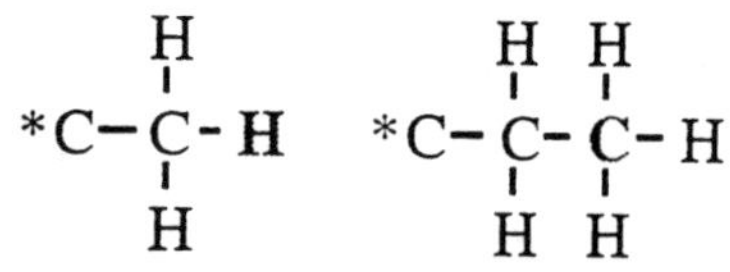

Chapter 13: Organic Chemistry.

$$*C-\underset{\underset{H}{|}}{\overset{\overset{H}{|}}{C}}-H \qquad *C-\underset{\underset{H}{|}}{\overset{\overset{H}{|}}{C}}-\underset{\underset{H}{|}}{\overset{\overset{H}{|}}{C}}-H$$

We should clearly see that the first point of difference has been bolded in the diagram above. In the methyl, it is a hydrogen, whereas in the ethyl it is a carbon. Since the atomic number of carbon is higher than that of hydrogen, then the ethyl group has a higher priority.

Of course, the highest priority is still that for the chlorine and the lowest is that of the hydrogen. Therefore,

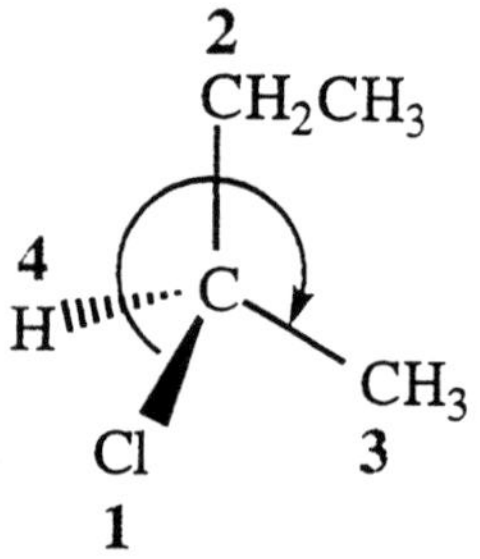

the name of this molecule is: (R)-2-Chlorobutane.

- Name the following molecule.

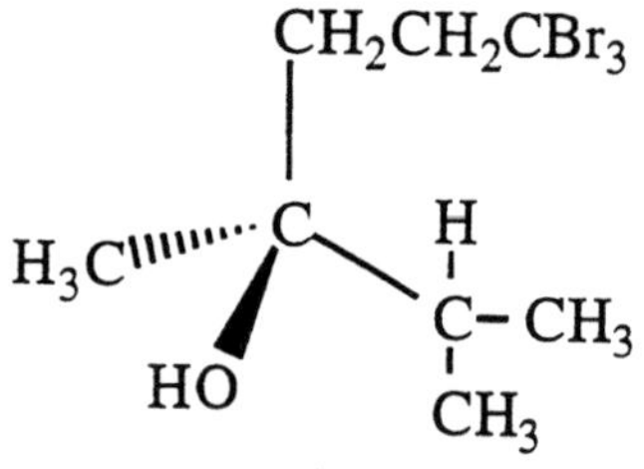

Once more, the first thing we notice is that the four groups bonded to the chiral center are different. This time, however, three of the groups bonded to the chiral center start with a carbon. Of course, the highest priority belongs to the oxygen. Let's analyze the other three, two at a time.

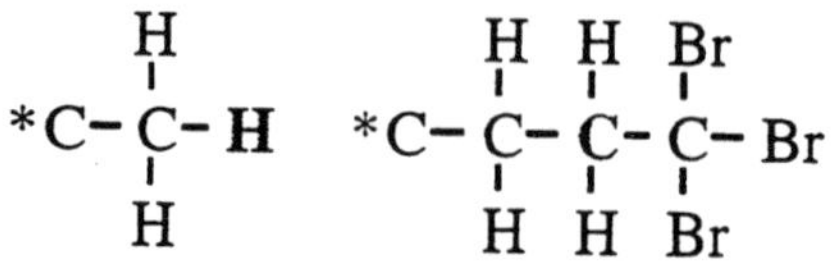

$$*C-\underset{\underset{H}{|}}{\overset{\overset{H}{|}}{C}}-H \qquad *C-\underset{\underset{H}{|}}{\overset{\overset{H}{|}}{C}}-\underset{\underset{H}{|}}{\overset{\overset{H}{|}}{C}}-\underset{\underset{Br}{|}}{\overset{\overset{Br}{|}}{C}}-Br$$

Comparing the methyl and the group with the bromines, it is clear that the one with the bromines has a higher priority because at the first point of difference (bolded atoms), the carbon has a higher priority than the hydrogen. Just to clarify what we mean by the first point of difference, we should ask ourselves the following question. What are the three other bonds to those carbons (the ones bonded to the chiral center)? Both of them have two hydrogens. Therefore, the first difference between those two groups is that the methyl has a third hydrogen whereas the other group has another carbon.

Now, let's compare the group with the three bromines with the isopropyl group.

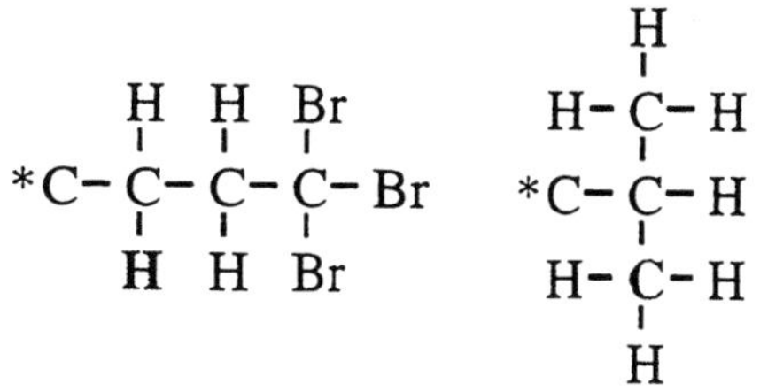

This time the difference occurs before. The carbon bonded to the chiral center in the left one is bonded to two hydrogens and another carbon. The one on the right is bonded to one hydrogen and two other carbons. Therefore, the difference is one hydrogen for one carbon. Therefore, the isopropyl group has the highest priority. This should, once and for all, clear up the misconception that it has something to do with the weight of the group. It has to do with the atomic number of the different atoms at the first point of difference. Finally,

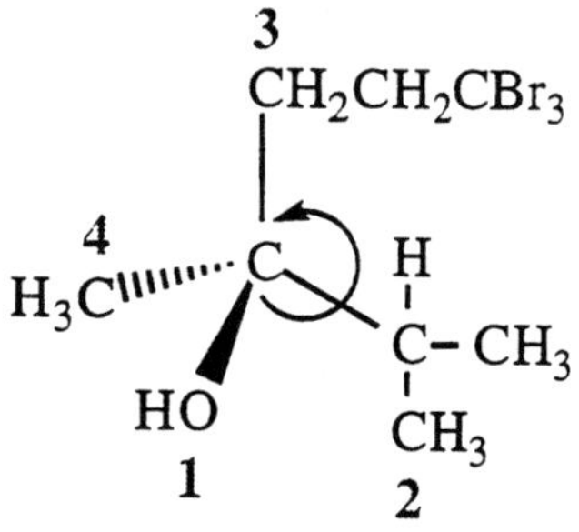

the name of this molecule is: (S)-6,6,6-Tribromo-2,3-dimethyl-3-hexanol. Can you see why? Remember that the OH takes precedence in the numbering of the carbon chain!

If a molecule is drawn in three dimensions and the group with the smallest priority is not in the dotted line (away from our view), then we MUST rotate the molecule until it is placed on the dotted line BEFORE we decide whether or not this compound is R or S. Let's look at one example.

- Name this molecule:

Chapter 13: Organic Chemistry.

$$\underset{H}{\overset{CH_2CH_3}{\underset{|}{H_3C^{\text{\tiny{III}}}}}}\overset{|}{\underset{}{C}}-OH$$

Looking at this molecule, we immediately realize that the hydrogen atom is the one with the lowest priority. Since it is NOT on the dotted line, we must rotate this molecule until the group with the lowest priority (the hydrogen atom) is located there. We do not have a choice about this particular problem. Of course, we can build the molecule with the molecular model sets, and then place the hydrogen away from us to decide. On paper, we perform the following rotation.

$$\underset{H}{\overset{CH_2CH_3}{H_3C^{\text{\tiny{III}}}C}}-OH \quad \xrightarrow{\text{rotation}} \quad \underset{HO}{\overset{CH_2CH_3}{H^{\text{\tiny{III}}}C}}-CH_3$$

Now we can easily name the molecule. The priorities are assigned as follows: 1) OH, 2) the ethyl group, 3) the methyl group, and 4) the hydrogen atom. When we trace a path from the OH to the ethyl to the methyl, we see that it is clockwise, therefore, the enantiomer in question is: (R)-2-Butanol.

We have stated many times that in order to talk about enantiomers (or to show them on paper), we must draw the molecule on three dimensions. So far, we have learned one way to draw a tetrahedral carbon. There is another way in which molecules can be drawn on paper in three dimensions, one that makes assigning the configuration a little easier at times. It is called a Fisher projection. We can take the normal drawing (where there are two bonds on the plane of the paper) and rotate it in such a way that there will be no bonds on the plane of the paper. In this scenario we will have two bonds coming at us (from the East and the West) and two bonds away from us (North and South)[1].

The Fisher projection is the one on the right and it is represented as a simple cross (we will have to remember that E and W are coming at us, while N and S are away from us). The good thing

---

[1] We should practice doing this with our molecular model sets.

about a Fisher projection is that we do NOT need (and SHOULD NEVER EVEN TRY IT) to rotate the molecule to decide its configuration. If the group with the smallest priority is away from us (N or S), then the configuration is normal (clockwise = **R**), BUT if the group with the smallest priority is coming at us (E or W), then the configuration is the opposite of what it seems (clockwise = **S**). This may sound a little confusing, but if we do a couple of examples, we will see how much simpler it can be.[1]

- Name this molecule.

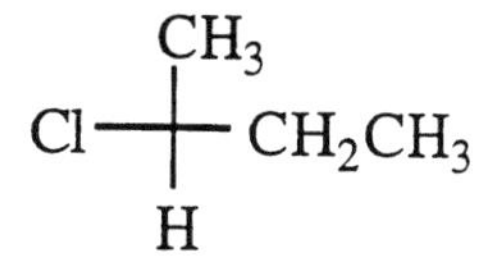

$$\text{Cl} \begin{array}{c} CH_3 \\ | \\ \!\!\!\!+\!\!\!\!\!\! \\ | \\ H \end{array} CH_2CH_3$$

We immediately notice that it does look like a cross. At the center of that cross is the chiral carbon. Notice that it does have four different groups attached to it: a hydrogen atom, a chlorine atom, a methyl group, and an ethyl group. Now, where is the hydrogen (which is, of course, the atom with the lowest priority)? This is the question we must always ask ourselves when using Fisher projections! It is in the south position, which happens to be going INTO the plane of the paper. That is exactly where we want it to be located (going INTO the plane of the paper, which is either the north or the south position). Therefore, this is a simple case: the same as before.

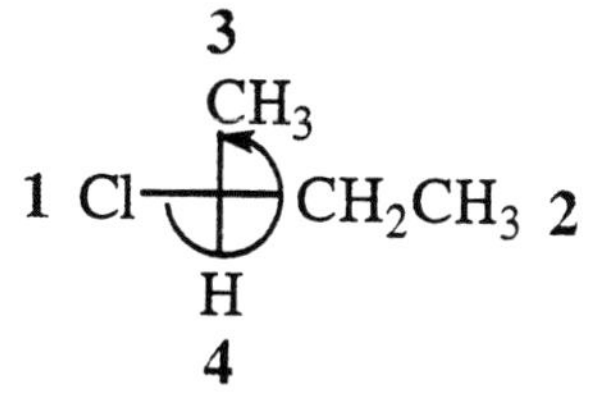

$$1\ \text{Cl} \begin{array}{c} 3 \\ CH_3 \\ | \\ \!\!\!\!+\!\!\!\!\!\! \\ | \\ H \\ 4 \end{array} CH_2CH_3\ 2$$

Therefore, the name of the molecule is: (S)-2-Chlorobutane[2].

- Name this molecule:

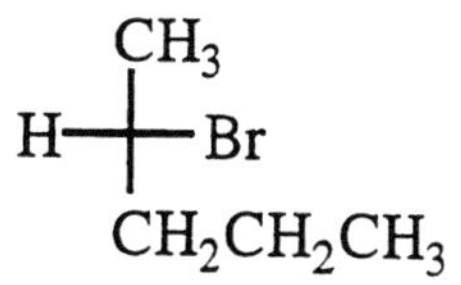

$$\text{H} \begin{array}{c} CH_3 \\ | \\ \!\!\!\!+\!\!\!\!\!\! \\ | \\ CH_2CH_2CH_3 \end{array} \text{Br}$$

---

[1] We should make it very clear that a Fisher projection CANNOT be rotated at all, whereas a normal projection can be rotated. This is the advantage: since it cannot be rotated, there is a "trick" to decide whether it is going to be the R or the S enantiomer.
[2] Because the rotation is counterclockwise.

This molecule has the hydrogen atom in the west position, which is COMING OUT of the paper. Therefore, since the hydrogen is in the wrong place, it will be the opposite of what it seems.

$$\begin{array}{c} 3 \\ CH_3 \\ 4\ H-\!\!\!\!-\!\!\!\!\!\!\!\!\!\!-\!\!\!\!\!\!\!\text{C}-Br\ \ 1 \\ CH_2CH_2CH_3\ \ 2 \end{array}$$

Looking at the clockwise rotation we would be tempted to say it is the R isomer. However, we must remember that the hydrogen is in the "wrong place", therefore, the enantiomer is actually the S enantiomer. Name: (S)-2-Bromo-2-pentanol.

13.11 Nucleophilic Substitution Reactions.

This are some of the most common organic reactions. We defined a **nucleophile** as a species that is looking for nuclei. Remember that the nucleus contains the positive charge in an atom. Therefore, since the nucleophile is attracted to it, it must be either negative or at least have a lone pair with which it can attract to that positive center. The molecule being attacked by the nucleophile is referred to as the **substrate**. In this section, we will consider organic substrates, which will turn out to be alkyl halides. In an alkyl halide, we always have somewhere a carbon atom bonded to a halogen. Since the halogen is more electronegative than the carbon, in every case, there will be a dipole, in which the partially positive end of the dipole will be at the carbon. When the nucleophile attacks the substrate, the lone pair in the nucleophile will try to form a bond with the partially positive carbon atom in the substrate. All carbons form four bonds, therefore, a fifth bond would be impossible. This means that the substrate must break free one of the atoms (or groups) attached to it. This is referred to as the **leaving group** (for us, it will be halogen always). The leaving group is displaced and will take both electrons of the bond with itself, thus it will leave with a negative charge (like Br^{-1}). Let us concentrate on the mechanism of this reaction, which we call nucleophilic substitution reactions. When we study all the possible reactions between alkyl halides with a nucleophile like OH^{-1}, the experimental results suggest that there are two different ways in which this reaction may take place. However, the bottom line, independent of the actual mechanism, is that one of the lone pairs in the hydroxide ion is attracted to the partially positive carbon in the alkyl halide. Let us look at each mechanism independently.

Chapter 13: Organic Chemistry.

13.12 Second Order Mechanism, the SN2 Mechanism.

Many of the reactions studied are observed to follow second order kinetics, where the rate equation is given by:

$$\text{rate} = k\,[\text{substrate}]\,[\text{nucleophile}] = k\,[\text{alkyl halide}]\,[OH^{-1}].$$

According to what we learned in Chapter 12, this means that the slow step of the mechanism involves the collision of one molecule of each, therefore, it should be a bimolecular collision between the substrate and the nucleophile. Let's look at one example.

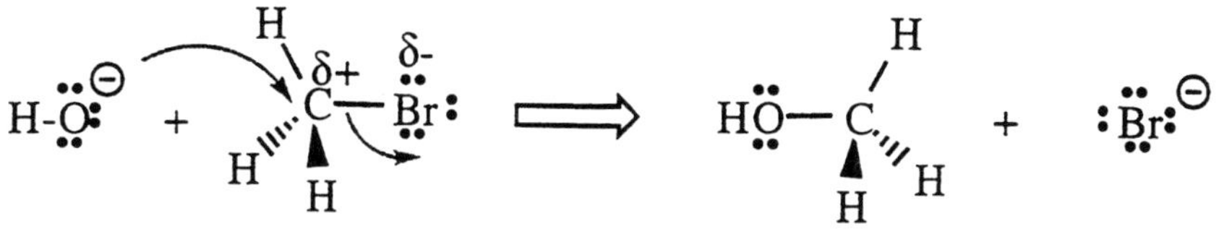

In this case, we see the reaction between the nucleophile (the hydroxide ion) and the substrate (methyl bromide) in a single, bimolecular collision. This yields directly the products of the reaction: methanol and the bromide ion. Notice that the reaction actually involves replacing the leaving group with the nucleophile. The leaving group also becomes negative once it leaves, which means that it itself is a nucleophile. Since the reaction occurs in the way written, this must mean that the hydroxide ion is a much better nucleophile than the bromide ion.

The mechanism we have just described is called nucleophilic substitution – second order, and it is symbolized by SN2. Therefore, the reaction between the hydroxide ion and methyl bromide follows the SN2 mechanism, which once again, is a single bimolecular collision between the hydroxide ion and the methyl bromide. One of the lone pairs in the oxygen of the hydroxide ion attacks the partially positive carbon from the back side, that is, 180° away from the carbon-bromine bond. The arrow in the diagram indicates the attack. This "attack arrow" also indicates the initiation of the formation of a bond between the oxygen and the carbon. Since the carbon cannot form five bonds, the bond between the carbon and the bromine is broken (the second arrow shown). If you look at the diagram carefully, you may notice that the carbon is inverted (flipped) during the attack, just like an umbrella is flipped by the action of a strong wind gust. In this particular case, this has no consequence whatsoever. However, if we use a substrate where the carbon being attacked is also chiral, then we will observe **inversion of configuration**. Let's look at such an example. Let's start with (R)-2-Bromobutane and react it with the hydroxide ion. The reaction follows exactly the same mechanism we described above (SN2).

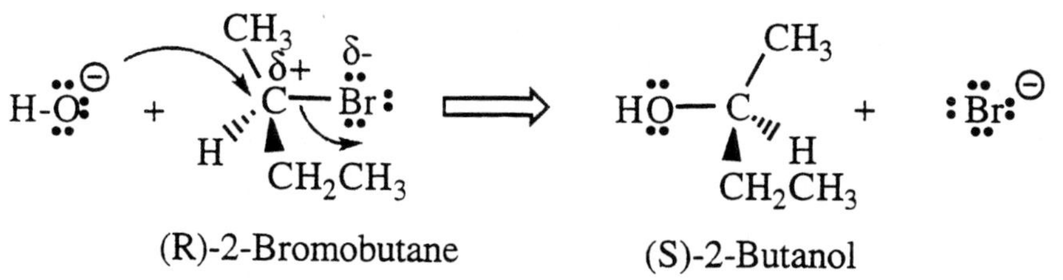

Chapter 13: Organic Chemistry.

$$H\text{-}\ddot{\underset{\cdot\cdot}{O}}{:}^{\ominus} + \overset{CH_3}{\underset{\underset{CH_2CH_3}{|}}{\overset{\delta+}{C}}}\!\!\!\!\!\!\!\!\!\!\!\!\!\!\!\!{}^{\delta-}_{}\ddot{\underset{\cdot\cdot}{Br}}{:} \implies H\ddot{\underset{\cdot\cdot}{O}}-\overset{CH_3}{\underset{\underset{CH_2CH_3}{|}}{C_{\cdots H}}} + {:}\ddot{\underset{\cdot\cdot}{Br}}{:}^{\ominus}$$

(R)-2-Bromobutane (S)-2-Butanol

We can immediately see that the effect of "flipping" the carbon during the attack ends up changing the configuration from R to S. This is what we mean by inversion of configuration. As the oxygen starts to form its bond with the carbon, the carbon–bromine bond starts to break. Of course, this must occur simultaneously since the reaction takes place in a single step. As this step progresses, the other three groups in the carbon (the methyl, the hydrogen, and the ethyl group) start to flatten out until eventually (at the end) they flip. We must understand that this step is very dynamic, and we can only see this clearly in a movie type demonstration. The diagram below will give us some "freeze frames" from this process. The frame marked **A** is the start of the process. The one marked **C** is the end whereas the one marked **B** is somewhere in between.

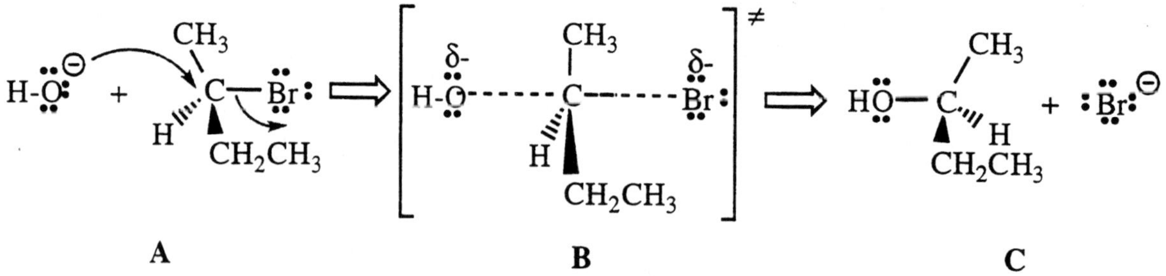

$$H\text{-}\ddot{\underset{\cdot\cdot}{O}}{:}^{\ominus} + \overset{CH_3}{\underset{\underset{CH_2CH_3}{|}}{C}}\!\!\!\!\!\!\!\!\!\!\!\!{-}\ddot{\underset{\cdot\cdot}{Br}}{:} \implies \left[H\text{-}\ddot{\underset{\cdot\cdot}{\overset{\delta-}{O}}}\text{-}\text{-}\text{-}\text{-}\overset{CH_3}{\underset{\underset{CH_2CH_3}{|}}{C}}\text{-}\text{-}\text{-}\text{-}\overset{\delta-}{\underset{\cdot\cdot}{\ddot{Br}}}{:}\right]^{\neq} \implies H\ddot{\underset{\cdot\cdot}{O}}-\overset{CH_3}{\underset{\underset{CH_2CH_3}{|}}{C_{\cdots H}}} + {:}\ddot{\underset{\cdot\cdot}{Br}}{:}^{\ominus}$$

A   **B**   **C**

This should remind us of the energy plots we studied in Section 12.10. These were plots of energy versus the reaction progress (the "freeze frames"). The frame marked **B** represents a high energy state, which we called the activated complex. The energy plot for this reaction can be seen below.

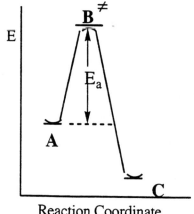

Reaction Coordinate

Chapter 13: Organic Chemistry.

- Consider the reaction of (R)-1-Bromo-2-methylbutane with hydroxide ions. Given that this reaction follows an SN2 mechanism, identify the organic product.

    Let's look at the compound first. It is clear that the reason the compound is R[1] is because of the second carbon, which has four different groups attached to it: a methyl, a hydrogen, an ethyl, and a bromomethyl.

$$\begin{array}{c} CH_2Br \\ H\!\!-\!\!\!\!\!\!|\!\!-\!\!CH_3 \\ CH_2CH_3 \end{array}$$

What is the situation in this problem? The bromine (the leaving group) is attached to carbon #1, which is NOT the chiral carbon. Therefore, the nucleophile is going to attack and replace the Br at that particular position. Since the chiral center is NOT ATTACKED, then there is no inversion of the carbon that matters and the product will still be R.

$$\begin{array}{c} CH_2OH \\ H\!\!-\!\!\!\!\!\!|\!\!-\!\!CH_3 \\ CH_2CH_3 \end{array}$$

(R)-2-Methyl-1-butanol

In that last problem, we were specifically told that this reaction followed an SN2 mechanism. Is there a way we can tell from just looking at the substrate that a reaction will follow an SN2 mechanism? In order for the reaction to follow the SN2 mechanism, the nucleophile (the OH$^-$) must be able to reach and attack the polarized carbon (the carbon bonded to the electronegative leaving group) in the substrate. This tells us right away that the only thing that should matter is whether this carbon is sterically hindered or not. In the case of CH_3-Br, we should be able to see that the nucleophile would have no problem whatsoever in reaching the carbon, since it has three hydrogens attached to it. Remember that H is the smallest atom, therefore, three of them do not pose much of a steric hindrance for the nucleophile. If we start substituting more and more alkyl groups, then we can see that it will keep getting harder and harder for the nucleophile to reach the polarized carbon in the substrate.

too crowded for the nucleophile to physically fit in the open space

---

[1] Looks like S, but the hydrogen is coming out at us, therefore, it is the opposite of what it seems.

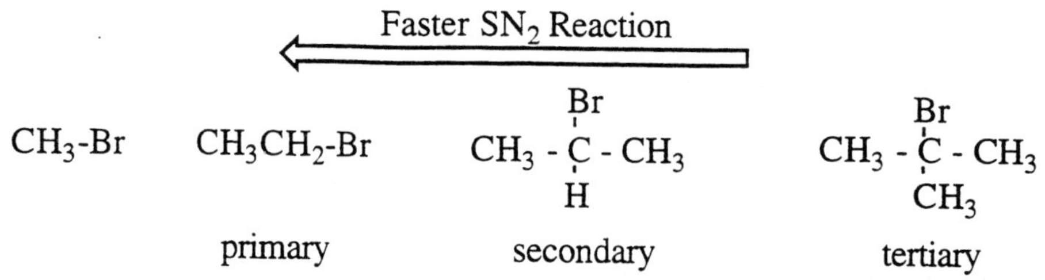

Chapter 13: Organic Chemistry.

⟵ Faster SN₂ Reaction

CH₃-Br   CH₃CH₂-Br   CH₃ - C(Br)(H) - CH₃   CH₃ - C(Br)(CH₃) - CH₃

primary   secondary   tertiary

In the figure above, we can see that the fastest and easiest reaction would be with the Methyl bromide (least amount of steric hindrance), whereas the hardest and slowest would be with the tertiary[1] substrate which has no hydrogens present on that carbon. When the carbon being attacked has no hydrogens bonded to it, we observe that the nucleophile is not able to penetrate close enough to initiate the attack. For the time being, let's say that the SN2 mechanisms will only be observed when our substrate is either methyl, primary or secondary, in other words, when we have at least one H in the carbon being attacked.

But what happens with the tertiary substrate? We have already mentioned that the OH⁻ cannot penetrate close enough to attack. Will we also observe substitution? Yes, we do, but it should be clear that is must follow a totally different mechanism.

13.13 First Order Mechanism, the SN1 Mechanism.

In order to understand this mechanism, we should try to follow one particular example. Let's take the case of the reaction of (S)-3-Bromo-2,3-dimethylpentane.

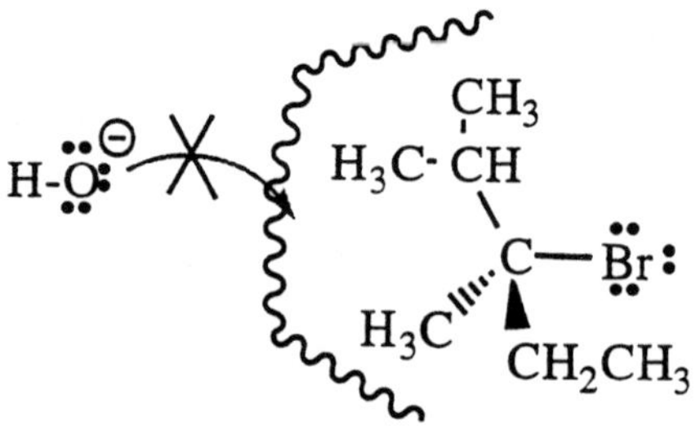

This diagram clearly shows the steric hindrance of the partially positive carbon (carbon #3), which is also the chiral carbon. The hydroxide ion cannot penetrate to attack the partially positive carbon due to the bulkiness of the three alkyl groups. Therefore, it has to wait. Wait for what? It has to wait for the bromine to leave on its own. At any temperature, we always have a

---

[1] The carbon with the bromine is bonded to THREE other carbons, therefore it is called tertiary. In the case of the secondary, that carbon is bonded to TWO other carbons, and in the case of primary, to only ONE other carbon.

wide distribution of molecular energies[1]. Some of the substrate molecules will have very little energy, most will have what we call the average energy, and some will have pretty large energies. Some of these high energy molecules will have enough energy to dissociate heterolytically. This means that the bromine will take BOTH electrons from the bond and leave as a bromide ion, leaving behind a carbocation. Of course, this is not going to be easy to do, therefore, this will be the slow step of the mechanism. Since in this step we only have the substrate as a reactant, then this reaction will follow first order kinetics.

First Step: Slow Step (RATE DETERMINING STEP)

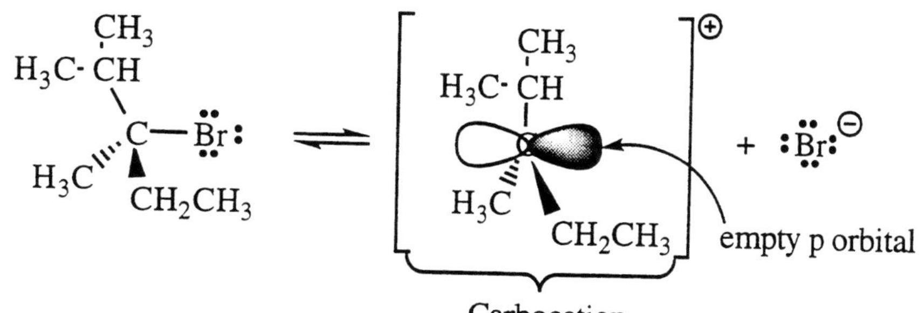

And what is a **carbocation**? It is an alkyl species that has a positive carbon. All carbocations are sp^2 in hybridization, therefore, the ion is completely flat right around the positive carbon.

The second step is very fast and it involves the collision of the OH⁻ ion with the carbocation. Since the carbocation is flat, the OH⁻ ion will have no problem at all reaching and attacking the fully positive carbon. However, this time the hydroxide ion has two clear positions into which it may attack the positive carbon (on either side). Therefore we will observe the formation of two products and the ratio of the two products will be 50:50 since the OH⁻ ion should have no preference for either side.

Second Step of the Mechanism

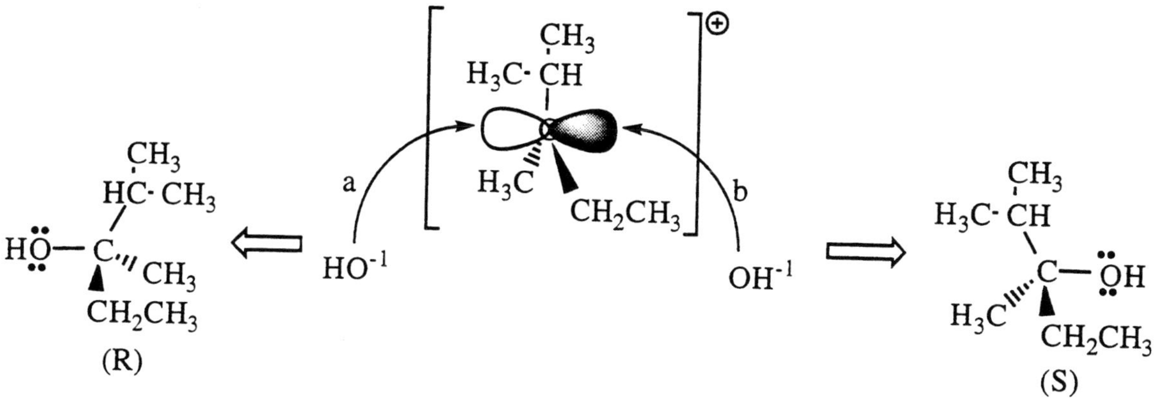

---

[1] Given by the Maxwell-Boltzmann expression.

This mixture of 50% **R** and 50% **S** will have no optical activity[1] because when we have two enantiomers in equal quantities, they offset each other.[2] This type of achiral[3] mixture, where we have equal amounts of the two enantiomers, are known as **racemic** mixtures. We will call this mechanism SN1, since it will follow first order kinetics (nucleophilic substitution, unimolecular). Clearly, the rate equation for this reaction can be seen from the rate equation.

$$\text{rate} = k\,[\text{substrate}]$$

This means that the rate of this reaction is totally independent of the concentration of the hydroxide ion. This should make sense, because if we were to increase the concentration of hydroxide ions with the hope of increasing the rate of the reaction, we should see that this is not going to have any effect whatsoever on the speed of the reaction, since the hydroxide ion CANNOT penetrate the molecule in order to attack the chiral center. Therefore, the entire issue is steric hindrance. If there is steric hindrance, the mechanism to be followed will be SN1; if there is no steric hindrance (or reduced hindrance, that is, at least one hydrogen on the carbon being attacked), then it will follow the SN2 mechanism.

Anytime a particular reaction follows the SN1 mechanism, it will lead to racemization of the chiral center, as long as the chiral center is the one that has the leaving group. Of course, if the chiral center is not involved in the reaction itself, then we will have retention of configuration.

We have explained this mechanism in terms of the difficulty of forming the carbocation in the first step (which is why it is the rate determining step). The energy profile for this mechanism can be seen below. We see two activation energies in the plot (two bumps), which corroborates the two steps of the mechanism. We also see that the first activation energy is much larger than the second one.

---

[1] It will not rotate a plane of polarized light at all.
[2] One rotates the light so many degrees to the right and the other one the exact same number of degrees to the left, so they cancel out.
[3] Not chiral; a substance that has lost its chirality.

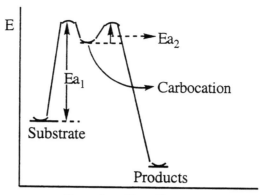

Reaction Coordinate

Why will this SN1 mechanism occur at all with tertiary substrates? It is clearly not an easy thing to do (to form a carbocation). We will see in the next chapter why tertiary carbocations are more stable, and therefore more likely to be formed, than any other kind.

General Trends for Nucleophilic Substitution.

We must distinguish between the Carbon being attacked (the one with the leaving group * the halogen * in it) and the chiral carbon, the one with the four different groups attached to it. These two carbons may be the same, but they do NOT have to be the same. If the chiral carbon is not attacked, then the stereochemistry of this Carbon is preserved, independent of the mechanism. At this point in time, it would be advantageous to state a few simple rules to follow in order to decide what happens to the stereochemistry when the nucleophile reacts with the substrate.

1) Circle the Carbon with the Bromine attached to it.

   a) If it has at least one hydrogen attached to it, it will be SN2.

   b) If it does not have any hydrogens attached to it, it will be SN1.

2) Draw a star by the chiral Carbon (the one with the four different groups attached to it).

   a) If the circled Carbon and the starred Carbon are one and the same, the stereochemistry **will** change.

       1. SN2 $\Rightarrow$ Inversion of configuration.
       2. SN1 $\Rightarrow$ Racemization.

   b) If the circled Carbon and the starred Carbon are not the same, then we have retention of configuration because the chiral Carbon was not attacked.

Chapter 14: Spectroscopy.

14.1 Experimental Determination of Structure.

At this point, we have had a short introduction to organic compounds and some of their properties and reactivities. There is one more aspect of organic chemistry that would be advantageous to learn at this point: Spectroscopy. Generally speaking, the term spectroscopy applies to the interaction of light with matter, and in most of these techniques, we take advantage of particularly that in order to elucidate the structure of some organic compounds. There are various techniques available for this purpose and most of them are very sophisticated and relatively new. We will now introduce the three main ones.

      a) Mass Spectroscopy (MS).
      b) Infrared Spectroscopy (IR).
      c) Nuclear Magnetic Resonance Spectroscopy (NMR).

The best way to end up figuring out the identity of the unknown compound is not by any one of these three techniques by themselves, but rather by the combination of all three. Mass spectroscopy will tell us the *mass* and the stability of molecular ions and fragments of the molecule. Infrared spectroscopy will tell us what types of bonds we have in the molecule (like O-H, C-H, C=O, etc.) and finally nuclear magnetic resonance spectroscopy will tell us how these atoms are bonded in the molecule.

14.2 Mass Spectroscopy.

Let us look at the schematic diagram of a mass spectrometer.

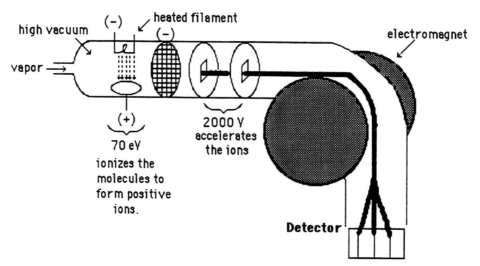

In order to determine the MS for a given molecule, we start by vaporizing a sample of the compound and then introduce it into the chamber. The molecules of the compound will be ionized when they pass by the 70 eV ionizer. The ions formed are then directed and accelerated through a region of 2000 V potential difference. This beam of accelerated molecular ions is then deflected by the presence of an external magnetic field. It can be shown that the heavier the ion (larger m/e ratio[1]), the smaller the angle of deflection. Therefore, ions of different masses will show different angles of deflection, *i.e..,* the degree or amount of deflection depends on the mass.

The implication behind the preceding paragraph is that we have more than one kind of cation formed in the gas phase. Why do we have more than one kind of ion to start with? Because once the molecular ion is formed, it will break up to try to achieve a more stable configuration. Remember that ions are not stable in the gas phase[2]. Therefore, as time goes by, the molecular ion is going to fragment into more relatively stable ions.

What does the instrument actually record at the end? What is it we call a mass spectrum? The mass spectrum is simply a series of peaks that let us know the mass of a certain cation (technically the mass to charge ratio, but since most of the ions are actually +1, then it is really the mass) and the relative abundance of each of these ions. This is the key to the whole MS experiment. The fragmentation pattern, including the relative stability of each, will be very useful in helping us to determine the identity of the unknown compound.

We have just mentioned again that ions are not stable in the gas phase, but he are making them anyway. That is exactly what happens in the apparatus. We have also stated that a given ion will fragment in order to form a smaller chunk that will be a little less unstable. What determines if a fragment is more stable than another? There are two main stabilizing effects: a) Inductive effect and b) Resonance effect. Both effects try to achieve the same purpose: to "spread out" the positive charge over a larger area, *i.e.., to delocalize the charge!* From the two, *resonance* imparts more stabilization (in general). Let's look at our first example. In here, we see a given carbocation, in which we happen to have the positive carbon directly connected to an atom with a lone pair (in this case the oxygen). How is the oxygen going to help the positive carbon? Well, the oxygen atom has two lone pairs, which if we remember it is the name we gave to non-bonding pairs. Of course, if the oxygen shared one of its two lone pairs with the carbon, it would really solve the problem for the carbon. This can be shown with the formation of an

---

[1] m/e means mass over charge.
[2] See Section 8.2.

extra bond between the carbon and the oxygen. We can see that now the carbon has a complete octet, and for that matter, so does the oxygen. Every atom ends up winning. However, an oxygen with three bonds and a complete octet carries a positive charge. This is the drawback. Oxygen is much more electronegative than carbon, but in this resonance form it carries a positive charge while the carbon remains neutral. We know that neither of these two forms are real and that instead what we achieve is electron delocalization between the two atoms, or in this case, a delocalization of the positive charge between the two atoms. Any time we have the potential of delocalizing a charge, it will always happen spontaneously because it will convey a lower energy.

$$H-\overset{\oplus}{\underset{H}{C}}-\overset{..}{\underset{..}{O}}-H \longleftrightarrow H-\underset{H}{C}=\overset{\oplus}{\underset{..}{O}}-H$$

Total Effect: $H-\overset{\overset{\oplus}{|||}}{\underset{H}{C}}-\overset{..}{\underset{..}{O}}-H$ } Charge is "spread out" over two atoms, ∴ more stable!

The inductive effect is normally a little weaker. This effect will be a lot easier to understand if we look at an example. Let's consider the ethyl carbocation:

$$H-\overset{\oplus}{\underset{H}{C}}-CH_3$$

We find out that the methyl group has a certain ability to donate electron density to the positive carbon, thus minimizing the positive charge. In other words, the electronic cloud of the CH_3 moves a little towards the positive carbon atom (*i.e.*, the CH_3 is polarized) and therefore takes a little bit of the positive charge away from the positive carbon. This creates a small stabilization.

$$H-\overset{\oplus\Leftarrow}{\underset{H}{C}}CH_3 \longrightarrow H-\overset{\delta+}{\underset{H}{C}}\overset{\delta\delta+}{CH_3}$$

Hydrogens do not do a good job at inducing charge, therefore, the only groups that have inductive ability are the alkyl groups (like methyls). Thus, the stabilization goes like this:

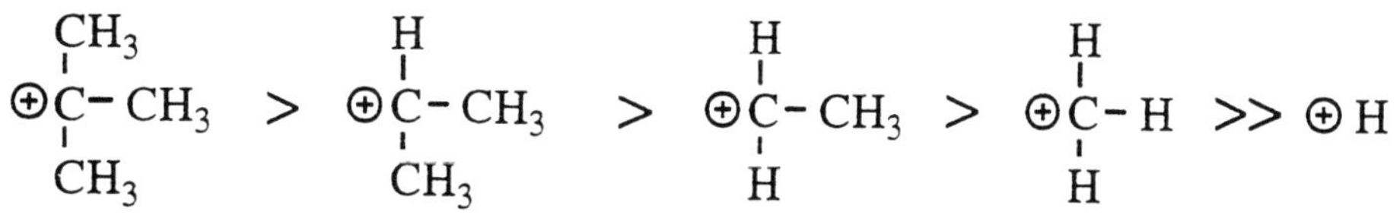

Chapter 14: Spectroscopy.

$$\underset{\text{tertiary}}{\overset{CH_3}{\underset{CH_3}{\oplus C-CH_3}}} > \underset{\text{secondary}}{\overset{H}{\underset{CH_3}{\oplus C-CH_3}}} > \underset{\text{primary}}{\overset{H}{\underset{H}{\oplus C-CH_3}}} > \underset{\text{methyl}}{\overset{H}{\underset{H}{\oplus C-H}}} \gg \underset{\text{hydrogen}}{\oplus H}$$

<u>Conclusion</u>: The more alkyl groups (like methyl) that we have connected to the positive carbon, the larger the inductive effect and the larger the stabilization. That is why tertiary (bonded to three alkyl groups) carbocations are more stable than secondary (bonded to two alkyl groups) carbocations, etc.

- Calculate the m/e values for the following ions: the methyl, ethyl, and isopropyl carbocations.

    We must remember that since they are all +1 ions, all we have to do is find the sum of their atomic weights and divide these by one (the charge). Therefore,

$$\overset{\oplus}{C}H_3 \qquad \overset{\oplus}{C}H_3CH_2 \qquad \overset{\oplus}{C}H_3\text{-}CH\text{-}CH_3$$

    m/e 15   m/e 29   m/e 43

We should now look at some mass spectra for some common molecules and see the type of information that we want or can get out of these.

- Methane gas, CH_4.

    The actual mass spectrum for methane gas is summarized below.

m/e	Relative Abundance
1	3.1
2	0.17
15	85.0
16	100.0
17	1.11
18	0.11

The actual mass spectrum looks somewhat like the one shown below.

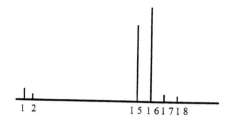

How are we going to interpret this spectrum? We should start at the very beginning. What happens to the methane molecule one it is ionized? One of the electrons in the molecule (one of the valence shell electrons) is lost. This has to be one of the electrons shared between the carbon and any of the four hydrogens. Once this molecular ion is formed (which would weigh m/e 16), then this ion could split into simpler ones. For example, it could break up by either forming the hydrogen ion (m/e 1) or the methyl ion (m/e 15). The spectrum tells us that these two ions come directly from the split of the original ions (the molecular ion, sometimes symbolized as the **M** ion). The breakage of **M** can therefore yield the **M – 1** or the **M – 15** ions. This can be seen below.

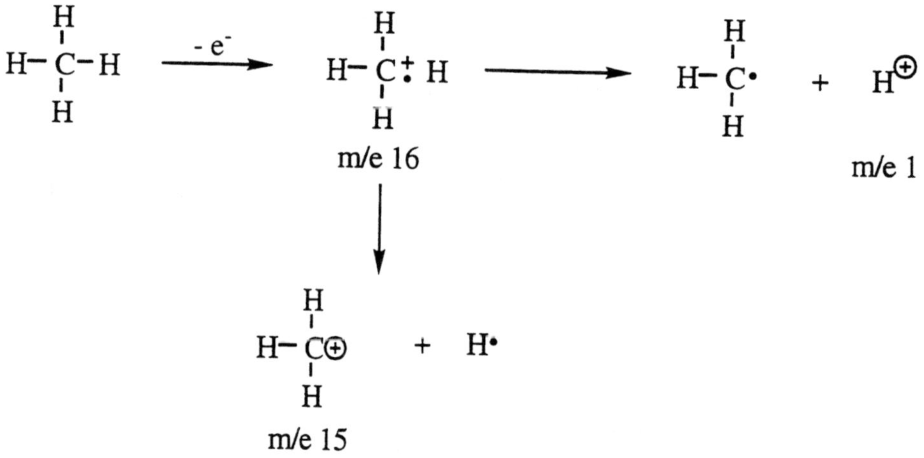

The fact that the spectrum shows that the m/e 15 ion is much stronger (more intense) than the m/e 1 ion, corroborates our idea of the relative stability of the positive ions. The one that is more stable is always easier to form.

How can we possibly explain the presence of peaks at m/e's of 17, 18, and at 2? These peaks are due to the isotopes that exist in nature. We know that we have present a certain amount of deuterium and also a certain amount of carbon-14. The interesting thing is that relative abundance of these isotopes in nature is clearly seen in the relative abundance of these ions in the mass spectrum.

Chapter 14: Spectroscopy.

$$^+D \quad\quad m/e\ 2$$

$$^+CH_3D \quad\quad m/e\ 17$$

$$^{+14}CH_4 \quad\quad m/e\ 18$$

Important Observations.

When a parent ion (**M**) breaks apart it forms two things: a carbocation and a radical.

$$R{:}X \longrightarrow R^\oplus + X^\bullet$$
$$\text{Carbocation} \quad \text{Radical}$$

The extent of the dissociation or breakage (*i.e..,* how many parent ions will follow that path) depends on two things, in the following order of importance:

1. The stability of the carbocation.
2. The stability of the radical formed.

How can we tell which radical is more stable? The relative stability of radicals follows the same identical pattern as that of carbocations:

Resonance > Tertiary > Secondary > Primary > Methyl > Hydrogen radical

We should keep in mind that the radicals formed are not "visible" in the mass spectrum. These radicals have no charge and therefore, they are NOT deflected by the electromagnet and they continue traveling in a straight path.[1] That does not mean that we do not have to pay attention to the radicals themselves. The most stable fragmentation is that one that yields the most stable carbocation AND the most stable radical possible. Remember that the only thing we can actually *detect* in the mass spectrum are the positive ions present in the gas phase during the experiment.

Let's do a few other examples, initially looking at the spectra, and later on we will flat out guess as to what parts of the spectra will look like if we are told the structure.

- Methanol.

    Remember that the general formula for methanol is $CH_3\text{-}OH$.

---

[1] See the general diagram for the mass spectrometer on the first page of this chapter.

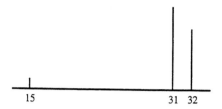

The first thing we notice from the spectrum is that the main peak in the mass spectrum of methanol occurs at m/e 31. This is actually the (**M – 1**) ion. This immediately tells us that the main peak occurs when we lose a hydrogen. Of course, the real question is: which hydrogen is lost? Of course, any one of them. However, when we look at all three possibilities[1], we can ascertain that the most probable is the third one (see the figure below), because it can place the positive charge on the carbon next to the oxygen with the two lone pairs. This will allow for stabilization by resonance. It is this resonance stabilization that is going to explain the preference for the fragmentation observed.

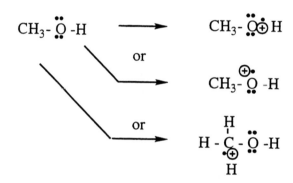

Therefore, the main peak of the mass spectrum for methanol can be rationalized through the following mechanism.

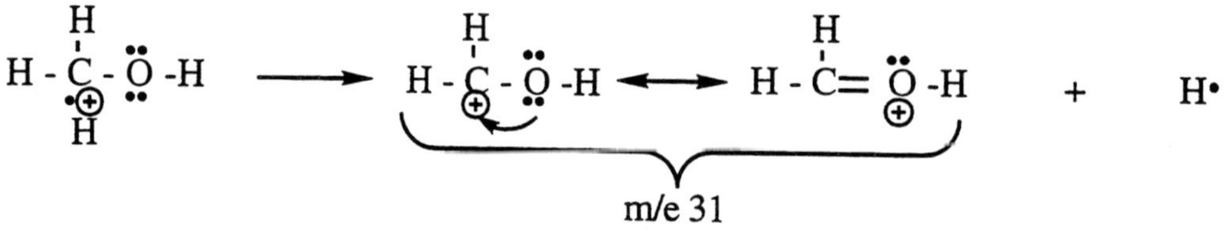

This is the best possibility or the most stabilizing fragmentation even though we formed a hydrogen radical (which is very unstable). In general, it is very difficult to lose a hydrogen radical and the only time that this happens is when the generated carbocation is stabilized by resonance. It should be clear that the peak at 15 is due to the methyl carbocation, whose

---

[1] The removal of an electron from the oxygen-hydrogen bond, the removal of an electron from one of the lone pairs in the oxygen, or the removal of one of the electrons from any of the carbon-hydrogen bonds.

Chapter 14: Spectroscopy. 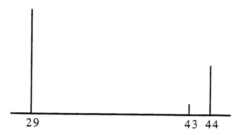 Ocasio - 465

formation is accompanied by the release of the hydroxide radical.

- Ethanal.

    This is an aldehyde, therefore, its molecular formula is CH_3CHO. The mass spectrum for this compound is seen below.

At this point, we should be able to see that the best possible scenario will be to create the charge on the carbon next to the oxygen (which has a lone pair and will therefore be able to stabilize it by resonance). Let's look at the molecule to make sure what are the distinct possibilities that would yield the kind of fragmentation that places the charge on that carbon atom.

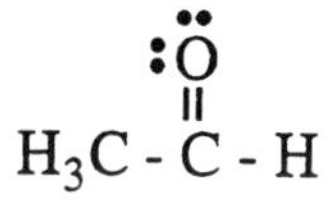

$$\overset{\overset{\displaystyle :\ddot{O}:}{\|}}{H_3C - C - H}$$

To achieve a positive ion at carbon #1, one of two things must occur:

1. A hydrogen radical is lost.
2. A methyl radical is lost.

The loss of either one of these two radicals will yield an equally "stable" carbocation, one that is stabilized from resonance with one of the lone pairs on the oxygen atom. Since the methyl radical is more stable than the hydrogen radical, we should expect the main peak to be 15 units smaller (case A below) than the molecular peak (**M – 15**) and if anything, a very small peak at one unit less (case B below) than the molecular peak (**M – 1**). As we can see from the spectrum above, this is exactly what we get.

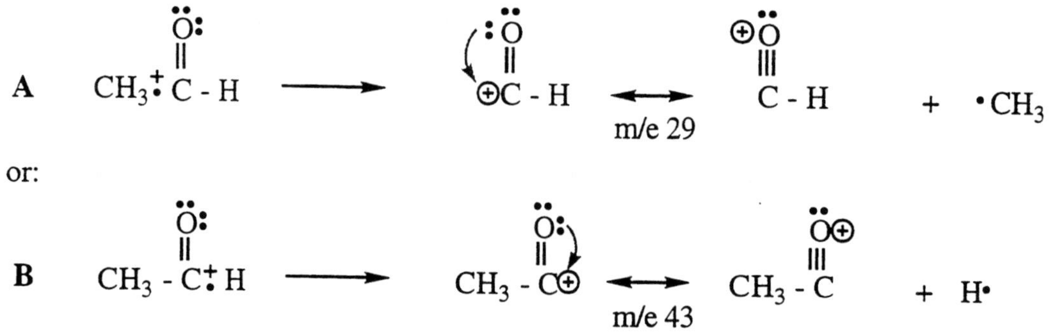

Chapter 14: Spectroscopy.

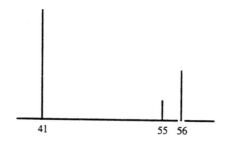

We can see that the two routes lead to the formation of a resonance stabilized carbocation, however, since the first route yields the most stable radical, then that will be the preferred pathway. This is why the main peak in the MS is at m/e 29.

Are there any other ways we can achieve resonance by stabilization other than delocalization with an electron pair from an oxygen (or a bromine, nitrogen, or any other atoms with lone pairs)? Of course. One of the most important is stabilization by the pi bonds of alkenes.

RULE: *To generate resonance on a carbocation, the charge must be on the must be on a carbon next to a carbon-carbon double bond (C=C).* This carbon is referred to as the **allylic** carbon.

The easiest way to see this stabilization is by looking at one such example.

- 1-Butene.

    The molecule is: $H_2C=CH-CH_2CH_3$, and its mass spectrum is shown below.

We can immediately see that the main peak of this spectrum occurs at m/e 41. We will concentrate our efforts not in explaining every peak in the MS from now on, but rather on explaining the main peak in the mass spectrum. This is also called the most prominent peak.

Chapter 14: Spectroscopy.

Applying the rule mentioned above we are going to generate a resonance stabilized ion by creating a charge on the carbon *next* to the carbon-carbon double bond (which is carbon #3, the allylic carbon). However, there are two distinct possibilities that yield an equally stable carbocation.

1. Remove a hydrogen radical.
2. Remove a methyl radical.

Just as in the previous example, we will observe that the peak at **M – 1** is much smaller than the peak at **M – 15** since the methyl radical is more stable than the hydrogen radical. Therefore, the ion responsible for the peak at m/e 41 would be:

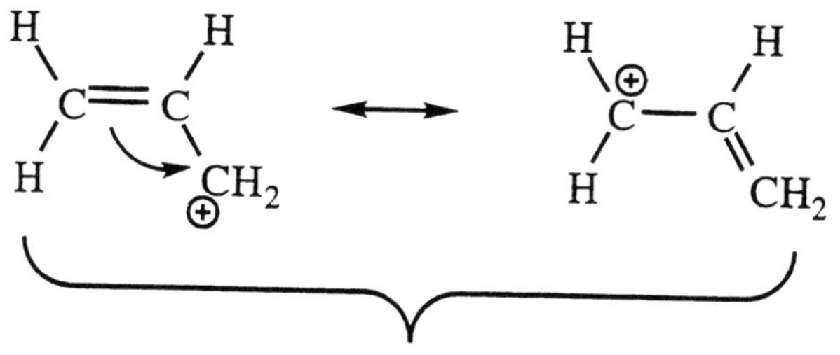

$$\underbrace{\begin{array}{c}H_2C=CH-\overset{+}{C}H_2 \quad \longleftrightarrow \quad \overset{+}{C}H_2-CH=CH_2\end{array}}_{m/e\ 41}$$

This ion gives rise to the most prominent peak in the MS for this compound because it is stabilized by resonance and it forms the most stable radical possible for this system.

- Determine the main peak in the MS for the following compound.

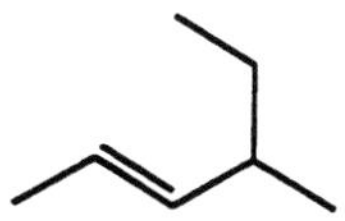

Once more, as soon as we immediately see the carbon-carbon double bond, we want to place the charge in the allylic carbon, which are carbons numbers 1 or number 4. The last place we would want to place this charge is in carbons #2 or #3. These are referred to as **vinylic** positions, and form, by far, the most unstable carbocations. The most stable radical possibly formed would be the ethyl radical, which is primary. Keep in mind that this is not a random choice. We are trying to remove a group (fragmentation) that is bonded to either carbons #1 or #4. Carbon #1 only has hydrogens (which form very unstable carbocations). Carbon #4 gives us three different choices: a hydrogen, a methyl, or an ethyl. Of course, the ethyl group is the most substituted and therefore will make the most stable radical.

Chapter 14: Spectroscopy.

$$\text{CH}_2=\text{CH}-\text{CH}_2-\overset{\oplus}{\text{CH}}_2 \longleftrightarrow \overset{\oplus}{\text{CH}}_2-\text{CH}_2-\text{CH}=\text{CH}_2 \quad \} \text{ m/e 69}$$

- Propanone.

    To find the most prominent peak in the spectrum of propanone, we must draw the molecule first and try to first find the place where the charge would be stabilized the most.

$$\underset{\text{H}_3\text{C} - \text{C} - \text{CH}_3}{\overset{\overset{..}{\text{O}}:}{\|}}$$

    We immediately should see that the positive charge should be in carbon #2, because that would place it right next to an atom with a lone pair (the oxygen). There is only one clear choice: to remove the methyl as a radical in order to achieve this. Therefore, the answer should be: m/e 43.

$$\underset{\text{H}_3\text{C} - \overset{\oplus}{\text{C}}}{\overset{\overset{..}{\text{O}}:}{\|}} \longleftrightarrow \underset{\text{H}_3\text{C} - \text{C}}{\overset{\overset{..}{\text{O}}\oplus}{\||\|}}$$

- Predict the main peak in the MS of 2-Methylpropane.

$$\text{H}_3\text{C}-\underset{\underset{\text{H}}{|}}{\overset{\overset{\text{CH}_3}{|}}{\text{C}}}-\text{CH}_3$$

    The first thing we do is look for the possibility of generating resonance but since there are no atoms with lone pairs *nor* doubly bonded carbons (π-electrons), then we will not get that kind of stabilization. Therefore, the only thing to look for is inductive stabilization. We might be tempted to expect the main peak at **M - 1** since that would generate a tertiary carbocation. However, that will not occur to a major extent. Remember that it is very difficult to remove a hydrogen radical (it is very unstable) and we will only be able to do it to an appreciable extent only if the ion formed is stabilized by resonance. Therefore, we will observe that a methyl radical is lost instead, thus generating an ion with m/e 43. This corresponds to a secondary carbocation.

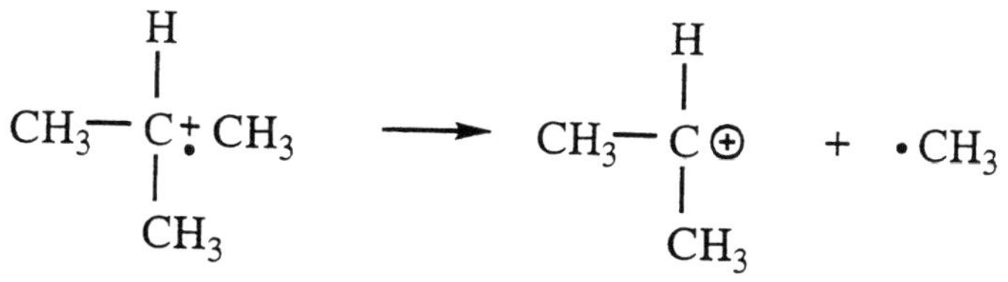

Chapter 14: Spectroscopy.

$$CH_3-\underset{\underset{CH_3}{|}}{\overset{\overset{H}{|}}{C}}\!\cdot\!CH_3 \quad \longrightarrow \quad CH_3-\underset{\underset{CH_3}{|}}{\overset{\overset{H}{|}}{C}}\!\oplus \quad + \quad \cdot CH_3$$

m/e 58   m/e 43

- Where will the main peak occur in the MS of 2,2,4-Trimethylpentane?

Once more, there is no possible stabilization by resonance so we must rely exclusively on inductive stabilization. There are two obvious places where the molecular ion could fragment to generate a tertiary carbocation.

$$CH_3-\underset{\underset{CH_3}{|}}{\overset{\overset{CH_3}{|}}{C}}\overset{1}{\cdot}\overset{\cdot}{\underset{2}{\cdot}}CH_2-\underset{\underset{CH_3}{|}}{CH}-CH_3$$

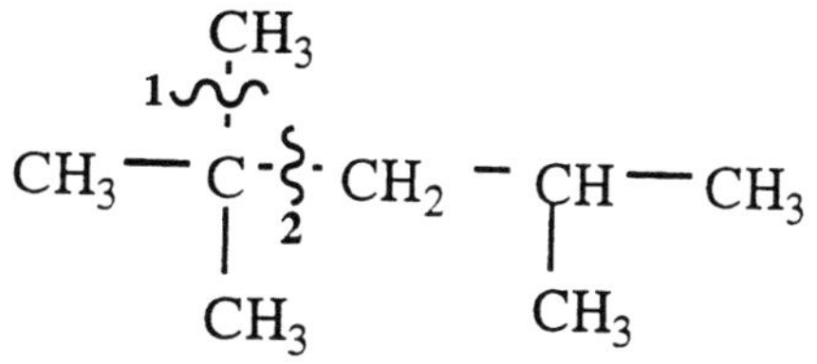

If we fragment it at **1** we would get a tertiary carbocation and a methyl radical:

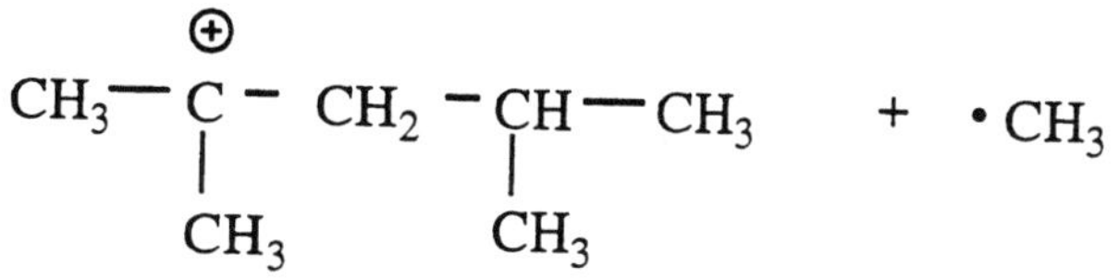

$$CH_3-\underset{\underset{CH_3}{|}}{\overset{\oplus}{C}}-CH_2-\underset{\underset{CH_3}{|}}{CH}-CH_3 \quad + \quad \cdot CH_3$$

m/e 99

Fragmentation at **2** would yield a tertiary carbocation also *and* a primary radical:

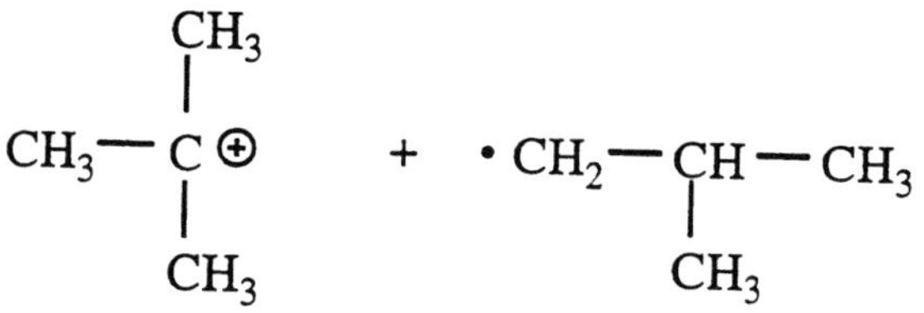

$$CH_3-\underset{\underset{CH_3}{|}}{\overset{\overset{CH_3}{|}}{C}}\!\oplus \quad + \quad \cdot CH_2-\underset{\underset{CH_3}{|}}{CH}-CH_3$$

m/e 57

Since the primary radical is more stable than the methyl radical, then fragmentation at **2** will be preferable. Therefore, the main peak will occur at *m/e 57*.

- Where will be the main MS peak for Ethyl Benzene?

In this case we can achieve stabilization by resonance if we place the positive charge on the carbon NEXT to a carbon-carbon double bond. Which carbon is that? The carbon right next to the phenyl ring. This particular carbon is referred to as **benzylic** and it yields a relatively "stable" carbocation. We should point out that just like there was a gigantic difference between allylic and vinylic carbocations, there is a huge difference between creating a charge on the carbon next to a phenyl group and generating it directly into one of the benzene carbons. The latter one is referred to as a **phenylic** position and it is one of the most unstable carbocations.

What are our choices in order to generate a charge on the benzylic carbon? We can either remove a hydrogen radical or a methyl radical. We already know the answer to that one. Since we are going to generate stabilization by resonance independent of which radical we remove, the path of lowest energy is that of the removal of the most stable radical: the methyl radical. Therefore, the main peak in the MS for Ethyl benzene occurs at m/e 91.

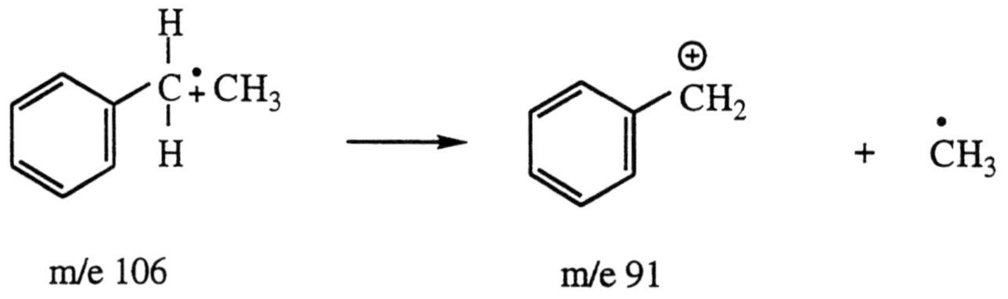

m/e 106   m/e 91

Why is this ion relatively stable? It is due to resonance. Actually, the positive charge will be helped by the closest double bond inside of the ring. However, once we draw the next resonance form, we can see that the new carbocation is going to be allylic to the next double bond inside that ring. Therefore, we will have extensive delocalization in this system. Let's look at all the resonant forms for this benzylic ion.

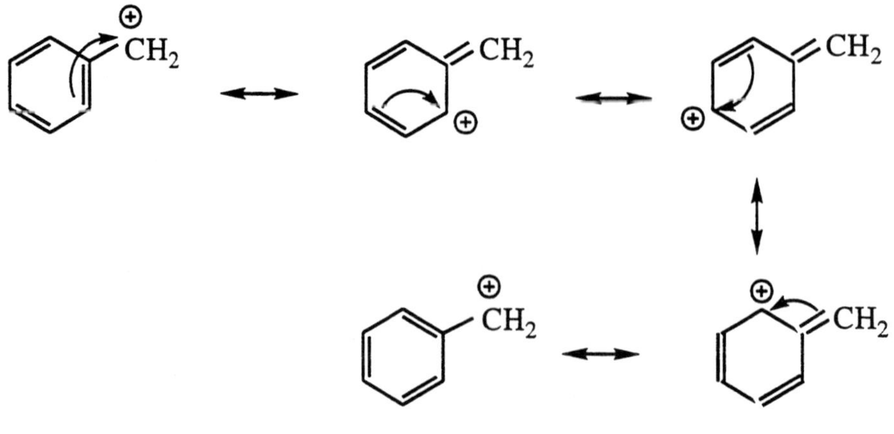

Are any of these forms real? Do any of these forms give us a true representation of the benzylic carbocation? Of course not. We have already learned that the true form is the resonance hybrid. What we have here is total delocalization (and therefore stabilization) of the positive charge into the ring. This can be seen in the diagram below.

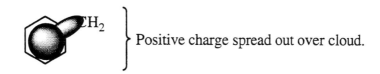

Positive charge spread out over cloud.

14.3 Infrared Spectroscopy.

So far we have seen one of these three techniques and it provided us with various important things: the molecular weight of the compound (from the parent peak) and some ideas as to certain possible groups (like methyl, isopropyl, or tert-butyl groups) present in the molecule (from the fragmentation patterns). Let's now concentrate on the second of these techniques, one that will give us great clues as to the types of bonds that we have present in our compound (like OH, carbonyl, COOH, CHO, etc.). We can imagine how useful this will be when we are trying to elucidate the structure of an unknown compound.

Spectroscopy in general can be defined as the interaction of light with atoms and molecules. Back in Section 6.5, we considered light to have wave properties such as wavelength (λ) and frequency (ν). These are related through the expression seen below.

$$\nu = c / \lambda$$

where: $c = 3.0 \times 10^8$ m/s or 3.0×10^{10} cm/s.

Furthermore, in Section 6.6, we said that Einstein proposed the quantization of light into little packets of energy called *photons*. The energy of these photons of light is proportional to the frequency and not the intensity.

$$E = h\nu, \quad \nu = 6.626 \times 10^{-34} \text{ J s}.$$

Spectroscopists like to work with what they call **wave number** (ω). By definition,

$$\omega = 1/\lambda, \quad \text{units} = \text{cm}^-$$

and then: $E = h c (1/\lambda) = \underline{h c \omega}$ ✓

If we take a beam of light and pass it through a prism, we know from our own personal experiences that a rainbow will be formed. This rainbow will have all possible colors that are visible to us, from red (lowest energy) to violet (highest energy). The prism is also able to separate out other waves that are not visible to the naked eye[1]. Those that are higher in energy than the violet are called the ultraviolet, and those that are lower in energy are called the infrared. If we place a thermometer in the path of the infrared radiation, we observe that the temperature increases. Furthermore, if we insert a sample of water in the path of the infrared radiation, we will see that the thermometer will not rise to quite as high a temperature. This is because the water is absorbing some of this infrared radiation, so that not as much radiation is making it to the thermometer. Not only that, but if we were to change the wavelength of infrared light that goes through the water, we will see that only some wavelengths in particular are absorbed. Why is the water absorbing some particular infrared wavelengths?

This should immediately remind us of the electronic energy of atoms, or the Bohr atom. Back in Section 6.7, we saw that the electronic energy of hydrogen (and hydrogen-like atoms) is quantized. The electron in the hydrogen atom is normally in the ground state (n = 1, or 1s). We can excite the electron and it will jump into a higher level (let's say 2p) **if** we add exactly the energy difference between the 1s and the 2p subshells in the form of a photon of light.

$$\Delta E = h \nu = h c \omega$$

Therefore, the electron in the hydrogen atom cannot just have any energy value possible, nor can it just absorb any amount of energy we supply. Only discrete values of energy are absorbed. Actually, the only reason spectroscopy exists is because of the quantization of energy, the generation of the energy levels, and the fact that light can induce jumping from one energy level to another.

The energy of a *molecule*, however, is made up of at least four components: translational, rotational, vibrational, and electronic energies. Translational energy is the kinetic energy atoms or molecules possess due to motion in space. Quantum restrictions are not important in the consideration of translational energy levels (essentially any energy is possible).

---

[1] Please, look at a program called "IR Tutor", by Charles B. Abrams from Columbia University in collaboration with Perkin Elmer (1992-93). It can be found in CWRU Software Library under the Chemistry Department.

The same is not true for the other kinds of energy.

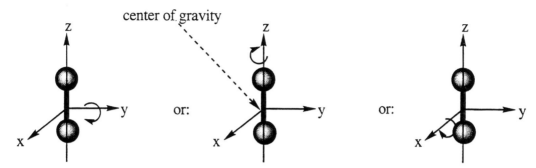

Rotational (see diagram above) energy is the kinetic energy molecules possess due to rotation about an axis through their center of gravity. Rotational energy **is** quantized and gives rise to absorption spectra in the microwave region of the electromagnetic spectrum. And what do we mean by absorption spectra? Just like we discussed, this type of spectrum occurs when light is absorbed to jump from one rotational energy level to another rotational energy level. We must keep in mind that one photon of light will be absorbed by one molecule and that the energy of this photon must correspond **exactly** to the energy difference between the two levels.

Vibrational energy is the potential and kinetic energy molecules possess due to vibrational motion. The atoms in a molecule can be considered as point masses held together by bonds acting like small springs. Because molecules are not rigid, their flexibility results in vibrational motion. The force constant (k) of a spring is a measure of the energy required to stretch the spring. Of course, different bonds will have different spring constants, because they will have different strengths.

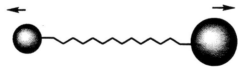

We should remember that when we compare bonds between two atoms, they will differ in various factors: the masses of the two atoms, the electronegativity differences between the two atoms, the ionization energy and electron affinity of the two atoms, etc. All of these factors will have an effect on the strength of that bond, therefore, they will require either more or less energy to get them excited vibrationally.

It should be clear from our discussion that vibrational energy is quantized and gives rise to absorption spectra in the <u>infrared region</u> of the electromagnetic spectrum. The relative spacing between vibrational energy levels increases with increasing strength of the chemical bonds between the atoms. Therefore, stronger bonds (like O-H, N-H, etc.) should require more energy in order to excite them. This means that they will require to absorb infrared energy of high wave number (ω). On the other hand, weaker bonds (like C=O, the pi part of the bond) will require less energy to excite. This would imply that it would require infrared radiation of smaller wave number. We can hopefully see the advantages of this particular fact. Different bonds absorb in different regions of the infrared region of the electromagnetic spectrum. We can use this information to determine which types of bonds are present in an organic compound.

We should finally point out that electronic energy (the fourth type) is the energy molecules and atoms possess due to the potential and kinetic energy of their electrons. The kinetic energy is the result of motion, and potential energy arises from the interaction of the electron with nuclei and other electrons. Within an atom there are many energy levels available for the electrons. These energy levels are designated by the four quantum numbers: **n, ℓ, m,** and **s**. Obviously, electronic energy is quantized and gives rise to absorption spectra in the visible and ultraviolet regions of the electromagnetic spectrum.

The following diagram illustrates the relationship between all these different energy levels. In this diagram, the translational levels have not been shown because they are too close to one another and they will essentially form a continuum.

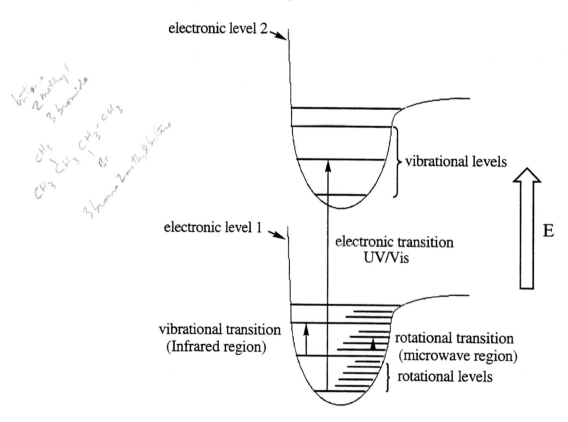

14.4 General Diagram of an IR Spectrometer.

How does an infrared spectrometer works? We should have at least an ideal of what happens when we place the sample of our organic compound in the spectrometer. In the following figure, we can see the general diagram for a simple infrared spectrometer. Of course, the first thing we should have is a source of infrared radiation. These sources will generate all possible infrared radiation wave numbers from between 4000 cm^{-1} (the highest wave number, or the highest energy) and about 600 cm^{-1} (the lowest wave number, or the lowest energy). This generated beam of infrared light is then passed through a wavelength selector which effectively allows us to select the passage of only one particular wave number. In other words, we will select only one wavelength to go through from that point on. This new beam is referred to as a monochromatic beam. Once the wavelength has been selected, the instrument proceeds to split this beam into two identical monochromatic beams through the uses of mirrors.

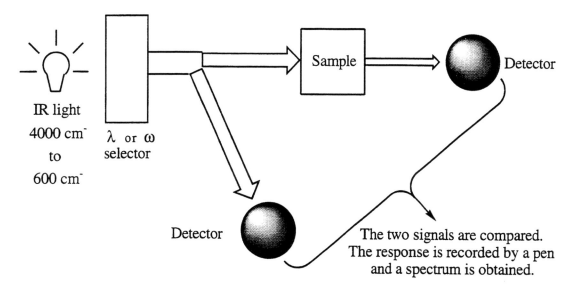

One of these beams will go through the sample, which will absorb some of the light if a certain type of bond in the molecule gets excited by this radiation, and the other will go directly to the detector. Both beams will then be compared. If some light was absorbed by the sample, it will be detected and properly recorded in the spectrum. An infrared spectrum is a record of the percent of light transmitted vs. the wave number. If the light was 100% transmitted by the solution, then nothing was absorbed, which means that there is no bond in the molecule that absorbs that particular wave number. However, at some point, a given bond will absorb a particular wave number, and the percent transmittance will be lower than 100%. This simply means that there is a bond somewhere in that molecule that is absorbing that particular wave number and is vibrationally excited.

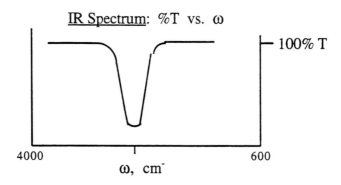

There is something else we should mention, although for a course like this, it is not as important. In order for a vibrational transition to be infrared active[1], the dipole moment (μ) of the molecule must change during the vibration. Let's look at a couple of examples to see what this means.

- Water.

There are three kinds or modes of vibration for the water molecule. These are the symmetrical stretch, the unsymmetrical stretch, and bending. We can see a representation for these three modes below.

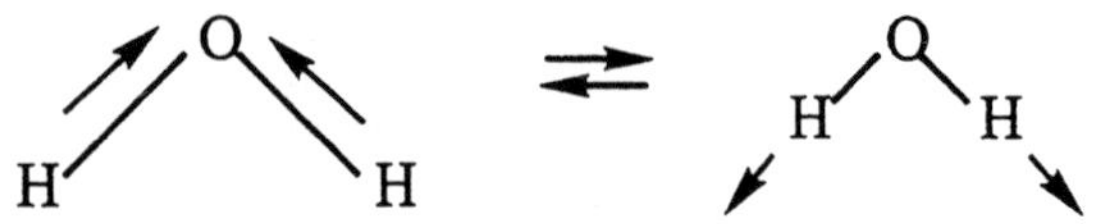

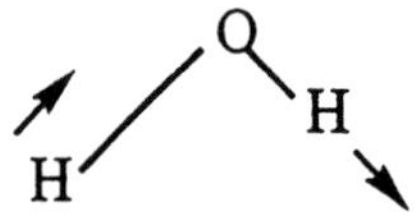

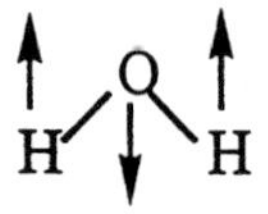

Symmetrical Stretch

Unsymmetrical Stretch

Bending

From these views, we should be able to tell that all the vibrations actually change the dipole moment of the water molecule, so we should expect to see all three present in the IR spectrum. It takes more energy to stretch a bond than to bent it, and it takes even a little bit more to stretch it asymmetrically than to stretch it symmetrically. An approximate IR

---

[1] An infrared active transition is one that shows an absorption peak in its spectrum.

spectrum for water is shown below.

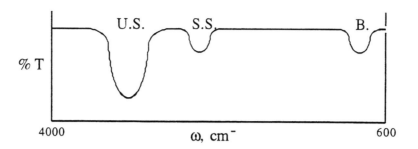

- Carbon Dioxide.

Once more, there are three vibrational modes. Let's look at them once more and let's analyze what happens to the dipole moment of the molecule as the vibration takes place. Being a triatomic molecule (like water), it will have the exact same three possibly vibrational modes. Let's consider first the case for the symmetrical stretching.

$$\overset{\leftarrow\cdots}{O}=C=\overset{\cdots\rightarrow}{O} \quad \rightleftarrows \quad \overset{\cdots\rightarrow}{O}=C=\overset{\leftarrow\cdots}{O}$$

Symmetrical Stretch

This stretching maintains at all times the dipole moment of the molecule equal to zero. Therefore, since the dipole moment does not change during the vibration, then this vibrational mode is **not** IR active (*i.e..,* we will not observe a peak in the IR corresponding to this vibrational mode).

The other two, the unsymmetrical stretch and the bending both indeed change the dipole of the molecule during the vibration, therefore these vibrations will be IR active and will show as bands or peaks in the IR spectrum. In conclusion, the IR spectrum for carbon dioxide will consist of two bands, the one for the unsymmetrical stretch and the one for the bending, as seen below.

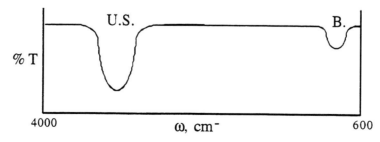

In conclusion, identification of characteristic absorption bands caused by different functional groups is the bases for the interpretation of infrared spectra. Let us look at some typical bonds in organic molecules and see where their characteristic absorption bands occur. On the next page, we will find a table of infrared characteristic group absorption frequencies that will indicate the approximate wave number of a few typical bonds that most organic molecules have. Please note that the intensity is also given. For the intensity column, **s** indicates strong absorption, **w** indicates weak absorption, and **m** indicates moderate absorption. This is not a complete table and we do not have to memorize it. We can find a much more complete table in any Organic Chemistry textbook or in the CRC Handbook of Chemistry and Physics.

IR Characteristic Group Absorption Frequencies

Band	Group	Intensity	Approx. ω (cm^{-})
C-H stretch	R-CH$_3$, alkanes	s	2960 and 2870
C-H stretch	R-CH$_2$-R, alkane	s	2930 and 2850
C-H stretch	R$_3$CH, alkane	w	2890
C-H bend	R-CH$_3$, alkane	s	1380-1360
C-H bend	R-CH$_2$-R, alkane	s	1470-1400
C-H stretch	=C-H, alkene	m	3080-3140
C-H stretch	≡C-H, alkyne	s	3300
C-H bend	≡C-H, alkyne	s	600-700
C-H stretch	H-C=O, aldehyde	m	2720
O-H stretch	in alcohols and acids	s	3600-3300
N-H stretch		s	3300-3500
Ar-H stretch	H in benzene (Ar) ring	m	3030
Ar-H bend	monosubstituted benzene	m/s	750 and 700
Ar-H bend	1,2-disubstituted benzene	m/s	750
Ar-H bend	1,3-disubstituted benzene	m/s	780-810
Ar-H bend	1,4-disubstituted benzene	m/s	850-800
C-H wag	monosubstituted benzenes	m	1600, 1500
C-H wag	(CH$_2$)$_n$ skeletal wag for n ≥ 3	m	≈ 750
C-O stretch	ether, alcohols, etc.	s	1070-1150
C=O stretch	ketones, aldehydes, etc.	s	1700-1800
C=C stretch		m	1645
C≡C stretch		m	2100-2200
C≡N stretch		s	2260-1690

14.5 Infrared Spectra Analysis.

This table is crucial in the analysis of an infrared spectrum. Actually, the worst possible thing we can do is to try to identify every single peak in the spectrum. It turns out that there is a region, somewhere between 600 cm^{-1} and 1100 cm^{-1}, that is many times different for every compound. This would imply that it would make the identification of these peaks almost impossible. This is referred to as the fingerprint region of the IR spectrum. So how are we supposed to attack this? We should look at some selected peaks (the ones highlighted in the table) and see if they are present or absent in your spectrum. These are so important that we should MEMORIZE those (the highlighted ones). Let's talk about some of these.

1) C-H stretch, as in Methyl groups.

The C-H bond in a methyl group can do the three things we spoke of before. It can stretch asymmetrically, it can stretch symmetrically, and it can bend. The stretches occur at 2960 and 2870 cm^{-1}. The first is the asymmetric stretch and the latter is the symmetric stretch. We should keep in mind that these numbers are only approximate. Depending on the proximity of other groups in the molecule, they will be shifted slightly upwards or downwards. The important thing is to identify their presence in that area. As an example, let's see a portion of the spectrum for Hexane.[1]

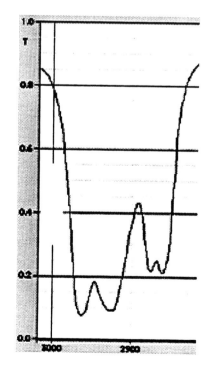

---

[1] Abrams, Charles B., "IR Tutor", Columbia University in collaboration with Perkin Elmer (1992-93).

We can clearly see four peaks in here. Two of these are for the methyl and the other two are for the CH_2's in the molecule. We call these methylenes. The one at the extreme left is the asymmetric stretch for the methyl, while the next one is for the asymmetric stretch for the methylene. The two on the right are for the symmetric stretches (the one on the furthest right is for methylene) of these two groups.

2) C=O stretch.

This is one of the most prominent peaks in the IR spectrum. It is fairly sharp and very strong, so once more, it is easy to spot. In the spectrum for 3-Heptanone, we can see this peak right around 1700 cm^{-1}.

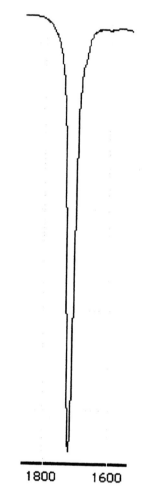

This peak will be present in compounds like ketones, aldehydes, carboxylic acids, and esters. Of course, in order to distinguish between those four, we would have to look at other parts of the spectrum. For example, to distinguish between a ketone and an aldehyde, we would need to look for a small difference, which is the hydrogen of the aldehyde (the H in CHO). This hydrogen would moderately absorb around 2720 cm⁻¹. Let's look at the following example. This is a portion of the IR spectrum for Heptanal.

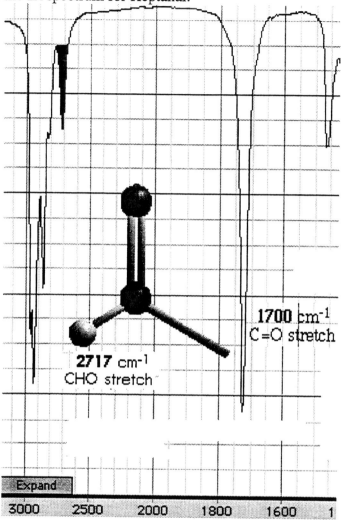

Here we see the small peak at around 2720 cm⁻¹ for the hydrogen in the aldehyde group, which coupled with the peak at 1700 cm⁻¹ (for the carbonyl stretch), strengthens our case for an aldehyde.

3) O-H stretch.

Any time we have an O-H present in a compound, this bond will show up as it stretches asymmetrically somewhere around 3300 cm⁻¹. Examples of such compounds are alcohols and carboxylic acids. In the case of an alcohol, the peak is very strong and wide. Of course, we could easily distinguish between alcohols and acids because the carboxylic acid would also have a very strong band at around 1700 cm⁻¹, which corresponds to the carbonyl stretch (C=O).

Chapter 14: Spectroscopy.

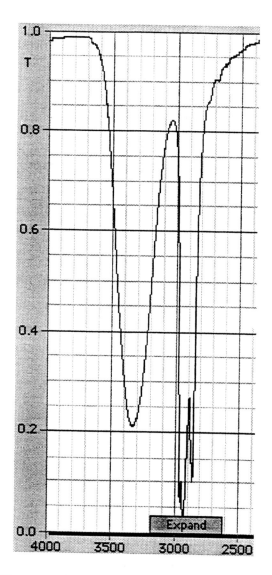

This is a portion of the IR spectrum for 1-Hexanol that shows the absorption of the O-H bond. In this spectrum we can indeed see that the peak for the OH is wide and very strong. This peak is centered at 3334 cm^{-1}. This is not easy to read off the bottom scale, but it is certainly clear to see that it is around 3300 cm^{-1}.

How does this band differ from the O-H stretch for a carboxylic acid? In a carboxylic acid, the band is much more wide due to the fact that two of these acid molecules will form a dimer with hydrogen bonds. This will weaken the bond slightly, thus it will move closer to the methyls and methylenes. Of course, there will be some overlap and at first glance, it looks more like a shoulder to the methyl band. Below, we can see that portion of the IR spectrum for Heptanoic Acid.

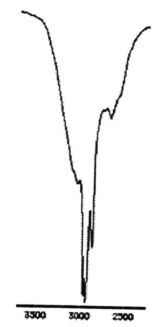

We can see that the OH band has moved very close to the CH_3 groups (close to 3300 cm^{-1}).

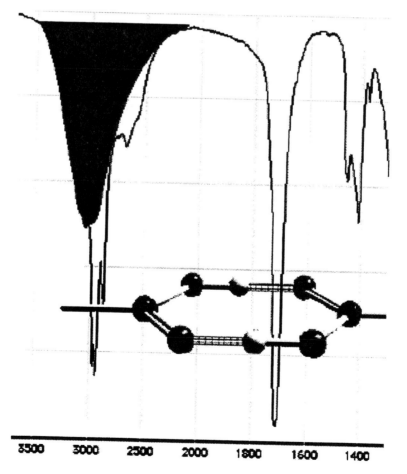

Hopefully at this point we can tell that given an unknown organic compound, the IR spectrum can provide a lot of clues that might help us to solve the puzzle and figure out what is the structure of the compound. Let's use the table of characteristic frequencies for IR absorption in order to predict the spectrum for the following examples.

- Which IR bands would we expect to see in the IR spectrum for 1-Butyne?

In order to answer this question, we need to look at the structure of the molecule: $HC\equiv CCH_2CH_2CH_3$. What type of bonds do we have here? We have an H bonded to a $C\equiv C$, the actual $C\equiv C$ bond, we have CH_2's and a CH_3. We should expect to observe bands in the following places (plus many more[1] that we will not discuss).

1) A strong band at about 600-700 cm^- for the H bonded to the $C\equiv C$.
2) A moderate band at about 2100-2200 cm^- for the $C\equiv C$ stretch.
3) A strong band at 2960 and 2870 cm^- for the CH_3 stretch.
4) A strong band at 1380-1360 cm^- for the CH_3 bend.
5) A strong band at 2930 and 2850 cm^- for the CH_2 stretch.
6) A strong band at 1470-1400 cm^- for the CH_2 bend.

This will only give us an idea of what bands should be present in the IR spectrum of this compound. Remember that when we are trying to identify an unknown, the last thing we would do would be to try to identify every single band in the spectrum. There is even the added problem that some similar bonds absorb in similar regions. The presence of a band is not proof that we have a certain type of bond, but the absence of a band is certainly a proof that we do NOT have that type of bond.

- Is there any way we can distinguish between the two isomers of C_2H_6O using infrared spectroscopy only?

The two isomers of this formula are Dimethyl ether and Ethanol. Clearly, one of them has the O-H bond (in Ethanol), whereas the other does not. Therefore, the easiest way to distinguish between them would be to look for the presence or the absence of a strong band at about 3300 cm^{-1}.

---

[1] Mostly from the fingerprint region.

Chapter 14: Spectroscopy.

- A compound has a molecular formula C_8H_8O and the IR spectrum shown below. Identify the compound based exclusively on its IR spectrum.

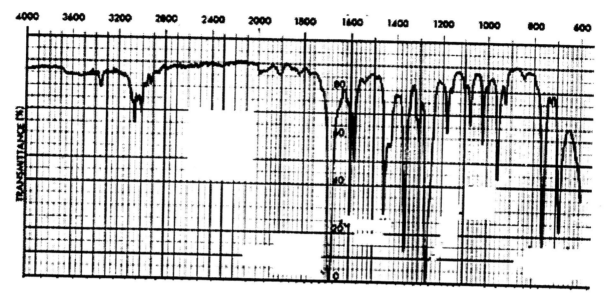

The first thing we should notice is that there is a strong band at 1700 cm^{-1}. This strongly suggests the presence of a carbonyl group. Since there is only one oxygen in the compound, then we either have a ketone or an aldehyde. How do we distinguish between these two? By the presence or the absence of the absorption of the hydrogen in the aldehyde (2720 cm^{-1}). The lack of a band there tells us that we do not have an aldehyde. Therefore, our unknown compound has to be a ketone.

The band at 3030 cm^{-1} suggests an aromatic ring. Of course, it is not easy to read 3030 from that spectrum, but it should be clear that it is higher than 3000 cm^{-1}. Had it been lower than 3000, then we would have had to assume that it was due to the methyl hydrogens. We actually have small bands after and before 3000 cm^{-1}. This means that we probably have either methyls or methylene hydrogens in the molecule. Finally, the two bands at 750 and 700 cm^{-1} indicate a monosubstituted benzene. This is further confirmed by the bands at 1600 and 1500 cm^{-1}.

With this information, and the fact that we know the molecular formula, we can figure out what the compound is going to be. The compound is: Phenyl methyl ketone.

This problem would be much easier if we would have known from the start that the unknown had five degrees of unsaturation. Since the phenyl group has four by itself, any time we have four or more degrees of unsaturation, the first guess we should try is to consider the possibility that there is a benzene ring somewhere in the molecule. We could have then seen that the fifth degree would have to be the carbonyl group. Therefore, it would be advantageous to be able to figure this information out BEFORE we try to determine the identity of an unknown compound.

14.6 Degrees of Unsaturation in Organic Compounds.

Since we know how many bonds are formed by the different atoms that bond with carbon in organic compounds, we can easily determine an equation that enable us to determine the number of degrees of unsaturation (Ω)[1]. As we have previously seen, the degrees of unsaturation will measure how many hydrogens[2] are "missing" in order to have a saturated compound. We will calculate Ω using the equation shown below.

$$\Omega = \frac{2 \times (\#C) + 2 - \#H - \#X + \#N}{2}$$

In this equation, #C represents the number of carbon atoms in the compound, #H is the number of hydrogens, #X is the number of halogens, and #N is the number of nitrogens. The oxygens have no effect on the degrees of unsaturation.

- Determine the number of degrees of unsaturation for the following formulas: C_2H_6O, C_4H_7Br, C_6H_6, and C_8H_8O.

    Applying the equation above, we can get the following results.

    a) C_2H_6O $[2(2) + 2 - 6 - 0 + 0]/2 =$ 0

    b) C_4H_7Br $[2(4) + 2 - 7 - 1 + 0]/2 =$ 1

    c) C_6H_6 (benzene) $[2(6) + 2 - 6 - 0 + 0]/2 =$ 4

    d) C_8H_8O $[2(8) + 2 - 8 - 0 + 0]/2 =$ 5

If we are told that a compound has a molecular formula of C_5H_4NBr, just by looking at the formula we can determine that it has 4 degrees of unsaturation and that, by itself, is very useful when we try to determine its structure. It should be clear that this is not enough

---

[1] Remember that a degree of unsaturation indicates either a double bond or a cyclic compound.
[2] Or univalent atoms.

information to decide what it actually is, but it is an important clue to add to the MS or the IR data.

14.7 Nuclear Magnetic Resonance Spectroscopy.

As we have seen so far, we can get a good idea of what type of bonds (or functional groups) are present in a molecule by a quick analysis of the infrared (IR) spectrum. However, this information only gives us a very general idea of the type of compound we are dealing with. The vast majority of an organic molecule is usually its hydrocarbon skeleton. It would be ideal if we had an experimental technique that would allow us to determine the structure of this skeleton. It turns out that there is such a technique available and it is due to the magnetic properties of the hydrogen nuclei along the skeleton. These magnetic properties are observed in the **nuclear magnetic resonance (NMR)** spectra of organic compounds.

The hydrogen nucleus (or proton) may be regarded as a spinning, positively charged unit and so, like *any* rotating electrical charge, it will generate a tiny magnetic field (**H'**) along its spinning axis.

Hydrogen nucleus

Under normal conditions, this small magnetic field (**H'**) can point anywhere (*i.e.*, it will not have any preferred orientation). However, if this nucleus is placed under the influence of an external magnetic field (**H°**), it will adopt one of only two possible orientations: it will either be aligned (or parallel) to the external magnetic field (**H°**), or it will be opposed (or anti parallel) to **H°**, as shown below. It should be pretty clear that the aligned orientation (or parallel) is of lower energy (*more stable*) than the opposed orientation.

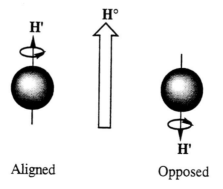

Aligned   Opposed

If we were to look at the relative energies of these two states, we could make a plot that shows that the aligned orientation is lower in energy than the opposed one. The following plot depicts just this fact.

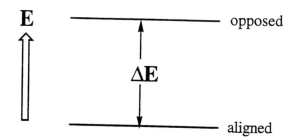

It is also true that the stronger we make the applied magnetic field (**H°**), the more difficult it will be to place the proton in the opposed orientation, because it will be opposing a stronger magnetic field. Therefore, we can see that the energy separation (ΔE) between these two possible orientations is going to be dependent on the strength of the applied magnetic field, **H°** (that is, ΔE α **H°**).

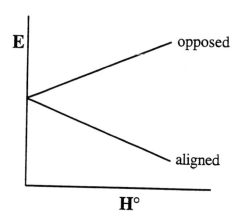

The energy difference (ΔE) between the two states will be absorbed or emitted as the nucleus flips from one orientation to the other. As we have seen in the past, the proton can either absorb or emit <u>exactly</u> the energy difference between the two states in the form of a photon of electromagnetic radiation (hν). Therefore, ΔE is related to both the applied magnetic field and also to a radiation frequency (ν). This <u>very small</u> energy difference (ΔE is about 0.01 cal/mol) is in the <u>radio</u> frequency of the electromagnetic spectrum (much <u>longer wavelengths</u> than the ΔE observed in the vibrational excitations, which are in the infrared region of the electromagnetic spectrum; see the diagram below).

Chapter 14: Spectroscopy.

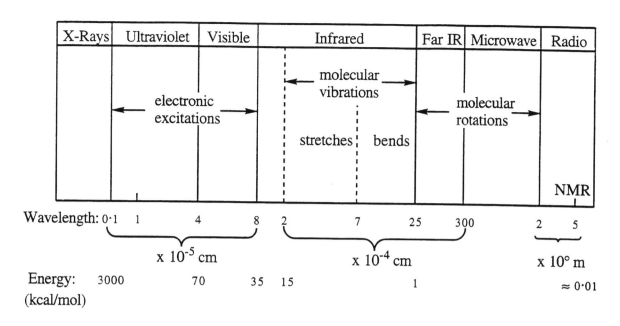

This difference in energy is so tiny that at room temperature the molecules possess enough thermal energy to flip the nuclei into the high energy state even without the introduction of a photon of electromagnetic radiation. Therefore, under normal experimental conditions, both energy states are almost identically populated[1] (as opposed to the other kinds of spectroscopy in which the lower energy level is always almost 100% populated). At a fixed frequency of 60 MHz[2] it would take an external field ($H°$) of 14,092 G[3] to reach the point where we would cause a <u>net</u> flip of protons from the low to the high energy state. When this condition is met we say that the magnetic component of the radiation frequency is in <u>resonance</u> with the magnetic field at the nucleus, the nuclei flip and we observe a peak in the NMR spectrum.

If all hydrogens in every single organic molecule were identical in their environment, then we would observe a peak in the NMR spectrum every time we applied a magnetic field of 14,092 G[4]. This would make the NMR experiment useless, because the only thing that we would achieve is to confirm that we have hydrogens in the molecule. Of course, that is something we already know. Well, it turns out that what matters is the effective magnetic field felt by the spinning hydrogen at its own nucleus. We will call this H_{eff}. So, as long as we are using a 60 MHz NMR machine, then we will observe an NMR signal anytime the effective field <u>at the nucleus</u> is 14,092 G. From the plot below, we can see that if we fix the radio frequency at 60

---

[1] Of course, there is a small difference between the two. It has been estimated that for every million protons in the higher energy state, there are one million and six in the lower energy state.
[2] One Mega Hertz (MHz) is a million Hertz (a million cycles per second).
[3] G = Gauss. This is a unit of magnetic field intensity.
[4] Assuming that we are using a 60 MHz NMR.

MHz, and slowly change the applied magnetic field, then the signal or peak will be observed when the magnetic field felt at the nucleus is 14,092 G, since at that point, the difference in energy between the two states will be equivalent to a photon of 60 MHz.

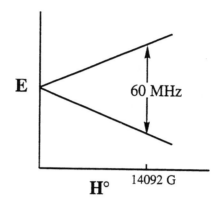

How are hydrogens different? Of course, they are all the same, after all. The differences we are referring to in this section deal with the chemical environment in which they are found. Let's try to follow very slowly what happens at the nucleus as this external magnetic field is applied. In order to do this, we should approach it one step at a time. Therefore, we will apply two approximations in our presentation.

The first approximation is to stop considering an isolated proton, as we have done so far, and instead, see what effect the electrons in its environment have on the proton (the hydrogen nucleus). Therefore, let's say that the effective magnetic field felt at the nucleus is equal to that of the applied external magnetic field minus the effect of the surrounding electrons. We will call this microscopically small effect, H_{shield}.

$$H_{eff} = H° - H_{shield}$$

These surrounding electrons, which are the ones between the hydrogen and the atom it is bonded to, will try to create a "shield" from this huge perturbation caused by the external magnetic field. Since electrons also spin on their own axis, they also create a tiny magnetic field that can be aligned or opposed to the magnetic field. These are also quantized, but their energy differences are much different than those of the protons, and they are not excited at these magnetic fields we are talking about. Therefore, under these conditions, these electrons will try to oppose the applied magnetic field to "help diffuse" the problem for the nuclei. Clearly, they will not be able to do much, but it is this tiny difference that is going to make the NMR experiment useful. Therefore, these shielding electrons will make the field felt at the nucleus a little smaller than the applied magnetic field. We may still be wondering how this is going to make a difference in the long run. Let's consider two very simple examples. In a normal C-H bond of an alkane (let's say methane), the electrons on the C-H bond can shield the proton from the applied magnetic field because there are no electronegative atoms to pull the electrons away. Let's compare that to chloroform ($CHCl_3$). The C-H bond here is very different from the one in the methane molecule. This is a very polar molecule, and the dipole is pointing away from the C-H bond. This means that the electrons in that bond are drawn towards the chlorines on the other side of the tetrahedron. Therefore, without as many electrons to shield it, the hydrogen nucleus is more exposed to the applied magnetic field. We say that it has lost a great part of its shield, and we use the terminology that they are more "deshielded" than the previous ones.

Is there a large difference between the magnetic field felt by those two hydrogens? The answer is NO. Remember that the effect of the shielding electrons is very small. It turns out that the difference is about 8 ppm (parts per million)[1]. Actually, this creates a small problem. Since in our experiment we are actually changing the applied magnetic field ever so slowly until a signal is observed, then in order to resolve[2] these two, we would need an instrument with the capacity to do this. Actually, most instruments have the capacity to resolve at least down to 0.01 ppm. However, in order for us to get an idea of the task of separating these peaks, it would be equivalent to having a telescope that is powerful enough not only to detect a dog sitting on the surface of the moon, but to differentiate between its right and left leg.

Experimentally, it is very difficult to determine the absolute value of the applied magnetic field ($H°$), to the required accuracy of one part in a hundred million[3]. Therefore, rather than trying to measure the exact applied magnetic field at any point in time, we will add a few drops of a liquid compound that has the most shielded protons, as compared to most organic compounds. This reference standard is called Tetramethylsilane (TMS). Because of silicon's low electronegativity, the shielding of the protons in TMS is greater than in essentially all other organic molecules. The separation (in ppm) between the signal for the TMS and the one for a particular type of hydrogen in an organic molecule is what we call its **chemical shift**.

---

[1] This would be equivalent to saying (although these numbers are not real) that one is excited at 14,091.88 G and the other at 14,092.00 G.
[2] The term resolution applies to the separation into different and distinct peaks in the spectrum.
[3] This is the required accuracy in order to resolve signals that are up to 0.01 ppm apart.

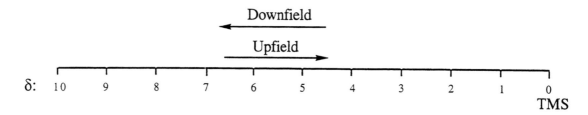

TMS

As a result, essentially all NMR signals for organic molecules appear in the same direction from the TMS signal: *downfield*.

```
                    Downfield
                  ←─────────
                         Upfield
                         ─────────→
δ:   10   9   8   7   6   5   4   3   2   1   0
                                                TMS
```

We will use a relative unit for the measurement of the chemical shift, which is one millionth of the magnitude of the external magnetic field, 1 ppm. This fractional unit (δ) sets the position of the highly shielded TMS hydrogens at a value of zero and gives the less shielded protons in organic molecules more positive values. Before we proceed, let's make sure we understand why the TMS hydrogens are far upfield. Since these are **the most shielded** hydrogens, they are the ones best protected from the external magnetic field. Therefore, we will have to apply a stronger magnetic field in order to overcome it, *i.e.*, they are upfield compared to anything else.

Since we have basically determined the NMR spectrum of most organic molecules, we have had the opportunity to establish a pattern as to where hydrogens absorb, depending on their chemical environment. For example, the hydrogens of a methyl group which is part of an alkane, will absorb relatively close to the TMS signal, at about 0.9 δ. These are referred to as primary hydrogens. It has also been observed that methylene hydrogens (also called secondary hydrogens) in an alkane absorb slightly downfield from the primary ones, at about 1.3 δ. On the other hand, the tertiary ones (for example, the C-H in isobutane), absorbs at around 1.5 δ. In sharp contrast, the methyl hydrogens attached to an oxygen atom (in the case of OCH_3) have a chemical shift of about 3.7 δ. This is because the electronegative oxygen will pull the electrons away from the protons, or "deshield" them). Below is a schematic diagram that shows an approximate delta (δ) for some functional groups.

Chapter 14: Spectroscopy.

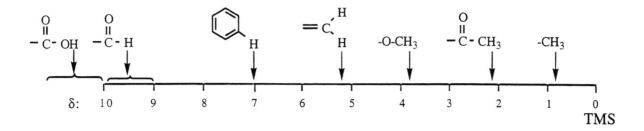

The actual chemical shift for different groups is not as exact as we may think from looking at the above diagram. Below is a small table[1] that gives us an idea of these shifts for some common groups. It is also true that the presence of neighboring groups on both sides will have an additive effect on the actual shift of the proton.

Type of Proton		Chemical Shift δ, in ppm
Primary	R-CH$_3$	0.9
Secondary	R$_2$CH$_2$	1.3
Tertiary	R$_3$C-H	1.5
Vinylic	C=C-H	4.6 - 5.9
Acetylenic	C≡C-H	2.0 – 3.0
Aromatic	Ph-H	6.0 - 8.5
Benzylic	Ph-C-H	2.2 – 3.0
Allylic	C=C-C-H	1.7
Fluorides	F-C-H	4.0 - 4.5
Chlorides	Cl-C-H	3.0 – 4.0
Bromides	Br-C-H	2.5 – 4.0
Iodides	I-C-H	2.0 – 4.0
Nitro	H-C-NO$_2$	2.8 – 5.0
Alcohols	H-C-OH	3.4 – 4.0
Ethers	H-C-OR	3.3 – 4.0
Esters	RCOO-C-H	3.7 - 4.1
Esters	H-C-COOR	2.0 - 2.2
Acids	H-C-COOH	2.0 - 2.6
Carbonyl	H-C-C=O	2.0 - 2.7
Aldehydic	R-CHO	9.0 – 10.0
Hydroxylic	RO-H	1.0 - 5.5
Phenolic	Ph-OH	4.0 – 12.0
Carboxylic	RCOO-H	10.5 – 12.0

A second aspect of the NMR spectrum that is extremely valuable is the **total area** under a band. The total area of one of these bands is proportional to the number of protons that are causing that particular absorption. This is in contrast to IR spectroscopy, where the bands have

---

[1] Taken in part from Morrison, Robert, and Robert Boyd. <u>Organic Chemistry</u>. Sixth Edition. Prentice Hall, 1992.

dramatically different strengths (we saw strong, moderate, and weak bands in the IR spectra). However, the relative strength of these bands do NOT tell us the relative number of types of bonds in the molecules. There is a much more complicated reason for the strength of these infrared bands, but there is no way to relate the size and strength of the IR bands to the molecules themselves. However, the same is not true for the NMR. Once we measure the relative areas under each of the bands, we will know the relative number of hydrogens absorbing in each band. This is an incredible piece of information we have suddenly acquired, and it will be extremely useful when we are trying to elucidate the structure of an unknown compound.

What have we learned so far? We have seen that different kinds of hydrogens absorb in different regions. Just remember, the more deshielded the hydrogens are, the further away they will be from the TMS signal. All we have to do is decide where the electrons are with respect to the hydrogen nucleus. If a highly electronegative atom (or group) is bonded to this hydrogen, then its electrons will move towards the electronegative atom (away from the hydrogen), leaving the hydrogen atom *deshielded*. We also learned that the area under the peak is proportional to the number of hydrogens absorbing.

Let's look at a few examples to apply all of this. We will draw some organic molecules and determine how many types of hydrogens each one has and where they absorb relative to the TMS.

a) Butane.

Butane is $CH_3CH_2CH_2CH_3$. At first glance, we may say that all the hydrogens are equivalent because there are no electronegative groups to deshield them. This is not true. Although they are very similar, they are not the same. We already saw that primary hydrogens (CH_3's) are a bit more shielded than secondary hydrogens (CH_2's). Therefore, there are two types of hydrogens in this molecule (labeled **a** and **b**).

$$\underbrace{CH_3}_{a} - \underbrace{CH_2 - CH_2}_{b} - \underbrace{CH_3}_{a}$$

The two end methyls are equivalent because of the symmetry of the molecule. Now, back on the chemical shift table, we see that the protons labeled **a** absorb at about 0.9 δ, whereas the ones labeled **b** absorb at around 1.3 δ. Finally, we have six protons labeled **a** and four labeled **b**, therefore, the **a** band should have a larger area than the **b** band.

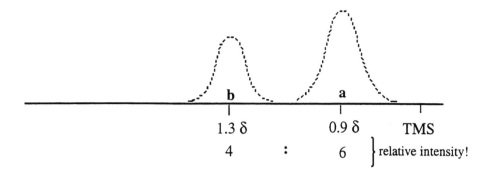

b) 2-Methylpropane.

As always, we should start by drawing the molecule and identifying how many different types of hydrogens we have. In this case, because of the symmetry of the molecule, there are two different types of hydrogens, as seen below.

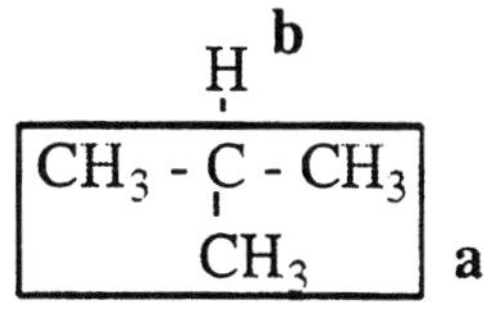

$$\boxed{\begin{array}{c} \text{H} \ ^{\mathbf{b}} \\ \text{CH}_3 - \overset{|}{\text{C}} - \text{CH}_3 \\ \underset{|}{\text{CH}_3} \end{array}} \ \mathbf{a}$$

Where do these absorb relative to the TMS? Once more, there are no electronegative groups in here (it is simply an alkane), therefore, the primary ones (**a**) will be around 0.9 δ, whereas the tertiary one (**b**) will be around 1.5 δ. Finally, the relative intensities of the two bands should be one (**b**) to nine (**a**).

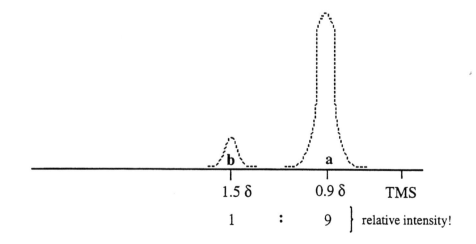

When the carbonyl group (C=O) is also singly bonded to an oxygen in a molecule (as we find in acids and esters), we observe much larger chemical shifts.

c) Ethanoic acid.

The hydrogen of the COOH group is extremely deshielded, and it absorbs at delta values greater than 10. This is mainly due to the delocalization of the pi electrons between the two oxygens in the COOH group.

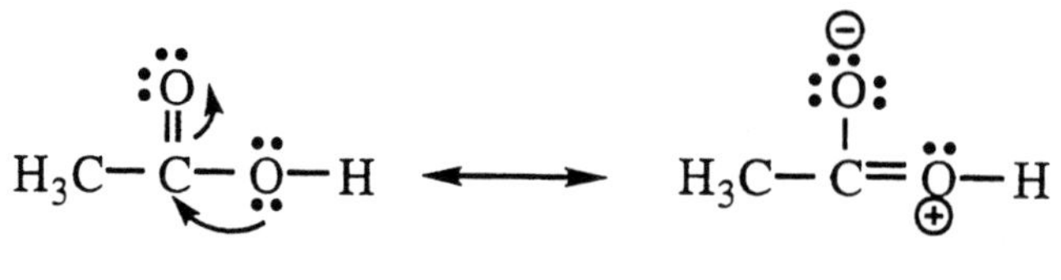

$$H_3C-\underset{\underset{\ddot{O}:}{\|}}{C}-\ddot{O}-H \longleftrightarrow H_3C-\underset{\underset{\overset{\ominus}{:\ddot{O}:}}{|}}{C}=\overset{\oplus}{\ddot{O}}-H$$

Since the oxygen bonded to the hydrogen in the COOH group has a partial positive character, it will really pull the electrons towards itself and out of the hydrogen. This is why that particular hydrogen is so deshielded.

Let's do the NMR analysis. We see two types of hydrogens in this molecule: the methyl hydrogens and the carboxylic hydrogen. We would expect, therefore, to see two bands in the NMR spectrum for this compound. One of these bands will be observed at a delta value that is higher than 10.5 δ (the table states that it absorbs somewhere in between 10.5 and 12 δ). How about the methyl hydrogens? The first thing one would think is that these are simply primary hydrogens, therefore, they should absorb at about 0.9 δ. However, this is not the case. These hydrogens are labeled in the table as shown below.

        Acids   **H**-C-COOH   2.0 - 2.6

Notice that the bold hydrogen is the one we are interested in finding. It is a hydrogen bonded to the carbon right next to the COOH group. Therefore, it will absorb somewhere between 2.0 and 2.6 δ. It is difficult to predict exactly where it will absorb, but a good rule of thumb is that since these hydrogens are primary, they will absorb in the lower end of that range. Therefore, these hydrogens will absorb around 2.0 δ, in a band that is three times more intense than the COOH band. If these hydrogens would have been secondary, they would have absorbed somewhere in the middle (around 2.3 δ), whereas if the hydrogen was tertiary, it would have been closer to 2.6 δ.

There is one more group we should mention now: the benzene molecule, or the phenyl group. In benzene we have six identical hydrogens and the table states that they absorb way downfield (the aromatic hydrogens absorb somewhere between 6.0 and 8.5 δ). How can we account for this? Unlike before, there are no electronegative atoms in here to have a deshielding effect on these hydrogens. Let's base our explanation on the diagram below.

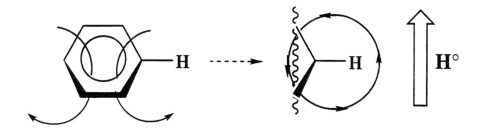

We should remember that the pi electrons are completely delocalized in this ring. This generates a ring current, and anytime we have a current, there is a perpendicular magnetic field associated with it. If we try to visualize this drawing as one of a flat ring that is perpendicular to the paper, then the magnetic field is in the direction shown. We probably learned this in physics: the right hand rule. If we curl our fingers on our right hand in the direction of the ring current and leave our thumb up, the thumb points in the direction of the magnetic field. Anyway, let's concentrate on the right side of the diagram. The first thing we have to contend with are the electrons in the actual bond between the carbon and the hydrogen (the sigma bond). These are trying to help in their usual way and they will not have much of an effect, as expected, since there are no electronegative groups in this molecule. However, the pi electrons have a very large effect in this case due to the ring current. We see that the new magnetic field generated by the resonating electrons is in the same direction as that of the applied magnetic field ($H°$). Therefore, in order for that hydrogen to feel a field of 14,092 G at its nucleus, the applied magnetic field will have to be LOWER since it already feels the ring current's magnetic field. Therefore, the ring current ends up deshielding the aromatic hydrogens.

d) Ethyl benzoate.

Once more, let's start by drawing the molecule and deciding how many types of hydrogens we have, their approximate chemical shift, and the relative intensities of the bands.

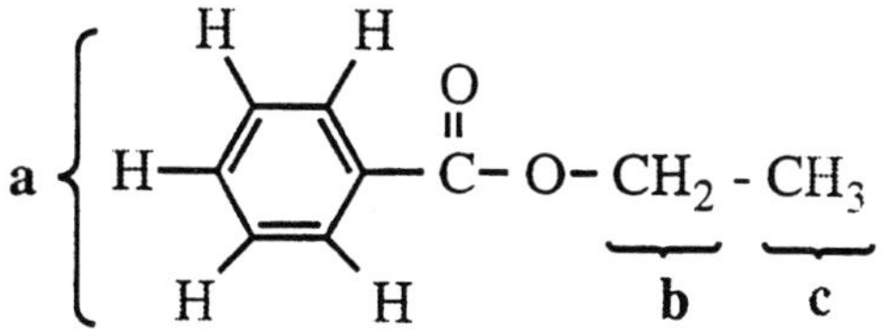

We see that this molecule has three different types of hydrogens, mainly the aromatic hydrogens (**a** - from the phenyl ring), the ones bonded on one end of the ester group (**b**), and the methyl group (**c**). From the drawing, we see that the relative intensities are 5 : 2 : 3. How about the chemical shifts? The aromatic ones absorb around 7 δ (we normally call it seven, although it really is anywhere between 6.5 and 8.0 δ). The methyl is simply a little larger than 0.9 δ, because of the proximity of the ester (COO) group. Finally, the other one is connected directly to the ester group. Let's analyze this one carefully. This is the information we get from the table.

| Esters | RCOO-C-**H** | 3.7 - 4.1 |
| Esters | **H**-C-COOR | 2.0 - 2.2 |

We have to be careful with the esters, because the C-H bond in question is either bonded to the carbonyl (the second case) or to the oxygen directly (the first case). The one bonded to the oxygen directly is much more deshielded than the other one, basically for the same reason as the hydrogen in the COOH. Therefore, our hydrogens labeled **b**, must be somewhere between 3.7 and 4.1 δ. Since these are secondary hydrogens, it will be close to the middle of the range, approximately 3.9 δ. The basic scheme for the NMR spectrum can be seen below.

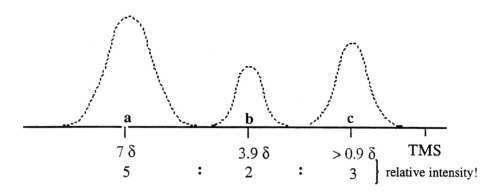

14.8 <u>Spin-Spin Coupling</u>.

To this point we have assumed that the magnetic field experienced by the proton (**H**$_{eff}$) depends only on the external field and the electronic environment around the proton (which gives rise to the chemical shift). However, there is an additional factor that we will consider in this course called <u>spin-spin coupling</u> which will affect the magnetic field experienced by a proton.

The NMR spectrum of 1,1,2,2-Tetrachloroethane consists of a single peak at approximately 6δ. The fact that the spectrum consists of only one peak (a **singlet**) is consistent with our understanding that the two protons in this molecule experience the **same** molecular environment.

However, that will not be the case for 1,1-Dibromo-2,2-dichloroethane. From our discussion so far, we should expect the NMR spectrum for this molecule to consist of two bands, one for each hydrogen.

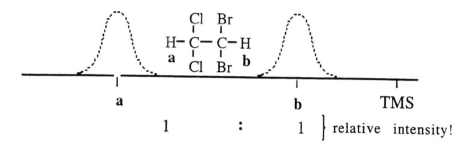

The hydrogen labeled **a** (connected to the CCl_2) is more deshielded than the one labeled **b** (connected to the CBr_2). This is because the chlorine atom is more electronegative than the bromine atom. This much makes sense, but the surprise comes when we look at the actual spectrum for this compound and see that each of the two bands is split into two peaks: **doublet's**. Let's try to explain this observation.

Let us look at the hydrogen connected to the CBr_2 first. During resonance, this hydrogen feels the strong magnetic field ($H°$), the weak magnetic field generated by the electrons (H_{shield}) AND the magnetic field of its neighboring hydrogen (the one connected to the CBr_2). That neighboring hydrogen is NOT absorbing at the time, therefore, there is a 50% chance that it is aligned with the external magnetic field and a 50% chance that it is opposed to the external magnetic field ($H°$). This effect of the neighboring proton may be expressed as:

$$H_{eff} = H° - H_{shield} + H_{cpl} \quad \text{for 50\% of molecules}$$

and:

$$H_{eff} = H° - H_{shield} - H_{cpl} \quad \text{for 50\% of molecules}$$

Thus, instead of getting one singlet, we observe that it splits into two peaks (a doublet) since half of the molecules feel a slightly larger magnetic field (larger by H_{cpl}) and the other half feel a slightly lower magnetic field (lower by the same amount, H_{cpl}). The same reasoning will explain the splitting for the other band.

Chapter 14: Spectroscopy.

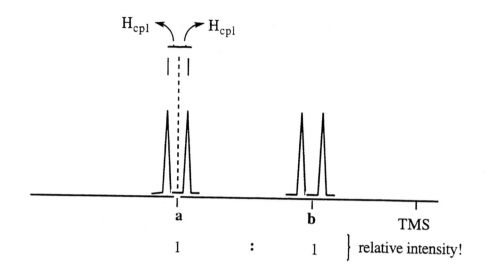

What will happen if instead of one neighboring proton we have more than one? Exactly the same thing! In general, a band will split if you have different kinds of H's (Hydrogens of different chemical environment or chemical shift) in the near vicinity (up to three bonds away). The number of peaks (**N**) that a band will split into is equal to the number of neighbors (**n**) plus one, or:

$$N = n + 1.$$

The figure below also shows the effect of having two and three neighbors (triplet and quartet, respectively).

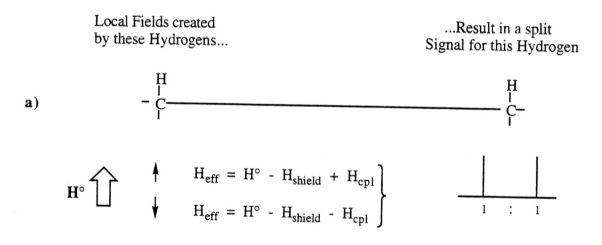

Chapter 14: Spectroscopy.

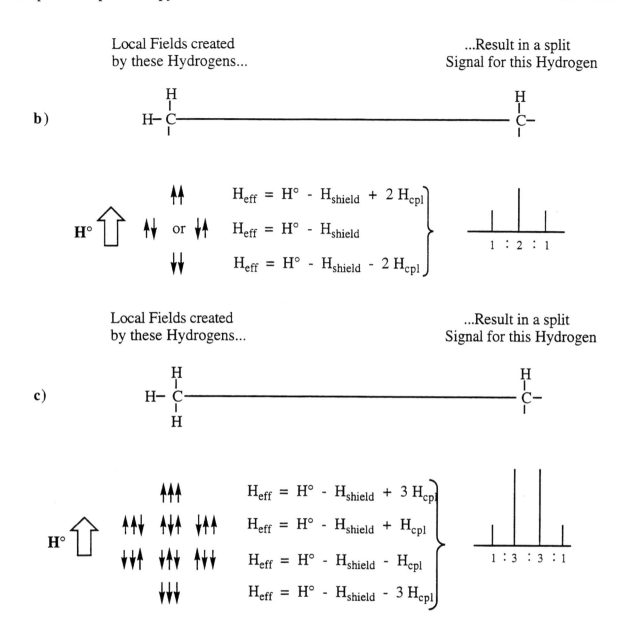

In general, in order to know the intensity of the peaks in a multiplet all we need is Pascal's Triangle. This will allow us to predict the relative intensities of all the peaks in any multiplet.

Neighbors

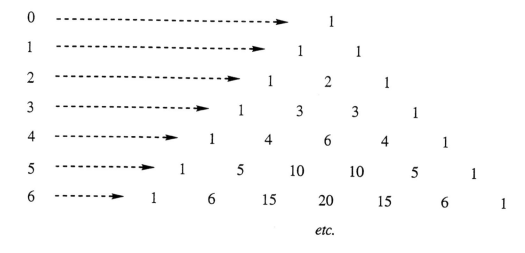

etc.

- What are the relative intensities for a sextet? A sextet is observed when we have a hydrogen in a given molecule with five neighbors of different chemical environment.

We simply read directly off Pascal's triangle and find that the sextet is going to have relative intensities of: 1 : 5 : 10 : 10 : 5 : 1. Since the two central peaks are ten times larger than the outside peaks in the sextet, there is a pretty good chance that the outside peaks will not be visible in the spectra. This fact will become clear as we work some examples later in this chapter.

Let us summarize the rules for the splitting pattern in the following way:

1) Splitting of a proton signal is caused only by neighboring protons of different chemical environments (different δ's).

2) Splitting of a proton signal by a proton separated by more than two atoms is uncommon. For our purposes, <u>only</u> the case of protons on <u>adjacent</u> carbons (or other atoms) needs to be considered (referred to as vicinal protons). In other words, we will see splitting from different kinds of hydrogens if they are up to <u>three</u> bonds away as a maximum.

3) The number of peaks (**N**) that a band will split into is given by: **N = n + 1.**

4) The intensity of the peaks within a band is given by the Pascal Triangle.

14.9 <u>NMR Spectra Analysis</u>.

Let's apply what we have learned and try to do two different things. First, let us practice drawing some simple NMR spectra for some compounds. Then we will reach our goal: to identify unknown organic compounds based on NMR data. In order to predict what the NMR spectrum of a given compound will look like you should always follow these steps.

1) Identify the different kinds of hydrogens.

2) Find the relative chemical shifts (δ) first and then estimate their approximate values.

3) Draw an approximate curve (including relative intensities) for the bands very lightly.

4) Count the neighbors of different chemical environment for each and decide what the splitting pattern should be.

5) Draw the peaks.

The following examples should clear up the use of these rules.

- Ethyl bromide.

The formula for this compound is CH_3CH_2Br. Since the bromine is electronegative, it will make the methylene hydrogens much more deshielded than the methyl hydrogens. Therefore, we have two distinct types of hydrogens in this molecule. Let's label the methyl hydrogens **a** and the methylene hydrogens **b**. Where do these hydrogens absorb? Well, the ones labeled **b** are the bromide type. Therefore, looking at the table we see the following.

Bromides   Br-C-**H**   2.5 – 4.0 δ

This means that these hydrogens should absorb somewhere between 2.5 and 4.0 δ. How about the methyl hydrogens, **a**? These are just primary hydrogens, therefore, all we know is that they normally absorb at approximately 0.9 δ. However, the proximity of the bromine on the other carbon will deshield those hydrogens to a certain extent. We did not go over the rules that help us to predict the amount of shifting from electronegative groups if they are further removed. It will suffice to say that it decreases rapidly with distance. Therefore, in this case it will be somewhat significant, but if there was one more carbon in between them, the effect would be negligible.

How about the relative size of the two bands? Clearly, the methyl hydrogens, **a**, will exhibit a larger band (more area) than the others, since the ratio of the two is 3 : 2. Finally, let's check the splitting for these bands. The methyl hydrogens have two neighbors of different chemical environment, therefore, they will be split into three peaks (**triplet**). The methylene hydrogens have three neighbors of different chemical environment, therefore this signal will be split into four peaks (**quartet**). The spectrum may be seen below.

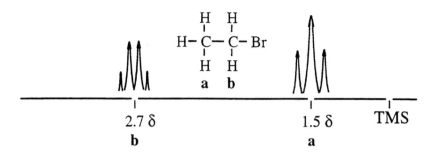

- Diethyl ether.

    The formula for this compound is $CH_3CH_2 - O - CH_2CH_3$. We immediately notice that there are two different types of hydrogens. The methylenes (CH_2's) are bonded to an oxygen, therefore they are "ether hydrogens". However, the methyls are not bonded to the oxygen. Clearly, because the oxygen is electronegative, the ether hydrogens will be a lot more deshielded than the methyl hydrogens. According to the table, the "ether hydrogens" should absorb around 3.6 δ since they are secondary hydrogens.

    Ethers   **H**-C-OR   3.3 – 4.0 δ

    The methyl hydrogens will show an effect due to the presence of a nearby oxygen, but once more, all we can say is that they are a little larger than 0.9 δ. The relative band area for these two types of hydrogens is the same as it was in the previous example, because we have six methyl hydrogens to four methylene hydrogens (ratio is still 3 : 2).

    How about the neighbors? We should notice that this is a symmetrical molecule, therefore, the methylene hydrogens (**b**) have three neighbors of different chemical environment while the methyl hydrogens (**a**) have two neighbors of different chemical environment. Therefore, the spectrum is exactly the same as it was before, but the difference is simply the chemical shifts, which are slightly different[1]. The NMR spectrum based on this data can be seen below.

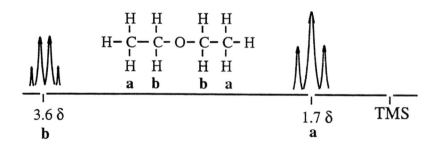

---

[1] Due to the difference in electronegativity between the bromine and the oxygen atoms.

Chapter 14: Spectroscopy.

- Dimethyl ether.

    The formula for this compound is $CH_3 - O - CH_3$. The symmetry of the molecule lets us see that all six hydrogens are completely equivalent, therefore, the NMR spectrum will consist of only a singlet. The only question we may have is about the chemical shift. Once more, being ether hydrogens, they should absorb somewhere between 3.3 and 4.0 δ. Since they are primary hydrogens, it should be close to 3.3 δ. The NMR spectrum can be seen below.

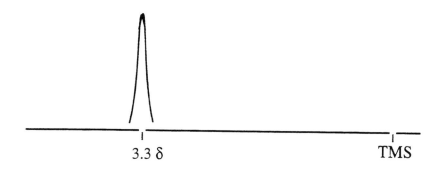

- 1-Chloropropane.

    The formula for this compound is: $CH_3CH_2CH_2-Cl$. There are clearly three types of hydrogens in the molecule. The chlorine has the largest effect on the methylene it is directly bonded to. It will have a minor effect on the methylene that follows, and it should have an essentially negligible effect of the methyl group (too far away). What is the effect of a chlorine atom on the hydrogens directly bonded to the same carbon? Looking at the table, we see the following.

    Chlorides   Cl-C-**H**   3.0 – 4.0 δ

Since those hydrogens are technically primary, it should be close to 3.0 δ. The other two are simply larger than their respective normal absorptions. In the case of the methylene, it is larger than 1.3 δ and the methyl only slightly larger than 0.9 δ. Of course, the area of the peaks will depend on the number of hydrogens, therefore, the relative area of the three bands should be: 3 : 2 : 2.

How about the neighbors? The methyl hydrogens (**a**) have two neighbors of different chemical environment, therefore, its band should split into a triplet. The methylene hydrogens bonded to the chlorine directly (**c**) also have two neighbors of different chemical environment, therefore they will form a smaller triplet (only for 2 H's). Finally, the central methylene (**b**) has five neighbors of different chemical environment, therefore, these should form a **sextet**[1]. The NMR for this compound can be seen below.

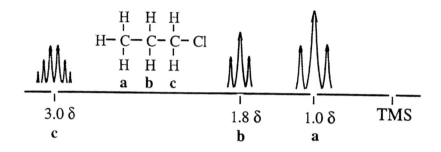

- 4-Methyl toluene.

    This compound has only two types of hydrogens. There are four aromatic hydrogens, which absorb around 7.0 δ, and six methyl hydrogens which are on the carbon next to the aromatic ring (benzylic). The absorption of benzylic hydrogens can be found in the table.

    Benzylic   Ph-C-**H**   2.2 – 3.0 δ

Since they are primary, they should be very close to 2.2 δ. The relative intensities are clearly given by the number of hydrogens, 4 : 6. Since neither has any neighbors, the spectrum will consist of two singlets.

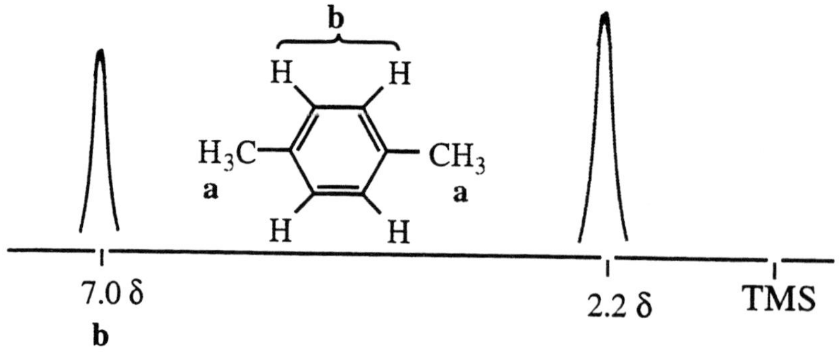

- 2-Bromopropane.

    The formula for this compound is: CH_3-$CHBr$-CH_3. Due to symmetry, there are only two types of hydrogen in this molecule. The methyls should be close to 0.9 δ, but a little deshielded by the vicinity of the bromine. On the other hand, the hydrogen on the second carbon will be much more deshielded by the bromine atom (appears as a bromide in

---

[1] Six peaks in the multiplet.

the table). This one should absorb at approximately 2.8 δ (in the middle of the range since it is secondary. As always, the relative areas corresponds to the number of hydrogens and would be: 6 : 1.

The splitting is easy to predict. The methyls (**a**) have one neighbor of different chemical environment, therefore, they form a large doublet. On the other hand, the CH (**b**) has six neighbors, therefore, it will show up as seven peaks. The NMR can be seen below.

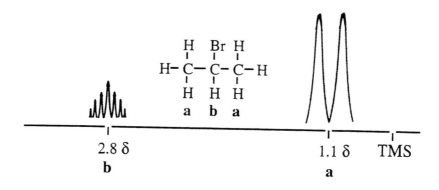

Let's try the reverse analysis, that is, reading the NMR spectrum itself, and from it, determining the structure of the compound. Well, this is not as hard as it seems at first, as long as we approach it in an orderly manner. Let's look at a few examples to see what we mean by this. Let's say that we have an unknown with the molecular formula of $C_2H_3Br_3$, and the NMR spectrum shown below.

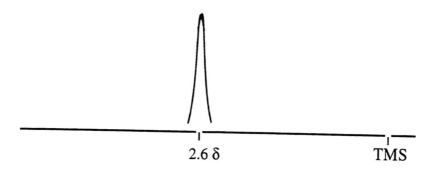

Can we identify the compound based on that alone? The answer is YES. The first thing we do is determine the number of degrees of unsaturation in the molecule. In this case, it happens to be zero, meaning that there are no double bonds or rings in the molecule. When we look at the spectrum there is one thing that becomes evident. There is only <u>one type of Hydrogen</u> in the compound. This means that all three hydrogens in the molecule are identical. At this point we ask ourselves, what can three identical hydrogens be? More likely than not, it would be a methyl group (CH_3). The fact that it is a singlet implies that it has no neighbors, therefore, the only possible compound would be 1,1,1-Tribromoethane. This explains one type of H in the molecule and why the signal is at 2.8 δ, because of the proximity of the three Br's.

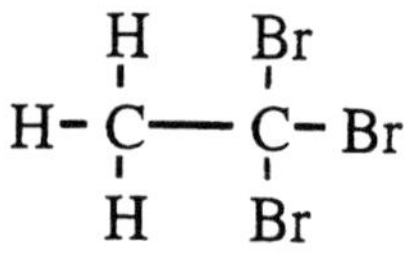

$$\begin{array}{c} H Br \\ | | \\ H-C-C-Br \\ | | \\ H Br \end{array}$$

This example was pretty straight forward but it serves one important purpose: to illustrate the analytical method of deciphering these spectra. One word of advice. At the point where we are supposed to make an educated guess as to what a certain number of hydrogens may be, the following table might prove useful. This by no means tells us what a given number of identical hydrogens will be, but rather an idea of what they MIGHT be. Don't be stubborn and insist that whatever we read out of this table is exactly what it is. Use it simply as a guide (or a first idea). That is why we call them educated guesses.

Educated Guesses

Number of Equivalent H's	Guess
3	CH_3
6	2 CH_3's, probably bonded to the same C. 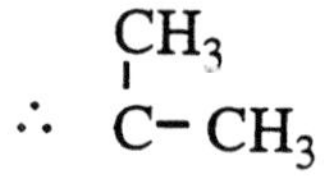 ∴ CH_3-C-CH_3
9	3 CH_3's, probably bonded to the same C. 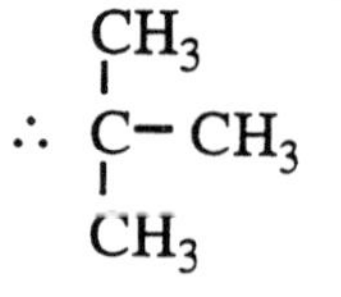 ∴ $CH_3-C(CH_3)-CH_3$ (tert-butyl)
2	CH_2
4	2 CH_2's, probably bonded to the same C.
1	CH

Let's look at another example. The molecular formula of a certain unknown compound is $C_4H_7Br_3$. Its NMR spectrum can be seen below.

Chapter 14: Spectroscopy.

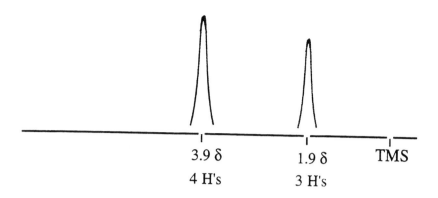

3.9 δ   1.9 δ   TMS
4 H's   3 H's

As always, the first thing we do is determine the degrees of unsaturation. In this case there are none, therefore, there are no π bonds or rings. Should we start to draw all possible isomers and see which one fits? NO! That would be the biggest mistake ever! All we must do is analyze it slowly. To start with, we see that there are two types of H's in the molecule. One type gives a singlet at about 1.9 δ while the other gives a singlet at about 3.9 δ. The fact that they are both singlets tells us that they have <u>no neighbors</u> of different chemical environment. Now, under the singlets there are numbers written: these numbers represent the number of hydrogens of similar environment absorbing at that particular chemical shift. This can be easily obtained experimentally by integrating the spectrum[1]. Anyway, in our example, the one furthest away from the TMS is for 4 hydrogens while the other one is for 3 hydrogens. At this point we should be ready to make educated guesses. Having three identical hydrogens in the molecule more likely than not indicates the presence of a methyl group. How about 4 identical hydrogens? More likely than not, these will be due to two identical methylenes, CH_2's. In order to make these methylenes identical, they should probably be connected to the same carbon. Let's try to build the molecule with this information.

We will start with the methylenes. Since they are the most deshielded, they must be bonded to bromine atoms.

$$\begin{array}{c} CH_2Br \\ | \\ C-CH_2Br \end{array}$$

Since the signal for these hydrogens constitutes a singlet, then these hydrogens cannot have any neighbors. What does that mean? That the carbon atom they are bonded to CANNOT have any hydrogens in it, because if it did, they would split the signal of the methylenes. Therefore, this carbon atom must be bonded to the remaining methyl and the remaining bromine atom. That

---

[1] Modern instruments do this automatically, and gives us the relative areas under the peaks – which is what we would need.

way, neither of them would have neighbors and our spectrum would consist of two singlet's. Therefore, our compound must be: 1,2,3-Tribromo-2-methylpropane.

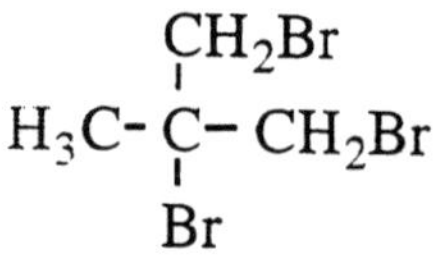

$$\text{H}_3\text{C}-\underset{\underset{\text{Br}}{|}}{\overset{\overset{\text{CH}_2\text{Br}}{|}}{\text{C}}}-\text{CH}_2\text{Br}$$

14.10 Additive Effect.

There is one more thing that we should know. What happens when we have a situation where more than one group directly affects the chemical shift of a given hydrogen? Let's take as an example the case of Bromomethyl benzene, $C_6H_5\text{-}CH_2Br$.

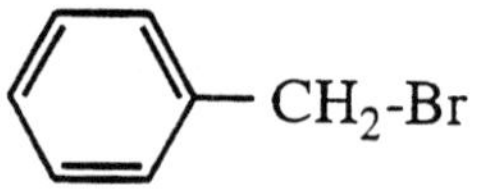

$$\text{Ph}-\text{CH}_2-\text{Br}$$

There are two kinds of Hydrogens in this molecule, and neither has any neighbors, therefore, we should observe two singlets. One of them should be around 7 δ, for the five hydrogens in the phenyl group, while the other should be for the methylene hydrogens outside of the ring. However, when we try to decide where these hydrogens will appear in the spectrum, we encounter a small problem. What type of hydrogens are these? We know that they are benzylic (bonded to the C directly bonded to the ring), but they are also bonded to the carbon with the bromine atom. Let's go back to the table once more. In there we see the following items of importance for this molecule:

Type of Proton		Chemical Shift δ, in ppm
Primary	R-CH$_3$	0.9
Benzylic	Ph-C-**H**	2.2 – 3.0
Bromides	Br-C-**H**	2.5 - 4.0

Now, since this hydrogen is affected by both the benzylic and the bromide effects, then we should **add** these together to figure out where the methylene will absorb. This would give us the following:

	Minimum		Maximum
Benzylic	2.2		3
Bromide	2.5		4
Total:	4.7 δ	⇔	7 δ

Is this the final answer? No! The reason for this is we must think carefully about the meaning of the two ranges given. These ranges represent the minimum and maximum deltas where these H's will absorb. These ranges already include the type of H they actually are (primary, secondary or tertiary : in our example it is primary!). Since these numbers already include this fact, by simply adding them, we will be including this fact twice. Therefore, we must subtract the normal delta for the primary hydrogens once, to get the true range of absorption of our hydrogens. In our case, that would be:

	Minimum	Maximum
Benzylic	2.2	3.0
Bromide	2.5	4.0
Primary H's	-0.9	-0.9
Total:	3.8 ⇔	6.1

Therefore, the methylene band will occur somewhere between 3.8 and 6.1, probably closer to the lower end, since the hydrogens are primary.

Let's make an analogy. Imagine that a large regular cheese pizza at Ocasio's Pizza Emporium costs $4. The menu also states that a large double cheese pizza is $6 and that a large pepperoni pizza is worth $5. What would be the price of a large pepperoni, double cheese pizza at this place? Well, using the additive effect, we can imagine that it would be the cost of the pepperoni plus the cost of the double cheese ($11). However, since we are only buying ONE pizza, not two, we must subtract the prize of the single pizza (remember that those prices include the basic pizza). Therefore, the pizza would cost: $11 - $4 = $7. Certainly, we could have simply said, the double cheese made it $2 more expensive and the pepperoni made it $1 more expensive. Therefore, the combination pizza will be the cost of the regular pizza PLUS the $3 for the extra items.

- The following two NMR's are for two isomers of $C_4H_8O_2$, and both of them are esters. One of them is Methyl propanoate and the other is Ethyl ethanoate. Match the spectra to the molecules.

Chapter 14: Spectroscopy.

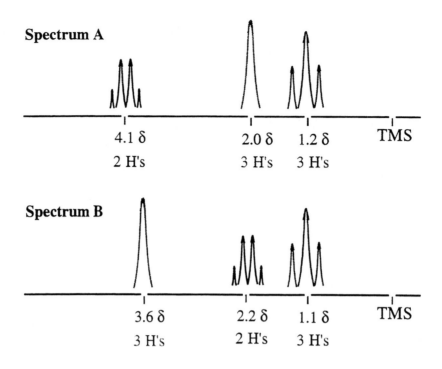

We should draw the two molecules so that we can make an informed decision.

Methyl propanoate   Ethyl ethanoate

The first thing we should notice is that both spectra consist of the same number of peaks and the same splitting: A singlet for 3 H's, a triplet for 3 H's and a quartet for 2 H's. The splitting of the triplet and the quartet is indicative of an ethyl group. Anyway, how can we tell by just looking at the spectra as to which spectrum corresponds to which isomer?

Well, in order to do this, we must look back at the table. Let's check out the part that deals with esters.

Type of Proton		Chemical Shift δ, in ppm
Esters	RCOO-C-**H**	3.7 - 4.1
Esters	**H**-C-COOR	2.0 - 2.2

It should be clear that the hydrogens bonded to the carbon that is connected directly to the oxygen are more deshielded (3.7 - 4.1) than those bonded to the carbonyl side (2.0 - 2.2). Based on that information, we can see that the first spectrum (**A**) is that of Ethyl ethanoate. The methylene is directly bonded to the oxygen, so these two hydrogens should be somewhere between 3.7 and 4.1 (indeed, they are around 4.1 δ). In the case of the other isomer (**B**), it is the lone methyl group that is bonded to the oxygen, so these three hydrogens should be between 3.7 and 4.1 (found at around 3.7 δ).

- Chloromethyl ethyl ether. Predict the NMR spectrum for this molecule using the additive effect.

The formula for this compound is: CH_3CH_2-O-CH_2Cl. We should be able to see that there are three different types of hydrogens in this molecule. Therefore, its NMR should consist of three bands: a singlet for two H's, a doublet for another two H's and a triplet for the remaining 3 H's. We should concentrate on the singlet, because it is the one influenced by the additive effect. Why is that so? Because those hydrogens are bonded to an oxygen on one side and a chlorine on the other side. These are the two entries from the table that are pertinent for those hydrogens.

Type of Proton		Chemical Shift δ, in ppm
Primary	R-**CH₃**	0.9
Chlorides	Cl-C-**H**	3.0 – 4.0
Ethers	**H**-C-OR	3.3 – 4.0

Remember that in order to use this table, we need to recall that these hydrogens are primary and because of the additive effect, the table will be used twice. Therefore,

	Minimum		Maximum
Chloride	3.0		4.0
Ethers	3.3		4.0
Primary H's	-0.9		-0.9
Total:	5.4	⇔	7.1

Since they are primary hydrogens, we expect this singlet to be around 5.4 δ. Therefore, the NMR spectrum for this molecule should look something like this:

Chapter 14: Spectroscopy.

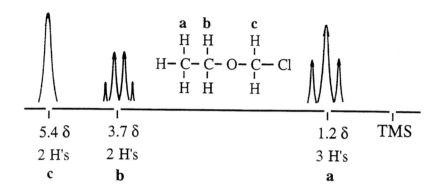

- Draw the NMR for 2,2-Dibromoethanoic acid.

The purpose of this example is to see the additive effect when there are three things affecting a given type of hydrogen. Let's start in the same place, drawing the molecule itself.

$$\text{Br} - \underset{\underset{\text{H}}{|}}{\overset{\overset{\text{Br}}{|}}{\text{C}}} - \overset{\overset{\text{O}}{\|}}{\text{C}} - \text{O} - \text{H}$$
$$\quad\quad\quad\text{a} \quad\quad\quad\quad\text{b}$$

We see that there are two types of hydrogens in this molecule. The one labeled **b** is for the COOH hydrogen, which we already know absorbs beyond 10.5 δ. However, the one labeled **a** is affected by three groups: the COOH and two bromine atoms.

Type of Proton		Chemical Shift δ, in ppm
Primary	R-CH$_3$	0.9
Bromides	Br-C-**H**	2.5 – 4.0
Acids	**H**-C-COOH	2.0 - 2.6

Looking at the table, and realizing that the carbon itself is primary, we can do the following calculation for the range:

	Minimum		Maximum
Bromide	2.5		4.0
Bromide	2.5		4.0
Acid	2.0		2.6
Primary H's	-0.9		-0.9
Primary H's	-0.9		-0.9
Total:	5.2	⇔	8.8

The reason we subtracted the primary hydrogens twice is because we added three entries from the table and each assumes a given type of hydrogen. We expect the absorption to be near 5.2 δ since it is a primary carbon. Since neither of the hydrogens has neighbors, it should be clear that the spectrum will consist of two singlets. The NMR spectrum for this compound can be seen below.

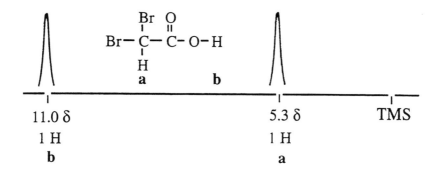

- An isomer of $C_3H_6Cl_2$ has the NMR spectrum shown below. Identify this isomer.

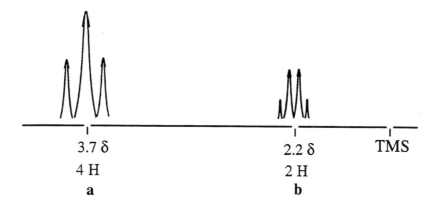

From the NMR spectrum, we see that there are two types of hydrogens. The band at 3.7 δ is for four hydrogens with two neighbors. The band at 2.2 δ is for two hydrogens with three neighbors. Since there are no more hydrogens left in the molecule, it should be clear that these hydrogens are neighbors to each other (that is, the four are neighbors to the two, and the two are neighbors to the four). Normally, we should start the drawing with the band that has the most splitting (in this case, the quartet). Therefore, these two hydrogens (**b**) have four neighbors which more likely than not are two methylenes.

$$\begin{array}{ccc} H & H & H \\ | & | & | \\ C - & C - & C \\ | & | & | \\ H & H & H \\ a & b & a \end{array}$$

What is left? Only to place the chlorines. Of course, they must go in the only place left which will give us 1,3-Dichloropropane for an answer. The reason the hydrogens labeled **a** are further downfield is because of the electronegative chlorines.

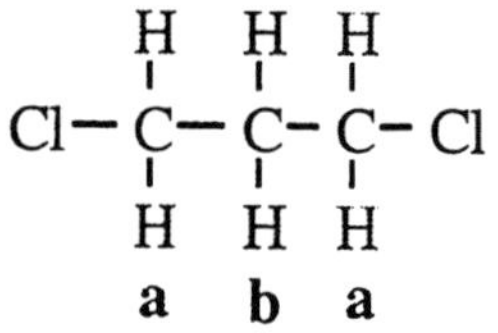

```
      H   H   H
      |   |   |
  Cl– C – C – C – Cl
      |   |   |
      H   H   H
      a   b   a
```

- An isomer of C_3H_7Br has the NMR spectrum shown below. Identify this isomer.

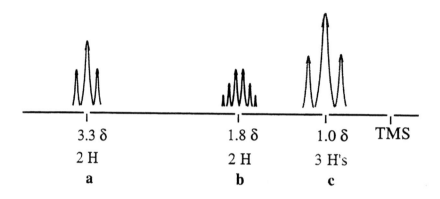

As always, we make a quick calculation and determine that there are no degrees of unsaturation in this molecule. We then notice that there are three types of hydrogens in this molecule, which we have proceeded to label **a**, **b**, and **c**. The spectrum has been integrated and we are told that there are two hydrogens in the triplet labeled **a**, which is probably a CH_2 bonded to the bromine. We make this assumption because the band is the most deshielded from all three. Secondly, we can see that this methylene has two neighbors. What is our educated guess for two neighbors? Probably another CH_2. The very next peak happens to be for two hydrogens at 1.8 δ and it is a sextet. This means that this particular methylene has five neighbors of different chemical environment. Therefore, it is probably bonded to a CH_2 and to a CH_3. This will immediately tell us the identity of the compound: 1-Bromopropane.

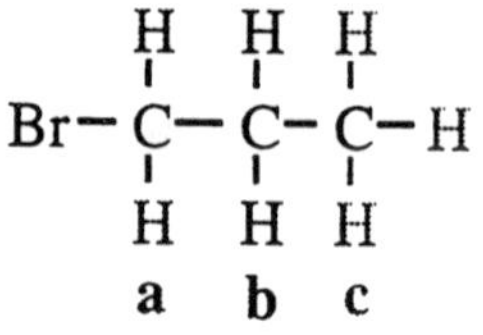

```
      H   H   H
      |   |   |
  Br– C – C – C – H
      |   |   |
      H   H   H
      a   b   c
```

Of course, the methyl group can easily be explained to give rise to the larger triplet at approximately 1.0 δ.

Chapter 14: Spectroscopy.

- An isomer of $C_4H_8Cl_2$ has the NMR spectrum shown below. Identify this isomer.

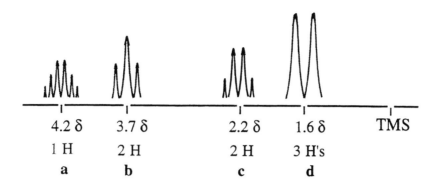

4.2 δ	3.7 δ	2.2 δ	1.6 δ	TMS
1 H	2 H	2 H	3 H's	
a	b	c	d	

Once more, we notice two immediate things: that there are no degrees of unsaturation and that there are four types of hydrogens in this molecule. We have labeled these **a, b, c,** and **d**, from left to right in the spectrum. Further analysis shows that the **a** band is for one hydrogen with five neighbors of different chemical environment. This is most likely a CH that is bonded to a CH_2 and a CH_3. Since it is the furthest one downfield, it must have one of the chlorines bonded to it. Therefore, we can start to build the molecule based on that fact only.

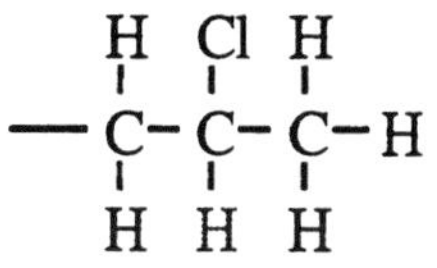

$$\begin{array}{c} \text{H} \quad \text{Cl} \quad \text{H} \\ | \quad | \quad | \\ -\text{C}-\text{C}-\text{C}-\text{H} \\ | \quad | \quad | \\ \text{H} \quad \text{H} \quad \text{H} \end{array}$$

This also explains the band for three hydrogens (labeled **d**). Clearly, these methyl hydrogens only have one neighbor, therefore, it should split into a doublet. What is left to complete the molecule? Another CH_2 and a chlorine. This must go in the remaining bond, to give the following compound: 1,3-Dichlorobutane.

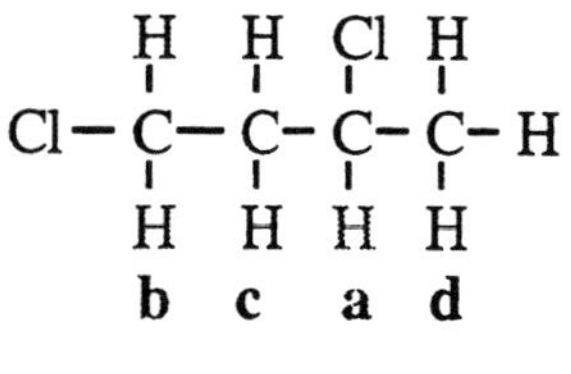

$$\begin{array}{c} \quad\ \text{H} \quad \text{H} \quad \text{Cl} \quad \text{H} \\ \quad\ | \quad | \quad | \quad | \\ \text{Cl}-\text{C}-\text{C}-\text{C}-\text{C}-\text{H} \\ \quad\ | \quad | \quad | \quad | \\ \quad\ \text{H} \quad \text{H} \quad \text{H} \quad \text{H} \\ \quad\ \text{b} \quad \text{c} \quad \text{a} \quad \text{d} \end{array}$$

For confirmation purposes, we can see that the **b** band is the second one from the left (the second most deshielded one[1], since it is also bonded to the chlorine) and that it is for two hydrogens with two neighbors, hence the triplet. Finally, the **c** band is for two hydrogens

---

[1] The reason the **a** band was more deshielded than the **b** band, even though they were both bonded to one chlorine, is because the **a** band has a tertiary hydrogen, whereas the **b** band has secondary hydrogens.

with three neighbors, which gives rise to a quartet at 2.2 δ.

- An isomer of C_5H_{10} has the NMR spectrum shown below. Identify this isomer.

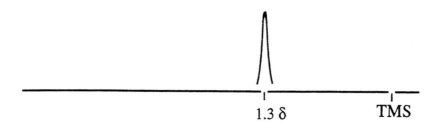

This time we notice that there is ONE degree of unsaturation. This is going to be invaluable to determine the structure of this compound. We also can clearly see that all ten hydrogens are completely equivalent. Let's put these two pieces of information together to elucidate the structure of the compound. What can cause a degree of unsaturation? One possibility is the presence of a pi bond or a ring (a cyclic compound), as we have previously seen.[1] Ten hydrogens is most likely five CH_2's, and in order to make them equivalent, we simply have to bond the two end ones together in order to form a ring. This also will explain the degree of unsaturation present in the molecule. Therefore, the compound must be: Cyclopentane. Remember that in this cyclic structure, every carbon is bonded to two hydrogens.

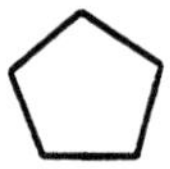

This can also be drawn as follows.

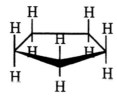

- An isomer of $C_5H_{10}O$ has the NMR spectrum shown below. Identify this isomer.

---

[1] See Section 8.8.

Chapter 14: Spectroscopy.

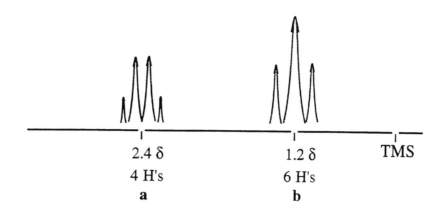

We notice that this isomer also has one degree of unsaturation, which means that it either forms a pi bond (like a carbon-carbon double bond, or a carbon-oxygen double bond) OR it forms a cyclic structure (a ring). That splitting we see above is suggestive of an ethyl group, as we have seen in previous examples. Given that there are 4 hydrogens to 6 hydrogens in these two groups, there must be two ethyl groups. Assuming this is correct, the only thing missing would be a carbonyl group (C=O), therefore, the compound could be 3-Pentanone. In order to corroborate this structure, let's analyze the possible chemical shifts of the compound. In the structure we proposed, there are two methylenes bonded directly to the carbonyl group. These four hydrogens would have to be the ones we labeled **a** in the spectrum above and that absorb at 2.4 δ. These type of hydrogens are described in the table as carbonyl hydrogens, and their absorption can be seen below.

       Carbonyl   **H-C-C=O**   2.0 - 2.7

Actually, since the carbonyl hydrogens in the proposed structure are secondary, we expect them to be somewhere in the middle of that range. This is exactly the case (they absorb at 2.4 δ)[1].

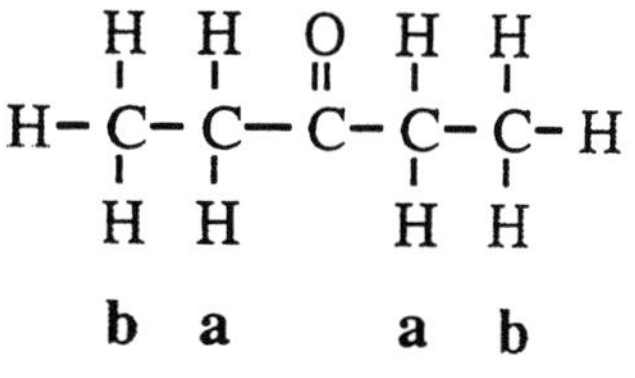

$$\begin{array}{cccccc}
 & H & H & O & H & H \\
 & | & | & \| & | & | \\
H- & C- & C- & C- & C- & C-H \\
 & | & | & & | & | \\
 & H & H & & H & H \\
 & b & a & & a & b
\end{array}$$

- An isomer of $C_6H_{12}O_2$ has the NMR spectrum shown below. Identify this isomer.

---

[1] See the spectrum for confirmation.

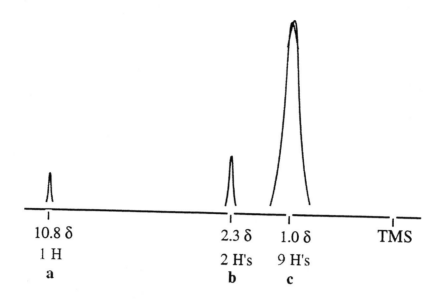

We quickly calculate the number of degrees of unsaturation and see that there is one. The presence of the two oxygens makes us suspect that we have either an ester or a carboxylic acid. Of course, the presence of a band for one hydrogen at 10.8 δ is an indication that it is indeed an acid. We also notice that there are three types of hydrogens and that none of them have neighbors. These have been labeled **a**, **b**, and **c** in the spectrum. We have already identified **a**, since this is an unique peak (the H in the COOH). Let's focus our attention to the other two singlets. The one at 1.0 δ is for nine hydrogens. An educated guess is to say that this is a tert-butyl group, whereas the two hydrogens for the **b** singlet must be a **CH₂**. Therefore, the unknown compound is more likely than not: 3,3-Dimethylbutanoic acid.

We should prove it by checking the chemical shift for the methylene hydrogens. Why? Because these hydrogens are next to the carbonyl of the acid group and they have a specific range in which they absorb. From the table, we get the following range.

    Acids   **H-C-COOH**   2.0 - 2.6 δ

The spectrum corroborates this. These secondary hydrogens absorb at 2.3 δ, which is somewhere in between those two numbers.

$$\begin{array}{c}
H_3CHO \\
H_3C-C-C-C-O-H \\
H_3CH \\
cba
\end{array}$$

Chapter 14: Spectroscopy.

14.11 Organic Reactions Revisited.

Before we proceed any further, it would be a good idea to review the organic reactions we discussed in the previous chapter. The main idea is to add the concepts that we learned in mass spectroscopy, mainly the stability of radicals and carbocations, to these reactions to see if this has an effect in the formation of the organic products. Let's separate this final section into three main categories and discuss each one separately.

1) Addition to Alkenes.

We saw that the pi electrons in the carbon-carbon double bond is what makes these molecules reactive (or functional). These electrons are sticking out of the molecule right around the sp^2 moiety, and since electrons are negative, they attract positive species (or electron deficient species which are called electrophiles). Back in Section 13.5, we mentioned one kind of reaction: addition reactions. In the addition reactions we are going to discuss, the alkene molecule reacts with some diatomic molecule (like Br-Br or H-Br) and in each case, the pi bond between the two carbons is broken and instead, each carbon atom forms a bond with one of the two atoms of the diatomic molecule, as seen below.

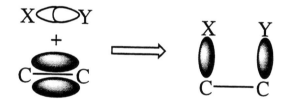

The question we should ask ourselves, therefore, is which atom (X or Y) will attach to which carbon? Does it even matter? Which one adds first and why? Do they both add simultaneously? All of these questions pertain to what we called in our last unit the mechanism of the reaction.

Let's start with the first example: the addition of Br_2 to an alkene, in particular Propene. The reaction itself is shown below:

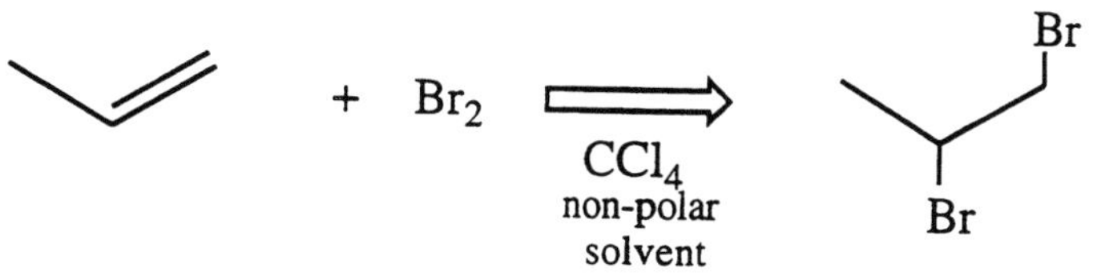

$$\text{CH}_3\text{CH=CH}_2 + Br_2 \xrightarrow{\text{CCl}_4 \text{ non-polar solvent}} \text{CH}_3\text{CHBrCH}_2\text{Br}$$

Notice that the CCl_4 is not a reagent, but rather the solvent. This reaction exemplifies the concept we mentioned above: both Br atoms are added to the pi bond to form 1,2-

Chapter 14: Spectroscopy.

Dibromopropane. In this case, there is no problem with which Br atom added to which carbon, because both atoms are the same. However, we did mention in that section that the stereochemistry of the addition was **anti**. This is because the mechanism involves the formation of a bromonium ion which completely blocks one face of the molecule, therefore forcing the Bromide ion (Br⁻) to attack from the other side. This leads to the addition of each Br from a different face of the molecule.

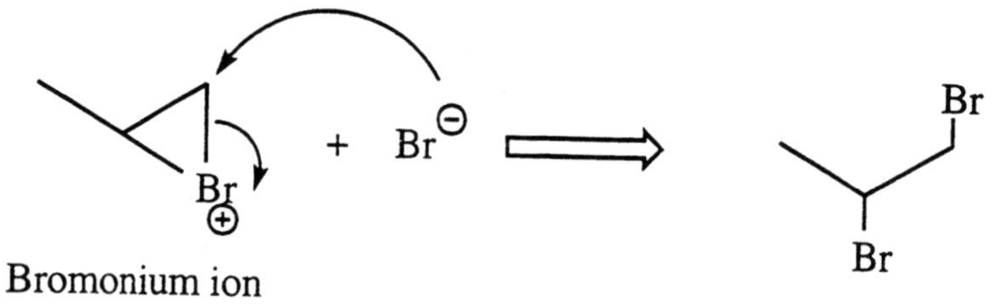

Bromonium ion

As a second example, let's look at the reaction of Propene with HBr. HBr happens to be a strong acid! Being a strong acid, we can think of it as a combination of two ions: H⁺ and Br⁻. Of course, this is not true, since it is not ionic, but it will help in the understanding of the reaction itself[1]. It should be clear by now that one of the carbons is going to accept the H whereas the other carbon will accept the Br. The question is, which carbon will take the H and which one will take the Br!

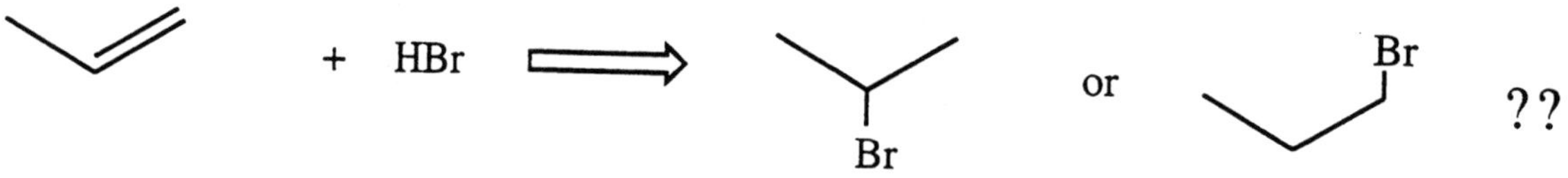

Let's think in terms of the mechanism of the reaction. Remember that mechanisms view the molecules themselves as they collide with one another in a series of steps that lead to the formation of the product. What happens when the two reagent molecules collide? The pi electrons will be attracted to the positive H, thus will "attack it" (form a bond between one of the carbons and the H). These are the two possibilities we just mentioned:

a) 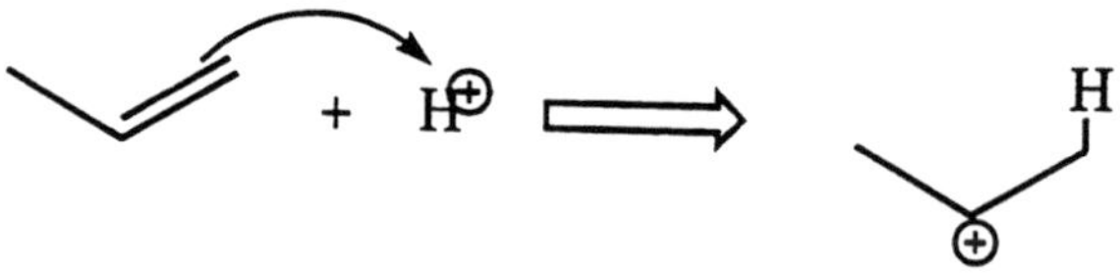 Secondary carbocation

OR:

---

[1] It is just a very polar bond with large partial charges on the H and the Br.

Chapter 14: Spectroscopy.

b) [alkene] + H⁺ ⇒ [carbocation with H] Primary carbocation

Back on Section 14.2, we learned that secondary carbocations are more stable than primary carbocations (due to induction), thus the first possibility has a lower activation energy and will occur more readily. When provided with the choice, we will follow the path of least resistance, that is, the one with the lower activation energy, or the one that forms the more stable carbocation. Remember that:

Tertiary carbocations >> Secondary >> Primary >> Methyl

What happens with the secondary carbocation we just formed? The Br⁻ ion will attach itself to that particular carbon (the positive one), thus yielding the product we observe (2-Bromopropane). In reality, we see about 98% of the 2-Bromopropane and only about 2% of the 1-Bromopropane.

Br⁻ + [secondary carbocation] ⇒ [2-bromopropane]

Is there anyway or any reaction that will actually yield the 1-Bromopropane as its main product? That is, is there anyway that we can add the two atoms reversed? The answer is yes. The reactants would be the same (since we are adding the same two atoms (H and Br), but there has to be an obvious difference. That would be the presence of peroxides (RO-OR). What is a peroxide bond? It is an oxygen-oxygen single bond as part of a chain in a molecule. This is a relatively weak bond; one that can be broken easily with light. We learned that in the presence of light, some bonds can absorb that photon of light and break homolytically. This means that a certain bond in a molecule will split in the presence of an appropriate photon of light to form radicals. In this case, the bond that breaks with the light is the oxygen-oxygen single bond in the peroxide. That means that this mechanism will differ from the previous one because instead of involving ions, it will involve radicals. The whole question then becomes: What do these radicals do once they are formed? The explanation that follows is what we call the mechanism of this reaction. We can see a series of elementary steps that start with the formation of the radicals (RO•) themselves. These radicals then proceed to abstract a hydrogen from an H-Br molecule, thus forming an alcohol and a bromine radical. Therefore, in the second step of this mechanism, we successfully convert one radical into another one. Anytime we convert one particular radical into a different one, we have generated a propagation step in the mechanism.

$$RO\text{-}OR \xrightarrow{light} 2\ RO\cdot \qquad \text{initiation}$$

$$RO\cdot\ +\ H\text{-}Br \implies ROH\ +\ Br\cdot \qquad \text{propagation \#1}$$

[alkene] + Br• ⟹ [secondary radical with Br] propagation #2

[secondary radical with Br] + H-Br ⟹ [product with Br and H] + Br• propagation #3

The bromine radical then reacts with the pi electrons of the double bond and will therefore initiate the addition. One of the two electrons in the pi bond will flip over and attack the incoming bromine radical, leaving the other carbon with an unpaired electron (the new radical). Once more we are phased with the dilemma: which carbon bonds with the Br and which one becomes the radical? The answer is the same: we form the most stable radical, which in this case is the secondary radical.

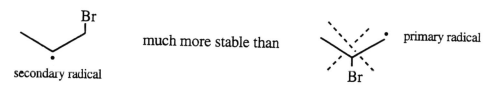

Finally, the carbon that has the unpaired electron is the one that reacts with another HBr molecule to form the final product: 1-Bromopropane.

The most important thing to remember is which atom adds first to the double bond: the H (in the form of H⁺) for the reaction that adds HBr in the dark and without peroxides and the Br (in the form of Br•) for the reaction that adds HBr in the presence of light and peroxides. It will always add in such a way that the carbon that forms the most stable species (a carbocation for the first reaction and a radical for the latter). Let's look at one example that illustrates the difference between these two reactions.

- Consider the reaction of 2-Methyl-2-butene with HBr in the presence and in the absence of peroxides. What is the name of the products formed in each case?

We should draw the molecule so that we can decide which way these two atoms will end up adding to the double bond. This can only be done if we think carefully about the mechanism of the reaction itself.

In the absence of peroxides, the proton from the acid adds first. In order to form the most stable carbocation, the proton will add to the third carbon, thus generating a tertiary carbocation (on the second carbon). This is where the bromide ion will subsequently attach. Therefore, the HBr reaction yields mostly: 2-Bromo-2-methylbutane.

However, in the presence of peroxides, the bromine radical (which is generated when the RO radicals attack an HBr molecule) will attach itself to the double bond at carbon number three, because this way it will generate the most stable radical (once more, the tertiary radical, which occurs at the second carbon). This is where the hydrogen will eventually attach. Therefore, the final product will be 2-Bromo-2-methylbutane.

2) Free Radical Substitution in Alkanes or Alkyl Groups.

This is a reaction we saw earlier in Section 13.2 with the example of CH_4 reacting with Br_2 in the presence of light. The mechanism was via free radicals and it led to the eventual substitution of one of the hydrogens for a bromine atom. This reaction is also observed with all alkanes, not just methane. Let's take the case of Propane. The mechanism involves the abstraction of a hydrogen from the alkane by a bromine radical in order to form an alkyl radical. Since this is a more complex molecule, does it matter which hydrogen is abstracted by the bromine radical? Yes, it does! There are two type of hydrogens in Propane, the primary hydrogens (**a**, which are the ones in the methyl groups) and the secondary hydrogens (**b**, the ones in the methylene group). The abstraction of a primary hydrogen will form a primary radical, whereas the abstraction of a secondary hydrogen will form a secondary radical. Since a secondary radical is the most stable of the two possibilities, there is no doubt that at least the vast majority of the bromine radicals are going to abstract a secondary hydrogen preferentially.

Therefore, the product we should expect to see is 2-Bromopropane, and not 1-Bromopropane. If we carry out this reaction in the laboratory, we will see that actually both compounds are formed during the reaction, however, the post-reaction mixture contains 98% of the 2-Bromopropane and only 2% of the 1-Bromopropane.

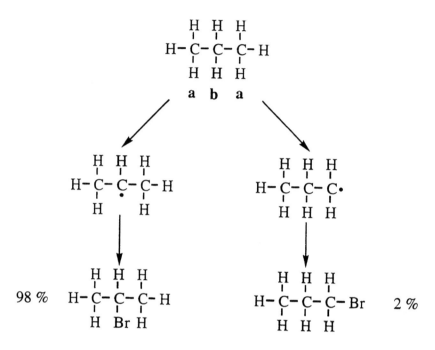

A quick overview of the general mechanism can be seen below.

a) First, the bromine radicals are formed when the bromine molecule absorbs a photon of light.
b) The bromine radical abstracts a hydrogen from the molecule, leaving behind a carbon radical. The hydrogen from the most substituted carbon will be easiest to abstract, since it will form the most stable radical.
  c) The carbon radical will abstract a bromine atom from another bromine molecule, forming another bromine radical, which will then proceed to repeat step **b** once again.
  d) OVERALL: Remove one hydrogen from the carbon that forms the most stable radical, and substitute it with a bromine atom.

Let's look at a few examples.

- What is the main organic product in the reaction of 2-Methylbutane with bromine in the presence of light?

    We should start by drawing the molecule itself, so that we can recognize how many

Chapter 14: Spectroscopy.

different types of hydrogens are present. Remember that the bromine is very selective, so we need to be aware of the different options present.

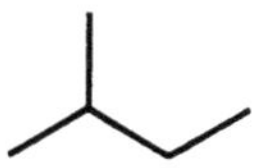

The most substituted carbon is carbon number two, which means that is where the most stable radical can be formed. Therefore, the main product will be: 2-Bromo-2-methylbutane.

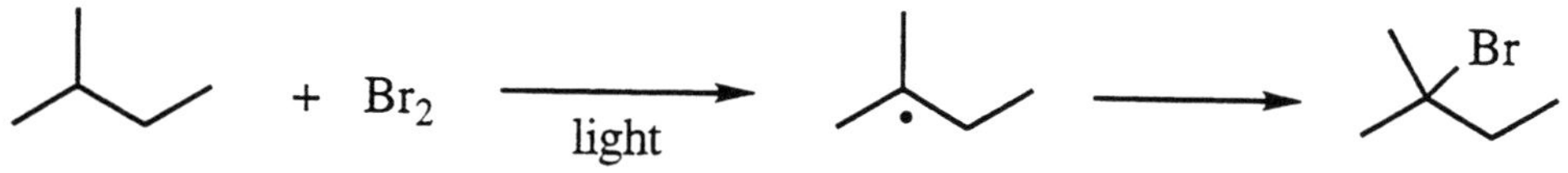

- What is the main organic product in the reaction of 1-Phenyl-2-methylpropane with bromine in the presence of light?

    The molecule itself can be seen below.

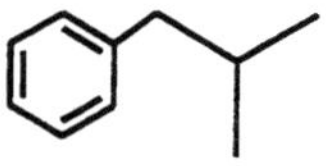

The first thing we notice is that there is a tertiary hydrogen in carbon number two. This is appealing because it forms a tertiary radical. However, if the bromine radical were to remove one of the two hydrogens from carbon number one, then we would form a benzylic radical. As we saw in Section 14.2, these are the most stable radicals we have seen so far. Therefore, the main organic product of this reaction will be 1-Bromo-1-phenyl-2-methylpropane.

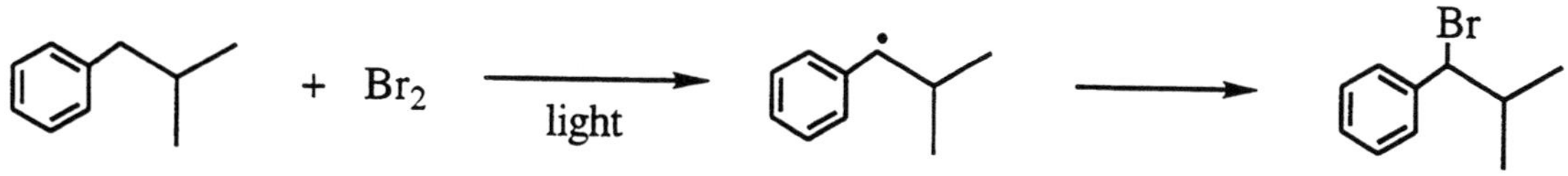

3) <u>Nucleophilic Substitution Reactions</u>.

We have already seen these type of reactions[1], so we will just reiterate the general reaction:

---

[1] See Section 13.11.

Chapter 14: Spectroscopy.

$$R - X \quad + \quad OH^{\ominus} \quad \rightleftharpoons \quad R - OH \quad + \quad X^{\ominus}$$

The key to these reactions is the mechanism. If the nucleophile (OH⁻) has direct access to the carbon with the leaving group (the one with the halogen, X, attached to it), then it will attack directly from the back side, leading to an inversion of the carbon. We called this mechanism SN2. What will allow us to decide whether the nucleophile will be able to attack the carbon directly or not? The presence or absence of hydrogen atoms on the carbon with the leaving group. Hydrogens are very small and they pose essentially no steric hindrance to the attack by the hydroxide ion. However, the absence of at least one hydrogen, will lead to the impossibility of an attack by the hydroxide ion. It will have to wait until the bromine (in the form of bromide) dissociates from the substrate, thus forming a carbocation. We pointed out that the structure of the carbocation is planar because the positive carbon is sp^2 in hybridization, therefore, not only is it planar, but it has an empty "p" orbital (where the positive charge is located). Of course, the hydroxide ion will immediately attack this carbocation from either side to form the alcohol. We called this mechanism SN1.

- Determine the main organic product in the reaction of (S)-2-Bromobutane with hydroxide ions.

    The molecule can be drawn using Fisher projections, as seen below.

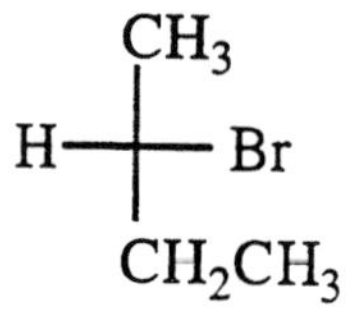

$$\begin{array}{c} CH_3 \\ H - \!+\! - Br \\ CH_2CH_3 \end{array}$$

Since the carbon with the bromine is the chiral one, and it has one hydrogen attached to it, the mechanism will be SN2 (direct attack from the back side). This will lead to inversion of configuration.

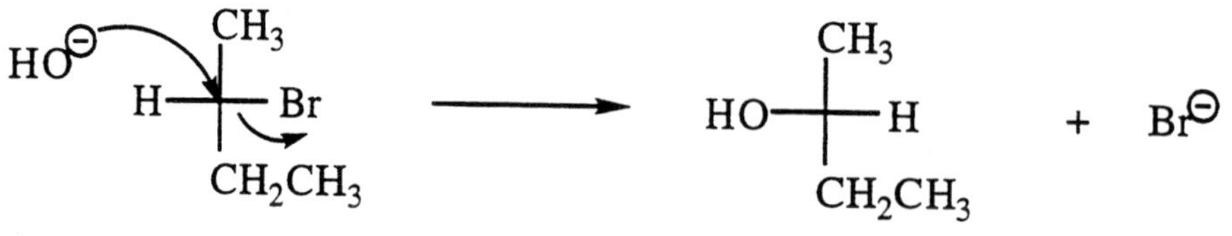

$$HO^{\ominus} \quad \begin{array}{c} CH_3 \\ H - \!\!\!\!\!\!\!\!\!\!\!\!\!+\!\!\!\!\!\!\!\!\!\!\!- Br \\ CH_2CH_3 \end{array} \quad \longrightarrow \quad \begin{array}{c} CH_3 \\ HO - \!\!\!\!\!\!\!\!\!+\!\!\!\!\!\!\!\!\!- H \\ CH_2CH_3 \end{array} \quad + \quad Br^{\ominus}$$

- Determine the main organic product in the reaction of (R)-1-Bromo-2-methylbutane with hydroxide ions.

    The molecule can be seen below.

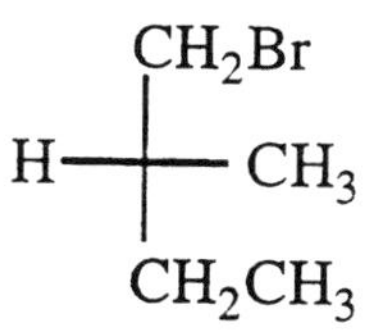

Chapter 14: Spectroscopy.

$$\begin{array}{c} CH_2Br \\ | \\ H\!-\!\!\!-\!\!\!-\!CH_3 \\ | \\ CH_2CH_3 \end{array}$$

There is one thing that we should notice immediately. The molecule is chiral because of carbon number two (which has four different groups attached to it), however, the leaving group is NOT attached to the chiral carbon: it is attached to carbon number 1. Therefore, the reaction will have retention of configuration independent of the mechanism since we are not touching the chiral carbon during the reaction itself. Finally, since the carbon with the leaving group (number 1) has two hydrogens in it, the hydroxide ion will be able to attack it directly, thus the mechanism will be SN2 and the main product will be (R)-2-Methyl-1-butanol.

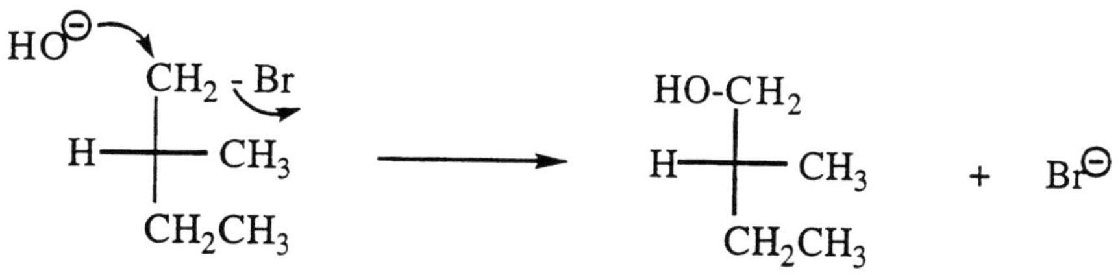

$$HO^\ominus \curvearrowright \begin{array}{c} CH_2\!-\!Br \\ | \\ H\!-\!\!\!-\!\!\!-\!CH_3 \\ | \\ CH_2CH_3 \end{array} \longrightarrow \begin{array}{c} HO\!-\!CH_2 \\ | \\ H\!-\!\!\!-\!\!\!-\!CH_3 \\ | \\ CH_2CH_3 \end{array} + Br^\ominus$$

- Determine the main organic product in the reaction of (S)-3-Bromo-3-methylhexane with hydroxide ions.

    The molecule can be seen below.

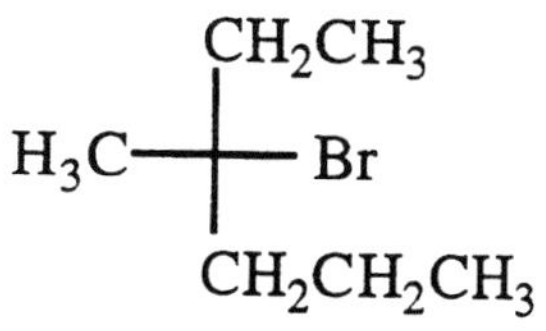

$$\begin{array}{c} CH_2CH_3 \\ | \\ H_3C\!-\!\!\!-\!\!\!-\!Br \\ | \\ CH_2CH_2CH_3 \end{array}$$

We can see that the chiral carbon and the carbon with the leaving group are one and the same (carbon number three). Therefore, we know that something is going to happen at that particular site, so the product will no longer be S. Since there are no hydrogens on that carbon, then the hydroxide ion cannot attack that carbon directly (steric hindrance), so it will have to wait for the dissociation of the bromide ion. This leaves behind a flat carbocation, which can be attacked on either side by the hydroxide. Therefore, the main product will be a racemic mixture of 3-Methyl-3-hexanol.

There are some reactions that follow an SN1 mechanism preferentially. Such a reaction is the substitution of an OH group from an alcohol with a halogen by the action of the strong acid

Chapter 14: Spectroscopy.

HX (where X is a halogen like Cl, Br, or I). The general reaction is seen below.

$$R\text{-}OH + H\text{-}X \Rightarrow R\text{-}X + H_2O.$$

The first thing that happens in the mechanism of this reaction is the protonation of the oxygen by the positive hydrogen in the acid. This forms the following species.

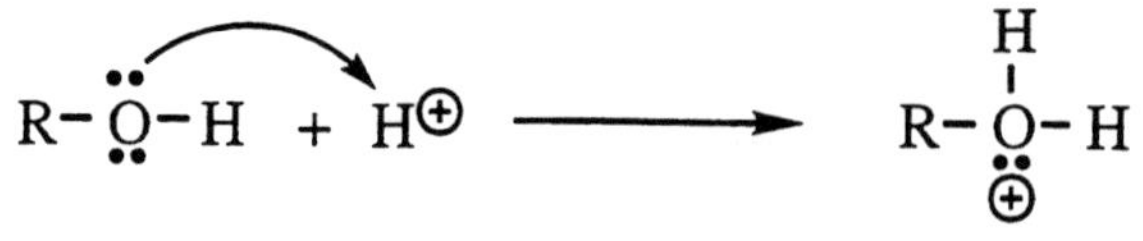

$$R\text{-}\ddot{\underset{..}{O}}\text{-}H + H^\oplus \longrightarrow R\text{-}\underset{\oplus}{\underset{..}{O}}\text{-}H \quad \text{(with extra H on O)}$$

We should mention that just like the bromine is a good leaving group, the hydroxyl group (OH) is a horrible leaving group (because it is such a strong base). However, the protonated hydroxyl group is an excellent leaving group. What do these excellent leaving groups end up doing? They leave and in the process of leaving, a flat, sp² carbocation is formed.

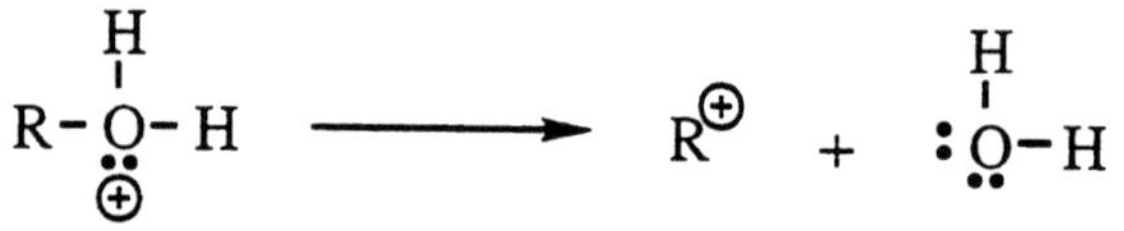

$$R\text{-}\underset{\oplus}{\underset{..}{O}}(H)\text{-}H \longrightarrow R^\oplus + :\ddot{\underset{..}{O}}\text{-}H$$

At this point, the halide ion will attack the carbocation to form the alkyl halide.

$$R^+ + X^- \Rightarrow R\text{-}X$$

The only thing we must be careful of is that carbocations are flat and therefore can be attacked on either side by the halide ion during that last step.

- Consider the reaction of (R)-2,3-Dimethyl-3-pentanol with the strong acid HCl. Name the product of this reaction.

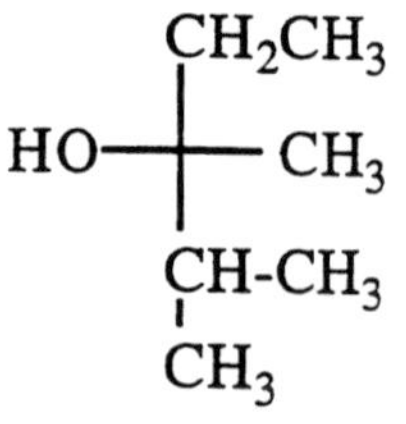

$$\begin{array}{c} CH_2CH_3 \\ | \\ HO\text{—}C\text{—}CH_3 \\ | \\ CH\text{-}CH_3 \\ | \\ CH_3 \end{array}$$

The first step in the mechanism of this reaction is the protonation of the hydroxyl group, which will be followed by the formation of the carbocation. The hydroxyl group is on the chiral carbon (carbon number three is bonded to four different groups – the methyl, the ethyl, the isopropyl, and the hydroxyl groups), therefore, the stereochemistry of that carbon will be affected by the reaction itself.

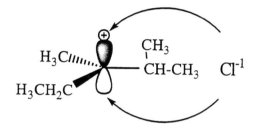

Since this carbocation is flat, it can be attacked on either side by the chloride ion.

Therefore, the final product will be a racemic mixture of 3-Chloro-2,3-dimethylpentane.

Chapter 15: Thermodynamics.

15.1 Introduction.

All the chemical transformations we have mentioned so far as well as physical and biological transformations, involve energy changes. In this chapter, we are going to concentrate our main focus on the study of these energy changes. The science of thermodynamics was developed during the Industrial Revolution, at a time when people were trying to maximize the performance of steam engines. Keep in mind that those days preceded Dalton's atomic theory, therefore, this science is not based on an atomic concept.

We can define **thermodynamics** as the branch of physical chemistry that studies processes in which there is a transfer of heat and a performance of work. From the definition, we can see the interest that this discipline has generated into many different fields. How much heat can we apply to a system in order to force it to do so much work for us? How can this be accomplished? Can we create an engine that will do work for us without any input of heat or any other form of energy? These are some of the questions that we will try to answer in this chapter.

We have mentioned the interconversion of heat and work in our introduction. These two quantities are different forms of energy and energy, in any one of its forms, may be interconverted. Some of the different types of energy are: heat (symbolized by q), light energy (in the form of photons of light, $h\nu$), mechanical energy (like expansion work), electrical energy (generated by the flow of electrons), and chemical energy (trapped in bonds, etc.).

There are some terms that we will be using constantly, and we should be familiar with their meaning and how to use them appropriately. The term **system** applies to the portion of matter under investigation. Therefore, it all depends on how the boundaries of the system are defined. For example, we can define a system to be a flask with 100 mL of water. This is what we call an **open** system. However, if this flask is sealed, then we refer to it as a **closed system**. This is because the water will not have the ability to escape the flask. Therefore, the system is not able to exchange matter with the surroundings. In contrast, imagine instead a perfectly sealed thermos, which would be called an **isolated** system. In this type of system, we have no exchange of mass nor of energy. It should be clear that the surroundings will correspond to anything around the system that has the ability to interact with the system itself.

The conditions under which a system exists at any point in time is referred to as the **state of the system**. In order for us to know which particular state the system is in at a particular time, we need to know everything about it, that is, its mass, its composition, its temperature, its volume, and its pressure. Once we know all the variables of the system, there is nothing left to know and we can easily envision how the system looks and behaves. For example, if we mention 25.0 g of pure water at 25.0°C under one atmosphere pressure, we know that it is a liquid. Failure to indicate any one of these would have resulted in a doubt as to the state of the system. An equation that relates all the variables for the system is referred to as an **equation of state**. An obvious example of an equation of state is that for an ideal gas:

$$PV = nRT$$

The variables for the system are the pressure, the volume, the temperature and the number of moles. Once three of those four are known, we know everything there is to know about this gas. The variables that are part of an equation of state are properties of the system and we refer to them as **state functions**. In other words, a system **has** a pressure, a volume, and a temperature in any particular state. When a system undergoes a transformation, any of its state function may change in the process. However, in order for us to know the change in this state function, all we need to know is the value of that function in the initial state and the value of that function in the final state, so that:

$$\Delta X = X_{\text{final state}} - X_{\text{initial state}}.$$

For example, let's say that during a particular stepwise transformation, the volume of a system changes from 10.0 L to 100.0 L, then back to 50.0 L, and finally it settles at 30.0 L. What is the change in volume for the transformation? Of course, the ΔV for this transformation is simply given as the difference between the final volume and the initial volume, which in this case it would be: $\Delta V = 30.0\,L - 10.0\,L = 20.0\,L$ ✓

Are there other variables that are also a property of the system? We have used the term energy throughout our initial discussion. What do we mean by **energy**? It has been defined by many as the capacity to do work. Back in Section 4.6, we saw that a gas exhibits an increase in temperature when its average translational kinetic energy is increased. We know that **kinetic energy** is related to the motion of the molecules. We will define the **internal** energy of a system as the combination of the translational energy of its molecules, together with other types of energy like electronic, vibrational, rotational, as well as nuclear.

What happens when we place a rod of red hot iron in contact with a bucket of very cold water? We know that the iron is going to cool down and that the water is going to warm up once they are in contact. This is because the high energy molecules of the hot iron will come in contact with the low energy molecules of water and upon their collision, energy will be transferred from the hot substance to the cold one. This transfer of energy will continue until the temperature of the two are the same. This type of energy transfer is known as **thermal energy**. It seems that any body has the ability to transfer some of its internal energy to another body, as long as they exist at different temperatures. The transfer of thermal energy from one body to another due to a temperature difference is what we call **heat**. On the other hand, if a system expands, as it pushes the boundaries against an opposing force, it will perform **work**. In a situation where a sufficient force is applied on a body to make it move, mechanical work can be defined as the product of the force applied times the distance of the displacement for that body.

15.2 The First Law of Thermodynamics.

The development of thermodynamics has brought forward three universal laws. The first of these laws is the law of conservation of energy. It basically states that energy may be transformed from one form to another, but it cannot be created nor destroyed. Another way of expressing the first law of thermodynamics is based on the concept that the universe is an isolated system. Of course, if we cannot exchange energy nor mass beyond the boundaries of the universe, then we can state that the energy of the universe is constant.

Experimentally, we have observed that every time a system absorbs heat from the surroundings, the internal energy of the system increases. It has also been observed that every time work is done on a system by the surroundings, the internal energy of the system increases. Mathematically, the first law of thermodynamics can be expressed with the equation seen below.

$$\Delta E = q + W$$

In this equation, q represents the heat absorbed or released by the system, and W represents the work done by the system or done on the system.

The internal energy is a state function and for a given transformation, the difference in energy is given by:

$$\Delta E = E_{final} - E_{initial}.$$

However, heat and work are not properties of the system. A system does not have heat at a given state, and certainly a system does not have work at a given state. These are forms of energy that are only observed along the boundary of the system during a transformation. These are NOT state functions. We shall focus our attention on work in a subsequent section.

Chapter 15: Thermodynamics

It seems odd that the sum of two variables that are not properties of the system will lead to the change in one of the actual properties of the system. This is based on a mathematical property which states that the sum of two inexact differentials will lead to an exact differential. Therefore, the sum of two variables that are not state functions will lead the change in one that is a state function.

The equation for the first law of thermodynamics gives us the ability to calculate ΔE for a transformation. It would be ideal to be able to determine the initial and the final energy of the system, because since the internal energy is a state function, we would not have to do much other than to find the difference ($\Delta E = E_{final} - E_{initial}$). However, it is pretty much impossible to determine the actual internal energy of a system. There are too many variables involved in the internal energy (electronic, rotational, vibrational, translational, nuclear energies, etc.). The first law ($q + W$) provides us with the tool necessary to calculate the change in the internal energy without ever actually knowing the values of the initial nor the final energy.

We have mentioned that neither work nor heat are state functions. This means that they are not properties of the system. Since they appear at the boundary of the system during the transformation itself, we need to know what happened as we went from the initial to the final state. In other words, we need to know the path followed by the reaction. For example, going from the first floor to the seventh floor in Crawford Hall would require much more work using the stairs than using the elevator. Of course, we will use a lot more of our internal energy if we take the stairs. Therefore, the amount of work done depends on the path.

Since the first law relates heat and work to the change in the internal energy, all we must do is keep track of these two in order to calculate ΔE. For example, a system may absorb a certain amount of heat. This will increase its internal energy by so many Joules. If the system uses some of that energy to expand, and therefore push the walls of the container, then as the expansion takes place, the system's internal energy will decrease. This particular case can be seen below.

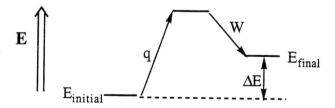

There is a sign convention for heat and work and we should be aware of it at all times. In general, when heat is absorbed by the system, we refer to this as a positive quantity. Similarly, when heat is released by the system, then it is a negative quantity. In terms of work, the work is said to be negative when it is done by the system, whereas it is said to be positive when work is done on the system by the surroundings.

Summary of Sign Convention.

Heat (**q**)	Positive	System *absorbs* heat.
	Negative	System *releases* heat.
Work (**W**)	Negative	Done *by* the system.
	Positive	Done *on* the system.

Since ΔE is a state function, it is <u>independent of the path</u>. If we know the change in internal energy between the initial and final state, then we can repeat the same process in a variety of ways (following different paths) and the ΔE will be the same value always as long as the initial and final state remain the same. On the other hand, heat and work are not state functions, so in order to know their values we must be told what was the actual path followed as the system went from the initial state to the final state.

The following example illustrates this concept. We are going to warm up a beaker that contains 100.0 g of water from 25.0°C to 35.0°C. This will be achieved by two totally different paths.

Path #1: <u>Addition of pure heat</u>. If we heat up the beaker with a burner, as the water absorbs the heat, it will eventually reach the desired temperature.

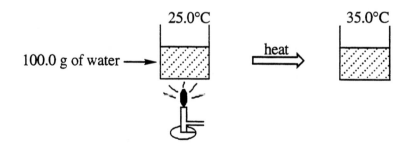

This represents the addition of pure heat (no work is done). From the first law of thermodynamics:

$$\Delta E = q + W = q \text{ (because there is no work done).}$$

However, in Section 5.2, we learned that for a physical process that does not involve a phase change, the heat absorbed is given by:

$$q = mc\Delta T.$$

Therefore,

$$q = (100.0 \text{ g})(4.184 \text{ J/gdeg})(10.0 \text{ deg})$$

So, $\quad\quad\quad\quad \Delta E = q = 4,180 \text{ J}^1$ ✓

Path #2: <u>Performing pure work</u>. Let's repeat the experiment but this time, instead of using a burner to heat up the sample, let us create a system of weights and pulleys as shown below. As the mass **M** drops, it will turn the paddle, which in turn will convert all that work energy into internal energy of the system. This will be manifested as an increase in the temperature.

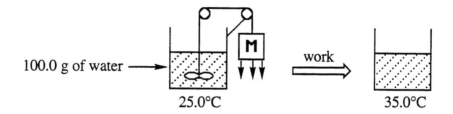

100.0 g of water → [25.0°C] — work ⇒ [35.0°C]

How do we determine the amount of work done on the system? We can go through the trouble of looking for the equation that would allow us to do this, but it is pointless. This is because we already know the change in internal energy for this transformation. The first example told us that ΔE was 1,480 J and since we are starting in exactly the same place (100.0 g of water at 25.0°C) and we are stopping the transformation at exactly the same place as the previous example (100.0 g of water at 35.0°C), then it must be true that ΔE must still be 1,480 J.

$$\Delta E = q + W = W \text{ (because there is no heat involved).}$$

Therefore, $\quad\quad\quad\quad W = 1,480 \text{ J}.$ ✓

A typical misconception is that as the temperature of the water (in the previous examples) is going up, the water must be absorbing heat. This is true only if you are heating up the system, but as we can see from the second case, the temperature can go up by just doing work on the system (in which case q = 0). Therefore, we must know the path between the initial and the final

---

[1] To three significant figures.

states in order to calculate q and W. Keep in mind that the temperature is a property of the system that reflects the random motion of the molecules of that substance. On the other hand, heat involves the transfer of thermal energy between two bodies due to a temperature difference.

15.3 Physical and Chemical Transformations.

We must learn to distinguish between a **physical** and a **chemical** transformation. A chemical transformation is one where there is a clear reaction taking place, where a particular substance is transformed into another different substance. As an example, consider the combustion of methane gas.

$$CH_{4(g)} + 2\, O_{2(g)} \Rightarrow CO_{2(g)} + 2\, H_2O_{(\ell)}$$

If instead we are referring to a change in temperature for a given substance, without a chemical transformation involved, then we are talking about a physical change. As an example, consider the process of warming up 25.0 g of methane gas from 25.0°C to 50.0°C. It was methane before the change and it is still methane after the change. All thermodynamic equations and properties apply exactly the same for either process.

Let's look at one example of a physical process. Imagine we have a cylinder, fitted with a movable piston, that has a certain amount of an ideal gas trapped inside. Let's also say that we will prevent the piston from moving by placing two stoppers right above it, as shown in the diagram below.

What would happen if we were to increase the temperature of the piston? It should be clear that the gas molecules will also get warmer because do to the temperature difference, they will absorb some of the energy from the piston (warmer) during the collisions, and upon doing so, their average kinetic energy will increase (thus, their temperature will also increase). This will continue until the temperature of both (the piston and the gas) are the same. Due to the stoppers, the piston did not move. Therefore, the energy that the piston lost must have been gained by the gas molecules. Overall, we can see that the energy simply was transferred from the warmer to the colder body. We can also see that the energy change (ΔE) was manifested as a temperature increase for the gas. In this case, there is no work done by either the system nor on the system, since there was no expansion. Let's look at a different example, where the gas will expand, and upon doing so, it will perform some mechanical (expansion) work. Actually, expansion work only applies to gases, since the amount of expansion or contraction for a liquid or a solid is negligible.

Consider a cylinder with a movable piston attached to it. This is very similar to the previous example, except this time the piston will have the ability to move (no stoppers). If there is an expansion, this means that the gas molecules must be pushing this piston away, therefore, the gas must be doing work.

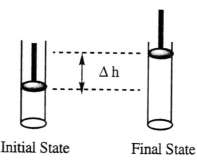

Initial State   Final State

How difficult is it for the gas to push that piston? It depends on the force that the piston is exerting on the gas. We will call the force the piston exerts on the gas (per unit area) the external pressure. Since work is defined as the force applied over the distance of displacement, then,

$$|\text{Work}| = |\text{Force} \times \text{distance}| = |F \times \Delta h|$$

but $F = P_{ext} \times A$, so:
$$|\text{Work}| = |P_{ext} \times A \times \Delta h|$$

But, $\Delta V = A \times \Delta h$, therefore,
$$|W| = |P_{ext} \Delta V|$$

This expression will give us the magnitude of the work. However, since our sign convention stated that work is a negative quantity when it is done by the system, then:

$$W = -P_{ext} \Delta V$$

In a case like the one we mentioned above, the gas pushes the piston as it expands, and thus uses some of its internal energy to do work. Therefore, the change in internal energy will depend on how much work the system will do. We should point out that the expression above is NOT the definition of expansion work. Work is not a state function, therefore it depends on the path. This equation only applies to the particular situation described above: an expansion of an ideal gas AGAINST a constant external pressure (this being the path). We will talk more about this in the next section.

- A certain ideal gas expands <u>against a constant external pressure</u> of 1.20 atm from a volume of 20.0 L to a volume of 50.0 L. During this process, the gas absorbs 2,500 J of heat. Determine the ΔE for this physical process.

The change in internal energy is given by the first law of thermodynamics.

$$\Delta E = q + W.$$

We need to determine both q and W in order to calculate ΔE. The heat was already given (2,500 J), therefore, only W must be calculated.

$$W = -P_{ext} \Delta V = -1.20 \text{ atm} (50.0 \text{ L} - 20.0 \text{ L}) = -36.0 \text{ Latm}.$$

These are *not* SI units of energy. Liters atmospheres are indeed units of energy, but not the SI units. Therefore, we must convert units into Joules. In order to do this, we can use the two values we have previously used for the gas constant, R.

$$1 Latm \times \frac{8.314 \frac{J}{Kmol}}{0.0821 \frac{Latm}{Kmol}} = 101.3 J$$

The conversion factor is: 1 Latm = 101.3 J

$$W = -36.0 \text{ Latm} \times \frac{101.3 \text{ J}}{\text{Latm}} = -3,650 \text{ J}$$

Finally,
$$\Delta E = q + W = +2,500 \text{ J} + (-3,650 \text{ J})$$

$$\Delta E = -1,150 J \quad \checkmark$$

Could we have done this problem if we had not mentioned that the transformation took place against a constant external pressure? NO! We need to know the path followed by the system in order to be able to determine the work involved in the transformation. Remember that both heat and work are not state functions, therefore we need to know their path. In the next section, we will deal with work to a greater detail, so that we can convince ourselves that different paths lead to different ways of determining the amount of work involved in the process.

15.4 Expansion Work.

As we mentioned in the previous section, at this point in time we are going to concentrate our efforts in one type of work only, expansion work. This is the work done by a gas as it expands and in the process pushes the walls of the container (the boundary of the system). Mathematically, we define expansion work as follows.[1]

$$W = -\int_{V_1}^{V_2} P_{ext} dV$$

Since work is not a state function, its path during the transformation needs to be defined. The gas is pushing against the walls of the container as it expands, so the information needed is actually the opposing pressure. The stronger the opposing pressure, the harder the gas will have to push against it and therefore, the more work it will have to do in order to achieve the expansion.

Some Defined Paths.

Rather than doing a pure mathematical definition of the paths, it would be easier to follow these with the example below. Notice that different amounts of work can be done by the gas as it expands from the initial to the final state through these different paths.

- Consider the isothermal[2] expansion of one mole of an ideal gas at 25°C as it expands from an initial volume of 1.00 L to a final volume of 10.00 L under the following conditions listed below.

    a) *Against a vacuum.*

    What do we mean by a vacuum? It is a system in which there is no opposing pressure (it is equal to zero). Therefore, the external pressure is zero. From the definition of work, it follows that there is no work done by the system during this expansion, therefore, W = 0.

    b) *Against a constant opposing pressure of 1.00 atm.*

    When the expansion occurs against a constant external pressure, then we can take

---

[1] The reason for the negative sign in the equation is to maintain the sign convention we presented in Section 15.2.
[2] The term isothermal means constant temperature, or dT = 0.

the external pressure out of the integral[1].

$$W = -\int P_{ext} dV = -P_{ext} \int dV = -P_{ext} \Delta V$$

Substituting the values we get:

$$W = -1.00 \text{ atm} (10.00 \text{ L} - 1.00 \text{ L}) = -9.00 \text{ Latm}.$$

This is the amount of work done by the system: 9.00 Latm. However, these are not appropriate units of energy, as we saw in Section 15.3. Therefore, we must convert it to SI units (Joules).

$$-9.00 Latm \times 101.3 \frac{J}{Latm} = -911 J \text{ [2]}$$

c) *Reversibly.*

A reversible process is one that can be performed in either direction with a simple infinitesimal change. In other words, if we increase the opposing pressure the smallest amount, it will start to contract, whereas if we decrease the opposing pressure the smallest amount, it will start to expand instead. Of course, in order for this to be true, the internal pressure (that of the gas) and the opposing pressure must be exactly equal to one another. We could even say that the two pressures are at equilibrium with one another. For a process to be reversible it must be true that all intermediate states along the pathway from the initial to the final state be equilibrium states. Therefore, in order for the gas to expand reversibly, its pressure must equal the opposing pressure, not only initially, but every step along the way. Therefore,

$$P_{gas} = P_{external} \text{ always.}$$

The implications of this pathway are such, that if we think about this for a while we will realize that this process would take an infinite amount of time. Let us imagine one simple scenario. We have a gas trapped in a piston and the pressure of the gas is equal to the pressure exerted by the mass of the piston on top. In order to expand this gas, we must decrease the opposing pressure infinitesimally, therefore, we will remove ONE METAL ATOM from the piston. This will make the mass infinitesimally smaller and

---

[1] Constants can be taken out of an integral before it is evaluated.
[2] Remember that any time we want to convert from Latm to Joules, all we need to do is multiply by the conversion factor: 101.3 J/Kmol.

therefore, the gas will expand a miniscule amount, dV. At this point, its pressure will also decrease by dP and then both pressures will be equal once more. Now we would remove a second atom from the piston, and wait for the expansion and equilibration to take place before we remove the third atom. We can imagine that if we want the volume to increase any appreciable amount, this process will take forever. Essentially, no process in nature is truly reversible. Nevertheless, if we take the process to be reversible, then:

$$W = -\int_{V_1}^{V_2} P_{ext}\, dV$$

The condition of reversibility means that: $P_{ext} = P_{gas}$, therefore,

$$W = -\int_{V_1}^{V_2} P_{gas}\, dV$$

Since the gas is ideal, then the pressure of the gas is given by the ideal gas law ($P_{gas} = nRT/V$). Substituting this above, we get the following expression.

$$W = -\int_{V_1}^{V_2} \frac{nRT}{V}\, dV$$

Since nRT are constants[1], they can come out of the integral.

$$W = -nRT \int_{V_1}^{V_2} \frac{dV}{V}$$

Knowing that: $\int dV/V = \ln V$, we then can solve for W.

$$W = -nRT\ln\left(\frac{V_2}{V_1}\right)$$

Now that we have derived the equation, let's go back to the problem. The reversible work would be:

---

[1] The temperature is constant because it is an isothermal process!

$$W = -(1.00 \text{ mol})(8.314 \text{ J/mol K})(298 \text{ K}) \ln(10.0/1.00)$$

$$\therefore W_{rev} = W_{max} = -5710 \text{ J} \quad \checkmark$$

The work we just calculated in the last part of the problem, corresponds to the maximum work that the system is capable of doing. Let's look at the following graph to try to understand this concept better. Work is actually the area under a curve of *external pressure vs. volume*. Most people get confused when they see the graph because they forget that we plot in the same graph not only the external pressure but also the pressure of the gas itself. Remember, all we need to follow is how the <u>external</u> pressure changes as the volume of the system changes. It is the area under that curve that gives us the work! In an isothermal expansion, the pressure *of the gas* times the volume is equal to a constant[1] ($PV = k$, Boyle's Law) and its graph is a curve.

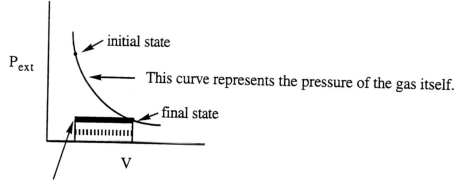

The dark line <u>is</u> the opposing pressure and it is the area under this line that represents the work done by the system as it goes from the initial state to the final state.

As a first example let's consider the expansion against a constant external pressure. Even though the pressure of the gas changed from P_1 to P_2 (initial to final states), the opposing pressure remained constant from the beginning to the end (dark line). Therefore, the work is *the area under that dark line*. We should be able to see that if we were to increase the opposing pressure, the area under the curve would continue to get larger, which means that the amount of work done by the gas would also increase. Is there a limit? The opposing pressure cannot be greater than the pressure of the gas. If that were the case, then we would not have an expansion and instead, we would observe a contraction. Therefore, the limit is when the opposing pressure is exactly equal to the pressure of the gas at all times. Then the dark line would be the same as the curve that describes the pressure of the gas as it expands. This is what we call the reversible work and from the plot, we can see that this reversible work also represents the maximum work that a

---

[1] $PV = k$, which is Boyle's Law.

system can do for us.

The plot below represents the case of a reversible isothermal expansion. The pressure of the gas is given by the solid curve (which is called an isotherm) and the opposing pressure is the solid dark line. Notice that the area under that dark line represents the maximum work as we move from V_1 to V_2.

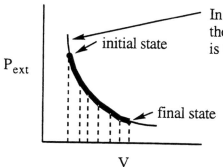

In this case, the actual pressure of the gas is exactly the same as the external pressure (dark line). This is what we call a *reversible expansion*.

15.5 Energy and Enthalpy.

In Section 5.4, we dealt with the enthalpy changes for chemical reactions (ΔH). Is there a difference between ΔH and ΔE? Let's look at the definitions.

ΔE - It is the amount of heat released or absorber during a transformation under conditions of constant volume. Therefore, by this definition, we can write the following.

$$\Delta E = q_v$$

The definition itself gives us a method to determine the ΔE for a process directly in the laboratory. It states that it is the HEAT associated with the **isochoric** process. An isochoric process is one in which the volume remains constant ($dV = 0$). Let us see why this is the case. Going back to the first law of thermodynamics, we get:

$$\Delta E = q + W$$

However, it the process is isochoric, then $q = q_v$, and W would have to be zero.

Therefore, $$\Delta E = q_v + 0$$

Or: $$\Delta E = q_v \checkmark$$

However, many times it is more convenient to measure instead the amount of heat

Chapter 15: Thermodynamics

released or absorbed during a transformation under conditions of constant pressure, not only externally, but also internally. These are referred to as **isobaric** conditions. Therefore, the isobaric conditions can be summarized as:

$$P_{ext} = P_{gas} = \text{constant}$$

Under isobaric conditions, $q = q_p$, and then the first law becomes:

$$\Delta E = q_p + W$$

So,

$$q_p = \Delta E - W = \Delta E - (-P_{ext}\Delta V)$$

Since it is isobaric,

$$P_{ext} = P_{gas} = P$$

$$\therefore q_p = \Delta E + P\Delta V$$

Many processes that we measure or observe take place under isobaric conditions, therefore, this quantity ($\Delta E + P\Delta V$) is constantly surfacing and people got tired of talking about $\Delta E + P\Delta V$, so they decided to give it a special name: ΔH --> Enthalpy.

ΔH - amount of heat released or absorbed during a transformation under conditions of constant pressure --> internal and external are equal and constant (isobaric conditions).

$$\therefore \Delta H = q_p$$

Thermodynamically, enthalpy is defined as:

$$\mathbf{H = E + PV}$$

Since enthalpy is a state function (as well as the internal energy), then during a change:

$$\Delta H = \Delta E + \Delta(PV)$$

and under isobaric conditions,

$$\Delta H = \Delta E + P\Delta V$$

Back in Section 4.6, we saw that the average kinetic energy for an ideal gas depends only on the absolute temperature.

$$KE_{avg} = 3/2\ RT$$

This equation is true as long as the gas is monoatomic and ideal. If the gas were polyatomic, then there would be other contributions to the average kinetic energy, like the vibrational and rotational energies. Of course, if we only have a single atom, it cannot rotate nor vibrate (at least not relative to another atom). Therefore, let's concentrate our discussion on monoatomic ideal gases.

If we wanted to change the temperature of one mole of a monoatomic ideal gas by one degree, its average kinetic energy would have to increase by 3/2 R.

$$\Delta KE^1 = 3/2\ R\ (301\ K) - 3/2\ R\ (300\ K) = 3/2\ R$$

We will define the **molar heat capacity** as the amount of energy required to raise the temperature of one mole of a substance by 1 K. It should be clear that for a monoatomic ideal gas, the molar heat capacity must be 3/2 R. However, as long as we are dealing with gases, we must be very careful to specify what conditions are used to measure its molar heat capacity. Under conditions of constant volume (C_v), the system does not have the ability to expand, therefore all the heat added to the system goes directly into increasing the thermal energy of the molecules. This is what we have called the average kinetic energy for the gas, therefore C_v would be 3/2 R. However, if we measure the heat capacity under conditions of constant pressure (C_p), then the system has the capacity to expand, and if it does, it will do work. We know that when a system does work its internal energy decreases, therefore, if we were to add the same amount of energy (3/2 R), the temperature would not rise an entire degree since some of it would be used to do work. How can we determine the amount of energy required to do that work so that we can add it up to the 3/2 R? Since it is isobaric, the amount of work done is PΔV. But what is PΔV? For an ideal gas:

$$P\Delta V = nR\Delta T = R\Delta T \text{ (per mole)}$$

Therefore, the total energy that would raise the temperature of one mole of a monoatomic ideal gas by one degree under isobaric conditions, would be:

$$3/2\ R\Delta T + R\Delta T = (3/2\ R + R)\ \Delta T$$

So,
$$C_p = 5/2\ R\ \checkmark$$

We should be able to see and corroborate that the difference between the two values for the molar heat capacity (isobaric and isochoric) is given by the universal gas constant, R.

---

[1] Using an increase from 300 K to 301 K, for example.

Chapter 15: Thermodynamics

$$C_p - C_v = R \quad \checkmark$$

We should point out that the experimental heat capacities for monoatomic gases (which would be the noble gases) are indeed 3/2 R and 5/2 R for C_v and C_p, respectively. However, for larger molecules[1], they are different. However, the difference between the two values is still equal to R.

Let's go back to the ideal monoatomic gas. Remember that for this case, the average translational energy is given by 3/2 RT.

$$\therefore E = 3/2 \, RT$$

If we change the temperature, then:

$$\Delta E = 3/2 \, R \Delta T$$

Since 3/2 R corresponds to the molar heat capacity under conditions of constant volume, then:

$$\Delta E = n \, C_v \, \Delta T. \quad \text{for } \mathbf{n} \text{ moles of gas.}$$

Let's derive a similar expression for ΔH. We should recall that

$$\Delta H = \Delta E + n R \Delta T$$

$$\Delta H = n \, C_v \, \Delta T + n \, R \, \Delta T = n \, (C_v + R) \, \Delta T$$

Finally:

$$\Delta H = n \, C_p \, \Delta T.$$

These two equations will allow us to determine the change in internal energy and the change in enthalpy for a certain amount of an ideal gas[2] as it undergoes a <u>physical transformation</u> that does not involve a phase change. Notice that according to these equations, the only way to tell when the internal energy for a system has changed during these type of physical transformations is to check the temperature before and after. If it remains constant, then the changes in both of these state functions would be zero.

- Consider 1.00 mole of nitrogen gas (C_p = 29.0 J/mol K) at STP. Let us add heat to the system while allowing it to expand isobarically against an external pressure of 1.00 atm until

---

[1] For example, the C_v and C_p for nitrogen gas are 20.71 and 29.03 J/molK, respectively.
[2] Although it was derived for a monoatomic idea gas, it works for all gases. The only problem would be to actually look up the values of the heat capacities for polyatomic gases.

the temperature of the system reaches 373 K.

a) Determine the final volume of the system.

We notice immediately that this is an isobaric process. This means that the initial pressure of the gas is equal to the final pressure and also to the constant external pressure. Therefore, all the pressures are 1.00 atm. We know this because we are assuming that the nitrogen gas is an ideal gas and one mole of any ideal gas at STP[1] occupies a volume of 22.4 L at a pressure of 1.00 atm and a temperature of 273 K.

Since the gas is ideal, then we can calculate the final volume from the final conditions, which we already know, using PV = nRT. Therefore,

$$V_2 = \frac{nRT_2}{P_2}$$

$$V_2 = \frac{1.00 mol \times 0.0821 \frac{Latm}{Kmol} \times 373 K}{1.00 atm}$$

So, $V_2 = 30.6 L$ ✓

b) Determine the ΔH for this transformation.

The ΔH for this process can be determined with the following equation.

$$\Delta H = n\, C_p\, \Delta T$$

Substitution of the values will give us:

$$\Delta H = 1.00 \text{ mol } (29.0 \text{ J/molK}) (373K - 273K)$$

$$\therefore \Delta H = +2,900 \text{ J} \checkmark$$

We should point out that the value of ΔH is also equal to q_p, and since this process is truly isobaric, then the heat absorbed by the process is also equal to +2,900 J, that is: q = q_p = ΔH.

c) Determine the ΔE for the transformation.

---

[1] See Section 4.2.

There are at least two different ways in which we can solve this problem. The easiest is to substitute directly into this equation, which applies to physical processes.

$$\Delta E = n\, C_v\, \Delta T$$

and substituting, we get:

$$\Delta E = 1.00 \text{ mol } (20.7 \text{ J/molK}) (100 \text{ K})$$

$$\Delta E = +2{,}070 \text{ J} \quad \checkmark$$

We can use the other way to solve this problem as a proof that this is accurate. Since we already know q (remember it is the same as ΔH since it is isobaric), let's determine W, so that we can add them up to find ΔE.

$$W = -P_{ext}\, \Delta V$$

and substituting, we get:

$$W = -1.00 \text{ atm } (30.6 \text{ L} - 22.4 \text{ L}) \times 101.3 \text{ J/Latm}$$

$$W = -830.4 \text{ J} \quad \checkmark$$

Since $q = q_p = +2{,}900$ J, and $\Delta E = q + W$, then:

$$\Delta E = +2{,}900 \text{ J} - 830 \text{ J}$$

$$\therefore \Delta E = +2{,}070 \text{ J} \quad \checkmark \qquad \text{QED}[1]$$

15.6 Chemical Transformations.

Let's apply all of this to chemical transformations. In a chemical transformation, we will be converting chemical energy (trapped in the bonds) into heat. We recognize the reactants as the initial state, and the products as the final state. Therefore, any change is visualized as a transformation from the reactants to the products. Just as before, the first law of thermodynamics tells us that $\Delta E = q + W$. How do we observe expansion work (PV work) during a chemical transformation?

$$\Delta H = \Delta E + \Delta(PV)$$

The only substances that have an appreciable compressibility factor are gases, so we will only consider gases in the term $\Delta(PV)$. If we assume these gases to be ideal, which is an excellent approximation under normal conditions, then $PV = nRT$ and we can rewrite the last equation as:

---

[1] Abbreviation for the Latin phrase: *"Quod Erat Demonstrandum"* which means, which has been demonstrated.

$$\Delta H = \Delta E + \Delta(nRT)$$

Since chemical reactions take place at constant temperature[1], then we can take R and T out of the parenthesis and we get the equation in the final form seen below:

$$\Delta H = \Delta E + RT\Delta n_{gases}$$

In this equation, Δn is the moles of gases in the product side minus the moles of gases in the reactant side.

- Consider the following reaction which is the combustion of liquid benzene. Given the following: $\Delta H° = -3269$ kJ, determine the $\Delta E°$ for this reaction.

$$C_6H_{6(\ell)} + 15/2\ O_{2(g)} \Rightarrow 2\ H_2O_{(\ell)} + 6\ CO_{2(g)}$$

We are told that the heat of combustion for benzene is $\Delta H° = -3269$ kJ. This number could be calculated[2] as we learned back in Section 5.7. Since both ΔE and ΔH are state functions, we can find one if we know the other with the equation derived above. It is a typical misconception to think that we cannot calculate $\Delta E°$ unless we carry out the process under conditions of constant volume (isochoric). From the definition, it is true that to *measure* $\Delta E°$, we need to carry out the reaction in an isochoric calorimeter, but to calculate its value, all we need it to know is the initial and the final states (the reactants and the products).

$$\Delta H° = \Delta E° + RT\Delta n_{gases}$$

And what is Δn for the gases? From the balanced chemical reaction, we can see that:

$$\Delta n_{gases} = 6 - 15/2 = -3/2\ mol[3]$$

$$\therefore \Delta E° = \Delta H° - RT\Delta n_{gases}$$

and: $\qquad \Delta E° = -3269 \times 10^3 J - (8.314\ J/mol\ K)(298K)(-3/2\ mol)$

$$\therefore \Delta E° = -3265\ kJ\ \checkmark$$

---

[1] As bonds are broken or formed, energy is released or absorbed. Once the reaction is over, it is indeed possible for the substances present to absorb any amount of heat that was released and end up at a higher temperature. However, the reaction itself took place at a constant temperature.

[2] Using the heats of formation of the products and of the reactants.

[3] The negative sign implies that there was a contraction as we went from reactants to products.

As we can see, there is not a very large difference between the $\Delta E°$ and the $\Delta H°$ for this reaction and this is almost always true. The work of expansion (which is basically the difference between $\Delta E°$ and $\Delta H°$) is normally small compared to the usually much larger values for $\Delta H°$. However, this would not be the case if the $\Delta H°$ for a reaction were close to zero or not much more than a few kJ/mol.

15.7 <u>Entropy</u>.

The first law of thermodynamics talks only about the amount of heat liberated or absorbed by the system during a transformation and the work involved in that same process, but it will not tell us whether the process will actually occur or not. Let's say we have a system that consists of a container with two identical halves. One of these halves is filled with a given gas at a given pressure, whereas the other half has a vacuum (it is completely empty). This system is immersed in a constant temperature bath. What would happed if we were to open the partition that separate the two halves? Everyone knows that the gas will spontaneously distribute evenly in between the two halves.

Why did that happen? Let's see if the first law of thermodynamics provides either an answer. Keeping in mind that the expansion took place under conditions of constant temperature, we can then calculate all of the thermodynamic properties from the first law.

$$\Delta E = n\, C_v\, \Delta T = 0$$

$$\Delta H = n\, C_p\, \Delta T = 0$$

W = 0, since it expanded against a vacuum.

Since $\Delta E = q + W$, then q = 0.

According to the first law, there is no need for this expansion to take place, since all of these parameters did not change at all. However, we know that the process still occurs spontaneously. What is the driving force for this process to take place? Certainly, none of the parameters from the first law has anything to do with it, since none of them changed. There must be some other property of the system that will explain why this occurs. As we shall see, this new property is related somehow to probability.

Chapter 15: Thermodynamics

Consider two identical flasks that are connected and imagine that we have only one molecule in these flasks. Where is this molecule most likely found? Well, let's try to visualize all the possible arrangements (also called microstates[1]) for the molecule (let's call it A). There are two possible microstates, as seen in the diagram below. What is the probability of finding the molecule in the left flask? It is certainly 50% (or 1/2).

Now, let's imagine that we have two molecules in this system. What are all the possible microstates for this system? The diagram below shows that there are four possibilities in which we can arrange these two molecules (A and B). Please, notice that we have two identical microstates that describe having one molecule in each flask. What is the probability of finding both molecules in the left flask? It is one possible microstate out of four, therefore, the probability is 1/4 (1/4 = 1/2 x 1/2), or 25 %.

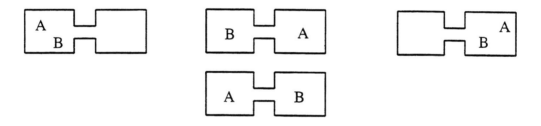

Let's extend this to one more molecule in the system, therefore, consider having three molecules (let's call them A, B, and C) in this system. This time there are eight possible microstates, so therefore, the probability of finding all three molecules in the left flask is one in eight (1/8 = 1/2 x 1/2 x 1/2), or 12.5 %.

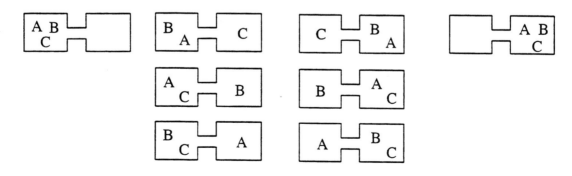

---

[1] Each possible but different arrangement is referred to as a **microstate**.

In general, the probability of finding all the particles in the left flask is given by the expression[1]:

$$\text{Probability for all molecules in the left flask} = \left(\frac{1}{2}\right)^n$$

where: n = number of molecules in the flask.

If we followed this, it should be obvious that any event[2] is in theory possible. However, the more molecules we have in a system, the **less likely** it is going to be for all the molecules to get "organized" into one side. Actually, from our discussion it follows that the more probable microstates are those that are more randomized (more disordered).

We will define **entropy** (symbolized with the letter **S**) as the property of the system that will help us explain why the original expansion took place. Entropy is a measure of the <u>degree of disorder or randomness</u> in the system. In nature we observe that disordered states are more favored than ordered states. The reason is probabilistic, as we just saw. This concept goes hand in hand with the theory of an expanding universe. As the universe expands, it gets more and more random, therefore, any event which directly or indirectly helps to increase the randomness of the universe will be favorable or *spontaneous*. If we were to do something that is not spontaneous, we would have to do work in order to achieve it and upon doing so, *we* would increase the disorder of the universe. For our purposes, if a reaction is not spontaneous, we could theoretically force it to occur by doing work.

Entropy is a state function, which means that it depends on the difference between the final and the initial state, and not on the path.

$$\therefore \Delta S = S_f - S_i$$

It should be clear that entropy must be related to heat because when you add heat to a system, molecules move faster and the randomness will therefore increase. However, entropy cannot just be equal to heat (q) because q is not a state function. This implies that there must be some restrictions on the heat (q) to make it a property of the system.

In order to originally define entropy for a physical process, let's consider a reversible and

---

[1] This expression is based on that example, that is, having only two chambers. If the container had three chambers, the equation would be the same, but instead of being 1/2, it would be 1/3 to the nth power.
[2] For example, having all the air molecules from the room go into one corner and stay there while we suffocate to death. There is a microstate that predicts this particular phenomenon, but it is one out so many that the probability for such an event is infinitesimally small.

isothermal expansion of an ideal gas. We have already seen that in this case, the work calculated happens to be the maximum work possible for this system and it was given by the following equation.

$$W_{max} = -nRT \ln[V_2/V_1]$$

Now, since the process we are considering is isothermal, then ΔE must be equal to zero. Therefore, $q = -W_{max}$. This quantity is referred to as the reversible heat for the process, and it is equal but opposite in sign to the maximum work: $q_{rev} = -W_{max}$.

This quantity (q_{rev}) is a property of the system and it represents the maximum amount of heat that the system can absorb as it goes from the initial to the final state and it corresponds to the one calculated during a reversible process. In reality, we will observe that most processes are not reversible, therefore, the system actually absorbs less heat.

$$q_{actual} \leq q_{rev}$$

We will <u>define</u> the ΔS for an *isothermal process* as:

$$\Delta S = q_{rev}/T$$

Keep in mind that this equation only works if the temperature is constant, otherwise we would not know what to substitute for the temperature.

- Assume that one (1.000) mole of an ideal gas expands reversibly from an initial pressure of 2.000 atm to a final pressure of 1.000 atm at a constant temperature of 273 K. The diagram below shows these facts. For this process, determine q, W, ΔE, ΔH, and ΔS.

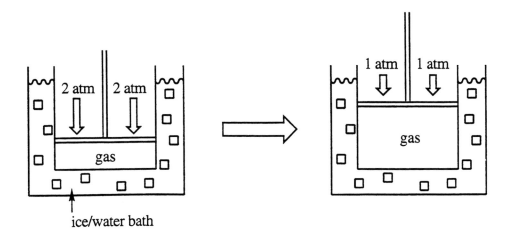

Since it is isothermal, then: $\Delta E = \Delta H = 0$. However, $\Delta E = q + W$, so $q = -W$. The fact that it is also reversible tells us that the work can be calculated with the equation shown below.

$$W = W_{max} = -nRT \ln[V_2/V_1]$$

There is one slight problem: we do not know the value for the volumes. Of course, we could determine these with the ideal gas equation of state, but it would be a lot easier if we noticed that since the temperature is constant and the gas is ideal, this gas should obey Boyle's Law: $P_1V_1 = P_2V_2$. Therefore, we should get the following expression instead.

$$W_{max} = -nRT \ln[P_1/P_2]$$

$$W_{max} = -(1.000 \text{ mol})(8.314 \text{ J/molK})(273 \text{ K}) \ln\{(2.00 \text{ atm})/(1.00 \text{ atm})\}$$

$$W_{max} = -1573 \text{ J}. \checkmark$$

Of course, since the heat is exactly the same as the work, except with opposite signs, then:

$$q_{rev} = +1573 \text{ J}. \checkmark$$

Finally, the calculation of ΔS yields:

$$\Delta S = q_{rev}/T = +1573 \text{ J} / 273 \text{ K}$$

$$\Delta S = +5.76 \text{ eu}[1] \checkmark$$

In general, the change in entropy for a transformation will be:

Positive: if there is an increase in disorder, which is what we observe when we have an expansion.

Negative: if there is a decrease in disorder, which is what we observe when we have a contraction.

Some processes in nature do resemble isothermal reversible cases and they are transformations that occur during a phase change[2]. This is true because in a phase change, we

---

[1] These is an abbreviation for entropy units. Technically, an eu corresponds to Joules per degree. Therefore, 1 eu = 1 J/K.
[2] For example: the melting process (solid to liquid), the boiling process (liquid to gas), and the sublimation process (solid to gas).

have an equilibrium between the two phases that are being interconverted, as long as it occurs at the temperature at which it is supposed to occur. For example, as long as we are melting ice at 0°C, we have the liquid and the solid in equilibrium with one another.

$$H_2O_{(s)} \rightleftharpoons H_2O_{(\ell)}$$

Since this is a truly reversible and isothermal process, then the heat we measure in the laboratory IS also the reversible heat (that is, $q_{measured} = q_{rev}$). We learned in Section 10.9 that the heat associated with this process is given by the following expression.

$$q = mass\ \Delta H_{fusion}$$

Therefore, due to the reversibility of the fusion process, we can now rewrite this equation as:

$$q_{rev} = mass\ \Delta H_{fusion}$$

- Consider the conversion of 2.00 grams of ice into 2.00 grams of liquid water at 0°C. Calculate the ΔS for this process.

$$\Delta S = q_{rev} / T$$

but: $\quad q_{rev} = mass\ \Delta H_{fusion}$

$$= (2.00\ g)(333\ J/g) = 666\ J$$

$$\Delta S = 666\ J / 273\ K = +2.44\ eu\ \checkmark$$

a) The answer to that problem was positive. Is that what we should have expected?

Of course! The liquid state is much more disordered than the solid state for any substance, therefore, we knew that there had to be an increase in the entropy of the system if we were going to melt the solid.

15.8 Entropy Changes for Other Physical Processes.

⇒ Reversible, Isothermal Expansion (IG)

The change in entropy is really defined as:

$$\Delta S = \int \frac{dq_{rev}}{T}$$

What is dq_{rev}? Since the process is isothermal, it follows that $\Delta E = 0$ ($nC_v\Delta T$). Therefore, $q_{rev} = -W_{max}$. But in the past we have seen what this expression is equal to.

$$W_{max} = -\int \frac{nRT}{V} dV$$

So,

$$dW_{max} = -\frac{nRT}{V} dV$$

Or,

$$dq_{rev} = +\frac{nRT}{V} dV$$

Finally, substituting we get:

$$\Delta S = \int \frac{nRT}{V\,T} dV$$

and:

$$\Delta S = nR\ln(V_2/V_1) \quad \checkmark$$

$\Rightarrow$ **General Expressions for ΔS (Physical).**

The equation above is too restrictive, because it only works if the transformation is isothermal. It would be convenient to have a more general equation that would work always, as long as there is no phase change involved. The two equations below, consider the possibility that not only the volume (or pressure) changes, but also the temperature. Notice that when the transformation is isothermal, the first term on either equation disappears, and in the case of the first one, it actually ends up with the same equation we derived above.

$$\Delta S = nC_v \ln(T_2/T_1) + nR\ln(V_2/V_1)$$

$$\Delta S = nC_p \ln(T_2/T_1) - nR\ln(P_2/P_1)$$

Since ΔS is a state function, we can use either one of these two equations in order to determine the ΔS for a physical process, as long as it does not involve a phase transformation. Don't forget that if we are talking about a phase transformation, ΔS is given by:

$$\Delta S_{transformation} = \frac{\Delta H_{transformation}}{T}$$

Chapter 15: Thermodynamics

15.9 Second Law of Thermodynamics.

We have already seen that the actual heat absorbed or released for a real process is always less than, or at best equal to, the reversible heat.

$$q_{actual} \leq q_{rev}$$

$$\{q_{actual}/T\} \leq \{q_{rev}/T\} = \Delta S$$

$$\therefore \Delta S \geq q_{actual}/T$$

If we consider an isolated system (no exchange of matter nor energy), then q = 0, and so:

$$\therefore \Delta S \geq 0 \checkmark$$

This is a statement of the second law of thermodynamics. *The entropy of an isolated system can never decrease.* Another way of stating this law would be, *in every possible process, the entropy of the universe increases.*

The second law can will allow us to predict if a process is going to be spontaneous[1] or not. If the ΔS of the universe is calculated to be a positive quantity, then it will be spontaneous. However, if the ΔS of the universe is calculated to be negative, then the process is not spontaneous. Therefore, the whole idea is to be able to determine the sign of the ΔS for the universe in order to decide whether a process will be spontaneous or not. This is not the easiest thing in the world to do, however.

It would be ideal to find another thermodynamic parameter for the system that would allow us to do just that, so that we would not have to deal with the calculation of a parameter for the universe. Let us look at the following scenario, which is going to allow us to define this new parameter.

Consider the following isolated system (like a small universe on its own). Notice that in this particular setting, we have the isolated system (the universe) subdivided into two parts: the smaller system (where a reaction may take place) and the surroundings, which in this case happen to be an ice/water constant temperature bath.

---

[1] A spontaneous process is one that occurs without any intervention on our part.

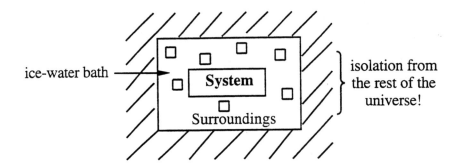

For any isolated system, the sum of the ΔS for the system and the one for the surroundings would have to be equal to the total ΔS (that of the "universe").

$$\Delta S_{tot} = \Delta S_{system} + \Delta S_{surroundings}$$

However, in this particular case, the surroundings happen to be a constant temperature ice/water bath (T = 273 K).

$$\Delta S_{tot} = \Delta S_{system} + \Delta S_{bath}$$

The ΔS for the bath can be calculated, since it involved in a reversible, isothermal process.

$$\Delta S_{bath} = q_{rev}/T = q_{actual}/T = \Delta H_{bath}/T$$

This whole thing is isolated, so if the bath gains so many Joules of heat, the system must have given off the exact same number of Joules of heat, and vice versa. Therefore,

$$q_{bath} = -q_{system}$$

which can be rewritten as:

$$\Delta H_{bath} = -\Delta H_{system}$$

Substituting this into the expression for the ΔS for the bath, we get:

$$\Delta S_{bath} = -\Delta H_{system}/T$$

and finally,

$$\Delta S_{tot} = \Delta S_{system} - \Delta H_{system}/T$$

Let's analyze what we have accomplished so far. We have written an equation that allows us to calculate the total ΔS for this isolated system, which for us it would translate to the

ΔS for the universe, in terms of parameters of the system itself. In order to continue with this derivation, let's drop the subscript system and multiply on both sides by (-T).

$$-T\Delta S_{tot} = -T\Delta S + \Delta H$$

We will define a new thermodynamic parameter: the Gibbs Free Energy (ΔG) as:

$$\Delta G = -T\Delta S_{tot}$$

and therefore,

$$\Delta G = \Delta H - T\Delta S.$$

This is the equation we were looking for from the start. It defines the thermodynamic parameter we wanted, ΔG, which will let us determine whether a process is going to occur spontaneously or not. The Gibbs Free Energy is a property of the system itself, but is defined in terms of (or is related to) the ΔS for the universe.

To determine if a reaction is spontaneous or not, all we need to do is determine the sign of ΔG. If ΔG is negative, that means that the ΔS of the universe is positive and therefore, the reaction will be spontaneous. If ΔG is positive, then the ΔS of the universe is negative and therefore, the reaction is not spontaneous in the direction written. Finally, if the ΔG is zero, that tells us that the ΔS of the universe is equal to zero and that implies an equilibrium condition.

The ΔG will also tell us the maximum work that we will ever obtain from the system. We call it the free energy available to do useful work and since no system is really reversible, we will never get that much work out of our system.

- One mole of an ideal gas at a temperature of 25°C expands isothermally and irreversibly against a constant pressure of 1.00 atm from a volume of 5.0 L to a volume of 20.0 L. Calculate W, q, ΔE, ΔH, ΔS and ΔG. What is the difference between W and ΔG; what does it mean?

    Calculating ΔE and ΔH is simple because the process is isothermal ($\Delta T = 0$). Therefore, both ΔE and ΔH are equal to zero. Therefore,

$$\Delta E = q + W = 0$$

$$\therefore W = -q = -P_{ext}\Delta V$$

$$W = -1.00 \text{ atm } (15.0 \text{ L}) (101.3 \text{ J/Latm})$$

$$W = -1520 \, J \quad \checkmark$$

Since q = - W, then:
$$q = +1520 \, J \quad \checkmark$$

In order to determine the ΔS, we can use of the two equations we proposed at the end of Section 15.9, and in this case, since we know both volumes, we would use this one.

$$\Delta S = n \, C_v \ln (T_2/T_1) + n R \ln (V_2/V_1)$$

In an isothermal process, $T_2 = T_1$, therefore, their ratio is equal to one, and the natural logarithm of one is equal to zero. This equation, therefore, can be simplified.

$$\Delta S = n R \ln (V_2/V_1)$$

$$\Delta S = 1.00 \text{ mol} \, (8.314 \, J/molK) \ln \{20.0L/5.0L\}$$

$$\Delta S = 11.5 \, eu. \quad \checkmark$$

Finally,
$$\Delta G = \Delta H - T\Delta S$$

And since ΔH is equal to zero, then:

$$\Delta G = - T\Delta S$$

$$= - 298 \, K \, (11.5 \, J/K)$$

$$\Delta G = - 3430 \, J \quad \checkmark$$

What kind of information can we extract out of that number? It is telling us, first of all, that the process is spontaneous because it is negative. The number itself means that the system **has 3430 J** of free energy available to do work. However, we know that the system **only used up 1520 J** of energy in order to expand (as it pushed the surrounding atmosphere, W). We might wonder, what happened to the other 1910 J (3430 - 1520) of free energy? It was lost to the surroundings. The free energy that is not used up by the system is lost forever and the universe uses it in order to expand. The only way that the universe could take that amount of free energy that it has trapped in the form of its own entropy and give it back would be for the universe to contract. This means that as the universe is expanding, it is constantly trapping some of the available energy of the universe in the form of entropy. As this occurs, the universe continues to get colder and colder and eventually, it will die. Of course, this will take a very long time, so it is not something we should be concerned about.

Chapter 15: Thermodynamics

15.10 Entropy Changes in Chemical Processes.

From now on, let's concentrate on chemical reactions rather than on physical processes. In order to determine if a chemical reaction is spontaneous or not, we need to calculate ΔG (its sign). However, to calculate ΔG, we need to know the values for ΔH and ΔS. How do we determine the $\Delta H°$ for a chemical reaction? We actually learned how to do this in Section 5.7, mainly:

$$\Delta H° = \Sigma \Delta H°_{f(prod)} - \Sigma \Delta H°_{f(react)}$$

Actually, for any thermodynamic state function, X:

$$\Delta X° = \Sigma \Delta X°_{f(prod)} - \Sigma \Delta X°_{f(react)}$$

However, we do not find a table of $\Delta S°_f$ in the thermodynamic tables. WHY? Because there is a way to determine the absolute value of the entropy. This is the only thermodynamic property for which this is possible. We can do this because of the Third Law of Thermodynamics, which states:

The entropy of a perfect crystal of a pure substance at 0 K is equal to zero.

The third law is reasonable because at 0 K, the lowest attainable temperature, the pure crystal is as ordered as it will ever be, therefore it should have the lowest possible entropy: zero.[1] A table of selected thermodynamic parameters can be found in Appendix #1. In there we will see the heats of formation, the free energies of formation, and the absolute entropies of many substances at 25°C and 1 atm pressure.

There are two factors that contribute to the magnitude of the absolute entropy, as reported on that appendix. These are:

1) <u>State of Matter</u>. The magnitude of the ΔS will depend on whether it is a solid, a liquid or a gas. Of course, the most ordered of all three phases is the solid phase, so it would have the lowest entropy, whereas the least ordered phase of all three is the gas phase. Therefore, the gas phase should have the highest entropy. We can see from the table that this is true. The absolute entropy of the gases are larger than those of liquids or solids.

2) <u>Size or Molecular Weight</u>. The heavier or bigger the molecules are, the larger the value for

---

[1] There is no such thing as a perfectly ordered crystal, so the definition simply states that it is as ordered as it will ever be for that particular system. In reality, the crystal will have, at the very least, some vacancies (spots in the lattice where atoms or molecules are actually missing). These are observed in every single crystal.

the absolute entropy. This is not as easy to explain at this level, so we will only use an analogy. Imagine we design an experiment where we can administer a certain dosage of a given drug to different animals that will make them go totally wild for an exact period of 5 minutes. Let us carry this analogy into a small glass figurines shop, where there are literally hundreds of these very small and delicate glass ornaments. Now, let's inject a small field mouse with the appropriate dosage and allow it to run wild and free for the 5 minutes, after which it will completely stop and we will retrieve it to assess the damage. Let's repeat the experiment, but this time we will use a ferocious bull and allow it run wild and free for the exact period of time. Which animal causes the greatest disorder in the shop? Clearly, the larger one: the bull.

Many times it could be confusing to try to decide which one has a higher entropy based on size and or molecular weight. Many times, the complexity of the molecule overshadows the molecular weight, as long as they are not tremendously different. For example, the absolute entropies of carbon dioxide (MW = 44.0) and ethane (MW = 30.0) gases are: 213.7 and 229.5 J/Kmol, respectively. The absolute entropy for ethane is larger simply because it has many more atoms (8 atoms) than the carbon dioxide molecule (3 atoms), while their molecular weights are not very different. Another example is the case of hydrogen gas (2 atoms, MW = 2.02) as compared to helium gas (1 atom, AW = 4.00). Once more, the absolute entropy for the hydrogen is 131, while that of the helium is 126 J/Kmol. Clearly, the fact that the hydrogen is diatomic overtakes the small difference in molecular weights.

Once we know the absolute entropies from that table, we can easily determine the ΔS for any reaction.

$$\Delta S° = \Sigma S°_{(prod)} - \Sigma S°_{(react)}.$$

- Is the following reaction spontaneous or not at 298 K?

$$N_{2(g)} + 3 H_{2(g)} \Rightarrow 2 NH_{3(g)}$$

In order to answer this question, we need to know the sign for the $\Delta G°$ for the reaction. There are two ways to achieve this. First, we could determine the values of $\Delta H°$ and $\Delta S°$ for the reaction and then calculate $\Delta G°$ with the equation: $\Delta G° = \Delta H° - T\Delta S°$. However, a quicker way would be to determine the value of $\Delta G°$ directly by using the free energies of formation, which can also be found in Appendix #1. In this case:

$$\Delta G° = 2\,\Delta G°_{f(NH3)}$$

substituting the values, we get:
$$\Delta G° = 2(-17\text{ kJ/mol})$$

$$\therefore \Delta G° = -34\text{ kJ} \quad \checkmark$$

There is no doubt that this is a spontaneous reaction (negative sign for the change in free energy). Try to prove that the answer would be the same if we were to do it the other way.

- Determine whether iron (III) oxide (rust) will decompose spontaneously to form iron metal at 25°C and 1 atm pressure.

$$2\text{ Fe}_2\text{O}_{3(s)} \Rightarrow 4\text{ Fe}_{(s)} + 3\text{ O}_{2(g)}$$

In order to answer this question, once more all we need is to determine the sign for the $\Delta G°$ for this reaction. To do this, we go back Appendix #1.

$$\Delta G° = 4\,\Delta G°_{f(Fe)} + 3\,\Delta G°_{f(O2)} - 2\,\Delta G°_{f(Fe2O3)}.$$

The first two are zero, because they are elements in their standard states. Therefore, the calculation yields:

$$\Delta G° = -2\,\Delta G°_{f(Fe2O3)}.$$

$$\Delta G° = -2\,(-742.2\text{ kJ})$$

$$\Delta G° = +1{,}484.4\text{ kJ} \quad \checkmark$$

The positive sign on the free energy tells us that this is NOT a spontaneous reaction or that rust will not spontaneously revert back to iron and oxygen.

- Could we reduce the iron oxide into iron metal spontaneously by adding carbon graphite in the presence of O_2 gas? In other words, will this reaction be spontaneous under these conditions?

$$2\text{ Fe}_2\text{O}_{3(s)} + 3\text{ C}_{(s)} \Rightarrow 4\text{ Fe}_{(s)} + 3\text{ CO}_{2(g)}$$

In order to solve this problem, we determine the $\Delta G°$ for the reaction as follows:

$$\Delta G° = 4\,\Delta G°_{f\,Fe} + 3\,\Delta G°_{f\,CO2} - [\,2\,\Delta G°_{f\,Fe2O3} + 3\,\Delta G°_{f\,C}\,]$$

$$= 4\,(0) + 3\,(-394\text{ kJ}) - [\,2\,(-742\text{ kJ}) + 3\,(0)\,]$$

$$\therefore \Delta G° = +\,302\text{ kJ} \quad \checkmark$$

As we can see, it is still NOT spontaneous. Why is this then used as a viable industrial process? Because the term TΔS is large enough that it is easily influenced with an increase in temperature. Let's determine the minimum temperature at which this reaction can become "spontaneous". Assume that the $\Delta H°$ for the reaction is +468 kJ and that the $\Delta S°$ is 560 J/K, and also assume that these values are temperature independent.[1]

$$\Delta G° = \Delta H° - T\Delta S°$$

$$0 = +468\text{ kJ} - T\,(0.560\text{ kJ/K})$$

$$\therefore T = 836\text{ K} \text{ or } 563°C. \quad \checkmark$$

Therefore, this reaction can be made to favor the products if the temperature is higher than 563°C, and this is easily achieved in an industrial furnace.

15.11 Calculation of Absolute Entropies.

How can we calculate the absolute entropies that we find in Appendix #1? All we know is that the absolute entropy of anything that can be made pure is exactly equal to zero at the absolute zero of temperature. Of course, the table is NOT at 0 K, but at 298 K. This means that we should be able, in theory, to determine the INCREASE in entropy for any substance as we go from 0 K all the way up to 298 K. In order to do this, we must think very carefully about what is happening to the this substance as the temperature increases. Remember that if the substance is a gas at 298 K, for example, then it must undergo a transformation from solid, to liquid, to gas. This cannot be done at random, but the phase conversions must occur at the normal temperature at which these substances melt and boil. Basically, we can divide the calculations into two major areas: the entropy increase as you warm up of a pure substance without a phase change, and the entropy increase when the phase changes. For the first case, we can use the equation:

$$\Delta S = n\,C_p \ln(T_2/T_1) - n\,R \ln(P_2/P_1)$$

and for the latter case, we can use the equation:

$$\Delta S = \Delta H_{\text{transformation}}/T$$

---

[1] This is not exactly true, but it is not a bad assumption either.

Chapter 15: Thermodynamics

If we limit ourselves to a constant pressure of 1 atm, then the two equations become:

$$\Delta S = n\, C_p \ln(T_2/T_1) \quad \text{and} \quad \Delta S = \Delta H_{transformation}/T.$$

- Explain how we would calculate the molar (for one mole) absolute entropy for Chlorine gas, given the following information.

Heat Capacities: Solid = 28.8 J/molK   Liquid = 67.7 J/molK   Gas = 34.2 J/molK

Transformations: Melting point = 172 K   Boiling point = 239 K

   Heat of Fusion = 6.402 kJ/mol   Heat of Vaporization = 20.41 kJ/mol

At this point we can envision a process where the temperature will rise from 0 K all the way up to 298 K, going through two clear transformations (from solid to liquid at 172 K and from liquid to gas at 239 K). We can visualize this with the following graph, which shows the change in entropy as the temperature increases. Remember that in the phase transformations mentioned above, the temperature remains constant until transformation is completed.

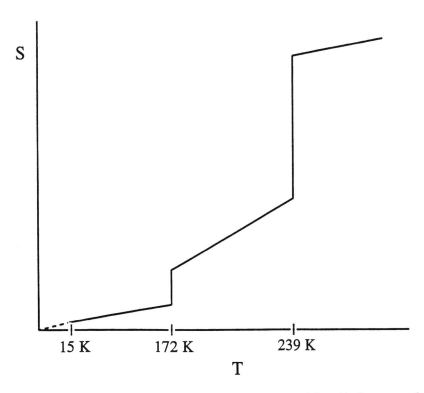

Notice that there is an extrapolation from about 15 K all the way down to zero Kelvin. The reason for this is simple: We cannot calculate ΔS from 0 K to 172 K using the equation below and using the initial temperature as 0 K.

$$\Delta S = n\, C_p \ln(T_2/T_1)$$

However, it has been confirmed that at very low temperatures, the heat capacity (C_p) follows the Debye "T-cubed" law accurately.

$$C_p = a\, T^3. \qquad a = \text{constant}$$

Thus, the ΔS at low temperatures can be calculated by the Debyes extrapolation. Therefore, we can divide the calculation into six parts.

1) The change in entropy from 0 K to 15 K given by the Debye extrapolation. This value will always be given to us, as we have not learned how to determine it. In this case, it is equal to 0.37 eu.[1]

2) The change in entropy from 15 K to 172 K.

3) The change in entropy at 172 K as we convert the solid into liquid.

4) The change in entropy from 172 K to 239 K.

5) The change in entropy at 239 K as we convert the liquid into a gas.

6) The change in entropy from 239 K to 298 K.

The calculations of each step yields the following results: Step #1: 0.37 eu (given). Step #2: 70.26 eu. Step #3: 37.22 eu. Step #4: 22.27 eu. Step #5: 85.40 eu. Step #6: 7.55 eu. The reported value for the absolute entropy for chlorine gas (the number that appears in the table) would therefore be the sum of all these parts: *223.07 eu.*

15.12 Dependence of Free Energy on Pressure.

The free energy does have a dependence on the pressure of the gas. Mathematically, this dependence is given by the following equation.

$$G = G^\circ + R\, T \ln(P)$$

In this equation, G° is the free energy when the pressure is standard (1 atm), R is the universal gas constant (8.314 J/molK), T is the absolute temperature, and G is the free energy when the pressure is equal to P (a pressure other than the standard pressure of 1 atm). At a pressure of one atmosphere, the equation becomes $G = G^\circ$, which should make sense, since that is the way we defined G°.

---

[1] This is always a very small correction, since the entropies must be very small at those low temperatures.

Chapter 15: Thermodynamics

Let's apply this equation to the chemical process shown below.

$$N_{2(g)} + 3\,H_{2(g)} \Rightarrow 2\,NH_{3(g)}.$$

Since ΔG is a state function, then it follows that:

$$\Delta G = \Sigma G_{products} - \Sigma G_{reactants}.$$

$$\Delta G = 2\,G_{NH3} - G_{N2} - 3\,G_{H2}.$$

But from what we just said:

$$G_{NH3} = G°_{NH3} + RT \ln(P_{NH3})$$

$$G_{N2} = G°_{N2} + RT \ln(P_{N2})$$

$$G_{H2} = G°_{H2} + RT \ln(P_{H2}).$$

So, substituting above we get the following expression.

$$\Delta G = 2[G°_{NH3} + RT \ln(P_{NH3})] - [G°_{N2} + RT \ln(P_{N2})] - 3[G°_{H2} + RT \ln(P_{H2})].$$

Rearranging the equation, we get:

$$\Delta G = 2\,G°_{NH3} - G°_{N2} - 3\,G°_{H2} + RT\,[2\ln(P_{NH3}) - \ln(P_{N2}) - 3\ln(P_{H2})]$$

Of course, the standard $\Delta G°$ must be equal to:

$$\Delta G° = 2\,G°_{NH3} - G°_{N2} - 3\,G°_{H2}$$

Therefore,

$$\Delta G = \Delta G°_{reaction} + RT\,[2\ln(P_{NH3}) - \ln(P_{N2}) - 3\ln(P_{H2})]$$

From elemental algebra, we know that: $2\ln(a) = \ln(a)^2$

and $\ln a + \ln b - \ln c = \ln\{(a \cdot b)/c\}$

Therefore,

$$\Delta G = \Delta G°_{reaction} + RT \ln Q$$

where:

$$Q = \frac{P^2_{NH_3}}{P_{N_2}\,P^3_{H_2}}$$

This ratio, Q, is known as the *reaction quotient* and it gives us an idea of where we are along in the reaction. If we start with pure reagents (nitrogen and hydrogen gases), then Q will be equal to zero. As the reaction progresses, the pressures of nitrogen and hydrogen gases start to decrease, while that of ammonia starts to increase. Therefore, Q starts to increase. If the reaction went 100 % to the right (all the way towards the formation of product – ammonia), then we would be dividing by zero, which although it is not defined, has a limit of infinity. Therefore, the equation highlighted above is basically telling us how the free energy changes as the reaction progresses. Let's investigate this a little further.

15.13 Equilibrium.

Consider the following equilibrium reaction:

$$N_2O_{4(g)} \rightleftharpoons 2\,NO_{2(g)} \qquad \rightleftharpoons \text{ means equilibrium}$$

If we start the reaction by placing pure N_2O_4 in a flask, the reaction will not go to completion (all the way to form pure NO_2); in fact, we only get about 16% NO_2. We can calculate $\Delta G°$ for this reaction and find that it is a small positive number. This would imply that it is a non-spontaneous reaction in the direction written. However, we already know that we will get 16% NO_2, therefore the reaction must go spontaneously to the right until we form about 16% product. Below we can see a plot of free energy as a function of the reaction progress (Q), which is what we call the reaction quotient.

$$Q = \Pi\,(P_{prod})^{coeff} \,/\, \Pi\,(P_{react})^{coeff}$$

where: Π = multiplication.

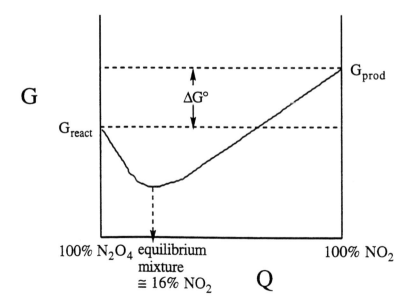

Chapter 15: Thermodynamics

$$Q = \frac{P_{NO_2}^2}{P_{N_2O_4}}$$

From the plot we can see that G decreases initially and reaches a minimum at a point when we have approximately 16% NO_2. The slope of that graph gives us the ΔG (which in turn will tell us the spontaneity of the reaction towards the formation of the products) for a particular mixture. Notice that the slope of the graph is negative (spontaneous) up to the minimum, at which point the slope is equal to zero. This is called the equilibrium point. Past that point, the slope of the graph becomes positive and therefore, the reaction is no longer spontaneous towards the formation of product. In other words, if we place pure NO_2 in a flask, we will observe that the reaction proceeds spontaneously until we reach that equilibrium mixture. It is also true that if we start with a pure mixture of N_2O_4, we will also go spontaneously up to a reaction mixture that contains approximately 16% NO_2.

Conclusion: *No matter where we start from, every system will go spontaneously towards equilibrium.*

What is the meaning behind $\Delta G°$? $\Delta G°$ is the change in free energy starting at a point when we have a standard mixture of reactants and products (the pressures of each being equal to one atmosphere) to the final state of 100% products. Therefore, a positive $\Delta G°$ implies that from that point on (when Q = 1), the reaction is not spontaneous toward the formation of products. In other words, a positive $\Delta G°$ implies that the minimum G (which we called the equilibrium point) is smaller than Q = 1 whereas a negative $\Delta G°$ implies that the equilibrium point is larger than Q = 1.

Mathematically, the equation that describes the preceding graph is:

$$\Delta G = \Delta G° + RT \ln Q$$

where: ΔG = slope of the graph at any Q
$\Delta G°$ = constant
Q = reaction quotient.

What happens at equilibrium? At equilibrium, the rate of the forward reaction is equal to the rate of the reverse reaction, therefore, the partial pressures of the reactants and the products do not change as a function of time. We can see that at equilibrium, the pressures of the reactants and the products stop changing as a function of time and therefore remain constant. This means that the reaction quotient (Q) becomes a constant and we will call it K_p (referred to as the equilibrium constant). From the graph it is evident that at equilibrium, $\Delta G = 0$, therefore:

$$0 = \Delta G° + RT \ln K_p$$

$$\therefore \Delta G° = -RT \ln K_p \quad \checkmark$$

This equation is extremely important since it will give us a mathematical way to determine the equilibrium constant for any reaction from thermodynamic parameters (in other words, using the tables, or Appendix #1). And what is the numerical value of the equilibrium constant telling us anyway? It tells us how far a reaction will proceed spontaneously. If K_p is very large (which we will get if $\Delta G°$ is a large negative number), then the reaction goes almost 100% towards the products before it reaches equilibrium. If K_p is very small (which we will get if $\Delta G°$ is a large positive number), then the reaction barely forms products by the time it reaches equilibrium. Two such examples can be seen below, both at 298 K.[1] The first example is one for a very large equilibrium constant. In this case, the $\Delta G°$ is -55.3 kJ.[2] In the second example, we have a relatively small value for the equilibrium constant, and this gives a positive value for the $\Delta G°$ (of $+19.3$ kJ).

$$CO_{(g)} + Cl_{2(g)} \rightleftharpoons COCl_{2(g)} \qquad K_p = 4.57 \times 10^{+9} \qquad \Delta G° = -55.3 \text{ kJ}$$

$$N_{2(g)} + O_{2(g)} \rightleftharpoons 2\, NO_{(g)} \qquad K_p = 4.08 \times 10^{-4} \qquad \Delta G° = +19.3 \text{ kJ}$$

- Consider the reaction shown below at 298 K.

$$3\, A_{(g)} + 2\, B_{(g)} \rightleftharpoons 4\, C_{(g)} + 2\, D_{(s)}$$

The heat **given off** for this reaction was measured in a constant pressure calorimeter to be 6,000 J/mol of A and the $\Delta S°$ for this reaction is determined to be -50.0 J/K. With this information, determine the equilibrium constant for this reaction.

A constant pressure calorimeter measures what we call q_p, which is by definition the $\Delta H°$ of the reaction. This means that the $\Delta H°$ for the reaction must be $-6,000$ J/mol of A, therefore,

$$\Delta H° = 3 \text{ mol A } (-6,000 \text{ J/mol A}) = -18,000 \text{ J} \quad \checkmark$$

---

[1] When the temperature is not specified, we will always assume that it is the standard temperature, which is 298 K. The same goes for the pressure. When it is not specified, the assumption is that it is one atmosphere.

[2] This value is calculated with $\Delta G° = -RT \ln K_p$.

Now, the ΔG° can be determined with $\Delta G° = \Delta H° - T \Delta S°$, so:

$$\Delta G° = -18{,}000 \text{ J} - 298 \text{ K}(-50.0 \text{ J/K})$$

$$\therefore \Delta G° = -3{,}100 \text{ J} \checkmark$$

Finally, to get K, we use $\Delta G° = -RT \ln K_p$.

$$-3{,}100 \text{ J} = -(8.314 \text{ J/mol K})(298 \text{ K}) \ln K_p$$

$$\therefore K_p = 3.50 \checkmark$$

Chapter 16: Chemical Equilibrium.

16.1 <u>Introduction</u>.

In the preceding chapter, we introduced equilibrium from a thermodynamic standpoint. In this chapter we are going to concentrate our efforts on understanding the properties of the equilibrium constant and properties of the equilibrium process in general. The methodology we will develop for problem solving will be very important, not only in the cases of chemical equilibria in the gas phase (Chapter 16), but also in equilibria for acids and bases (Chapter 17), as well as equilibria for the solubility process (Chapter 18).

16.2 <u>Le Chatelier's Principle</u>.

One of the most important properties of the equilibrium constant is the fact that any system at equilibrium will always stay at equilibrium unless a stress is applied. Whenever that happens, the system will counteract the stress in order to reach equilibrium once again. This is due to the fact that the equilibrium point is the point of lowest free energy, and every system goes spontaneously to the point of lowest free energy. This is something we saw in our last chapter.

Le Chatelier's principle[1] basically states this fact. <u>If a stress is applied to a system at equilibrium, the position of the equilibrium will shift to reduce the stress.</u> As an example, if we were to introduce an excess of a given reagent into a flask, once the reaction has reached equilibrium, the reaction will initially shift to the product side, because upon doing so, it would partially destroy the excess, and it will continue shifting in that direction until the system reaches equilibrium once again (that is, until $Q = K_p$). The easiest way to follow this principle is to apply it to a particular example. Let's consider the following reaction already at equilibrium. This reaction has a $\Delta H°$ (at 25°C and 1 atm pressure) of: -8.80 kJ.

$$2 SO_{2(g)} + O_{2(g)} \rightleftharpoons 2 SO_{3(g)} + 8.80 \text{ kJ}$$

We have written the heat loss for the reaction as a product of the reaction. This is the implication of a negative $\Delta H°$. Of course, had it been positive, then it would have been a heat gain, therefore, it would have been written as a reactant. The mathematical expression for the K_p for this reaction is given by:

---

[1] Henry Le Chatelier was a French chemist who lived from 1850 until 1936.

Chapter 16: Chemical Equilibrium.

$$K_p = (P_{SO3})^2 / (P_{SO2})^2 (P_{O2}).$$

In this equation, these pressures specifically represent the equilibrium partial pressures of all three gases. That is, once we reach equilibrium, the rate of the forward reaction is equal to the rate of the reverse reaction, and we will observe that the pressures of these three gases will remain constant as a function of time. Therefore, this ratio we used to call Q has suddenly become a constant, which we now call K_p. Let's apply some stresses to see what effect they have on the partial pressures and the position of equilibrium.

a) Addition of some SO_2 gas to the equilibrium mixture.

Before we add the sulfur dioxide gas, we have an equilibrium mixture, which means that the reaction quotient for the mixture is equal to K_p. As soon as we add some of the sulfur dioxide gas, the reaction quotient will no longer be equal to K_p, since we have an excess of sulfur dioxide. How does the new Q compare to K_p? Since the pressure of sulfur dioxide is in the denominator, the Q will be smaller than K_p. Although it is true that we do not know the actual value for the equilibrium constant, we will assume an arbitrary value. Consider the plot below. Here we are tracking the value of Q as the reaction moves from the reactants to the products (left to right). We have seen something similar to this in the previous chapter.[1] The dashed line marked K_p represents the normal position of equilibrium. After the stress, Q became smaller than K_p. Therefore, the reaction will have to shift to the right until once more Q is equal to K_p.

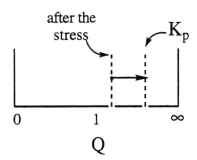

Shifting to the right simply means that the partial pressures of the reactants will become smaller and the pressure of the product will become larger. This will mathematically increase the value of Q until it is equal to K_p. Another way to look at this is by saying that when we add sulfur dioxide gas into the equilibrium mixture, we create a stress and the only way to counteract that stress is by reducing the amount of sulfur dioxide in the mixture.

---

[1] See Section 15.14.

Therefore, the reaction will shift to the right which is the direction in which the sulfur dioxide is consumed.

b) Removal of some O_2 from the equilibrium mixture.

Since we have partially depleted the amount of oxygen gas in the equilibrium mixture, the reaction will shift in the direction that will create some more oxygen gas. This means that the reaction will shift to the left until equilibrium is re-established.

c) Increasing the Temperature of the equilibrium mixture.

We should consider the heat released (it is exothermic) by this reaction to be a product of the reaction. Therefore, the addition of heat to the system (which is one way in which we can increase the temperature), will shift the equilibrium to the **left** in order to destroy the excess heat. This is an oversimplification, because in reality what happens is that the actual value of the equilibrium constant changes with temperature. The change is dependent on the ΔH of the reaction. In this case, since we concluded that the equilibrium shifted to the left, this would imply that the equilibrium constant decreased at the higher temperature. The equation below relates the magnitude of the two constants with the temperature. In that equation, R is the gas constant in units of J/molK, and the temperature are in Kelvin.

$$\ln\left(\frac{K_2}{K_1}\right) = -\frac{\Delta H}{R}\left(\frac{1}{T_2} - \frac{1}{T_1}\right)$$

- Determine the ratio of the equilibrium constants if the temperature for the reaction discussed above is increased from 25°C to 50°C.

Since we know the ΔH for the reaction, we can easily evaluate this ratio.

$$\ln\left(\frac{K_2}{K_1}\right) = -\frac{-8{,}800 \text{ J/mol}}{8.314 \text{ J/molK}}\left(\frac{1}{348 \text{ K}} - \frac{1}{298 \text{ K}}\right)$$

Therefore,

$$\left(\frac{K_2}{K_1}\right) = 0.600$$

d) Decreasing the volume of the system.

Changing the volume of the system has an effect on the position of equilibrium only if the number of moles of gases change from the reactant to the product. We have previously labeled this quantity as: Δn_g.[1] In this particular example, we are starting with three moles of gases and end up with two moles of gases. Therefore, we have a net contraction as we move from the reactants to the products. This should be clear, since three moles of gases occupy a larger volume than two moles of gases. Therefore, if we decrease the volume of the system, the reaction will shift to the **right**, where it occupies less volume. We should point out that if the Δn_g would have been zero, as in the reaction shown below, there would have been no effect whatsoever upon changing the volume of the system.

$$H_{2(g)} + Cl_{2(g)} \rightleftharpoons 2\,HCl_{(g)}$$

e) Adding some Neon gas (an inert gas) without changing the volume of the system.

If the volume is not changed, then the presence of the neon gas has absolutely no effect on the position of equilibrium for this reaction. Remember that the neon gas does not take part in the reaction itself, therefore, it is not part of K_p.

f) Adding some Neon gas and changing the volume in the process (achieved at constant pressure).

If the addition of Ne gas increases the volume of the container (in order to allow the pressure to remain constant), then the reaction will no longer be at equilibrium and it will shift to the **left** in order to release some stress. Remember that at a larger volume, the reaction will shift to the side that occupies more volume (left side has three moles of gas while the right side has only two).

16.3 Properties of K.

In thermodynamics we learned that Hess' Law applied to all the properties of a system, meaning that if we were to add two reactions, the total $\Delta H°$ would be equal to the sum of the $\Delta H°$ for each reaction. The same was true for $\Delta G°$, $\Delta E°$, and $\Delta S°$. However, the same is not true for K_p since K_p is defined logarithmically in terms of $\Delta G°$.

---

[1] See Section 15.7.

Chapter 16: Chemical Equilibrium.

$$A_{(g)} + 2B_{(g)} \Leftrightarrow C_{(g)} \qquad K_{p1} : \Delta G°_1$$

$$D_{(g)} \Leftrightarrow 2B_{(g)} \qquad K_{p2} : \Delta G°_2$$

$$\overline{A_{(g)} + D_{(g)} \Leftrightarrow C_{(g)}} \qquad \overline{K_{p\,tot} : \Delta G°_{tot}}$$

In the example above, we see that the addition of the $\Delta G°$ for these two reactions would give us the total $\Delta G°$ for the overall reaction, that is: $\Delta G°_{total} = \Delta G°_1 + \Delta G°_2$. But,

$$\Delta G°_1 + \Delta G°_2 = -RT \ln(K_{p_1}) + [-RT \ln(K_{p_2})]$$

$$\Delta G°_{total} = -RT [\ln(K_{p_1}) + \ln(K_{p_2})]$$

$$\Delta G°_{total} = -RT \ln(K_{p_1} \cdot K_{p_2}) = -RT \ln K_{p_{total}}$$

$$\therefore K_{p_{total}} = K_{p_1} \times K_{p_2}$$

1) In conclusion, when we add two equations, their respective equilibrium constants are multiplied, not added.

What is the relationship between the equilibrium constant of a given reaction and the equilibrium constant for the reverse reaction? Let's look at a simple example. Take the case of the decomposition reaction shown below. The mathematical expression for the equilibrium constant is also shown.

$$A_{2(g)} \rightleftharpoons 2A_{(g)} \qquad K_p = \frac{P_A^2}{P_{A_2}}$$

Let's write the reverse reaction, and determine also the mathematical expression for its equilibrium constant.

$$2A_{(g)} \rightleftharpoons A_{2(g)} \qquad K_p^* = \frac{P_{A_2}}{P_A^2}$$

We should notice immediately that the new equilibrium constant is exactly the inverse of the original one.

2) In conclusion, when a reaction is reversed, its equilibrium constant is simply the inverse of the initial one.

Chapter 16: Chemical Equilibrium.

- What is the equilibrium constant for the combination reaction seen above (2 A ⇌ A_2), if we are told that the decomposition of A_2 has an equilibrium constant of 0.025 ?

    Since these are cases of two reverse reactions, then the equilibrium constant for the combination reaction must be the inverse of 0.025, or: *40*. √

    Sometimes we prefer to work problems with concentrations, rather than pressures. It would be to our advantage then to work with an equilibrium constant that is in terms of concentrations. We shall call this equilibrium constant: Kc, to distinguish it from the one that is in terms of pressures (Kp). Since they are not one and the same, we need to derive an expression that will relate the numerical values of these two equilibrium constants.

    Consider the following reaction in the gas phase.

$$A + 2B \rightleftharpoons C$$

The mathematical expressions for the two proposed equilibrium constants can be seen below.

$$K_p = \frac{P_C}{P_A \cdot P_B^2} \qquad \text{and} \qquad K_c = \frac{[C]}{[A][B]^2}$$

$$\text{but:} \quad P_C = \frac{n_C RT}{V}$$

$$K_p = \frac{\dfrac{n_C RT}{V}}{\dfrac{n_A RT}{V} \left(\dfrac{n_B RT}{V}\right)^2} \qquad \text{but:} \quad \frac{n}{V} = \text{Molarity}$$

$$K_p = \frac{[C] RT}{[A] RT [B]^2 (RT)^2} = \frac{[C]}{[A][B]^2} \cdot (RT)^{[1-(1+2)]}$$

$$\text{but:} \quad [1 - (1+2)] = \Delta n_g$$

$$\therefore K_p = K_c (RT)^{\Delta n_{gases}}$$

- Assume that Kc for the above reaction is 25.0 @ 300 K. What would be the numerical value for ΔG°?

    Since ΔG° is related to Kp, we must determine Kp first.

$$Kp = Kc(RT)^{\Delta n_{gases}}$$

What is Δn_g for this reaction? It is determined as the change in number of moles of gases as you go from reactants to products (products minus reactants). In this case, it would be:

$$\Delta n_g = 1 - 3 = -2 \text{ moles.} \checkmark$$

Now the calculation of Kp (which although it is technically unitless must have all the quantities in atm) involves units of Latm, therefore we need the value of R (the gas constant) in Latm.

$$Kp = (25.0) [(0.0821 \text{ Latm/Kmol}) (300 \text{ K})]^{-2}$$

$$\therefore Kp = 0.0412 \checkmark$$

Technically, Kp is actually unitless[1], and we will discuss the reason below.

Finally, $\qquad \Delta G° = - R T \ln Kp$

$$\Delta G° = - (8.314 \text{ J/mol K}) (300 \text{ K}) \ln (0.0412)$$

$$\therefore \Delta G° = + 7,950 \text{ J/mol} \checkmark$$

3) In conclusion, Kp and Kc are related by the equation shown below, but the one that is obtained from the $\Delta G°$ equation is Kp.

$$Kp = Kc(RT)^{\Delta n_{gases}}$$

There is one more thing we have not mentioned yet. What happens when we have more than one particular phase in the equilibrium reaction? What happens when we involve gases with solids and liquids? How do we deal with these other phases?

Consider the following equilibrium.

$$CaCO_{3(s)} \rightleftharpoons CaO_{(s)} + CO_{2(g)}$$

For any reaction, the true <u>thermodynamic</u> expression for the equilibrium constant (the one that is obtained from $\Delta G° = - RT \ln K$) can be written as:

---

[1] See the last property below.

$$K = \frac{a_{CaO} \times a_{CO_2}}{a_{CaCO_3}}$$

In this equation, **a** is a quantity we call the activity. It is a measure of the real or the effective concentration of a particular substance.

How do we relate activities to things we can measure? Activities are actually unitless quantities; they are actually ratios that compare the actual value to some standard value, and like all ratios, are unitless. There are three extremely good approximations which we will be using here:

a. The activity of gas is approximately equal to its pressure in atmospheres but written without units. This is because the activity of a gas is defined as:

$$a_{gas} = \frac{P_{gas}}{P°}$$

In this equation, P° is exactly one atmosphere (the standard pressure). Since we are dividing by one atmosphere, then the actual activity is unitless!

b. The activity of an ion in solution is approximately equal to its molar concentration but again written without units. This is very similar to what we did for the pressure, except the standard concentration is exactly 1 M, therefore it yields a unitless ratio.

c. The activity of pure liquids and pure solids is exactly equal to one.

Based on these approximations, we can write the thermodynamic equilibrium constant for the reaction above as seen below. Please notice that it does not include the solids (activity of one), as they do not have an effect on the equilibrium process. This is an important point when thinking about Le Chatelier's principle. Solids will have no effect on the position of the equilibrium point, since they are not part of the mathematical equation for the equilibrium constant.

$$K = P_{CO_2}$$

4) In conclusion, pure solids and pure liquids do not form part of the mathematical expression for the equilibrium constants. This is independent of whether we are using K_p or K_c for any problem.

Chapter 16: Chemical Equilibrium.

16.4 Chemical Equilibrium Problems.

In this section, we will simply learn a method by which most chemical equilibria problems can be solved. We should start with a problem that will show to us how these equilibrium constants may be determined experimentally.

- One mole of SO_2 gas and 1.00 mol of O_2 gas were confined at 1000 K in a 5.00 L flask. At equilibrium, 0.840 moles of SO_3 gas was formed. Calculate the numerical value for Kc.

$$2\ SO_{2(g)} + O_{2(g)} \rightleftharpoons 2\ SO_{3(g)}$$

It should be clear that the mathematical expression for the equilibrium constant is given by:

$$Kc = \frac{[SO_3]^2}{[O_2] \times [SO_2]^2}$$

We need to know the equilibrium concentrations of all the species in the gas phase. Since molarity is moles per liter, we need to determine first the initial concentration of all species. There are two initial species present, the SO_2 and the O_2 gases, in each case we have one mole in five liters. This amount to a concentration of 0.200 M. How do these concentrations change as a function of time? It follows that both of them are going to get smaller as a function of time, whereas the concentration of SO_3 is going to increase as a function of time. We should also see that these concentrations will not change at the same rate. Since the SO_2 and the O_2 react in a 2:1 mole ratio, it is clear that the SO_2 is disappearing twice as fast as the O_2, and that the SO_3 is being formed at the same rate as the SO_2 is disappearing. In general, we can say that for every **x** moles of O_2 that are consumed, **2x** moles of SO_2 are consumed, and **2x** moles of SO_3 are formed. This is summarized below.

	2 SO_2	+ O_2	$\rightleftharpoons$ 2 SO_3
initial:	0.200 M	0.200 M	0 M
@ equilibrium:	0.2 - 2x	0.2 - x	2x

However, in this problem we are told that at equilibrium, the concentration of SO_3 is 0.840 moles in 5 L. This leads to a concentration of (0.840 mol/5.00 L) = 0.168 M for SO_3. What did we call the concentration of SO_3 at equilibrium? We called it **2x**. Therefore, **x** = 0.0840 M. It follows that:

$$[SO_2] = 0.200 - 0.168 = 0.032 \text{ M},$$

$$[O_2] = 0.200 - 0.084 = 0.116 \text{ M},$$

$$\text{and } [SO_3] = 0.168 \text{ M}.$$

With these numbers (remember NOT to include the units!) we can determine the numerical value for the equilibrium constant.

$$K = \frac{[0.168]^2}{[0.032]^2 [0.116]}$$

$$\therefore K_c = 240 \checkmark$$

- Consider the following equilibrium, whose equilibrium constant (K_c) is 4.0 at 800 K.

$$H_{2(g)} + Cl_{2(g)} \rightleftharpoons 2 HCl_{(g)}$$

If we mix initially 1.0 mol each of H_2 and Cl_2 gases in a 2.0 L vessel, what will be the molar concentration of all species at equilibrium?

In order to solve this problem, we should start by looking at the initial conditions and determining the direction in which the reaction is going to proceed. With the initial conditions, we will calculate the initial quotient (Q) so that we will be able to compare its value to the equilibrium constant. Remember that Q changes spontaneously until it is equal to K.

In this example, initially we have a gaseous mixture which happens to be 0.50 M in H_2 and 0.50 M in Cl_2 (remember that M = n/V). Therefore, initially Q is determined to be:

$$Q = \frac{[HCl]^2}{[H_2][Cl_2]} = \frac{0^2}{(0.50)^2} = 0$$

Since K = 4, then it is certain that Q will get larger as a function of time: until it is equal to four.

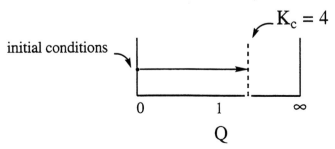

This is accomplished by making the numerator larger (products) at the expense of the denominator (reactants). The meaning of this is a shift to the right, towards the formation of products. We can see right above a pictorial representation of the shift towards equilibrium.

Let's summarize the initial and the equilibrium conditions (once we have shifted to the right until we reach equilibrium):

$$H_{2(g)} + Cl_{2(g)} \rightleftharpoons 2\,HCl_{(g)}$$

initial:	0.50 M	0.50 M	0 M
@ equilibrium:	0.5 - x	0.5 - x	2 x

Now,

$$K_c = \frac{[HCl]^2}{[H_2][Cl_2]} = 4.0$$

$$\frac{(2x)^2}{(0.50 - x)^2} = 4.0$$

Taking square roots on both sides, we get.

$$\frac{2x}{0.50 - x} = 2.0$$

and solving for x, we get: **x = 0.25 M.**

Finally, the concentration of all species at equilibrium would be:

$$[H_2] = [Cl_2] = 0.50\,M - 0.25\,M = \mathit{0.25\,M} \checkmark$$

$$[HCl] = 2\,(0.25\,M) = \mathit{0.50\,M} \checkmark$$

- Consider the following equilibrium, for which $K_c = 5.0 \times 10^{-6}$ at 300 K.

$$2\,A_{(g)} + B_{(g)} \rightleftharpoons C_{(g)}$$

If we mix initially in a <u>1.0 L</u> flask 1.0 mol of A and 2.0 moles of B, what will be the equilibrium molar concentration of the gas C?

Let's analyze the initial conditions and calculate Q with the initial conditions. We shall then compare this value with the known equilibrium constant to see the direction where the spontaneous shift will take place.

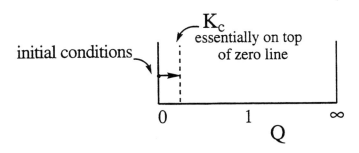

From the initial conditions (1.0 M for A and 2.0 M for B), we can tell that Q initially will be equal to zero (there is no C at the beginning). Therefore, the reaction will have to shift to the right. No reactions have an equilibrium constant equal to zero. They may be close, but it will not be zero. In this case, it is very close. The equilibrium constant is only 5.0×10^{-6}. This number is very close to zero, and if we were going to plot it above, we would not be able to tell it apart from the zero. However, for the practical purpose of being able to see where we are going, let us mark the equilibrium point to the right of the zero (which is where it is). What we learn from this plot is that we are essentially at equilibrium from the beginning (but technically we still have to shift to the right). Therefore, the amount of displacement (which we will call x) or shift, although to the right, is going to be minimal or negligible.

A summary of the initial and equilibrium conditions may be seen below.

$$2 A_{(g)} + B_{(g)} \rightleftharpoons C_{(g)}$$

	2 A$_{(g)}$	B$_{(g)}$	C$_{(g)}$
initial:	1.0 M	2.0 M	0 M
@ equilibrium:	1 - 2x	2 - x	x

Since we decided that the shift was minimal, then 2x is negligible compared to 1.0 M and x is negligible compared to 2.0 M. Therefore,

$$K_c = \frac{[C]}{[A]^2 [B]} = \frac{x}{1^2 (2)} = 5.0 \times 10^{-6}$$

$$\therefore x = [C] = 1.0 \times 10^{-5} M \quad \checkmark$$

Chapter 16: Chemical Equilibrium.

- Consider the equilibrium below, for which $K_p = 2.00$ at 298 K.

$$A_{(g)} + B_{(g)} \rightleftharpoons C_{(g)}$$

In a given flask at 298 K we initially introduce gases A, B, and C with initial pressures of 0.200, 0.300, and 0.400 atm, respectively. After equilibrium is established, what will be the partial pressures of all three gases?

Once more, let's analyze the initial conditions and calculate Q with these conditions. We shall then compare this value with the known equilibrium constant to see the direction where the spontaneous shift will take place. The calculation of Q with the initial pressures will give us a value of 6.67. This number is larger than the equilibrium constant, therefore, the reaction will shift to the left.

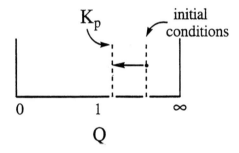

The summary can be seen below.

	$A_{(g)}$	+	$B_{(g)}$	$\rightleftharpoons$	$C_{(g)}$
initial:	0.200 atm		0.300 atm		0.400 atm
@ equilibrium:	0.2 + x		0.3 + x		0.4 - x

Finally,

$$K_p = \frac{P_C}{P_A P_B} = \frac{0.4 - x}{(0.2 + x)(0.3 + x)} = 2$$

Solving a quadratic equation, we get that **x = 0.124 atm**. Therefore,

$$P_A = 0.200 + 0.124 = 0.324 \text{ atm} \checkmark$$
$$P_B = 0.300 + 0.124 = 0.424 \text{ atm} \checkmark$$
$$P_C = 0.400 - 0.124 = 0.276 \text{ atm} \checkmark$$

Chapter 16: Chemical Equilibrium.

- If we mix 2.00 moles of A and 1.00 moles of B in a 1.0 L flask at 300 K, what will be the molar concentration of all species once equilibrium is established? The reaction below has an equilibrium constant (Kc) equal to 1.0×10^8 at 300 K.

$$A_{(g)} + 2 B_{(g)} \rightleftharpoons C_{(g)}$$

With the initial conditions, we can figure out that Q is equal to zero. This is very far from the value of Kc (Q << K).

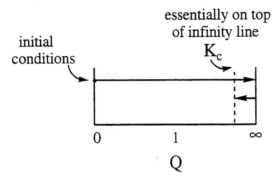

The easiest way to solve this problem is to shift the equilibrium 100% towards the formation of the products. Why? Because once we redefine the initial conditions at the extreme right, then we will be dangerously close to the equilibrium point. After that, we will shift backwards towards the formation of the reactants, but since we are so close to equilibrium now, the shift will be minimal and **x** will be negligible. In order to shift 100% to the right, we need to determine the limiting reagent between A and B. This LR will be consumed 100% and it will tell me how much product will be formed and how much of the other reagent will be consumed. Doing all this with this particular example gives us the following summary:

	$A_{(g)}$	+	$2 B_{(g)}$	$\rightleftharpoons$	$C_{(g)}$
initial:	2.00 M		1.00 M		0 M
100% shift: $\Longrightarrow$	1.50 M		0 M		0.50 M
@ equilibrium:	1.5 + ~~x~~'		2 x		0.5 - ~~x~~'

So,

$$K_c = \frac{[C]}{[A][B]^2} = \frac{0.50}{1.50 \, (2x)^2} = 1.0 \times 10^8$$

Chapter 16: Chemical Equilibrium.

$$\therefore 2x = [B] = 5.8 \times 10^{-5} M \checkmark$$

Conclusion.

1) It is convenient to approach equilibrium from the side that is closest to the equilibrium mixture.

2) A large Kc (10^5 or larger) implies that the reaction goes essentially to completion.

- Consider the following reaction at 25°C, for which Kc = 5.25 x 10^{+10}.

$$A_{(g)} + 2 B_{(g)} \rightleftharpoons 3 C_{(g)}$$

In a 1.00 L flask, we place 0.300 mol of gas A and 0.100 mol of gas B at 25°C. What will be the concentration of gas B at equilibrium?

Let's look at the initial conditions.

	A	+	2 B	⇌	3 C
initially:	0.300 M		0.100 M		0 M

The initial value of Q is zero. This is a lot smaller than Kc, therefore, we should redefine the initial conditions 100% to the right. At that point, we will be past the equilibrium point but very close to it. Let's determine the limiting reagent for the forward reaction.

$$R_A = 0.300/1 \qquad R_B = 0.100/2 = 0.050$$

Since 0.050 < 0.300, then **B** is the LR.

The limiting reagent (0.100 M of **B**) will allow us to determine the amount of C formed and the amount of A left over.

$$0.100 \text{ M } \cancel{B} \times \frac{3 \text{ C}}{2 \cancel{B}} = 0.150 \text{ M C formed}$$

$$0.100 \text{ M } \cancel{B} \times \frac{1 \text{ A}}{2 \cancel{B}} = 0.050 \text{ M A consumed}$$

$$\therefore [A]_{\text{left over}} = 0.300 \text{ M} - 0.050 \text{ M} = 0.250 \text{ M} \checkmark$$

A summary of the initial, the redefined initial conditions (100% to the right), and the equilibrium conditions appears next.

	A	+	2 B	⇌	3 C
initially:	0.300 M		0.100 M		0 M
100% shift: ⟹	0.250 M		0 M		0.150 M
@ equilibrium:	0.250 + x		2 x		0.150 - 3 x

Substituting the equilibrium concentrations into the mathematical expression for Kc, we get:

$$K_c = \frac{[C]^3}{[A][B]^2} = \frac{(0.150)^3}{(0.250)(2x)^2} = 5.25 \times 10^{10}$$

and finally: $[B] = 2x = 5.07 \times 10^{-7} M$ ✓

16.5 Strategy for Solving Chemical Equilibria Problems.

a. If the equilibrium constant is "normal" (10^{-4} to 10^{+4}), then **x** is never negligible! After the initial conditions, we just would go directly to the equilibrium conditions.

b. If the equilibrium constant is very large ($\geq 10^{+5}$), then

   1. **x** is negligible if Q = ∞. Once more, after we establish the initial conditions, we just would go directly to the equilibrium conditions.

   2. **x** is not negligible if Q < 100. Therefore, we would proceed to shift 100% to the right (⟹) and return back ever so slightly to the equilibrium conditions. The reason for this, as we saw in a few problems above, was in order to make **x** negligible.

c. If the equilibrium constant is very small ($\leq 10^{-5}$), then:

   1. **x** is negligible if Q = 0. After the initial conditions, we would go directly to the equilibrium conditions.

   2. **x** is not negligible if Q is between 1 and infinity. Therefore, we would shift 100% to the left (⟸) and return back ever so slightly to the equilibrium conditions. Once more, the reason for doing this is to make **x** negligible.

Chapter 17: Acids and Bases.

17.1 Brönsted Lowry Acid/Base Concept.

The simplest acid base concept was introduced by Arrhenius. In this concept, an acid is defined as a substance that increases the concentration of hydronium ions (H^+) in aqueous solutions, whereas a base is defined as a substance that increases the concentration of hydroxide ions (OH^-) in aqueous solution. This places serious restrictions to the acids and the bases. An acid must have a polarized hydrogen that will be identified as the acidic hydrogen. For example: HCl, H_2SO_4, and ethanoic acid. This last one can be written as: $HC_2H_3O_2$. The acidic hydrogens in these Arrhenius acids are always written first in the formula. From the four hydrogens in this molecule, only one is polarized enough to be classified as acidic.

$$\begin{array}{c} H \quad\; O \\ |\quad\;\; \| \\ H-C-C-O-H \\ | \\ H \end{array} \quad \text{Ethanoic acid}$$

The Arrhenius bases are easy to identify because they have the OH directly in the formula. We should be aware that only ionic hydroxides are basic, such as compounds like NaOH and $Ca(OH)_2$. These would therefore be classified as Arrhenius bases. On the other hand, methanol (CH_3OH) is not an Arrhenius base since it is not ionic.

The most widely used acid base concept is the Brönsted Lowry one. It basically removes the restriction of having water as a solvent, so the definitions change slightly.

Acid - Substance that donates a proton (H^+) to another substance. These are the same kinds of acids as in the Arrhenius concept, that is, the acid itself must contain acidic hydrogens (always written at the beginning of the formula): HCl, HNO_3, H_2SO_4, H_3PO_4, $HC_2H_3O_2$, etc.

Base - Substance that can accept a proton. This is a more encompassing definition than the one from the Arrhenius concept. What does a base need in order to be able to function as one (accepting an H^+)? A *lone pair!* The small proton, which is nothing more than a point charge, will be very strongly attracted to one of the lone pairs on the base to form a coordinate covalent bond (a bond where initially both electrons come from one atom).

Let's look at an example of a Brönsted Lowry base that is not classified as an Arrhenius base. Such is the case for ammonia (NH_3). This base does NOT contain the OH directly on the formula, but it does classify as a Brönsted Lowry base because of the lone pair on the nitrogen atom. The same thing is true for the water molecule itself. It is also a base because it can accept a proton into one of its lone pairs.

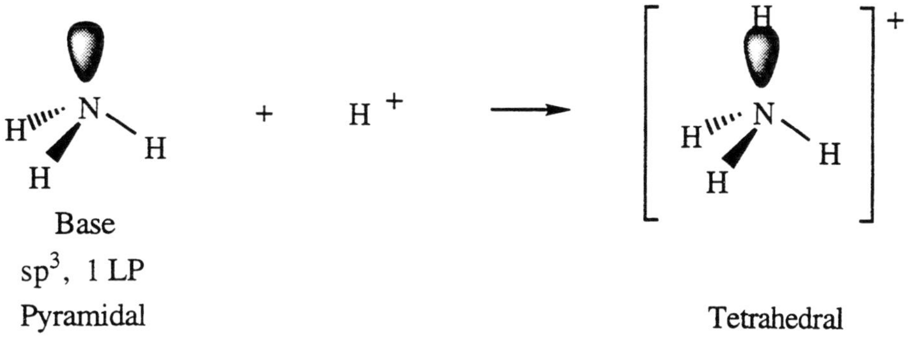

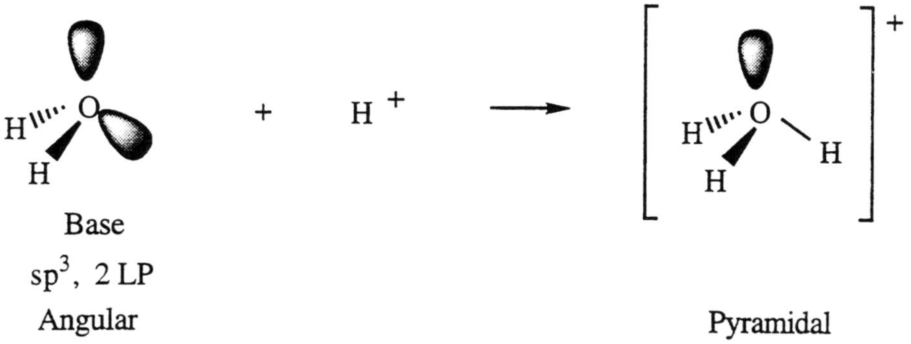

Base
sp³, 1 LP
Pyramidal

Tetrahedral

Base
sp³, 2 LP
Angular

Pyramidal

Actually, water has the ability to accept a proton (as we just saw), but it also has the ability to donate a proton (since it has two polarized hydrogens). This makes water an **amphoteric** or amphiprotic substance. This is a substance that can function both as an acid or as a base.

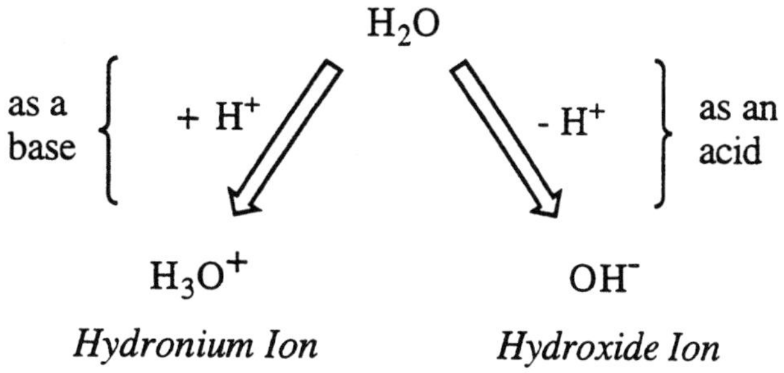

H_2O

as a base { $+H^+$ $-H^+$ } as an acid

H_3O^+ OH^-
Hydronium Ion   *Hydroxide Ion*

This shows that acid base reaction are competition reactions. When we dissolve a substance in water, the dilemma is whether the water has the strength to give a proton to the substance, or the opposite. This will depend on the relative strengths (donating abilities) of the two. The examples below illustrate this concept.

a) HCl is a stronger acid than water. Therefore, the HCl will donate the proton to the water molecule to form:

$$HCl + H_2O \rightleftharpoons H_3O^+ + Cl^-$$
Acid   Base   Acid   Base

b) NH_3 is a weaker acid than water. Therefore, the water will donate a proton to the ammonia instead to form:

$$NH_3 + H_2O \rightleftharpoons NH_4^+ + OH^-$$
Base   Acid   Acid   Base

How can we tell if a particular substance is a stronger or a weaker acid than water? Is there a way to tell by just looking at the molecular structure? The next section should answer this question.

17.2 Acid Strength and Molecular Geometry.

The acidity of a given hydrogen in a molecule depends on the degree of polarization of the hydrogen in the molecule and how easy it will be to break the bond itself. Furthermore, to explain these, we need to concentrate on two things: 1) the electronegativity difference between the two atoms bonded and 2) the size of the atom directly bonded to the hydrogen. It should be clear that the larger the electronegativity difference between the two atoms, the easier it will be to break the bond since the hydrogen will carry a larger partial positive charge and should have a stronger attraction for the base. This is typically the case when we are comparing acids where the size of the atoms bonded to the polarized hydrogen are relatively close. Not only that, but under these conditions the differences may be quite dramatic. For example: HF is a much stronger acid than H_2O, and H_2O is a much stronger acid than NH_3.

However, if we are comparing acids that vastly differ in the size of the atoms bonded to the polarized hydrogen, the reverse is true. In other words, the larger the atom bonded to the polarized hydrogen, the stronger the acid. As an example, the hydrides of the halogens have the following relative strengths:

$$HF < HCl < HBr < HI$$

Actually, HF is a weak acid, whereas the rest are all strong acids.

Chapter 17: Acids and Bases.

Based on these observations, let's arrange the following acids in order of increasing strength: HF, NH_3, HI, and PH_3. The acid HI is the strongest (since it is the only strong one of the group) and the weakest would be the NH_3. The correct order would be: $NH_3 < PH_3 < HF < HI$. The reason for phosphine to be a weaker acid than HF is that the strength of the acids increases slowly as we move down a column, but increases dramatically as we move from left to right.

There are many acids in which the polarized H is bonded to an oxygen which in turn is bonded to something else (H-O-X). We call these **oxyacids**. In these cases, the polarized H is always bonded to the same atom (an oxygen atom) therefore, size plays no role in determining the strength of the acid. The electronegativity rule is only important if the only variation from acid to acid is the atom bonded to the oxygen. For example: $HClO_2$ versus $HBrO_2$. Since Cl is more electronegative than the Br, then $HClO_2$ is a stronger acid than $HBrO_2$. Keep in mind that the structure of these oxyacids is H-O-X! In the case of $HClO_2$, it would be:

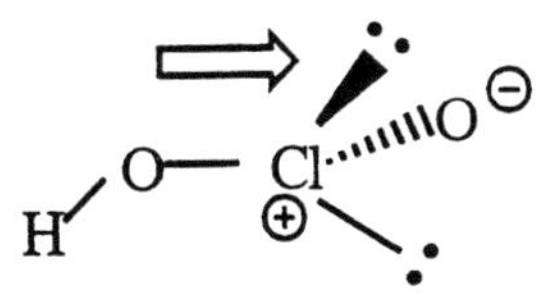

If we were to compare acidity for substances like HOCl, HOBr, HOI, and $HOCH_3$, the first thing we would do is ask ourselves, which is more electronegative: Cl, Br, I, or C? The answer is $Cl > Br > I > C$. Therefore we can tell instantly that the strongest acid from these four would be HOCl ($K_a = 4 \times 10^{-8}$), followed by HOBr ($K_a = 2 \times 10^{-9}$), HOI ($K_a = 2 \times 10^{-11}$) and finally $HOCH_3$ (whose Ka is so small that for all practical purposes it is zero --> it is about 10^{-16}). This should not surprise us since CH_3OH is an alcohol and we learned that alcohols are not acidic *per se*.

Another aspect of oxyacids is comparing oxyacids with varying number of oxygens in it (ex.: HClO, $HClO_2$, $HClO_3$, and $HClO_4$). We find that the more oxygens you have in the formula, the stronger the acids turns out to be. Therefore,

$$HClO < HClO_2 < HClO_3 < HClO_4$$

This is because as the number of oxygens increases in the formula of the acid, the formal charge of the central atom increases, therefore that increases the pull for the electrons from not only the oxygens but also from the central atom. The table in the next page illustrates this example. Notice how the formal charge on the Cl increases from $0 \Rightarrow +1 \Rightarrow +2 \Rightarrow +3$ as we go from

$HClO \Rightarrow HClO_2 \Rightarrow HClO_3 \Rightarrow HClO_4$. This causes a much greater polarization of the O-H bond and therefore increases the strength of the acidity.

		Ka
$HClO_4$	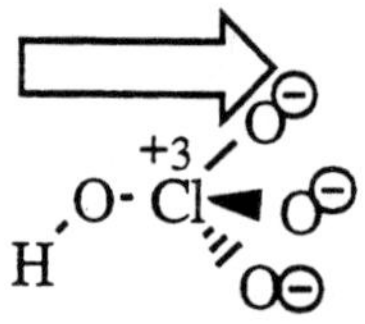 structure	Very Large
$HClO_3$	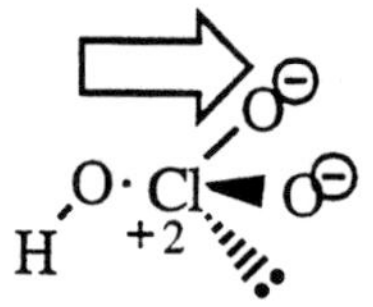 structure	$\cong 1$
$HClO_2$	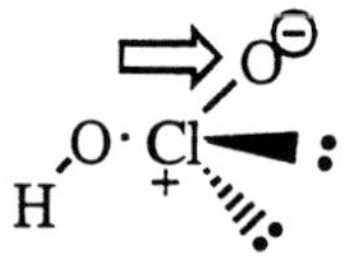 structure	1.2×10^{-2}
$HClO$	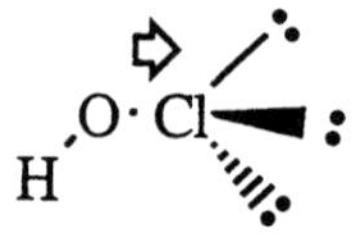 structure	3.5×10^{-8}

The numbers on the table above (Ka) represent a quantitative measurement of the strength of the acid. We will discuss the way these numbers are measured and their meaning in the next section but at this point it will suffice to say that the larger Ka, the stronger the acid so we will actually be using it to be able to see instantly which acid is stronger.

There is another kind of competition prevalent in acids and bases, which is that between the forward and the reverse reactions. Obviously, that will depend on the relative strengths of the acids and bases for both reactions (the reverse and the forward reactions).

Take the case of HCl dissociating in water. We find that HCl is a **strong acid** – which by definition means that it dissociates 100% in water. That must mean that its conjugate base (the Cl⁻ ion) must be very weak - it does not want an H⁺ very much.

$$\therefore \quad HCl + H_2O \rightleftharpoons H_3O^+ + Cl^-$$

We find that there is an inverse relationship between the strength of and acid and that of its conjugate base.

* THE STRONGER THE ACID, THE WEAKER THE CONJUGATE BASE.

How do we measure acid strength? All we need to do is to determine how much a

Chapter 17: Acids and Bases.

substance (acid) will dissociate in water. The more it dissociates, the stronger it is. However, all strong acids dissociate 100% in water so we cannot tell which is stronger than which directly. We call this the **leveling effect of the solvent**. In order to tell which of the strong acids shown below is stronger we would need to find the % dissociation in a different solvent. For example, in ethanol as a solvent, HCl does not dissociate 100% but HBr and HI do. That immediately tells us that HCl is a weaker acid that either HBr or HI.

List of **strong acids** in water (we must learn to recognize them).

Substance	K_a
HNO_3	125.1
HCl	$1.0 \times 10^{+7}$
HBr	$1.6 \times 10^{+7}$
H_2SO_4	$1.6 \times 10^{+9}$
HI	$1.0 \times 10^{+10}$
$HClO_4$	$1.0 \times 10^{+11}$

$HClO_4$ is the strongest acid in that table and it is actually one of the strongest acids known but don't forget that in water they are all equally strong, since they dissociate 100% and therefore are indistinguishable.

Now, how about the bases? Which are the strong bases? All the conjugate bases of the following very weak acids are by definition strong bases.

Substance		K_a	Conjugate Base
H-O-H	water	10^{-16}	OH^- (strongest)
CH_3O-H	R-OH, alcohols	10^{-18}	RO^-
CH_3-C≡C-H	R-C≡C-H, terminal alkynes	10^{-26}	$RC≡C^-$
H-H	hydrogen gas	10^{-35}	H^-
H-NH_2	ammonia	10^{-36}	NH_2^-
H-CH_3	R-H, alkanes	10^{-60}	R^- (weakest)

Their strong conjugate bases will follow the order for increasing basicity:

Chapter 17: Acids and Bases.

$$OH^- < RO^- < RC\equiv C^- < H^- < NH_2^- < R^-$$

This last base is extremely strong and we will call it: "super base". Let's predict if some organic reactions will occur or not just based on the strength of the bases we just discussed. An important concept is that when a strong base is used in a reaction, that reaction is pushed essentially 100% towards the formation of product. This implies that we **cannot form** a stronger base than the one we started with. For example:

$$CH_4 + OH^- \not\Rightarrow CH_3^- + H_2O$$

This reaction will NOT occur since we would be forming a stronger base (CH_3^-) than the one we started with (OH^-). If it did form, the CH_3^- would drive the reaction all the way back towards the reactants.

- Predict the product of the following reactions. If the reaction will not go, we should answer it with N.R. (for **no reaction**).

  a) $CH_3O^- + H_2O \Rightarrow$???

    This is not as difficult as it seems at first. We should remember that we are dealing with acid base reactions. This means that one of these species is going to donate a proton to the other one. In this reaction, the negative species is the base, therefore, the water would have to be the acid. This would yield the following reaction (after the water gives one of its hydrogens to the negative oxygen).

    $$CH_3O^- + H_2O \Rightarrow \mathbf{CH_3O\text{-}H + OH^{-1}}.$$

    Since the base formed (the hydroxide ion) is weaker than the initial base (CH_3O^{-1}), then this reaction does occur.

  b) $CH_3CH_3 + H^- \Rightarrow$???

    Once more, the negative ion must be the base. Since all that it takes is the transfer of a proton from the acid to the base, it is simple to predict the product for this or any acid base reaction.

    $$CH_3CH_3 + H^- \not\Rightarrow CH_3CH_2^{-1} + H_{2(g)}$$

Chapter 17: Acids and Bases.

The new base we formed is MUCH stronger than the hydride ion, therefore, this reaction will not occur. Conclusion: **NR**.

c) $CH_3CH_2^- + NH_3 \Rightarrow$???

The base is the negative ion, therefore, the ammonia must be the acid. The reaction would be:

$$CH_3CH_2^- + NH_3 \Rightarrow \mathbf{CH_3CH_3 + NH_2^{-1}}.$$

Since the base formed is weaker than the initial one, this reaction does take place.

d) $CH_3CH_2\text{-}C\equiv CH + CH_3CH_2\text{-}O^- \Rightarrow$???

The base is the negative ion, therefore, the terminal alkyne must be the acid. Therefore,

$$CH_3CH_2\text{-}C\equiv CH + CH_3CH_2\text{-}O^- \Rightarrow CH_3CH_2\text{-}C\equiv C^{-1} + CH_3CH_2\text{-}O\text{-}H$$

Since the base formed is stronger than the initial base, then this reaction will NOT occur. Therefore, **NR**.

Let's look now at some substituted benzenes that have groups that will make the molecule either acidic or basic. As we learned previously, carboxylic acids (RCOOH) are acidic (hence the name) and amines (RNH$_2$) are basic. Now we can tell why amines are basic since we can see a lone pair on the nitrogen of the amine and a hydride of nitrogen is less acidic than water (since the nitrogen is less electronegative than oxygen). It is not surprising, therefore, to find out that aniline (φ-NH$_2$, where φ is a phenyl group) is basic but it might be surprising to find out that NH$_3$ is about 100,000 times more basic than aniline. We must be able to explain this observation. This is evident if we think of what makes a base react as a base. It is the accessibility of the lone pair to attract the polarized hydrogen from the acid. If the lone pair on the base is localized (unable to move and rather sticking out) on a given atom (in this case on the nitrogen in ammonia), then it will be easy to attack. That is the case of NH$_3$.

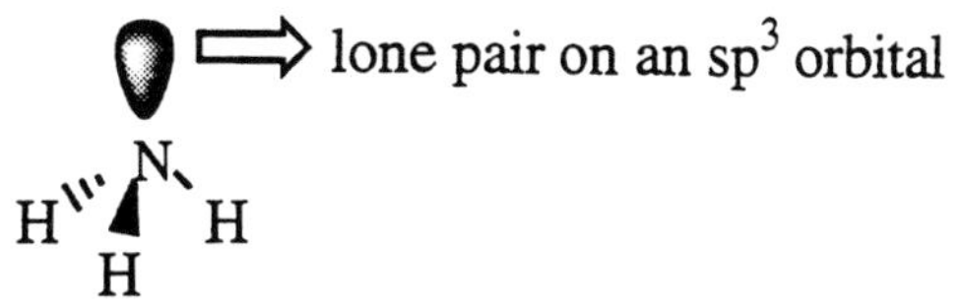

⇒ lone pair on an sp^3 orbital

We can see that the lone pair on the nitrogen is in an sp^3 orbital which is "sticking" out of the molecule and therefore easily accessible to the acid. On the other hand, the lone pair on the nitrogen in the aniline molecule is drawn into the ring in order to establish delocalization (or resonance), therefore lowering the energy of the system[1] (stabilization). This makes the lone pair a lot less accessible to the acid, since most if the time is not on the nitrogen but rather in the ring.

What happens when we place a substituent on the para position to the NH$_2$ group in the aniline molecule? It should have an effect on the basicity of the molecule. Let's compare the basicity of aniline to p-nitroaniline (p-NO$_2$-φ-NH$_2$). The nitro group is one of the strongest electron withdrawing groups we know of, so we would expect the NO$_2$ group to pull the lone pair on the nitrogen very strongly into the ring, therefore making it even less accessible to the acid than in the case of aniline itself. This is exactly what we observe, as aniline is about 100 times stronger as a base than p-nitroaniline. If instead we have p-methoxyaniline (p-CH$_3$O-φ-NH$_2$), we now would expect one of the lone pairs on the oxygen to participate in the delocalization with the benzene ring. This is exactly what happens. Now we have a competition for delocalization on the ring and independent of which atom wins, both of them are delocalized so the lone pair of the nitrogen is delocalized less often (since it is competing with the oxygen) and therefore it resides longer on the nitrogen itself. This makes it more accessible to the acid and therefore a stronger base. This prediction holds true, as the p-methoxyaniline is about 10 times stronger as a base than aniline itself.

17.3 Ka's and Kb's.

How do we quantitatively measure the strength of an acid? By measuring its degree of dissociation or the numerical value of the equilibrium constant. Let's look at the case of a weak acid dissociating in water: Acetic acid[2]. The bond between the oxygen and the hydrogen is very polarized and therefore will break heterolytically when it is dissolved in water.

---

[1] Electron delocalization always yields stabilization, or lower energy for the system.
[2] Common name for Ethanoic acid and main component of vinegar.

Chapter 17: Acids and Bases.

[Structural diagram showing acetic acid reacting with water to form hydronium ion and acetate ion, with the acetate ion shown with two resonance structures, labeled "Acetate ion is stabilized by resonance!"]

For the sake of simplicity, we will abbreviate acetic acid from now on as HAc, therefore we can rewrite that equilibrium as:

$$HAc + H_2O \rightleftharpoons H_3O^+ + Ac^-$$

What is the mathematical expression for K_c?

$$K_c = [H_3O^+][Ac^-] / [HAc][H_2O]$$

Notice that we used molar concentration of acetic acid and molar concentration of water. At first glance this should sound like we made a mistake because water and acetic acid are liquids, therefore we might be tempted to say that their activities are equal to one. However, these liquids are not pure since they are mixed with one another. Therefore, their molar concentrations must be expressed in the mathematical equation for K_c. Furthermore, since water is the solvent, there is lots of water in the flask relative to acetic acid. Therefore, the amount of water that actually reacts to form hydronium ions is very small. We may then say that the concentration of water does not change, that is, it remains constant as a function of time. We will therefore rewrite the equilibrium constant expression as:

$$K_c [H_2O] = K_a = [H_3O^+][Ac^-] / [HAc]$$

K_a's are measured experimentally at 25°C and then are tabulated. We can find a table of weak acid dissociation constants in Appendix #2.

For HAc, $K_a = 1.8 \times 10^{-5}$.

What is the meaning of such a small value for the K_a? This tells us that most of the acetic acid remains undissociated in solution at this temperature.

Chapter 17: Acids and Bases.

How about **bases**? We will take the same approach. The equilibrium for the dissociation of the base B in water is shown below.

$$B\!: + H_2O \rightleftharpoons HB^+ + OH^-$$

We can write the mathematical expression for the dissociation constant in exactly the same way as we did above and we get:

$$K_b = [HB^+][OH^-] / [B\!:]$$

K_b's are also tabulated at 25°C in Appendix #2.

- When 1.00 mol of HCN is dissolved in water to form 1.00-L of solution, the concentration of hydronium ions is found to be 2.00×10^{-5} M at equilibrium. What is the numerical value of K_a for HCN?

These problems are approached in the exact way as we did the equilibrium problems in the previous chapter. We should start by writing the equilibrium and writing the initial conditions, shift towards equilibrium and then write the equilibrium conditions in terms of x. Further substitution into the mathematical expression for the equilibrium constant should yield the numerical value of K_a.

$$HCN + H_2O \rightleftharpoons H_3O^+ + CN^-$$

	HCN	H₃O⁺	CN⁻
init:	1.00 M	≈ 0	0
@ equil:	1.00 - x	x	x

We were told the value for the concentration of hydronium ions at equilibrium, and that is what we just called x. Therefore,

$$K_a = \frac{[H_3O^+][CN^-]}{[HCN]} = \frac{x^2}{1.00 - x}$$

$$K_a = \frac{(2.00 \times 10^{-5})^2}{(1.00 - 2.00 \times 10^{-5})}$$

$$\therefore K_a = 4.00 \times 10^{-10} \quad \checkmark$$

17.4 Autoionization of Water.

Back in Section 17.2, we mentioned that water is an amphoteric substance. This means that it can either give a hydronium ion or take one at one of its lone pairs. Therefore, water can and will establish this particular equilibrium.

$$H_2O + H_2O \rightleftharpoons H_3O^+ + OH^-$$

The mathematical expression for the equilibrium constant can be seen below.

$$K_c = [H_3O^+][OH^-]/[H_2O]^2$$

Once more, since water is the solvent, its concentration will not change as some of its molecules autoionize. Therefore,

$$K_c [H_2O]^2 = K_w = [H_3O^+][OH^-] = 1.00 \times 10^{-14} \text{ @ 25°C.}$$

This equilibrium must always be satisfied. Therefore, in pure distilled water, only a small fraction of the water molecules dissociate to form identical amounts of hydronium and hydroxide ions. Anytime this reaction takes place is because one water molecule split apart to form an H^+ and an OH^- ion. This is why we have identical concentrations of these ions in pure distilled water.

$$\therefore [H_3O^+] = [OH^-] = 1.00 \times 10^{-7} M \checkmark$$

This is what is known as a neutral solution.

What happens when we dissolve an acid in water? Now we bring a foreign species (the acid) into the liquid. This species will split to form additional hydronium ions. However, it is always true that the multiplication of the concentration of hydronium and hydroxide ions must be equal to 1.00×10^{-14}, therefore, when the concentration of hydronium ions goes up, it must be at the expense of the hydroxide ion concentration, which must decrease. Let's take as an example a 0.100 M solution of HCl. What are the concentration of both hydronium and hydroxide ions? Since HCl is a strong acid, we do not have any HCl molecules in solution, and instead all of it is converted into H_3O^+ and Cl^-. Therefore, the concentration of H_3O^+ ions in solution has to be 0.100 M. This is to be added to the 1.00×10^{-7} M concentration of hydronium ions in water, but since 0.100 M is so much larger than the other, we can say that it is approximately equal to 0.100 M. The concentration of hydroxide ions must be 1.00×10^{-13} M, because the product of the two must equal 1.00×10^{-14}.

When we dissolve a **base** in water, we are actually increasing the concentration of hydroxide ions at the expense of the hydronium ions in solution.

- Calculate the concentration of hydronium and hydroxide ions for a 0.0100 M solution of NaOH.

NaOH is a strong base, therefore it is completely dissociated to form Na^+ and OH^- ions. Thus, the concentration of hydroxide ions must be 0.0100 M. The concentration of hydronium ions must be 1.00×10^{-12} M because at all times the product of the two must be equal to 1.00×10^{-14}.

17.5 pH, a Logarithmic Scale.

Many times it is inconvenient to work with exponentials and a lot easier to deal with logarithms.[1] Therefore, the pH scale was invented just to circumvent the small and always exponential notation for the concentration of the hydronium ions. We will define pH as:

$$pH = -\log [H_3O^{+1}].$$

Actually, the "p" is an operator that tells us to first find the logarithm (base 10) of the particular quantity, and then to change the sign. Therefore, if we were told to calculate the pCl, that would mean to take the logarithm (base 10) of the concentration of chloride ions and then to multiply it by minus one.

$$\therefore \quad p = -\log$$

We should remember the expression for Kw from Section 17.2,

$$K_w = [H_3O^+][OH^-]$$

Taking the negative logarithm on each side of the equation, we get the following expression.

$$-\log K_w = -\log [H_3O^+] - \log [OH^-]$$

or: $\quad pK_w = pH + pOH$

However, $\quad pK_w = -\log(1.00 \times 10^{-14}) = -(-14) = 14$

---

[1] This was particularly true in the days before the calculators.

Therefore, the equation takes the final form shown below.

$$14 = pH + pOH \quad \checkmark$$

This equation will enable us to quickly convert from the concentration of hydronium ions to the concentration of hydroxide ions, using the logarithmic notation.

From our previous discussion, a neutral aqueous solution has a concentration of hydronium ions equal to 1.00×10^{-7} M, therefore, such a solution would have a pH of 7.00. When an acid is dissolved in water, the concentration of hydronium ions will increase at the expense of the hydroxides ions, therefore, the pH of the solution will be lower than seven, while its pOH will be higher than seven. Finally, if a base is dissolved in water, the concentration of hydroxide ions will increase at the expense of the concentration of hydronium ions, therefore, the pOH of the solution will be lower than seven while the pH will be higher than seven.

- What is the pH of a 0.0100 M solution of HI?

Since HI is a strong acid, it is completely dissociated in water, thus forming hydronium ions and iodide ions. So, the concentration of the hydronium ions would be 0.0100 M plus the contribution from water (which happens to be negligible). Therefore, the pH of the solution would be:

$$pH = -\log(0.0100) = 2.000 \quad \checkmark$$

- What is the pH of a 2.50×10^{-3} M solution of NaOH?

Since NaOH is a strong base, it is completely dissociated into its ions (the sodium and the hydroxide ions). So, the concentration of hydroxide ions would be 1.00×10^{-3} M plus the contribution from the water (10^{-7} M) which would be negligible. Since we know the concentration of hydroxide ions and not that of the hydronium ions directly, we should calculate the pOH of the solution first.

$$pOH = -\log[OH^-] = -\log(1.00 \times 10^{-3}) = 3.000$$

Therefore, the pH would be:

$$pH = 14 - pOH = 11.000 \quad \checkmark$$

A graphic summary of the pH scale can be seen below. This diagram does not mean to imply that there are limits to the pH scale. The pH can be lower than two and higher than 12.

Chapter 17: Acids and Bases.

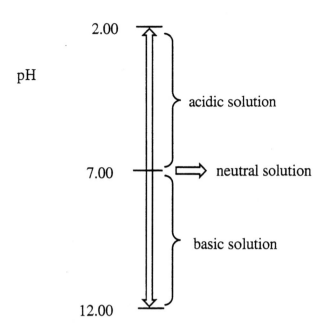

- What is the pH of a 10.0 M solution of HCl?

    HCl is a strong acid, therefore it is dissociated 100% in water. The concentration of hydronium ions in this solution is 10.0 M. Thus,

    $$pH = -\log(10.0) = -1.000 \checkmark$$

    This shows that the pH can even be a negative number!

- Would the pH of a 1.00×10^{-8} M solution of HBr be higher than seven, lower than seven, or equal to seven?

    A 1.00×10^{-8} M solution of HBr will dissociate 100% in water to form the ions, which are the hydronium and the bromide ions. This will make the concentration of hydronium ions 1.00×10^{-8} M plus the contribution from water. Since the contribution from water this time is NOT negligible (it is actually larger than the concentration provided by the HBr), then the concentration of hydronium ions will be slightly larger than 1.00×10^{-7} M. Therefore, the pH of the solution will be *slightly lower than seven*.

- Determine the pH of a 0.0500 M solution of Ca(OH)$_2$.

    Ca(OH)$_2$ is a strong base, meaning that it dissociates 100% in water. What happens when the calcium hydroxide dissolves in water? It dissociates into the ions.

    $$Ca(OH)_2 \Rightarrow Ca^{+2} + 2\,OH^-.$$

Since one mole of Ca(OH)$_2$ breaks up to form two moles of hydroxide ions, then the concentration of hydroxide ions will be 0.100 M. We need to get the pOH first (since we know the [OH$^-$] instead).

$$pOH = -\log[OH^-] = -\log(0.100) = 1.00$$

$$\therefore pH = 14 - pOH = 14 - 1.00 = \mathit{13.00} \checkmark$$

As long as we are determining the pH of a solution of a strong acid or a strong base, the problem is straight forward, since these dissociate completely. However, in the case of a weak acid or a weak base, we would be forced to solve the equilibrium first in order to determine the concentration of the hydronium (or the hydroxide) ions in solution.

- Determine the pH of a 0.100 M solution of HB. The Ka for HB is 1.00 x 10^{-7}.

Certainly, this is a little more involved than the examples above because this is a weak acid. In other words, we do not know the concentration of the hydronium ions to start with, so we will have to determine that first.

$$HB + H_2O \rightleftharpoons H_3O^+ + B^-$$

init: 0.100 M   0   0
@ equil: 0.1 - x   x   x

$$K_a = \frac{[H_3O^+][B^-]}{[HB]} = \frac{x^2}{0.100 - x} = 1.00 \times 10^{-7}$$

Solving for x, we get:

$$x = [H_3O^+] = 1.00 \times 10^{-4} \text{ M}$$

$$\therefore pH = -\log(1.00 \times 10^{-4}) = \mathit{4.00} \checkmark$$

- Determine the pH of a 1.00 M solution of the weak base B. Kb = 1.00 x 10^{-8}.

First, we must write the equilibrium and define the initial and equilibrium conditions.

Chapter 17: Acids and Bases.

$$B + H_2O \rightleftharpoons OH^- + HB^+$$

	B		OH$^-$	HB$^+$
init:	1.00 M		0	0
@ equil:	1.00 - x		x	x

$$K_b = \frac{[OH^-][HB^+]}{[B]} = \frac{x^2}{1.00} = 1.00 \times 10^{-8}$$

Solving for x will give us the concentration of OH$^-$ this time. This means that we must determine first the pOH and then the pH.

$$x = [OH^-] = 1.00 \times 10^{-4} \text{ M.}$$

$$\therefore \text{pOH} = -\log(1.00 \times 10^{-4}) = 4.00$$

and finally,

$$\text{pH} = 14 - \text{pOH} = 10.00 \checkmark$$

17.6 Hydrolysis.

In acid base chemistry, the term hydrolysis refers to the reaction (as an acid or as a base) of a particular ion with water. In order to predict if an ion will be able to hydrolyze in water, all we need to know is where did that ion came from, that is, is the conjugate base of that ion a strong acid or a weak acid? Remember that the conjugate base of a strong acid is very weak, therefore it will not react with water, that is, it will not hydrolyze.

Ex. $Cl^- + H_2O \xrightarrow{\times} HCl + OH^-$

However, let's consider the equilibrium of HCN. We have already seen that HCN is a weak acid. Will the cyanide ion (CN$^-$) hydrolyze in water? YES; the CN$^-$ will exhibit a certain power as a base since it comes from a weak acid.

$$\therefore CN^- + H_2O \rightleftharpoons HCN + OH^- \qquad K_b$$

In general, for a weak acid/conjugate base system:

$$\begin{array}{ll} HA + H_2O \rightleftharpoons H_3O^+ + A^- & K_a \\ A^- + H_2O \rightleftharpoons HA + OH^- & K_b \\ \hline 2 H_2O \rightleftharpoons H_3O^+ + OH^- & K_{tot} \end{array}$$

Chapter 17: Acids and Bases.

From the properties of K, we should remember that $K_{tot} = K_a \times K_b$. This is an important relationship since it will allow us to determine the numerical value for the equilibrium constants of ionic species when they dissociate in water as either an acid (K_a), or as a base (K_b).

Often, we will not find tables of dissociation constants for ionic species. Instead we find the dissociation constant for their respective conjugate acids or conjugate bases. Obviously, we can use the relationship that we just derived to calculate the one we really want.

- Calculate the K_b for the cyanide ion, CN^-. K_a for HCN is 4.00×10^{-10}.

$$K_b = K_w/K_a = 1.00 \times 10^{-14} / 4.00 \times 10^{-10}$$

$$\therefore K_b = 2.50 \times 10^{-5}. \checkmark$$

- Calculate the K_a for the ammonium ion, NH_4^+. K_b for NH_3 is 1.80×10^{-5}.

We should recognize the ammonium ion as the conjugate acid for ammonia. Since ammonia is a weak base, then the conjugate acid has a measurable strength as an acid.

$$K_a = K_w/K_b = 1.00 \times 10^{-14} / 1.80 \times 10^{-5}$$

$$\therefore K_a = 5.56 \times 10^{-10}. \checkmark$$

- Calculate the pH of a 1.00 M solution of NaCN. K_a for HCN is 4.00×10^{-10}.

We already saw that the cyanide ion is a base with a K_b of 2.50×10^{-5}. Therefore, we should determine the concentration of hydroxide ions in this solution.

	CN^-	+	H_2O	$\rightleftharpoons$	OH^-	+	HCN
init:	1.00 M				0		0
@ equil:	1.00 - x				x		x

$$K_b = \frac{[OH^-][HCN]}{[CN^-]} = \frac{x^2}{1.00} = 2.50 \times 10^{-5}$$

Therefore, $\quad x = [OH^-] = 5.00 \times 10^{-3}$ M.

and: $\quad pOH = -\log(5.00 \times 10^{-3}) = 2.30$

and finally, $\quad pH = 14 - 2.30 = 11.70$ ✓

- What would be the pH of a 0.30 M solution of NH_4Ac? The Ka for HAc is equal to the Kb for NH_3 (numerically they are both 1.8×10^{-5}).

    This is a salt composed of two ions: the ammonium ion and the acetate ion. We need to consider each ion separately, to see if they will hydrolyze or not. We have already seen that the ammonium ion hydrolyzes with an equilibrium constant (Ka) of 5.56×10^{-10}. How about the acetate ion? Well, since the dissociation constant of the acetic acid is exactly the same as that of ammonia, it will have a Kb of 5.56×10^{-10}. This means that both of these ions are equally strong, one as an acid and the other as a base. Therefore, the amounts of hydronium and hydroxide ions formed by the dissociation of each will be exactly the same, and these will neutralize each other. Therefore, the pH will be equal to seven.

17.7 Common Ion Effect.

The common ion effect is observed when we have both the weak acid and its conjugate base present initially or when we have initially present species from both sides of the equilibrium. It will further reduce the dissociation of the weak acid itself.

- Calculate the pH of a 0.50 M solution of HA which is also 0.10 M in NaA. The Ka for HA is 1.0×10^{-5}.

    We should start by thinking what is the equilibrium for this system. The presence of the weak acid and its conjugate base initially may be a bit confusing, because we may be tempted to try to force one to react with the other, as seen below.

$$HA + NaA \rightleftharpoons NaA + HA$$

Although this happens all the time, it is pointless and certainly does not lead toward the formation of either hydronium nor hydroxide ions. Therefore, this particular approach would never give us the pH. This means that the equilibrium should be the dissociation of either the weak acid or the weak base in water. These are the equilibria that will bring forth the generation of hydronium or hydroxide ions in the balanced equation.

$$HA + H_2O \rightleftharpoons H_3O^{+1} + A^{-1}.$$

or: $\quad\quad A^{-1} + H_2O \rightleftharpoons HA + OH^{-1}.$

Although either would give the same result, it is easier to work with the dissociation of the weak acid, since we already know Ka. If we chose to use the second equation, then we would need to calculate Kb first, solve for the concentration of hydroxide, calculate pOH, and then the pH.

$$HA + H_2O \rightleftharpoons H_3O^+ + A^-$$

init: 0.50 M   ≈ 0   0.10 M

@ equil: 0.50 - x   x   0.10 + x

$$K_a = \frac{[H_3O^+][A^-]}{[HA]} = \frac{(0.10 + \cancel{x})\, x}{0.50 - \cancel{x}} = 1.0 \times 10^{-5}$$

$$\therefore x = [H_3O^+] = 5.0 \times 10^{-5} \text{ M}$$

and pH = *4.30* ✓

17.8 Buffers.

The solution we just described in the previous problem is referred to as a **buffer** solution. In order to have a buffer solution we must have present initially a weak acid and its conjugate base (or a weak base and its conjugate acid). The relative concentrations of these species must be very close to one another. Actually, the closer they are, the better the buffer solution is going to be. As a matter of fact, the best buffer solution we can prepare with a given system is one in which the concentration of the conjugate acid and base are exactly the same, $[HA] = [A^-]$. We call this the optimum buffer for that particular acid base system.

The most important property of buffer solutions is that they will maintain a relatively constant pH upon the addition of small quantities of strong acids or strong bases. We say that buffer solutions are pH Resistant.

For example, let's say that we have some bacteria growing in a large 1.00 L biological dish and that the pH of this system happens to be seven. If we were to add 0.10 mol of HCl, that would cause the pH to drop from seven down to one. Clearly, the bacteria would die almost immediately, since all their biological reactions are sensitive not only to temperature but also to pH. However, the buffers in the system would only let the pH change slightly (let's say to 6.92!) with this addition of HCl.

Chapter 17: Acids and Bases.

In a good buffer, the ration of [A⁻] / [HA] is very close to one. The addition of small quantities of strong acids or bases would change this ration only slightly, thus the pH would remain essentially the same.

- Consider the weak acid HAc, whose Ka is 1.85×10^{-5}.

   a) What is the pH of a 1.00 M solution of HAc which is also 1.00 M in NaAc?

   We can see that this is simply a common ion effect problem.

	HAc	+	H₂O	⇌	H₃O⁺	+	Ac⁻
init:	1.00 M				≈ 0		1.00 M
@ equil:	1.00 − x				x		1.00 + x

   $$K_a = \frac{[H_3O^+][Ac^-]}{[HAc]} = \frac{(1.00)x}{1.00} = 1.85 \times 10^{-5}$$

   and $x = [H_3O^+] = 1.85 \times 10^{-5}$ M

   ∴ pH = 4.73 ✓

   b) If we were to add 0.10 mol of HCl to 1.00 L of this buffer, what will be the pH of the resulting solution?

   Initially, we have a solution with a pH of 4.73 and equal concentrations of HAc and Ac⁻ (1.00 M). What happens when we add 0.10 moles of HCl to this solution? We are really adding 0.10 moles of hydronium ions to the solution (which is a common ion). This will force the reaction to go backwards, and since Q will be so much larger than K, we will shift it 100% to the left and then allow it to return to equilibrium, as shown below.

	HAc	+	H₂O	⇌	H₃O⁺	+	Ac⁻
init:	1.00 M				1.85×10^{-5} M		1.00 M
after:	1.00 M				0.10 M		1.00 M
←100%:	1.10 M				0 M		0.90 M
@ equil:	1.10 − x				x		0.90 + x

Chapter 17: Acids and Bases.

$$K_a = \frac{[H_3O^+][Ac^-]}{[HAc]} = \frac{(0.90)x}{1.10} = 1.85 \times 10^{-5}$$

Solving for x, etc., we get a mere decrease of 0.08 pH units!!!

$$pH = 4.65 \checkmark$$

17.9 The Henderson-Hasselbach Equation.

Henderson and Hasselbach derived a logarithmic equation that is identical to Ka.

$$K_a = \frac{[H_3O^+][A^-]}{[HA]}$$

$$[H_3O^+] = \frac{[HA]}{[A^-]} K_a$$

Taking (-log) on both sides, we get:

$$-\log [H_3O^+] = -\log \frac{[HA]}{[A^-]} - \log K_a$$

$$pH = -\log \frac{[HA]}{[A^-]} + pK_a$$

or in the final form:

$$\mathbf{pH = pK_a + \log \frac{[A^-]}{[HA]}}$$

This is referred to as the **Henderson-Hasselbach equation**. It allows us to determine the pH of a buffer solution without having to actually write down the equilibrium.

- Determine the pH of the buffer solution that is 0.100 M in HAc and 0.200 M in NaAc. The Ka for HAc is 1.85×10^{-5}.

    The pKa for HAc = $-\log (1.85 \times 10^{-5})$ = 4.73.

Therefore, the pH of the solution is given by

$$pH = 4.73 + \log(0.200 \text{ M}/0.100 \text{ M})$$

$$\therefore \text{pH} = 4.73 + 0.301 = \mathit{5.03} \checkmark$$

A buffer solution only functions as such (pH resistant) if the ratio of base to acid is between 10 : 1 and 1 : 10. This means that if the ratio is any larger than that or any smaller than that, then this solution will no longer be pH resistant. It would simply be a solution that has a common ion present initially. The range of a buffer is defined as the pH's at which this solution will be pH resistant. We have already seen that the best possible buffer (optimum) is when we have equal concentrations of the weak acid and its conjugate base. Under these conditions, the Henderson Hasselbach equation becomes:

$$\text{pH} = \text{pKa} + \log(1)$$

or: $\underline{\text{pH} = \text{pKa}}$ ✓

What happens towards the end of this range? In the upper end, the concentration of the base is ten times larger than the concentration of the acid, therefore, the Henderson Hasselbach equation becomes:

$$\text{pH} = \text{pKa} + \log(10)$$

or: $\underline{\text{pH} = \text{pKa} + 1}$ ✓

At the lower end of the range, the concentration of the base is ten times smaller than that of the acid, therefore, the Henderson Hasselbach equation becomes:

$$\text{pH} = \text{pKa} + \log(0.10)$$

or: $\underline{\text{pH} = \text{pKa} - 1}$ ✓

Combining them, we get the general range for a buffer solution, which is:

$$\underline{\text{pH} = \text{pKa} \pm 1} \checkmark$$

- The pKa for acetic acid (HAc) is 4.73. Can we prepare a solution with acetic acid and sodium acetate (the conjugate base) at a pH of 6.00?

    With this system, we can prepare a buffered system of pH's ranging from 3.73 to 5.73. However, since the question did not mention the word buffer, the answer should be YES. We can prepare a solution of acetic acid and sodium acetate that will have a pH of 6, but this solution WILL NOT be a buffer solution. Therefore, it will not be pH resistant. If we were asked to describe how the solution would be prepared, we would have to perform a simple calculation.

$$pH = pKa + \log\{[Ac^-]/[HAc]\}$$

$$6.00 = 4.73 + \log\{[Ac^-]/[HAc]\}$$

$$\therefore [Ac^-]/[HAc] = \mathit{18.6} \checkmark$$

Since the concentration of the base is nineteen times larger than the concentration of the acid in this solution, it falls outside of the range in which a buffer has the potential to be effective (it is more than 10 times larger than the concentration of the acid).

There is one more aspect of buffers that we should mention. The **strength** of a buffer depends on the actual amounts of the components in this solution. The larger the amounts of the components, the stronger the buffer solution will be. Since we have viewed a buffer solution as one that can handle small amounts of strong acids and bases without changing the solution's pH, the larger the concentrations of the components, the larger the amounts of strong acids and bases that this solution will be able to handle without changing its pH considerably.

Finally, if we were asked to choose an acid base system to prepare a buffer solution, the best system will be the one whose pKa is closest to the desired pH. Of course, this presumes that all of the systems are going to be equally acceptable in other ways (like toxicity, etc.).

- Let's say that we want to prepare a buffer solution of pH 5.63 and we have available the following acid/base systems:

System	Ka	pKa
HA	2.50×10^{-5}	4.60
HB	3.33×10^{-6}	5.48
HD	1.11×10^{-7}	6.96

Which one should we use, and how should we prepare that solution?

The pKa that is closest to the desired pH is that of HB. Therefore, we should prepare a solution that contains both **HB and** its conjugate base, B^-, in proportions such that the pH of the solution ends up being 5.63. How do we determine the ratio in which to prepare this solution? By using the Henderson Hasselbach equation.

$$pH = pKa + \log\{[B^-]/[HB]\}$$

$$5.63 = 5.48 + \log\{[B^-]/[HB]\}$$

Solving for the ratio, we get:

$$[B^-]/[HB] = 1.41 \checkmark$$

This means that in order for us to prepare this solution, we must combine B^- and HB in such way that we have 1.41 times more moles of B^- (the base) than HB (the acid).

- What is the pH of a buffer prepared by dissolving 0.10 mol of HA and 0.20 mol of NaA in enough water to form 1-L of solution? K_a for HA is 2.5×10^{-6}.

    As soon as we read the word buffer, we immediately recognize that this problem can be solved with the Henderson Hasselbach equation.

$$pH = pKa + \log\{[A^-]/[HA]\}$$
$$pH = -\log(2.5 \times 10^{-6}) + \log\{0.20/0.10\} = 5.60 + 0.30$$
$$\therefore pH = 5.90 \checkmark$$

- How would you prepare 5.00 L of a buffer solution with a pH of 5.00 if you have available a 0.0350 M solution of HAc and solid NaAc? K_a for HAc is 1.85×10^{-5} and the MW of NaAc is 82.0 g/mol.

    This problem asks us to prepare a specific volume of a certain buffer solution. The availability of a solution of 0.0350 M solution of HAc defines the concentration of one of the species, therefore, we only have to figure out what is the molar concentration of its conjugate base (the sodium acetate). For this purpose, we use the Henderson Hasselbach equation.

$$pH = pKa + \log\{[Ac^-]/[HAc]\}$$
$$5.00 = 4.73 + \log\{x/0.0350 \text{ M}\}$$

Solving for **x**, we get:

$$x = [Ac^-] = 0.0652 \text{ M}$$

This means that we need to have in this buffer solution 0.0652 moles of NaAc in every liter of solution. Since we are preparing 5.00 L of solution, then:

$$5.00 \text{ L} \times \frac{0.0652 \text{ moles NaAc}}{1 \text{ L}} \times \frac{82.0 \text{ g NaAc}}{1 \text{ mole}}$$

Therefore, we need to add *2.67 g of NaAc* to the 5.00 L of 0.0350 M solution of acetic acid in order to obtain a buffer solution of pH 5.00.

- Let's say that we have 1-L of a 0.200 M solution of the weak acid HA ($K_a = 1.00 \times 10^{-6}$). If we are told that the strong base NaOH reacts quantitatively[1] with HA as shown below, how many grams of NaOH (MW = 40.0) are needed in order to form a buffer solution of pH 6.00?

$$NaOH_{(s)} + HA_{(aq)} \Rightarrow H_2O_{(\ell)} + NaA_{(aq)}$$

Once more, we can solve for the concentration of the conjugate base (A^-) using the Henderson Hasselbach equation. However, we should immediately recognize that the desired pH happens to be the pKa exactly. When the pH of the solution is equal to the pKa, then we should have equal amounts of the two components (the acid and its conjugate base). However, the difference between this problem and the previous one is that we are NOT adding the conjugate base from an external source. We are converting the acid into its conjugate base by the action of the strong base NaOH. From the balanced reaction, the NaOH and the HA react in a one to one mole ratio. We are starting with 0.200 moles of HA (calculated by multiplying the molarity times the volume). It will take exactly 0.100 moles of NaOH to convert half of the HA sample into A^-. Therefore, we should add *4.00 g of NaOH* (one tenth of a mole).

17.10 Fractions of the Conjugates Forms in a Solution of a Given pH.

We have mentioned that a good buffer is one that has concentrations of both the weak acid and its conjugate base that are close to one another, the optimum being exactly the same. For this solution, the pH of this buffer is equal to its pKa. Let's try to understand a little bit better what happens around the pKa region. In order to do that, we will derive an equation that allows us to calculate the fraction of one of the forms, let's say the basic form (A^-).

$$pH = pK_a + \log \frac{[A^-]}{[HA]}$$

$$\log \frac{[A^-]}{[HA]} = pH - pK_a$$

$$\therefore \frac{[A^-]}{[HA]} = 10^{(pH - pK_a)}$$

---

[1] The context of the word quantitative is that the reaction goes 100% towards the formation of products (or very close to it).

Remember that we are looking for an equation that gives us the fraction of [A⁻]. Such fraction would be given by:

$$\text{Fraction of A}^- = \frac{[A^-]}{[HA] + [A^-]}$$

What does the denominator represent? It represents the total amount of acid present in the solution at any time. [HA] is the acidic form of the acid, also called the undissociated form. [A⁻] is the basic form of the acid, also called the dissociated form. By dividing both the top and the bottom of the equation by [HA], we get:

$$\text{Fraction of A}^- = \frac{[A^-]/[HA]}{\frac{[HA]+[A^-]}{[HA]}} = \frac{[A^-]/[HA]}{1 + [A^-]/[HA]}$$

but: $\dfrac{[A^-]}{[HA]} = 10^{(pH - pKa)}$

$$\boxed{\text{Fraction of A}^- = \frac{10^{(pH-pKa)}}{1 + 10^{(pH-pKa)}}}$$

Previously, we said that a buffer system works only within a range of pH = pKa ± 1. We also saw that in order for a buffer system to function as such, the [A⁻] must be close to that of the [HA]. This implies that the fraction of A⁻ should be close to 0.500 for a buffer (the point when we have identical amounts of each, the A⁻ and the HA). It should follow that we will have identical amounts of A⁻ and HA only when the pH = pKa. It is also true that if the pH is not equal to the pKa, we will have different amounts of A⁻ and HA. Thus, we will say that the solution has an excess of the acidic form (HA) whenever the pH is lower than the pKa and that the solution will have an excess of the basic form (A⁻) when the pH is higher than the pKa.

A buffer solution whose pH is exactly equal to the pKa is said to be "neutral" with respect to its acid base components. Let's make the comparison and contrast to the pure water system. When the pH of water is 7, we have equal amounts of hydronium and hydroxide ions and we say that the aqueous solution is neutral. Similarly, at a pH lower than 7, we have an excess of hydronium ions and we state that the solution is acidic. At a pH higher than 7, we have an excess of hydroxide ions, and we state that the solution is basic.

Now, going back to our HA/A⁻ system, when the pH = pKa, we have equal amounts of

Chapter 17: Acids and Bases.

HA and A⁻, therefore, the solution is said to be "neutral" with respect to this system. When the pH is lower than the pKa, we have an excess of the acidic form (HA). Finally, when the pH is higher than the pKa, we have an excess of the basic form (A⁻). Let's prove this mathematically by calculating the fraction of A⁻ at both ends of the pH range for a buffer.

a) <u>When pH = pKa -1</u>.

Under these conditions, pH - pKa = -1. So, fraction of A⁻ = $10^{-1} / (1 + 10^{-1})$.

Therefore, Fraction of A⁻ = <u>0.0909</u>

As expected, at this pH, we have a lot more of the HA (acidic form) than we do of the A⁻ (basic form). In other words, when the pH = pKa - 1, 90.9% of all the molecules are in the acidic form (HA) and 9.09% of all the molecules are in the basic form (A⁻).

b) <u>When pH = pKa + 1</u>.

Under these conditions, pH - pKa = +1. So, fraction of A⁻ = $10^{1} / (1 + 10^{1})$.

Therefore, Fraction of A⁻ = <u>0.909</u>

As expected, at this pH, we have a lot more of the A⁻ (basic form) than we do of the HA (acidic form). Once more, this means that when the pH = pKa +1, then 9.09% of all the molecules exist in the acidic form (HA), and 90.9% of the molecules exist in the basic form (A⁻).

The graph on the next page shows the behavior of the fraction of A⁻ as a function of pH for an example of a weak acid, HA, whose Ka is 1.00×10^{-5}. Notice that we will be plotting the fraction of either A⁻ vs. pH or the fraction of HA (= 1 - fraction of A⁻) vs. pH. It should be clear that as the pH increases, the solution is becoming more and more basic, therefore, the fraction of A⁻ is increasing and the fraction of HA is decreasing.

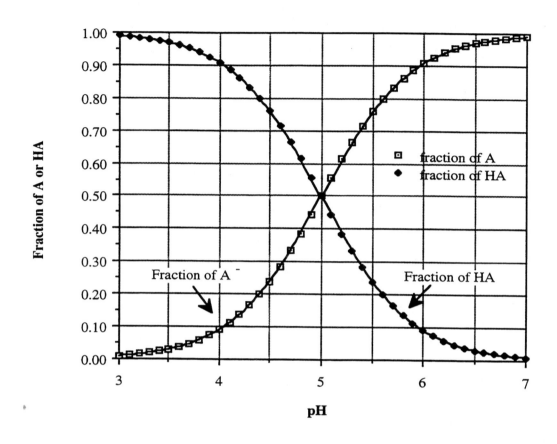

The graph indicates that the pH = pKa when the fractions of HA and A⁻ are equal to 0.500. It can also be concluded that for all practical purposes, when the pH is at least 2 units above or below the pKa, we only have one of the forms present. To be more specific, at two pH units below the pKa, we have approximately 100% HA and no A⁻. On the other hand, at two pH units above the pKa, we have approximately 100% A⁻ and no HA.

Conclusions:

1. A buffer works only when its pH is within one unit of its pKa.

2. At a pH one unit below the pKa, we have 91% of the acidic form.

3. At a pH one unit above the pKa, we have 91% of the basic form.

4. At a pH two units below the pKa, we only have the acidic form.

5. At a pH two units above the pKa, we only have the basic form.

17.11 Acid Base Titrations.

An acid base titration reaction represents a quantitative proton transfer from an acid to a base. As long as one of the two species (the base or the acid) is strong, the reaction will proceed to completion. We will discuss two particular examples: the case of a strong acid titrated with a strong base, and the case of a weak acid titrated with a strong base.

For our first example, let's consider the case of 50.0 mL 0.100 M solution of HCl in a flask being titrated with 0.100 M NaOH from a buret[1]. The initial pH of the HCl solution depends exclusively on the concentration of the HCl in that solution. HCl is a strong acid, therefore if the concentration of HCl is 0.100 M, then the pH must be 1.00. As we start to add some of the NaOH solution, the HCl reacts with it according the following reaction.

$$HCl + NaOH \Rightarrow H_2O + NaCl$$

Since we amount of HCl will be less after the reaction with a certain amount of NaOH, the concentration of HCl (left over) will be smaller than 0.100 M, therefore, the pH will be higher than 1.00. At the point where we have added 50.0 mL of the NaOH solution, we will have added exactly the same number of moles of NaOH as we had moles of HCl initially. This means that we have reached what we call the **equivalence point** of the titration ⇒ the point where we have run out completely of both, the acid (HCl) and the base (NaOH). What do we have present in solution at the equivalence point? We have two things: water and sodium chloride. What type of salt is NaCl? We know that it is composed of Na^+ and Cl^- ions. Do any of these ions hydrolyze in water? We can that the Na^+ cannot be an acid because it does not have a polarized hydrogen to donate. How about the Cl^- as a base? It does not hydrolyze either because it is the conjugate base of a strong acid. What did we say previously about the conjugate bases of strong acids? They are not bases, per se. This means that since neither is hydrolyzing, the pH of the solution at the equivalence point must be equal to seven.

Past the equivalence point, all we have is an excess of the NaOH, therefore the pH will be determined by the excess of the NaOH and the pH will be higher than seven. For example, consider the case in which we have added 60.0 mL of the 0.100 M solution of NaOH. We know that it took 50.0 mL to completely neutralize this solution. Therefore, there were 10.0 mL of the solution that did not react and are present in the solution. How many moles of NaOH are present in those 10.0 mL? We can determine this by the multiplication of the molarity times the volume.

---

[1] A buret is a uniform-bore glass tube with fine gradations and a stopcock at the bottom. It is used in laboratory procedures for accurate fluid dispensing and measurement.

Thus, we have 0.00100 moles of NaOH present in a total volume of 110.0 mL of solution. This is the total volume present after we mix 50.0 mL of the acid with 60.0 mL of the base. Therefore, the concentration of hydroxide ions in solution at this point would be: 0.00100 mol / 0.110 L, which is 0.00909 M. Finally, the pOH would be 2.041 and the pH 11.959.

The following plot shows the gradual increase of pH as a function of the volume of the NaOH added. We should point out that the equivalence point is easily detected from this type of plot because it is always observed at the point where the pH changes the fastest (the steepest slope) as a function of the volume added.

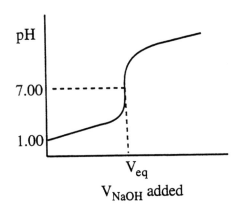

How about the case of a weak acid with a strong base? Let's take the specific case of 50.0 mL of a 0.100 M solution of aspirin (HAsp) with 0.100 M NaOH. Aspirin is a weak acid with a Ka of 3.02×10^{-4}. These two react according to the following reaction:

$$HAsp + NaOH \Rightarrow H_2O + NaAsp$$

It should be clear that the HAsp is the weak acid and that the NaAsp (or Asp$^-$) is what we call the conjugate base.

Before we add any NaOH from the buret, all we have is a 0.100 M solution of HAsp. Since this is a weak acid, the pH of this solution is determined as follows:

$$HAsp + H_2O \rightleftharpoons H_3O^+ + Asp^-$$

init:	0.100 M		0	0
@ equil:	0.100 - x		x	x

$$K_a = \frac{[H_3O^+][Asp^-]}{[HAsp]} = \frac{x^2}{0.100 - x} = 3.02 \times 10^{-4}$$

Using a quadratic equation, we can determine:

$$x = 0.00535 \text{ M} = [H_3O^+]$$

So, $$pH = -\log x = -\log(0.00535)$$

$$= 2.27 \checkmark$$

Let's determine the pH of the resulting solution after the addition of 10.0 mL of the NaOH solution. Remember that the molarity of the NaOH solution is 0.100 M. Therefore, in 10.0 mL of this solution we have (n = M V): 0.00100 moles of NaOH. Since we started with 50.0 mL of the aspirin, which is also 0.100 M, we had 0.00500 moles of HAsp in the flask initially. This reaction is 1:1 (mole ratio), therefore, the LR is the NaOH, which means that we will have 0.00400 moles of HAsp left unreacted and we will have 0.00100 moles of Asp⁻ formed. How do we determine the pH of this mixture of HAsp and Asp⁻? As soon as we read this question, we should think that there is a very simple way to do this: with the Henderson Hasselbach equation.

$$pH = pKa + \log\{[Asp^-]/[HAsp]\}$$

$$= -\log(3.02 \times 10^{-4}) + \log(0.00100/0.00400)$$

$$pH = 3.52 - 0.602 = 2.92 \checkmark$$

It should become clear that at exactly half-way through the titration, we will have identical amounts of HAsp and Asp⁻, therefore, the pH will be equal to the pKa because the ratio of Asp⁻ to HAsp will be equal to one (and the logarithm of one is equal to zero). Conclusion: After the addition of 25.0 mL of the NaOH solution, the pH of the solution will be **3.52**. Please, remember that when the pH = pKa, we have a buffer solution of this system, meaning that the solution will be pH resistant. This means that the pH will not be changing considerably when we are close to half-way through the titration.

At the equivalence point, the point where we have added a total of 50.0 mL of the NaOH solution, we find that all the HAsp is gone and all the NaOH is consumed also. Therefore, the only thing present at this point is the conjugate base of the aspirin, Asp⁻ and lots of water. Since the conjugate base of the aspirin is a base of measurable strength, the pH of this solution WILL NOT BE EQUAL TO SEVEN!! Indeed, it should be larger than seven and in order to determine the pH of this solution, we must work with the following equilibrium:

$$Asp^- + H_2O \rightleftharpoons HAsp + OH^-$$

Since we are not using the Henderson Hasselbach equation, we must determine the initial concentration of Asp^-, which means that we must determine the moles and divide them by the total volume at the equivalence point. We have 0.00500 moles of Asp^- and the volume is 100.0 mL. Therefore, the $[Asp^-]$ is 0.00500 mol / 0.1000 L = 0.0500 M. Now,

	Asp^-	+ H_2O	$\rightleftharpoons$	$HAsp$	+ OH^-
init:	0.0500 M			0	0
@ equil:	0.05 - x			x	x

What is the equilibrium constant that controls the equilibrium shown above? We call it K_b. The K_b for Asp^- must be calculated with the expression:

$$K_b = \frac{K_w}{K_a} = \frac{1.00 \times 10^{-14}}{3.02 \times 10^{-4}} = 3.31 \times 10^{-11}$$

$$K_b = \frac{[HAsp][OH^-]}{[Asp^-]} = \frac{x^2}{0.0500} = 3.31 \times 10^{-11}$$

and finally[1]: $x = [OH^-] = 1.29 \times 10^{-6}$ M

Therefore, $pOH = -\log x = 5.89$

and: $pH = 14 - 5.89 = 8.11$ ✓

Conclusion: At the equivalence point, the pH is higher than seven because of the presence of the weak base, Asp^-.

Past the equivalence point, we have excess base (NaOH), therefore the pH would be controlled by the excess of the NaOH, a strong base. This is completely analogous to the previous example (the reaction between NaOH and HCl). Below, we can see a pictorial representation of all the points that we have discussed for this titration. Let's contrast the two plots we have seen so far. In the case of the strong acid with the strong base, the increases were smooth up to the equivalence point. However, in the case of a weak acid with a strong base, we see that initially there is a sharp increase in the pH and that it kind of levels off half-way through

---

[1] x is negligible compared to 0.0500 M due to the very small size of K_b.

the titration. This is due to the formation of a buffered system which disappears once the pH rises one unit above the pKa. Then the pH continues to rise more sharply until we reach the equivalence point. As in the previous example, the pH increases very sharply at the equivalence point and this point can be easily identified as the one where the change in the pH is the largest.

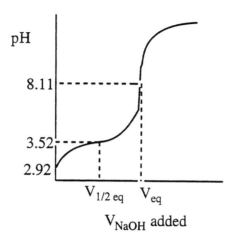

- Let us try to follow the titration of 30.00 mL of 1.00 M acetic acid (weak acid, HAc, with Ka = 1.85 x 10^{-5}) with a 1.00 M solution of the strong base NaOH. To get a good understanding of what is actually going on as we add the base slowly, let's observe the behavior of the pH as a function of the volume of NaOH added. We can divide this study into four parts: 1) before the addition of NaOH, 2) after addition of NaOH but before the equivalence point, 3) the equivalence point, and 4) after the equivalence point.

1. <u>Before the addition of NaOH.</u>

    At the very beginning of the titration, all we have is the weak acid (HAc) and lots of water, the solvent. How can the weak acid establish an equilibrium to maintain the pH in an aqueous solution? We have already seen that and we know that it depends on how much acid we have.

$$\begin{array}{cccccc}
 & HAc & + & H_2O & \rightleftharpoons & H_3O^+ & + & Ac^- \\
init: & 1.00\ M & & & & \cong 0 & & 0 \\
@\ eq: & 1.00 - x & & & & x & & x
\end{array}$$

$$K_a = \frac{[H_3O^+][Ac^-]}{[HAc]} = \frac{x^2}{1.00 - x} \cong \frac{x^2}{1.00} = 1.85 \times 10^{-5}$$

Chapter 17: Acids and Bases.

So, $\quad x = [H_3O^+] = 4.30 \times 10^{-3}$ M

$$pH = 2.37 \checkmark$$

2. <u>After addition of 5.00 mL of the NaOH solution.</u>

Once we start adding the base, we must think about what is actually happening. The NaOH is going to react with the HAc and this is a reaction that goes essentially 100% towards completion.

	HAc	+	NaOH	⇒	H_2O	+	NaAc
init:	30 mmol		5 mmol				0
100%:	25 mmol		0		⟹		5 mmol

What do we have present in the flask after this reaction is over? We have 25 mmoles of HAc, 5 mmol of NaAc (which is the same as Ac⁻) and lots of water, which is the solvent. The question we should ask ourselves is, how are these species going to establish an equilibrium that will in turn enable us to determine the pH of the solution? The equilibrium is simply the dissociation of the weak acid (HAc) in the presence of a common ion (Ac⁻). We have learned already that we can solve problems like these by either solving directly the equilibrium or by the use of the Henderson-Hasselbach equation.

$$pH = pK_a + \log\{[Ac^-]/[HAc]\}^1$$

$$= 4.73 + \log(5/25) = 4.73 - 0.70 = 4.03 \checkmark$$

3. <u>After addition of 15.00 mL of the NaOH solution.</u>

At this point it becomes a routine to perform these calculations. First, you calculate the mmoles of HAc and NaOH that are being added. Use then the titration reaction to figure out how many mmoles of HAc and Ac⁻ are left over after the reaction. Finally, use the Henderson Hasselbach equation to determine the pH of the resulting solution.

	HAc	+	NaOH	⇒	H_2O	+	NaAc
init:	30 mmol		15 mmol				0
100%:	15 mmol		0		⟹		15 mmol

---

[1] Since this is a ratio of concentrations, we may use just milli moles instead of molarity.

Using the Henderson Hasselbach equation, we get:

$$pH = pK_a + \log\{[Ac^-]/[HAc]\}$$

$$= 4.73 + \log(15/15) = 4.73 \checkmark$$

Please notice that at this point in the titration, the pH = pKa. This is always true halfway through a titration.

4. **After addition of 30.00 mL of the NaOH solution.**

It is fairly simple to notice that this is the equivalence point of the titration (the point where you have added exactly the same number of mmoles of NaOH as you had of HAc when you started). What do we have after the titration reaction is over? Only 30 mmoles of Ac$^-$ and lots of water, which is the solvent. The question is once more, how do we determine the pH of this reaction? We cannot use the Henderson-Hasselbach equation because we only have one of the two species, not both. We should recognize this as the hydrolysis of the Ac$^-$ ion.

	Ac$^-$	+	H$_2$O	⇌	HAc	+	OH$^-$
init:	30 mmol/60 mL				0		0
	0.500 M				0		0
@ eq:	0.500 - x				x		x

$$K_b = \frac{[HAc][OH^-]}{[Ac^-]} = K_w/K_a = 5.41 \times 10^{-10}$$

$$\frac{x^2}{0.500} = 5.41 \times 10^{-10}$$

So, $\quad x = [OH^-] = 1.64 \times 10^{-5}$ M.

Finally, $\quad pOH = 4.78$

and: $\quad pH = 9.22 \checkmark$

5. <u>After addition of 35.00 mL of the NaOH solution.</u>

At this point we are past the equivalence point and we have an excess of NaOH (same as OH^-). Actually, the excess if 5 mmoles of OH^- and it will be dissolved in 65 mL of solution. The pH is clearly determined by the excess of NaOH.

$$[OH^-] = 5 \text{ mmol} / 65 \text{ mL} = 0.0769 \text{ M}$$

$$\therefore \text{ pOH} = 1.11 \text{ and } \text{pH} = 12.89 \checkmark$$

Below, we can see a plot of the pH of the solution as a function of the volume of NaOH added for this last example. Once more, we can see that somewhere around halfway through the titration, we have the buffered region of the titration, a region where the pH does not change drastically upon the addition of the strong base NaOH.

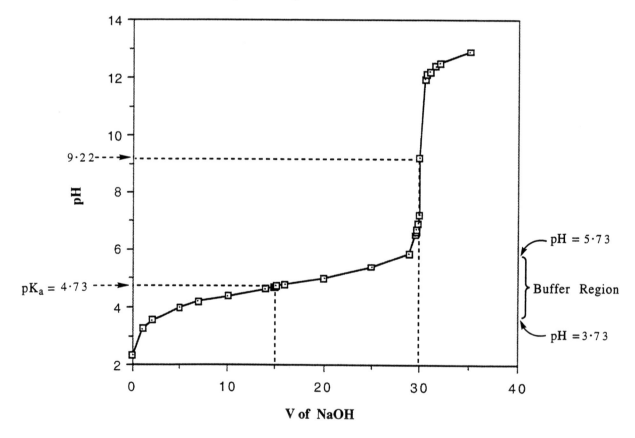

Titration of 30 mL of 1M HAc with 1M NaOH

Chapter 17: Acids and Bases.

17.12 Polyprotic Acids.

These acids contain more than one acidic hydrogen and they follow a stepwise dissociation. *Examples:* H_2S, H_2CO_3, H_2SO_4, etc. In the case of carbonic acid we have the following two equilibria:

$$H_2CO_3 + H_2O \rightleftharpoons H_3O^+ + HCO_3^- \qquad K_{a_1} = 4.20 \times 10^{-7}$$

$$HCO_3^- + H_2O \rightleftharpoons H_3O^+ + CO_3^{-2} \qquad K_{a_2} = 4.80 \times 10^{-11}$$

It should be obvious that $K_{a_2} < K_{a_1}$ since it should be harder to remove a positive proton from a species that is already negative (HCO_3^-). However, when K_{a2} is more than 1,000 times smaller than K_{a_1}, then the amount of hydronium ion formed in the second step is negligible compared to the amount of hydronium ion formed in the first step.

- Determine the pH of a 0.10 M solution of H_2CO_3.

The Ka for the first dissociation is much larger than the Ka for the second dissociation. Therefore, it is safe to say that all the hydronium ions are formed during the first step.

$$H_2CO_3 + H_2O \rightleftharpoons H_3O^+ + HCO_3^-$$

init: 0.10 M   ≈ 0   0

@ equil: 0.10 - x   x   x

$$K_a = \frac{[H_3O^+][HCO_3^-]}{[H_2CO_3]} = \frac{x^2}{0.10} = 4.20 \times 10^{-7}$$

$$\therefore x = [H_3O^+] = 2.0 \times 10^{-4} \text{ M}$$

and pH = *3.69* ✓

- What is the concentration of the carbonate ions for the solution in the previous problem?

In order to answer this question, we need to consider the second dissociation, because this is where the carbonate ion appears (look at the dissociations above).

$$HCO_3^- + H_2O \rightleftharpoons H_3O^+ + CO_3^{-2}$$

init: 2.0×10^{-4} M 2.0×10^{-4} M   0

@ equil: $2.0 \times 10^{-4} - x$ $2.0 \times 10^{-4} + x$ x

Chapter 17: Acids and Bases.

$$K_{a2} = \frac{[H_3O^+][CO_3^{-2}]}{[HCO_3^-]} = \frac{(2.0 \times 10^{-4}) x}{2.0 \times 10^{-4}} = 4.80 \times 10^{-11}$$

$$\therefore x = [CO_3^{-2}] = 4.80 \times 10^{-11} \, M \checkmark$$

- What is the pH of a 0.10 M solution of Na_2CO_3?

In this case, we only have the carbonate ion and lots of water initially, therefore, the reaction is between the carbonate ion and the water. The carbonate ion is a base, and it could take up to two protons (to form carbonic acid). However, the amount of hydroxide ions present in solution can be calculated with the first step only because the second one is more than 1000 times smaller than the first (just like the first example).

$$CO_3^{-2} + H_2O \rightleftharpoons HCO_3^- + OH^-$$

Let's calculate the first K_b for the carbonate ion.

$$K_{b_1} = K_w / K_{a_2}$$

This presents an interesting dilemma. How do we know that we need the second Ka and not the first one? Well, looking at the two equilibria on top of the previous page, we can see that the one that involves the carbonate AND the bicarbonate ions is the second equilibrium. These are the two species that are at equilibrium when the carbonate ion hydrolyzes in water.

$$\therefore K_b = 1.00 \times 10^{-14} / 4.80 \times 10^{-11} = 2.08 \times 10^{-4} \checkmark$$

Therefore,

	CO_3^{-2}	+ H_2O $\rightleftharpoons$	HCO_3^-	+ OH^-
init:	0.10 M		0	≈ 0
@ equil:	0.10 - x		x	x

Let's assume that x is negligible compared to 0.10. This is not a great assumption because the K_b is in the order of 10^{-4}, however, since we have only two significant figures, the error will be minimized. This way we can avoid solving a quadratic equation.

$$K_b = \frac{[OH^-][HCO_3^-]}{[CO_3^{-2}]} = \frac{x^2}{0.10} = 2.08 \times 10^{-4}$$

$$\therefore x = [OH^-] = 4.6 \times 10^{-3} \text{ M}$$

So, $\quad$ pOH = 2.34 and *pH = 11.66* ✓

Sometimes it is hard to tell whether a species is an acid or a base. This is particularly true for ions of polyprotic acids. An ion like HCO_3^-, the bicarbonate ion, presents this dilemma. Notice that this ion has a polarized, acidic hydrogen, but it also has a negative charge which technically gives it the potential to be a base. How can we tell? Let's look at the example below, which illustrates this concept and how the decision is reached.

- Calculate the pH of a 0.25 M solution of NaHS. For H_2S: $Ka_1 = 5.7 \times 10^{-8}$ and $Ka_2 = 1.2 \times 10^{-13}$.

This problem is a lot more complicated than it seems at first, because for the very first time, we are encountering a situation where we do not know initially if the ion present is an acid or a base. Remember that what we have initially here is HS^- ions and lots of water. There are also sodium ions but we should know that these are just spectator ions. The question we should ask ourselves is: What is this ion (HS^-), an acid or a base? It has the polarized hydrogen to donate and thus it qualifies as an acid. It also has the negative charge (thus the lone pairs) to qualify as a base. In order to solve this situation, we need to determine the equilibrium constants for both of these cases and then compare them and decide.

$$HS^- + H_2O \rightleftharpoons S^{-2} + H_3O^+ \qquad Ka_2 = 1.2 \times 10^{-13}.$$

$$HS^- + H_2O \rightleftharpoons H_2S + OH^- \qquad Kb = Kw/Ka_1.$$

We should be able to tell why we used Ka_1 for the calculation of Kb above. At any rate, the Kb = 1.8×10^{-7}. When we compare these two values, we can see that the HS^- ion is really a base. The ion is about one million times stronger as a base than it is as an acid (because Kb ≈ 10^6 larger than Ka). Another way of stating this is that for every molecule that acts as an acid, there are one million others that act as a base. Anytime that one of the constants is at least 1000 times larger than the other, then that will tell us which is the predominant function (acid or base).

Once we know it is a base, the rest of the problem should be trivial.

Chapter 17: Acids and Bases.

$$HS^- + H_2O \rightleftharpoons H_2S + OH^-$$

init: 0.25 M   0   ≈ 0
@ equil: 0.25 - x   x   x

$$K_b = \frac{[OH^-][H_2S]}{[HS^-]} = \frac{x^2}{0.25} = 1.8 \times 10^{-7}$$

$$\therefore x = [OH^-] = 2.1 \times 10^{-4} \text{ M}$$

So, pOH = 3.67 and

$$\therefore pH = 10.33 \checkmark$$

17.13 Lewis Acids and Bases.

This is the last concept of acids and bases that we will cover in this course. In this concept, we will look at everything from the point of view of the base: the substance with the lone pairs. This is by far the most comprehensive of all the theories of acids and bases that we have discussed.

Base - Substance that donates an electron pair to an acid. We refer to these substances as **nucleophiles**. Basically they are the same kind as the bases in the Brönsted-Lowry concept.

Acid - Substance that accepts an electron pair from a base. We refer to these substances as **electrophiles**. The acid concept is very different form the Brönsted-Lowry concept. Any positive ion or any electron deficient compound can act like a Lewis acid. For example, positive ions like Na^+, K^+, Ca^{+2}, Al^{+3} now qualify to be Lewis acids, since they have the ability to accept electrons. Of course, we would have never considered these to be acids before, since they do not have hydrogens in their formula. Not all Lewis acids are positive ions. We may also have electron deficient species, like BF_3, as possible Lewis acids. Actually, all trivalent[1] boron and aluminum compounds are by definition Lewis acids, since they are sp^2 hybridized and have an empty p orbital. Let's look at a diagram that explains why the boron trifluoride hybridizes sp^2 and ends up being a Lewis acid.

---

[1] The term trivalent technically means having a valence of three, or that it forms three bonds.

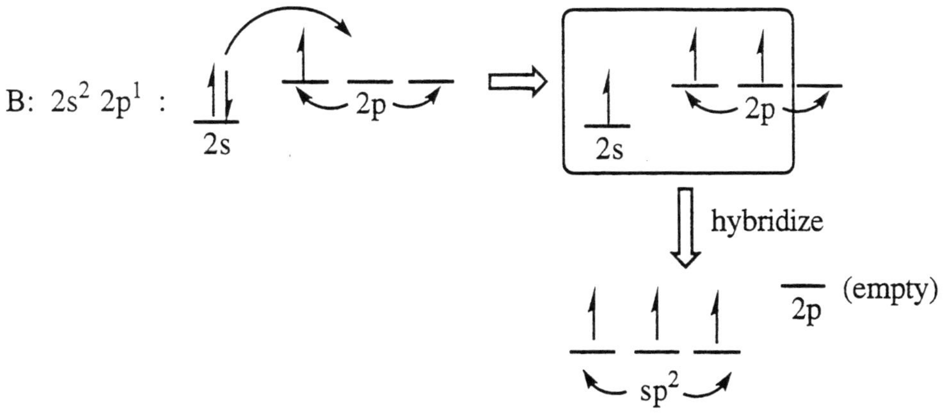

Chapter 17: Acids and Bases.

B: $2s^2\ 2p^1$:

[Orbital diagram showing ground state $2s^2\ 2p^1$ being promoted to $2s^1\ 2p^2$, then hybridized to three sp^2 orbitals with one electron each, plus an empty $2p$ orbital.]

We can see that the Boron will form the three bonds in a plane (trigonal planar, at 120° angles) and have an empty p orbital perpendicular to it. This empty p orbital is offering a "window" into the nucleus. This means that any negative species, called a nucleophile, will approach this molecule and try to form a bond with the Boron through this empty p orbital. As an example, let's have an ammonia molecule act as a base and attack the flat boron trifluoride molecule to form the acid/base complex.

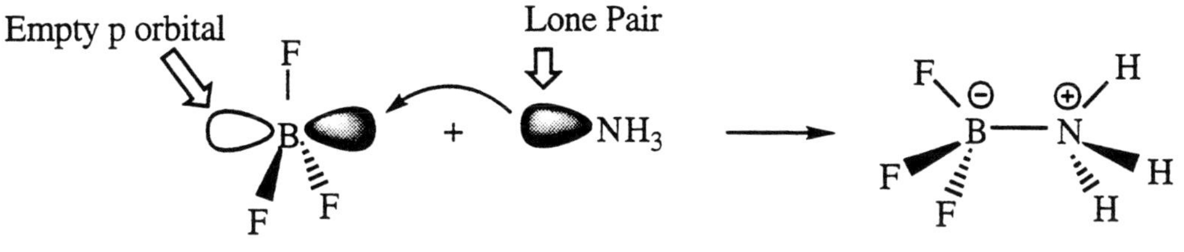

[Diagram showing BF₃ (Lewis Acid, trigonal planar) with empty p orbital reacting with NH₃ (Base, lone pair) to form the Acid/Base Complex F₃B—NH₃ with formal charges ⊖ on B and ⊕ on N.]

The term acid/base complex is very useful because it reminds us of the fact that this complex was formed from the reaction of the acid with the base, and the reaction could be reversible. This means that the acid/base complex could dissociate to form the Lewis acid and the base once more. In general, we will see in Organic chemistry that we can use tetravalent boron or aluminum compounds as Lewis bases donors. For example, if we were to use the ion BH_4^{-1} in a given reaction, we should be aware that this ion is really an acid/base complex, formed by the reaction of the Lewis acid BH_3 with the base H^{-1}. This means that the BH_4^{-1} ion (the acid/base complex) could react with some organic compound by donating a hydride ion (H^{-1}).

- Consider the following reaction:

$$BF_{3(aq)} + ClF_{3(aq)} \rightleftharpoons ClF_2^+{}_{(aq)} + BF_4^-{}_{(aq)}$$

Classify all the species in this reaction as either Lewis acids, Lewis bases, or acid/base complexes.

We should analyze the reaction first to decide what actually happened. The analysis let us see that we can break up the reaction into two basic steps.

$$BF_3 + F^- \rightleftharpoons BF_4^-$$

$$ClF_3 \rightleftharpoons ClF_2^+ + F^-$$

Let's look at the first step. The boron trifluoride is the Lewis acid, whereas the fluoride ion is the Lewis base. Of course, the ion they form is the acid base complex. How about the second step? In here we see that the ClF_3 breaks up to form a Lewis acid (ClF_2^+) and a Lewis base (the fluoride ion). Conclusion:

$$BF_{3(aq)} + ClF_{3(aq)} \rightleftharpoons ClF_2^+{}_{(aq)} + BF_4^-{}_{(aq)}$$

L Acid   A B Complex   L Acid   A B Complex

Chapter 18: Other Aqueous Equilibria: Ksp.

18.1 Slightly Soluble Salts.

In this chapter, we shall consider the equilibria established by slightly soluble salts as they dissolve in water. We will define a slightly soluble salt as one that has a solubility of less than one gram per liter of solution. An example of such a salt is silver bromide, a yellowish solid with a melting point of 432°C. Although it is true that only a minuscule amount of AgBr goes into solution, whatever amount dissolves will be totally dissociated into ions. This is because all ionic compounds are strong electrolytes.

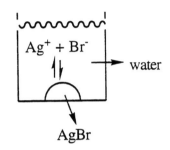

$$AgBr_{(s)} \rightleftharpoons AgBr_{(aq)} \rightleftharpoons Ag^+_{(aq)} + Br^-_{(aq)}$$

We will be dropping the subscript (aq), since we know it is solvated by water. Thus, we can write the following equilibrium:

$$AgBr_{(s)} \rightleftharpoons Ag^+ + Br^-$$

and the mathematical expression for the equilibrium constant is given by:

$$K_{sp} = [Ag^+][Br^-] \quad \text{solubility product.}$$

- Given that the solubility of AgBr is 7.11×10^{-7} M in water at 25°C, determine the equilibrium constant for the solubility equilibrium, Ksp.

    We should start this problem in the same way we started all the previous equilibria problems: writing the equilibrium together with the initial conditions, as well as the conditions once the system reaches equilibrium. What do we have initially in a solubility problem? In order to answer this question we need to think in terms of the solubility experiment itself. In order to dissolve the maximum amount of a given salt in water, we need to add a lot of the salt into a beaker with water, and wait for the solution to be saturated.

Once this happens, we will not observe any change in the amount of solid that is left in the bottom of the flask. How much of the solid did dissolve? Since we do not know, we shall call this amount **x**. Therefore, **x** will represent the maximum amount of salt dissolved in the water (per Liter of solution), and this is what we call solubility.

$$AgBr_{(s)} \rightleftharpoons Ag^+ + Br^-$$

	$AgBr_{(s)}$	Ag^+	Br^-
init:	lots	0	0
@ equil:	lots - x	x	x

but: $Ksp = [Ag^+][Br^-]$

∴ $Ksp = x^2$.

We just mentioned that **x** represents the solubility of the salt AgBr. Since the solubility was given, then $x = 7.11 \times 10^{-7}$ M.

∴ $Ksp = (7.11 \times 10^{-7})^2 =$ ***5.06 x 10⁻¹³*** ✓

The same can be done with any other salt in order to determine its Ksp. The values for the Ksp for different slightly soluble salts can be found in Appendix 3.

A typical misconception is that the Ksp is always equal to the solubility squared. This is only true for salts of the 1 : 1 type (same number of cations as anions, like AgBr, AgCl, etc.). However, if the salt is of a different type, then we MUST determine the relationship between the Ksp and the solubility. As an example, let's take the case of a 2 : 1 salt.

- Calculate the solubility of Ag_2S in water. $Ksp = 1.6 \times 10^{-49}$.

Once more, we start by writing the equilibrium, which will always be the solid in equilibrium with the ions in solution. Then, we should write the initial concentration of the ions, as well as the equilibrium concentration of these ions.

$$Ag_2S_{(s)} \rightleftharpoons 2\,Ag^+ + S^{-2}$$

	$Ag_2S_{(s)}$	$2\,Ag^+$	S^{-2}
init:	lots	0	0
@ equil:	lots - x	2x	x

$Ksp = [Ag^+]^2[S^{-2}]$

$$\therefore (2x)^2 x = 1.6 \times 10^{-49}$$

$$\text{or} \quad x = \text{Solubility} = 3.4 \times 10^{-17} M \checkmark$$

Therefore, we can see that for a salt of this type (1: 2), the relationship between solubility and the Ksp is given by:

$$Ksp = 4(\text{Solubility})^3.$$

- Calculate the solubility of $Ca(OH)_2$ in water. $Ksp = 1.3 \times 10^{-6}$.

$$Ca(OH)_{2(s)} \rightleftharpoons Ca^{+2} + 2\ OH^-$$

	$Ca(OH)_{2(s)}$	Ca^{+2}	OH^-
init:	lots	0	≈ 0
@ equil:	lots - x	x	2 x

In this case, we assumed that the initial concentration of hydroxide ions was zero. We know that this is not really the case, because in water, the concentration of hydroxide ions is 10^{-7} M. However, when the Ksp is that large (larger than about 10^{-19}), that is normally a safe assumption. What we should do is make that assumption and then look at the answer. The concentration of OH^- ions CANNOT be smaller than 10^{-7} M because if it is, then the original assumption (that it was zero) would not be appropriate.

$$Ksp = [Ca^{+2}][OH^-]^2$$

$$\therefore x(2x)^2 = 1.3 \times 10^{-6}$$

$$\therefore x = 6.9 \times 10^{-3} M = \text{Solubility}.$$

Hopefully we can see that the original assumption was a good one! We should look at an example where this is not the case.

- What is the molar solubility of $Al(OH)_3$ in pure distilled water? $Ksp = 2.1 \times 10^{-32}$.

Let's do the <u>incorrect</u> assumption first, so that we can see how the answer does not make sense.

$$Al(OH)_{3(s)} \rightleftharpoons Al^{+3} + 3\ OH^-$$

	$Al(OH)_{3(s)}$	Al^{+3}	OH^-
init:	lots	0	≈ 0
@ equil:	lots - x	x	3 x

$$K_{sp} = [Al^{+3}][OH^-]^3$$

$$x(3x)^3 = 27x^4 = 2.1 \times 10^{-32}$$

$$\therefore x = 5.3 \times 10^{-9} \text{ M}$$

If that were the correct answer, the concentration of hydroxide ions would be three times that amount, or 1.6×10^{-8} M. This is impossible, because water itself provides more hydroxide ions than that. Therefore, our initial approximation was not correct this time and we should do this problem the following way.

$$Al(OH)_{3(s)} \rightleftharpoons Al^{+3} + 3\,OH^-$$

init:	lots	0	1×10^{-7} M
@ equil:	lots - x	x	$1 \times 10^{-7} + 3x$

$$\therefore x(10^{-7})^3 = 2.1 \times 10^{-32}$$

$$\therefore x = \text{Solubility} = 2.1 \times 10^{-11} \text{ M} \checkmark$$

The common ion effect that we observed in the acid base equilibria is also observed in these equilibria. The presence of a common ion reduces the solubility of a slightly soluble salt even more.

- What is the molar solubility of AgCl in a 0.10 M solution of NaCl? $K_{sp} = 1.8 \times 10^{-10}$.

If the AgCl was dissolved in pure distilled water, then the solubility would be simply the square root of the K_{sp} (since it is a 1 : 1 salt). This would make the solubility for AgCl equal to 1.3×10^{-5} M. Now let's see the effect of the common ion, the chloride ion.

$$AgCl_{(s)} \rightleftharpoons Ag^+ + Cl^-$$

init:	lots	0	0.10 M
@ equil:	lots - x	x	0.10 + x

$$K_{sp} = [Ag^+][Cl^-] = x(0.10) = 1.8 \times 10^{-10}$$

$$\therefore x = \text{Solubility} = 1.8 \times 10^{-9} \text{ M} \checkmark$$

This proves our argument: the presence of the chloride ion in the solution, further reduces the solubility of the salt, from 1.3 x 10^{-5} M to 1.8 x 10^{-9} M.

18.2 Precipitation.

In this section, we will look at the equilibrium for the slightly soluble salts, but from a rather different perspective. Rather than placing a solid in the bottom of a flask and waiting to see what is the maximum amount of that solid that will actually dissolve in the solvent, we will mix two ionic solutions and wait to see if the solid will come out of solution. We use the term precipitation referring to this event: the formation of a slightly soluble solid from a clear solution.

It would be convenient if we could predict if precipitation is going to occur or not. Many times we can prevent the precipitation of a solid by making simple changes in the solutions (like altering the pH of the solutions). Previous knowledge of the problem can save us from many headaches. Therefore, the entire question for us would be to decide whether a solution is saturated[1] with respect to the salt or not.

$$MX_{(s)} \rightleftharpoons M^+ + X^-$$

Will $MX_{(s)}$ form?

In order to answer this question we will have to first calculate Q (the reaction quotient) and compare it to Ksp. It will depend on whether Q is equal to, larger than or smaller than Ksp. What happens when Q is larger than Ksp? Under these conditions, the solution is said to be supersaturated. Therefore, the equilibrium shifts spontaneously to the left to reach a minimum of free energy. Upon doing so, a precipitate will form. How far to the left will this equilibrium shift? As always every system will proceed spontaneously to a minimum of free energy, therefore, it will continue to go to the left until Q is equal to Ksp, that is, until the solution remains saturated.

If, however, Q is exactly equal to Ksp, we have a situation where we are already at the minimum of free energy and the system will not shift anywhere spontaneously. This is called a

---

[1] A **saturated** solution is one that has the maximum concentration of solute possible at that temperature.

Chapter 18: Other Aqueous Equilibria: Ksp.

saturated solution.

Finally, if Q is smaller than Ksp, the solution is said to be unsaturated. Therefore, we should observe a spontaneous shift to the right because that would lower the free energy of the system. However, nothing actually happens because we have no solid present initially, therefore, the process cannot occur. Below you may find a summary of what we have stated so far.

IF	Solution is...	Will it Precipitate?
Q < Ksp	Unsaturated	NO
Q = Ksp	Saturated	NO
Q > Ksp	Supersaturated	YES

Let's do a few problems that apply this concept. Remember that in precipitation problems, we have no solid present initially and what we are trying to figure out is if the initial concentration of the ions is enough to cause precipitation. In other words, in this type of problems we always need to calculate Q initially.

- A solution is prepared by mixing equal volumes of 0.010 M $MgCl_2$ and 0.020 M $Na_2C_2O_4$. Will the MgC_2O_4 precipitate? Ksp for MgC_2O_4 is 8.57×10^{-5}.

Some people get initially confused with precipitation and believe that they need to generate some new equilibrium where they will have to react the ions mixed originally. Of course, that is not the case. The equilibrium is simply the Ksp equilibrium. The only problem is to simply check the initial conditions carefully.

$$MgC_2O_{4(s)} \rightleftharpoons Mg^{+2} + C_2O_4^{-2}$$

Since it is a precipitation problem, there is absolutely no solid present initially. However, we have effectively prepared a solution that contains these two ions (the magnesium ion and the oxalate ion). Therefore, we need to know their initial concentrations in order to calculate Q. What happens when we mix equal volumes of two solutions? The concentration of each ion is reduced in half before they even react. This is due to the fact that we have the same number of moles, but the volume is twice as large ($M_1 V_1 = M_2 V_2$, remember?). Therefore, we are presented with the following initial scenario:

$$MgC_2O_{4(s)} \rightleftharpoons Mg^{+2} + C_2O_4^{-2}$$

init: 0   0.0050 M   0.010 M

Chapter 18: Other Aqueous Equilibria: Ksp.

$$\therefore Q = [Mg^{+2}][C_2O_4^{-2}] = 0.0050\,(0.010) = 5.0 \times 10^{-5}$$

Notice that: $\quad Q < Ksp \quad \checkmark$

Therefore, there is no precipitation observed, although the solution is almost saturated (but close is not enough).

- What will be the $[CO_3^{-2}]$ after the addition of 0.053 g of Na_2CO_3 (MW = 106) to 100 mL of 0.0100 M Ca^{+2} solution? Ksp for $CaCO_3$ is 1.0×10^{-8}.

We need to convert those grams to moles (5.0×10^{-4} moles) and then determine the initial concentration of carbonate ions (5.0×10^{-3} M) in solution. We should realize that we need to know the concentrations of the calcium and the carbonate ions initially.

$$CaCO_{3(s)} \rightleftharpoons CO_3^{-2} + Ca^{+2}$$

init: 0   0.0050 M   0.0100 M

When we calculate Q with these initial concentrations and get $Q = 5.0 \times 10^{-5}$. Remember that Q is calculated in the same way as we would calculate Ksp, except we use the initial concentrations of the ions, rather than the equilibrium concentrations. When we compare Q to Ksp, we find that Q > Ksp, therefore precipitation will indeed occur this time. However, Q is a lot larger than Ksp (actually, it is about 1000 times larger). Anytime Q is so much larger than Ksp, we can shift the equilibrium 100% to the left and to simplify the problem.

$$CaCO_{3(s)} \rightleftharpoons CO_3^{-2} + Ca^{+2}$$

init:	0	0.0050 M	0.010 M
←100%:	some	0	0.0050 M
@ equil:	some − x	x	0.0050 + x

$$Ksp = [Ca^{+2}][CO_3^{-2}]$$

$$(0.0050)\,x = 1.0 \times 10^{-8}$$

$$\therefore x = [CO_3^{-2}] = 2.0 \times 10^{-6}\,M \quad \checkmark$$

- A solution is prepared by mixing 200 mL of 0.200 M Pb(NO$_3$)$_2$ with 200 mL of 2.00 x 10^{-2} M NaI. What will be the concentration of I$^-$ ions after precipitation, if it occurs? Ksp for PbI$_2$ is 1.43 x 10^{-8}.

    Once more, we are mixing equal volumes of two solutions, therefore, the initial concentration of both solutes will be halved. Keep in mind that the two ions of interest are the Pb^{+2} (0.100 M) and the iodide (0.0100 M), since the others are spectator ions. From the initial concentrations, Q is calculated to be 1.00 x 10^{-5} {= 0.100 (0.0100)^2}. Since this number is about 1000 times larger than Ksp, we will shift 100% to the left.

	PbI$_{2(s)}$ ⇌	Pb^{+2}	+ 2 I$^-$
init:	0	0.100 M	0.010 M
←100%:	some	0.095 M	0
@ equil:	some - x	0.095 + x	2 x

    $$Ksp = [Pb^{+2}][I^-]^2$$

    $$(0.095)(2x)^2 = 1.43 \times 10^{-8}$$

    $$\therefore 2x = [I^-] = 3.88 \times 10^{-4} M \checkmark$$

    Just in case that you have forgotten how to determine the number 0.095 M for the concentration of Pb^{+2} after the shift, let us quickly review this.[1] We must start with the moles (or molarity) of the limiting reagent (in this case it is the I$^-$) and use the coefficients from the balanced reaction to determine how much Pb^{+2} will be **consumed**. Once we know this number, we can figure out how much is **left over** (by difference).

    $$0.010 M_{I^-} \times \frac{1 M_{Pb^{+2}}}{2 M_{I^-}} = 0.0050 M_{Pb^{+2}} \text{ consumed}$$

    $$\therefore [Pb^{+2}]_{left} = 0.100 M - 0.0050 M$$

    $$= 0.095 M \checkmark$$

- A solution is prepared to be 1 x 10^{-5} M in NaBr and 1 x 10^{-6} M in NaCl. Which of these ions (Cl$^-$ or Br$^-$) will precipitate first upon the addition of silver ions in the form of solid silver nitrate? Ksp for AgCl is 2 x 10^{-10} and Ksp for AgBr is 5 x 10^{-13}.

---

[1] See Section 2.8.

Chapter 18: Other Aqueous Equilibria: Ksp.

This is a multi step problem. We need to first figure out the concentration of silver ions that will saturate the solution with respect to each ion. Once we know that, then we can answer the question. How do we determine the concentration of silver ions that will saturate the solution with respect to the chloride ion? The word saturation implies an equilibrium situation, which is the case when Q = Ksp. For this salt (AgCl), Ksp = [Ag$^+$] [Cl$^-$]. Therefore,

$$2 \times 10^{-10} = [Ag^+] (1 \times 10^{-6})$$

$$\therefore [Ag^+] = 2 \times 10^{-4} \text{ M} \checkmark$$

This number represents the concentration of silver ions that will saturate the solution with respect to AgCl. In other words, the concentration of silver ions is zero initially and we will start adding some, therefore, its concentration will slowly start to rise. When it reaches 2×10^{-4} M, the solution will be saturated with respect to the chloride ion and the very next addition of silver ions will start the precipitation of the silver chloride.

Let's repeat this calculation for the silver bromide. Once more, the mathematical expression for Ksp is given by: Ksp = [Ag$^+$] [Br$^-$], therefore,

$$5 \times 10^{-13} = [Ag^+] (1 \times 10^{-5})$$

$$\therefore [Ag^+] = 5 \times 10^{-8} \text{ M} \checkmark$$

This number represents the concentration of silver ions that will saturate the solution with respect to the bromide ions. Since this number is smaller than the one for the chloride ions, it should be clear that we will reach the latter number first, therefore, *the silver bromide will precipitate first.*

- Silver ion in the form of AgNO$_3$ is added to a liter of a solution containing 0.10 mol of Cl$^-$ and 0.10 mol of C$_2$O$_4^{-2}$. AgCl is white and Ag$_2$C$_2$O$_4$ is red. Ksp for each is 1.7×10^{-10} and 1.0×10^{-12}, respectively.

a) Will the first precipitate formed be red or white?

The question here is whether the first precipitate will be the silver chloride or the silver oxalate. We need to do something very similar to what we did in the previous problem. That is, we need to determine the concentration of silver ions that will saturate the solution for each ion separately.

$$AgCl_{(s)} \rightleftharpoons Ag^{+1} + Cl^{-1}$$

$$Ksp = [Ag^+][Cl^-]$$

$$1.7 \times 10^{-10} = [Ag^+](0.10)$$

$$\therefore [Ag^+] = 1.7 \times 10^{-9} \text{ M} \checkmark$$

This number represents the concentration of silver ions that will saturate the solution with respect to the chloride ions.

$$Ag_2C_2O_{4(s)} \rightleftharpoons 2Ag^{+1} + C_2O_4^{-2}$$

$$Ksp = [Ag^+]^2[C_2O_4^{-2}]$$

$$1.0 \times 10^{-12} = [Ag^+]^2(0.10)$$

$$\therefore [Ag^+] = 3.2 \times 10^{-6} \text{ M} \checkmark$$

This number represents the concentration of silver ions that will saturate the solution with respect to the oxalate ions. Since this number is much larger than the one for the chloride ions, then it is evident that the solution will be saturated with respect to the silver chloride first and therefore, it will precipitate first. Since the silver chloride is white, then *the first precipitate will be white*.

b) What are the $[Cl^-]$ and $[C_2O_4^{-2}]$ at the point when a trace of the second precipitate starts to form?

We already know that the second precipitate will be the silver oxalate, or the red one. Let's try to understand what is going on here. Once the concentration of silver ions barely exceeds 1.7×10^{-9} M, the white precipitate will start to form. From that point on, most of the silver that we add will be precipitated as a white solid. However, we must keep in mind that an equilibrium has been established between the silver ions, the chloride ions and the solid silver chloride, which means that at all times, the product of the concentration of the silver ions and the chloride ions must be equal to 1.7×10^{-10} (the Ksp). So, as the concentration of chloride ions continues to decrease (because of the precipitation), the concentration of silver ions will slowly start to increase. Eventually it will reach the value of 3.2×10^{-6} M. What happens at this point? That is when the solution becomes saturated with respect to the oxalate ions. Any further increase of the silver ion concentration will lead to the appearance of the red precipitate. So, what is the concentration of the oxalate ions at this point? It is still 0.10 M because it has not precipitated yet. However, the concentration of the chloride ions will be a lot less, because most of it is found as a white solid in the bottom of the flask. We do know, however, that its concentration can be determined because it has the established equilibrium with the silver ions.

$$AgCl_{(s)} \rightleftharpoons Ag^{+1} + Cl^{-1}$$

$$K_{sp} = [Ag^+][Cl^-].$$

$$1.7 \times 10^{-10} = 3.2 \times 10^{-6} [Cl^-]$$

$$\therefore [Cl^-] = 5.3 \times 10^{-5} M \checkmark$$

- A solution is prepared to be 1.00×10^{-4} M in $Mg(NO_3)_2$, 0.123 M in NH_3, and 0.0321 M in the salt NH_4Cl. Will the $Mg(OH)_2$ precipitate? The K_{sp} for $Mg(OH)_2$ is 1.81×10^{-11} and K_b for NH_3 is 1.80×10^{-5}.

Once more, this is a precipitation problem. Therefore, we need to know the initial concentration of the ions in question, that is, the magnesium and the hydroxide ions. Since one of them was given, the whole problem centers around getting the concentration of the hydroxide ion. The [OH⁻] will depend on the solution itself since it contains a base (NH_3) dissolved in it.

	NH_3	+ H_2O $\rightleftharpoons$	NH_4^+	+ OH^-
init:	0.123 M		0.0321 M	≈ 0
@ equil:	0.123 − x		0.0321 + x	x

Solving for x (through K_b), we get:

$$x = [OH^-] = 6.90 \times 10^{-5} M \checkmark$$

The rest should be straight forward by now.

	$Mg(OH)_{2(s)} \rightleftharpoons$	Mg^{+2}	+ $2\,OH^-$
init:	0	1.00×10^{-4} M	6.90×10^{-5} M

$$\therefore Q = (1.00 \times 10^{-4})(6.90 \times 10^{-5})^2 = 4.76 \times 10^{-13} \checkmark$$

Since Q < K_{sp}, the solution is unsaturated with respect to the hydroxide ions. Therefore, *no precipitation will occur.*

Chapter 19: Electrochemistry.

19.1 <u>Redox Reactions</u>.

Almost all chemical reactions carried out for producing energy are **redox** reactions. *Examples:* Combustion of fuels, generation of electricity by cells (or batteries), metabolism of food. What are redox reactions? Let's look at one very simple example that will illustrate what actually happens in a redox reaction.

$$Fe + Cu^{+2} \rightleftharpoons Cu + Fe^{+2}.$$

When we think about what actually happened as these two substances reacted, we can see that a redox reaction simply involves the transfer of electrons from one species to another. The term redox is a combination of two other words: reduction and oxidation. The name redox is meant to stress a point: oxidation will not occur unless reduction takes place simultaneously. What do we mean by reduction and oxidation?

Any element which *increases* in oxidation number is said to be <u>oxidized</u>. Another way to state this is: A substance which loses electrons is said to be oxidized. On the other hand, any element which *decreases* in oxidation number is said to be <u>reduced</u>. Another way to state this is: A substance that gains electrons is said to be reduced. Let's look at another example so that we can point out the species being reduced as well as the one being oxidized.

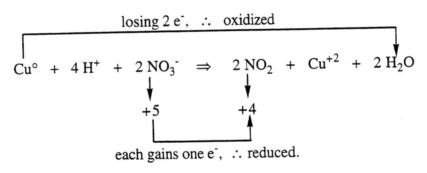

We can make the following observations.

- ◊ The $Cu°$ was oxidized since it lost 2 e^-.

- ◊ The NO_3^- was reduced since it gained 1 e^-.

- ◊ We can say that the $Cu°$ was oxidized because it forced the NO_3^- to be reduced. Therefore, we refer to the $Cu°$ as the <u>reducing agent</u>.

◊   We can say that the NO_3^- was reduced because it forced the $Cu°$ to be oxidized. Therefore, we refer to the NO_3^- as the <u>oxidizing agent</u>.

Therefore, species that are reduced are referred to as the oxidizing agents, whereas species that are oxidized are referred to as the reducing agents.

There are a few factors that we should be aware of while we deal with these types of reactions. Three of these factors can be seen below.

1. Concentration of reacting species.
2. Presence of acid or base to serve as catalysts.
3. Laws of conservation of mass and energy must hold!

19.2 Oxidation States.

The concept of oxidation states will help us to determine where the electrons go in a redox reaction. The rules for assigning oxidation states are shown below.

a) The oxidation state of an atom in an element is zero. For example, the oxidation state of each atom in the following cases is zero: O_2, S_8, K, etc.

b) The oxidation state of a monoatomic ion is the same as its charge. For example, the oxidation state for the Li^+ ion is +1.

c) Oxygen is assigned an oxidation state of -2 in its covalent compounds, such as CO, CO_2, SO_2, etc. The exceptions to this rule occur in the peroxides in which the oxygen atom is assigned an oxidation state of -1 (Ex. H_2O_2) and with the molecule OF_2, in which the oxidation state for the oxygen is +2.

d) In its covalent compounds with non-metals, hydrogen is assigned an oxidation state of +1. For example, in CH_4, H_2O, NH_3, etc., the oxidation state for the hydrogen is +1. When the hydrogen is bonded to metals, then it is -1. Example: NaH.

e) The sum of the oxidation states must be zero for an electrically neutral particle and must be equal to the overall charge in an ionic species. For example, NH_4^+ is evaluated this way. We know that hydrogen is +1. Since we have 4 hydrogens, they account for a charge of +4. Since the total charge of the ion is +1, then the nitrogen must be -3.

- Determine the oxidation state for Mn in the permanganate ion (MnO_4^{-1}).

    Since it is an ion, the sum of the oxidation state of all the elements involved must be equal to the charge of the particle, -1. Therefore,

$$OS\ of\ Mn\ +\ 4\ (OS\ of\ O)\ =\ -1$$

$$OS\ of\ Mn\ +\ 4\ (-2)\ =\ -1$$

$$\therefore\ OS\ of\ Mn\ =\ +7\ \checkmark$$

19.3 <u>Balancing Redox: Ion-Electron Method</u>.

We mentioned that these reactions are affected by the presence of acids or bases. Therefore, we will learn two separate rules to balance these reactions whether the solutions are acidic or basic. In an acidic solution, there are basically two things present in large quantities: water and hydronium ions (H^+). In a basic solution, there are basically two things present in large quantities: water and hydroxide ions (OH^-).

A. <u>Acidic Rules</u>.

1. Identify and separate the two half reactions (one will be oxidation and the other will be reduction).
    2. Balance each individually.
        a. Balance the central atom first (mass).
        b. Balance the **O** by adding H_2O.
        c. Balance the **H** by adding H^+.
        d. Balance the charge by adding electrons to the most positive side.
    3. Add the two half reactions making sure that the electrons cancel out.

Let's look at an example that will illustrate how to apply these rules in an acidic medium.

- Balance the following reaction using the ion-electron method in an acidic medium.

$$NO_3^-\ +\ Cu_2O\ \rightleftharpoons\ Cu^{+2}\ +\ NO_2$$

Let's separate the two half reactions first. The nitrate ion is going to react to form the nitrogen dioxide, whereas the cuprous oxide will react to form the cupric ions.

Chapter 19: Electrochemistry.

$$NO_3^- \rightleftharpoons NO_2.$$

$$Cu_2O \rightleftharpoons Cu^{+2}.$$

The procedure asks next for us to balance these reactions separately following the rules stated above. Let's start with the nitrate reaction. The first thing we notice is that we have the same number of nitrogens on both sides. Therefore, we must balance the oxygens by adding water molecules to the side of the equation that is missing the oxygens. Since there are three oxygens on the left and only two oxygens in the right, we need to add one water to the right side.

$$NO_3^- \rightleftharpoons NO_2 + H_2O$$

This introduces two hydrogens in the right side. Therefore, we need to add two hydronium ions on the left side of this equation.

$$2 H^+ + NO_3^- \rightleftharpoons NO_2 + H_2O$$

We should notice that the mass is now completely balanced. That is, we have the same number of nitrogens, hydrogens and oxygens on both sides of the equation. HOWEVER, the reaction IS NOT balanced. This is because there is a charge imbalance. The total charge on the left side is +1 (the result of +2 −1 = +1), whereas there is no charge on the right side. Therefore, we need to add one electron to the most positive side in order to get the charge balance.

$$1 e^- + 2 H^+ + NO_3^- \rightleftharpoons NO_2 + H_2O$$

Therefore, this half reaction turned out to be a reduction half reaction (an electron was gained by the nitrate ion). This must mean that the next one must be an oxidation. Let's balance the other half reaction.

The first thing we notice is that we must balance the copper first, since there are two Cu's on the left side, and only one on the right side.

$$Cu_2O \rightleftharpoons 2 Cu^{+2}$$

We immediately notice that the left side has one oxygen atom, therefore, we should add a water molecule on the right side of the equation to take care of that oxygen.

$$Cu_2O \rightleftharpoons 2\ Cu^{+2} + H_2O$$

Of course, this has introduced two hydrogens on the right side, which means that we should add two hydronium ions on the left side.

$$2\ H^+ + Cu_2O \rightleftharpoons 2\ Cu^{+2} + H_2O$$

How about the charge balance? There is a total charge of +2 on the left side, while there is a charge of +4 on the right side. This means that we should add two electrons on the right side.

$$2\ H^+ + Cu_2O \rightleftharpoons 2\ Cu^{+2} + H_2O + 2\ e^-$$

As expected, this second half reaction was an oxidation. We can see the cuprous ions lost electrons in order to form the cupric ions. At this point, we must add the two half reactions in order to get the overall reaction. It is imperative to cancel the electrons during this process, so that there are no leftover electrons in the overall reaction. Remember, an oxidation does not take place unless a reduction takes place simultaneously, therefore, the cuprous oxide cannot give away two electrons while the nitrate accepts only one, because that would imply that the cuprous oxide was oxidized by itself. Another way of looking at it would be to see that there is an electron floating in the aqueous solution. Therefore, the reduction reaction must be multiplied by two, so that we have the same number of electrons in each.

$$1\ e^- + 2\ H^+ + NO_3^- \rightleftharpoons NO_2 + H_2O$$
$$2\ H^+ + Cu_2O \rightleftharpoons 2\ Cu^{+2} + H_2O + 2\ e^-$$
$$\overline{6\ H^+ + Cu_2O + 2\ NO_3^- \rightleftharpoons 2\ Cu^{+2} + 3\ H_2O + 2\ NO_2}$$

Now let's look at the basic rules. Remember that, in a basic medium, we have lots of water and hydroxide ions.

B. <u>Basic Rules</u>.

1. Identify and separate the two half reactions.

2. Balance each individually.

    a. Balance the central atom first (mass).

    b. Balance the **O** by adding **twice** as many OH⁻.

Chapter 19: Electrochemistry.

    c. Balance the **H** by adding H_2O.

    d. Balance the charge by adding electrons to the most positive side.

3. Add the two half reactions making sure that the electrons cancel out.

Let's try the following example to apply this rules.

- Balance the following reaction in a basic medium.

$$CrO_4^{-2} + I^- \rightleftharpoons Cr^{+3} + IO_3^-$$

Once again, we separate the two half reactions first. Of course, the CrO_4^{-2} ion will react to form the Cr^{+3} while the iodide ion will react to form the iodate ion.

$$CrO_4^{-2} \rightleftharpoons Cr^{+3}.$$

$$I^- \rightleftharpoons IO_3^-.$$

Let us start with the first one. We notice that there are two chromium atoms in each side of the equation, therefore, all we need to do is balance the oxygens. However, being a basic medium, to balance the oxygens we will add twice as many hydroxide ions as we need oxygens. The left side has four oxygens, therefore, we will add eight hydroxide ions in the right side.

$$CrO_4^{-2} \rightleftharpoons Cr^{+3} + 8\,OH^-$$

There seems to be a problem. We have four oxygens in the left side of this equation, while there are eight in the right side. Apparently we did not succeed at balancing the oxygens. That is absolutely correct, because in this method, the oxygens are balanced at the same time we balance the hydrogens. Since there are eight hydrogens in the right side, we need to add four water molecules in the left side.

$$4\,H_2O + CrO_4^{-2} \rightleftharpoons Cr^{+3} + 8\,OH^-$$

This proves that upon balancing the hydrogens, we introduce the missing oxygens and now the mass is completely balanced for this reaction. The only thing left is to balance the charge. The total charge in the left side is –2, whereas the total charge in the right side is -5. This creates a difference of three between the two sides, therefore we should add three electrons to the most positive side (the left side):

$$3\,e^- + 4\,H_2O + CrO_4^{-2} \rightleftharpoons Cr^{+3} + 8\,OH^-$$

This turned out to be the reduction half reaction. This means that the CrO_4^{-2} ion was reduced and upon gaining those three electrons, it formed the Cr^{+3} ion. Therefore, the CrO_4^{-2} ion is the oxidizing agent.

Let us balance the other half reaction at this point. Once more, we see that we have the same number of iodines in both sides of the equation, therefore, we should start by balancing the oxygens. Since there are three oxygens in the right side of the equation and none in the left side, we should add six hydroxide ions to the left side.

$$6\,OH^- + I^- \rightleftharpoons IO_3^-.$$

To balance the hydrogens, we will add three water molecules to the right side of the equation.

$$6\,OH^- + I^- \rightleftharpoons IO_3^- + 3\,H_2O$$

We can see that the mass is completely balanced at this point. Now, we should balance the charge. The total charge in the left side is -7, whereas it is -1 in the right side. That makes for a difference of six, which means that we should add six electrons to the most positive side (the right side):

$$6\,OH^- + I^- \rightleftharpoons IO_3^- + 3\,H_2O + 6\,e^-.$$

This is the oxidation half reaction, which makes the iodide ion the reducing agent.

In order to add the two half reactions, we need to multiply the first reaction by two, in order to cancel the electrons.

$$\mathbf{2x}\ (3\,e^- + 4\,H_2O + CrO_4^{-2} \rightleftharpoons Cr^{+3} + 8\,OH^-)$$

$$6\,OH^- + I^- \rightleftharpoons IO_3^- + 3\,H_2O + 6\,e^-$$

$$\overline{}$$

$$5\,H_2O + 2\,CrO_4^{-2} + I^- \rightleftharpoons 2\,Cr^{+3} + IO_3^- + 10\,OH^-$$

We will do one more example. The reaction below is a **disproportionation** reaction. Disproportionation reactions are those in which the reagent molecules are both oxidized and reduced. This means that some of the molecules will be oxidized while others will be reduced.

Chapter 19: Electrochemistry.

- Balance the following disproportionation reaction in a basic medium.

$$Br_2 \rightleftharpoons BrO_3^- + Br^-$$

Since it is a disproportionation reaction, the two half reactions must be a bromine molecule reacting to form a bromate ion, while another bromine molecule reacts to form a bromide ion.

$$Br_2 \rightleftharpoons BrO_3^-$$

$$Br_2 \rightleftharpoons Br^-$$

The first reaction is balanced as follows. The bromine is balanced first since we have different number of bromine atoms in each side. At that point, there are six oxygens in the right side, therefore, we are forced to add twelve hydroxide ions in the left side. This also presumes that we will balance the hydrogens by adding six water molecules in the right side.

$$12\,OH^- + Br_2 \rightleftharpoons 2\,BrO_3^- + 6\,H_2O$$

This equation has a complete mass balance. However, in order to balance the charge, we need to add ten electrons to the right side.

$$12\,OH^- + Br_2 \rightleftharpoons 2\,BrO_3^- + 6\,H_2O + 10\,e^-$$

The second half reaction is trivial, since it does not contain oxygens nor hydrogens.

$$2\,e^- + Br_2 \rightleftharpoons 2\,Br^-$$

Adding the two half reactions, we get:

$$12\,OH^- + Br_2 \rightleftharpoons 2\,BrO_3^- + 6\,H_2O + 10\,e^-$$

$$\mathbf{5x}\,(2\,e^- + Br_2 \rightleftharpoons 2\,Br^-)$$

$$\overline{12\,OH^- + 6\,Br_2 \rightleftharpoons 10\,Br^- + 2\,BrO_3^- + 6\,H_2O}$$

There is one more situation that we should cover. If a formula contains **OH** directly in it, don't attempt to balance those OH's with the method we just discussed because it will not work. Just write them on the opposite side so that you don't have to balance them.

- Balance the following half-reaction in a basic medium.

$$Cr_2O_3 \rightleftharpoons Cr(OH)_2$$

We will balance the chromium atoms first.

$$Cr_2O_3 \rightleftharpoons 2\,Cr(OH)_2$$

We proceed to get rid of the OH's that are part of the formula by writing them in the left side.

$$4\,OH^- + Cr_2O_3 \rightleftharpoons 2\,Cr(OH)_2$$

Certainly, the only oxygens that are not balanced are those that are part of the Cr_2O_3 species. Therefore, we will balance these with the method we learned.

$$3\,H_2O + 4\,OH^- + Cr_2O_3 \rightleftharpoons 2\,Cr(OH)_2 + 6\,OH^-$$

We could combine the hydroxide ions from both sides, in order to simplify the equation. This would yield:

$$3\,H_2O + Cr_2O_3 \rightleftharpoons 2\,Cr(OH)_2 + 2\,OH^-$$

Notice that the mass is completely balanced at this point, therefore, the only thing missing is the charge.

$$2\,e^- + 3\,H_2O + Cr_2O_3 \rightleftharpoons 2\,Cr(OH)_2 + 2\,OH^-$$

19.4 Electrochemistry – Introduction.

In this unit we will be concerned with chemical transformations produced by the passage of an electric current. Let's say that we have an aqueous solution of NaCl. We know that nothing at all will ever happen to that solution, whose pH happens to be exactly seven. However, electricity (which is a flow of electrons) will have an effect on this solution. Suddenly, we will observe that some ions are accepting electrons, while others are giving them away. According to this observation, the passage of electricity through some solutions can force a redox reaction to take place. The reaction we just mentioned is not spontaneous ($\Delta G° > 0$), which is why nothing ever happened until we forced an electric current through the solution. Actually, we were forced to do electrical work (for example, connecting a battery that will force a flow of electricity through the solution) in order for a reaction to take place.

Electrochemistry is also concerned with the opposite situation. Electricity can be produced by means of chemical reactions (allows the storage of electrical energy in the form of chemical reagents). In this situation, we are assuming to have a spontaneous redox reaction ($\Delta G° < 0$). Therefore, the two reagents will combine directly, whether we like it or not, to transfer an electron from the reducing agent **B** (gives away electron) to the oxidizing agent **A** (takes the electrons). The important concept is what we choose to do at this point. We can allow the two reagents to come in direct contact with one another, as seen below.

   No useful work

In this particular case, **B** donates electrons to **A** directly and the reaction is over. This will happen anyway, whether we intervene or not (since it is spontaneous). However, we could try to carry the reaction in such a way as to get some kind of electrical work out of the system. If we separate the two reagents and connect them with a wire, we could force the electrons to be transferred through the wire. This generates a flow of electrons through the wire (electricity). If we place a light bulb along its path, the bulb will be lit and therefore, the reaction is doing electrical work for us. This is exactly what we see in the diagram below and it will be the primary goal[1] for these spontaneous reactions: to obtain useful electrical work out of these spontaneous redox reactions by the generation of an electrical current.

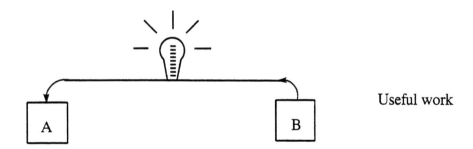   Useful work

One of the connections between these two reagents (**A** and **B**) is through a metal wire. How does the metal wire conduct electricity? We talked about this in Chapter 10 when we talked about metals and metallic conduction. As a review, let's mention the "electron-sea" model.

Let's consider a piece of wire made of metallic silver (Ag°).

---

[1] Of course, these reactions will not be as simple as the diagram seems to indicate. There are other factors to take into consideration, like the fact that we must complete a circuit, if we want for the current to be generated. This is something we will learn throughout the chapter, particularly in the section titled "Galvanic Cells".

Chapter 19: Electrochemistry.

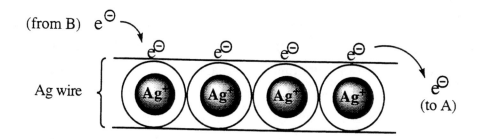

Metals have a low ionization potential, so we can visualize a metallic surface as consisting of essentially positive ions (Ag^+) and the corresponding electrons pretty loosely held. This "sea" of electrons will serve as a "glue" to keep all the Ag^+ ions together in the solid, otherwise the repulsion between them would break the solid structure.

If an electron is pushed in one of the ends of the wire (**B** is being oxidized, therefore it is giving the electron away), the whole electronic cloud on the surface of the wire will be pushed (look at the diagram above) and one electron will "pop out" on the opposite end of the wire where *simultaneously* a substance (**A**) will be reduced, i.e. will accept the electron. Remember that **B** will *not* be oxidized at all unless **A** is reduced simultaneously. Somehow, these loose electrons in the metal feel the potential that **B** has to lose electrons while simultaneously feel the potential that **A** has to accept electrons. This information is then transferred from **B** to **A**, even though they are not in direct contact with one another.

19.5 Units.

What are the units in which we measure electrical charge? These units are called Coulombs (C). It is the International System (SI) unit of charge. One Coulomb is the amount of charge carried by 6.241×10^{18} electrons. Please, notice that this is quite a significant number of electrons, actually, it is short of one mole by a ratio of only about 96,500!

The current[1] itself is measured in Amperes (A). One Ampere represents the passage of one Coulomb of charge in one second through a given point. As an example, a current of 2.52 A represents the passage of 2.52 C of charge through a given point every second.

$$\therefore A = C/s$$

---

[1] The word current refers to the actual **flow** of electrons thorough the wire.

19.6 Galvanic Cells.[1]

This is the name we give to the electrochemical cells that consist of spontaneous cell reactions. From our previous discussion, if we were to allow the two reactants to come in direct contact, the reaction will take place at the interface and therefore, no useful work (electrical) will be obtained. Let's look at one of these galvanic cells and see what we must do in order to get some useful work out of it.

Let's take the case of the spontaneous reaction between zinc metal and cupric ions. This particular cell is known as the Daniell Cell.

$$Zn° + Cu^{+2} \rightleftharpoons Cu° + Zn^{+2}$$

If we place a cupric solution in a beaker and place in it a bar of Zinc metal, we will observe immediately that a reaction is taking place and a shiny metal (the color of new pennies) is deposited on top of the zinc bar. The potential for electrical work is wasted in this case. If somehow we were able to physically separate the two reactants (the cupric ions and the zinc metal bar), then the flow of electrons from the zinc to the cupric ions can be utilized to do electrical work for us. Let's look at the diagram for the Daniell cell, set up in such a way that it will produce electrical work for us. We should point out that we are preparing a **standard** cell at this point in time. What is a standard cell? It is one that is kept at 25°C and 1 atm, but keep in mind that it is also one that has all the reagents and all the products present initially at unit activity. This is a concept we discussed previously when we did chemical thermodynamics (Chapter 15, Section 15.14). The implication of this is that the initial value of Q, the reaction quotient, will be exactly equal to one. In order to achieve this, we need to prepare two beakers, one where the half reaction of Zn metal to form Zn^{+2} ions will take place, and one where the half reaction of Cu^{+2} to form Cu metal will take place simultaneously. Since we want a standard cell, we will immerse a bar of Zn metal into a solution of zinc ions that is exactly 1-M (example: 1.00 M $ZnSO_{4(aq)}$). The other beaker will consist of a copper metal bar immersed into a solution of the cupric ions that is also exactly 1-M (for example: 1.00 M $CuSO_{4(aq)}$). If a wire is connected between the two metal bars, nothing actually happens. It is not until we connect a **salt bridge** that we observe a current. This is because until we do this, we have not completed a circuit, or in other words, if the reaction were to start, we would be forming more zinc ions in the left beaker, but we would not have enough sulfate ions to counterbalance them. This is an impossible situation. We cannot create ions of just one type. On the other hand, in the right beaker we

---

[1] We will use equilibrium arrows throughout this chapter to symbolize reversibility.

would be removing copper ions from solution (to form solid copper metal on top of the metal bar). This would leave an excess of sulfate ions in that beaker: also impossible. To solve that problem we add a salt bridge (typically a 1-M solution of the salt K^+Cl^-). As soon as we do this, the current starts to flow as indicated in the diagram below. This is because as soon as we start forming the zinc ions in the left beaker, chloride ions come out of the salt bridge to balance them and *simultaneously*, as the cupric ions disappear from the right beaker, the potassium ions come out to take their place. Therefore, the main purpose of the salt bridge is to maintain electrical (or charge) neutrality at all times.

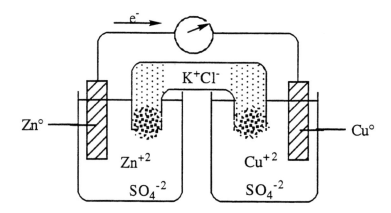

There is a terminology that we should be familiar with at this point. The solid metal bars that we placed in both beakers initially are called **electrodes**. These solid metallic surfaces provide a surface where the actual electron exchange takes place. The electrode in the left beaker, where the oxidation half reaction takes place (electrons are left behind by a zinc atom to form a zinc ion), is called the **anode**. The electrode in the right beaker, where the reduction half reaction takes place (electrons are picked up at this electrode by the cupric ions in order to form the solid copper), is called the **cathode**. There is a reason why there names were chosen. The positive ions (cations) are attracted to the cathode, whereas the negative ions (anions) are attracted to the anode. Remember that as we formed the zinc ions in the anodic compartment, the anions from the salt bridge came into that compartment, whereas as the cupric ions disappeared from the cathodic compartment, the cations from the salt bridge came out into that compartment.

Another way in which we can connect the two half cells without utilizing a cell bridge, is by connecting the two with a semi-permeable membrane. This membrane would have to allow the passage of the ions through it, but would have to impede the physical mixing of the two solutions. Remember that we do not want the cupric ions to come in contact with the zinc metal.

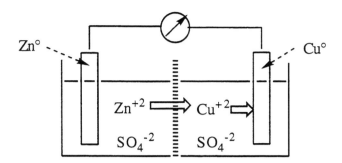

The positive ions (Zn^{+2} and Cu^{+2}) move toward the cathode, whereas the negative ions (SO_4^{-2}) are attracted to the anode! The porous membrane allows the passage of the ions through but we need not to fear for the Cu^{+2} ions to come in contact with the zinc bar because they are attracted to the cathode, not the anode!

As we can see, it is painstaking to have to write or draw a cell (like the one seen above) every time that we mention a galvanic cell. It would therefore be easier if we had a "shorthand" notation to write galvanic cells. There is a conventional notation that we will be following to represent galvanic cells.

19.7 Conventional Notation.

The rules for this "shorthand" conventional notation are seen below.

1. Anode is represented or written down first (on the left).

2. We will go from the anode to the cathode and write down everything that we see.

3. Slashes (/) represent phase boundaries.

4. Commas (,) separate species in the same phase.

5. Double slashes (//) represent a salt bridge.

6. The concentrations of all ions (or pressures of all gases) must be written in parenthesis <u>unless</u> it is a 1.00 M solution (or if the pressure is 1.00 atm).

7. A general expression for the cell notation is shown below.

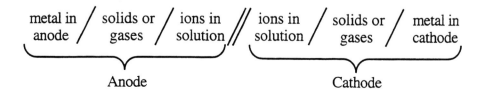

Chapter 19: Electrochemistry.

◊ Using this notation, the standard Daniell cell can be represented as follows:

$$Zn° / Zn^{+2} // Cu^{+2} / Cu°$$
(anode)　　　　　(cathode)

Keep in mind that the reason we did not write the concentration of the ions in parenthesis is because this is a standard cell (all the concentrations are 1 M).

- What is the cell reaction for the following galvanic cell?

$$Zn° / Zn(NO_3)_{2(aq)} // Sn(NO_3)_{2(aq)} / Sn°$$

It is worth noting that many salts (i.e., $Zn(NO_3)_2$ and $Sn(NO_3)_2$) dissociate completely in water. Therefore, since they both say aqueous in parenthesis, they really exist as Zn^{+2} and Sn^{+2}. The nitrate ions are spectator ions. It should be fairly straight forward to determine the two half reactions:

$$Zn° \rightleftharpoons Zn^{+2} + 2e^-$$
$$Sn^{+2} + 2e^- \rightleftharpoons Sn°$$

Adding them up (the electrons cancel out), we obtain the cell reaction:

$$Zn° + Sn^{+2} \rightleftharpoons Zn^{+2} + Sn°$$

- Consider the following galvanic cell:

$$Cu° / Cu^{+2}, Br^- / Br_{2(\ell)} / Pt°$$
(anode)　　　　　(cathode)

Determine the cell reaction.

In order to determine the cell reaction, we need the individual half-cell reactions. We should remember that the electrode on the left is the anode (where oxidation occurs) and the electrode on the right is the cathode (where reduction takes place). The anodic half reaction is easy to deduce from the conventional notation above. The cathodic one, however, is not as straightforward. First of all, the question comes up, why do we use platinum metal as the electrode on the cathode? This is because in order to make the standard cell, we cannot make a solid electrode with either of the two species that we have present in that beaker: liquid bromine and bromide ions. However, we do need to have a solid metallic surface so that the electron transfer will occur between these two species. Thus, we need to search for an inert metal: one that will not interfere with the reaction. Platinum (Pt°) serves this role very effectively. It does not have to be Pt° all the time, but it is convenient to remember that this metal will typically be inert. The two half reactions are seen below.

$$Cu° \rightleftharpoons Cu^{+2} + 2\,e^-$$
$$2\,e^- + Br_{2(\ell)} \rightleftharpoons 2\,Br^-$$

Secondly, we might question how we decided to go from Br_2 to Br^- instead of Br^- to Br_2. This is simple if we remember that at the cathode we have reduction, that is, we want to have the electrons on the left side of the reaction (as reactants - showing that they are being gained by the substance being reduced). Had we done the exact opposite, it would have also been an oxidation. We cannot have two oxidation half reactions. The actual cell looks like this:

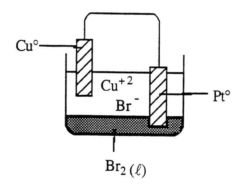

◊ Food for thought: How come there is no salt bridge in this cell?

19.8 Electromotive Force (EMF).

We will define the **electromotive force** as the potential difference (measured in volts, V) between the two electrodes in a galvanic cell. It will give us an idea of how spontaneous the reaction is, i.e., it gives us an idea of the driving force for the reaction. The superscript in $\varepsilon°$ implies *standard state* and that means that everything is in unit activity, that is, the concentration of all ions is exactly 1 M and the pressure of all gases is exactly 1 atm, as we discussed previously.

We can measure $\varepsilon°$ by using a *potentiometer*, which is a source of variable EMF. All we need to do is connect the spontaneous cell to the potentiometer and provide an increasingly larger opposing force until there is no current flowing. This measured potential is referred to as

Chapter 19: Electrochemistry.

the *reversible EMF* - related to the ideal *maximum* amount of electrical work that can be obtained from the chemical reaction taking place in the cell. It is called reversible because the slightest change in the opposing force will make the direction of the current flow go from one direction to the other.

In our definition of reversible EMF we used the word WORK. However, work has units of energy (J) whereas $\varepsilon°$ has units of volts (V). But:

$$1 J = 1 C \cdot 1 V$$

(One Joule is the amount of work done when we carry one Coulomb of charge through a potential difference of one Volt). Therefore, we must multiply the $\varepsilon°$ by the total number of coulombs of charge that went through the cell (which depends on the number of electrons that went through the cell). We should also recall that the thermodynamic parameter we called $\Delta G°$ gives us (or tells us) the maximum amount of work that we can get from a system.

$$\Delta G° = - nF\varepsilon°$$

where: n = number of moles of electrons
F = 96,500 C
$\varepsilon°$ = standard EMF in Volts.

It should be clear that if you have a galvanic cell, $\Delta G° < 0$ and the $\varepsilon° > 0$.

- Calculate the $\Delta G°$ for the reaction $Zn° + Cu^{+2} \rightleftharpoons Cu° + Zn^{+2}$ if the standard EMF ($\varepsilon°$) measured with a potentiometer is +1.10V.

In the actual redox reaction there is a net transfer of 2 moles of electrons from the zinc metal to the cupric ions. Therefore, n = 2.

$$\therefore \Delta G° = - (2)(96,500 \text{ C})(1.10 \text{V})$$

$$= - 212,000 \text{ J} = \textit{- 212 kJ} \quad \checkmark$$

The previous example represents a spontaneous reaction and the maximum amount of work that can be theoretically obtained from that cell is 212 kJ of work for every mole of Zinc metal that reacts. In reality, we will never get this much work, because the cell reaction will not proceed reversibly.

What would happen if we were to multiply the reaction of the preceding problem by two? If we double all the reagents and the products, then the reaction would be:

$$2\ Zn° + 2\ Cu^{+2} \rightleftharpoons 2\ Cu° + 2\ Zn^{+2}$$

and the $\Delta G°$ of the reaction would be twice as large (- 424 kJ). The question is: why is the $\Delta G°$ doubled? It is *not* because the $\varepsilon°$ increased by a factor of two (it is still +1.10 V), but rather because we have twice as many electrons flowing through the cell, $\therefore$ n = 4. Therefore, the *free energy change per coulomb* ($\varepsilon°$) is unchanged.

However, the reverse reaction is obviously non-spontaneous, so:

$$Cu° + Zn^{+2} \rightleftharpoons Zn° + Cu^{+2} \qquad \varepsilon° = -1.10\ V$$

19.9 <u>Half-Cell Potentials</u>.

It would be nice if we could have tables of half-cell potentials so that we could combine any two of them to make a cell and find the $\varepsilon°$ of the cell by just adding them up. However, a half-cell potential cannot be measured since it is impossible to have only an oxidation or a reduction half reaction occurring by itself. $\therefore$ We will choose a "standard" half-cell reaction and by convention say it has a potential of zero volts exactly.

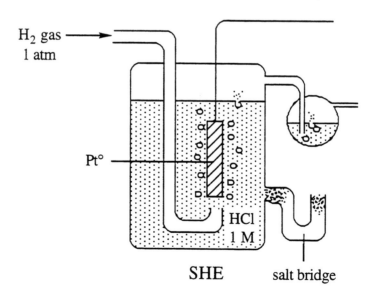

◊ reaction: $2\ H^+ + 2\ e^- \rightleftharpoons H_{2(g)}$

◊ by definition: $\varepsilon° = 0.00\ V$ (exactly).

At this point we will compare all other half-cells to our standard one to see how much easier or how much harder it is to reduce the unknown half-cell as compared to the standard one. We have chosen the standard **hydrogen** electrode (SHE) as the standard half-cell of zero potential.

If we coupled the SHE half-cell to a silver electrode in contact with a 1.00 M solution of $AgNO_3$ (same as Ag^+) and then measured the $\varepsilon°$ of the cell with a potentiometer we would find out that it is: + 0.7994 V. We could also observe that the electrons are coming out of the Pt° electrode and flowing into the silver electrode. This is the all the information needed in order for us to be able to determine the $\varepsilon°$ for that half reaction. Knowing that the electrons come out of the Pt° electrode tells us that the SHE is the anode (where oxidation occurs), therefore the silver electrode must be the cathode (where reduction occurs).

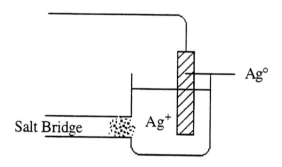

Therefore, the cell diagram would be:

$$Pt° / H_{2(g)} / H^+ // Ag^+ / Ag°$$

and the two half reactions would be:

$$H_{2(g)} \rightleftharpoons 2 H^+ + 2 e^- \qquad \varepsilon° = 0.00 \text{ V}$$

$$\underline{2 Ag^+ + 2 e^- \rightleftharpoons 2 Ag° \qquad \varepsilon° = x \text{ V}}$$

$$H_{2(g)} + 2 Ag^+ \rightleftharpoons 2 H^+ + 2 Ag° \qquad \varepsilon°_{cell} = 0 + x$$

but $\varepsilon°_{cell} = + 0.7994 \text{ V}$

$$\therefore + 0.7994 \text{ V} = 0 + x = x$$

Conclusion: The standard EMF for this half cell is seen below.

$$Ag^+ + 1 e^- \rightleftharpoons Ag° \qquad \varepsilon° = + 0.7994 \text{ V}$$

This is exactly the way that all the standard EMF's are measured experimentally. We

Chapter 19: Electrochemistry.

may find the results in the table of Standard Reduction Potentials in Appendix #4. We can now use these $\varepsilon°$ to calculate the $\varepsilon°_{cell}$ for any cell we want (as long as the two half-cells are included in the table in Appendix #4).

- Find the standard EMF ($\varepsilon°$) for the following cell:

$$Cu° / Cu^{+2} , Cl^- / Cl_{2(g)} / Pt°$$

Let's first determine the two half reactions before we look them up in the table. The two half reactions are:

$$Cu° \rightleftharpoons Cu^{+2} + 2\,e^- \qquad \varepsilon° = -(+0.3402\ V)$$
$$\underline{2\,e^- + Cl_{2(g)} \rightleftharpoons 2\,Cl^- \qquad \varepsilon° = +1.3583\ V}$$
$$Cu° + Cl_{2(g)} \rightleftharpoons Cu^{+2} + 2\,Cl^- \qquad \varepsilon°_{cell} = +1.0181\ V$$

We may wonder why the sign for the $\varepsilon°$ of the anodic cell was changed. This is because in the table, we find the potential for the reduction of cupric ions, *not* the oxidation of copper metal. Since the reaction was reversed (for the oxidation), the sign must be changed.

- Is the following reaction spontaneous at 25°C and under standard conditions (all ions at unit molarity)?

$$2\,Ag^+ + Cd° \rightleftharpoons 2\,Ag° + Cd^{+2}$$

In order to answer this question we must split this reaction into the two half reactions first and find the $\varepsilon°$ for a theoretical cell built with these two half-cells.

$$Cd° \rightleftharpoons Cd^{+2} + 2\,e^- \qquad \varepsilon° = -(-0.4026\ V)$$
$$\underline{2\,e^- + 2\,Ag^+ \rightleftharpoons 2\,Ag° \qquad \varepsilon° = +0.7994\ V}$$
$$Cd° + 2\,Ag^+ \rightleftharpoons Cd^{+2} + 2\,Ag° \qquad \varepsilon° = +1.2020\ V$$

Since the $\varepsilon°$ is positive, then the reaction is spontaneous in the direction written.

If we want to make a galvanic cell out of any two half reactions from Appendix #4, all that we must do is leave the half-cell with the most positive $\varepsilon°$ the way it is and reverse the other one so that when we combine them, the $\varepsilon°$ for the resulting cell will be positive. This will make the $\Delta G°$ negative (therefore it will be spontaneous or galvanic).

Before we proceed, we should think about the meaning of those numbers ($\varepsilon°$) from the appendix. The more positive the EMF for a given substance, the more likely it is for that substance to be reduced. Remember that this is a *reduction* potential table, so it measures the ease with which a species is reduced. That is why when we compare two species we can very easily determine which is more likely to be reduced (the one with the more positive EMF).

It is also important to realize that we can read the table backwards. If we were to look at the reverse reactions, we would be looking at an oxidation table instead. This means that we can decide which species are more likely to be oxidized by comparing those on the right hand side of the table. Let's look at the particular portion of the table shown below.

Reduction Half Reaction	EMF ($\varepsilon°$), V
$Cl_{2(g)} + 2\,e^- \rightleftharpoons 2\,Cl^-$	1.3583
$Cr_2O_7^{-2} + 14\,H^+ + 6\,e^- \rightleftharpoons 2\,Cr^{+3} + 7\,H_2O$	1.33
$O_{2(g)} + 4\,H^+ + 4\,e^- \rightleftharpoons 2\,H_2O$	1.23
$MnO_{2(s)} + 4\,H^+ + 2\,e^- \rightleftharpoons Mn^{+2} + 2\,H_2O$	1.224
$IO_3^- + 6\,H^+ + 5\,e^- \rightleftharpoons 0.5\,I_{2(s)} + 3\,H_2O$	1.195
$Br_{2(\ell)} + 2\,e^- \rightleftharpoons 2\,Br^-$	1.095

In such a table, the species on the left hand side (bold) are getting reduced and the species on the right hand side would be getting oxidized, if we were looking at the table backwards. Anyway, from that short table, the Cl_2 gas is the easiest substance to reduce (it has the most positive EMF). We could also state the Cl_2 is the strongest oxidizing agent. Remember that the oxidizing agent is reduced.

Let's reverse the table and convert it into an oxidation potential table. The reactions are exactly reversed and so are the EMF's. Notice that the least negative oxidation EMF is the one for the oxidation of the bromide ions, therefore, the easiest substance to oxidize in this short table would be the bromide ions.

Oxidation Half Reaction	EMF ($\varepsilon°$), V
$2\,Cl^- \rightleftharpoons Cl_{2(g)} + 2\,e^-$	-1.3583
$2\,Cr^{+3} + 7\,H_2O \rightleftharpoons Cr_2O_7^{-2} + 14\,H^+ + 6\,e^-$	-1.33
$2\,H_2O \rightleftharpoons O_{2(g)} + 4\,H^+ + 4\,e^-$	-1.23
$Mn^{+2} + 2\,H_2O \rightleftharpoons MnO_{2(s)} + 4\,H^+ + 2\,e^-$	-1.224
$0.5\,I_{2(s)} + 3\,H_2O \rightleftharpoons IO_3^- + 6\,H^+ + 5\,e^-$	-1.195
$2\,Br^- \rightleftharpoons Br_{2(\ell)} + 2\,e^-$	-1.095

Chapter 19: Electrochemistry.

Keep in mind, however, that the table at the end of the book (on Appendix 4) is exclusively a reduction potential table. We need to learn how to read this table by itself. This means that in such a table, the substances being reduced are on the left hand side of that table, whereas the things being oxidized are on the right hand side of the table. If someone were to ask us a question pertaining to the ease of oxidation for a particular substance, you would look for the substance on the right hand side (where the reducing agents are located). Conclusion: the easiest substance to be oxidized from the short reduction table in the bottom of the previous page, is the Br⁻ ion (the bromide ion), because it has the least positive reduction potential (or it would have the most "positive" potential as an oxidation).

This table should let us also estimate the feasibility of many redox reactions. For example, let's say that we want to know if chloride ions would react spontaneously with liquid bromine. This question could be stated as follows. *Are the chloride ions capable of reducing liquid bromine?* In order to answer this question we need to establish a possible reaction between these two substances and determine the EMF for this possible reaction. If it turns out to be positive (the EMF), that would mean that the $\Delta G°$ would be negative, and the reaction would be spontaneous.

$$Br_{2(\ell)} + 2\,e^- \rightleftharpoons 2\,Br^- \qquad\qquad 1.095\text{ V}$$

$$2\,Cl^- \rightleftharpoons Cl_{2(g)} + 2\,e^- \qquad\qquad -1.3583\text{ V}$$

The overall cell reaction would be:

$$Br_{2(\ell)} + 2\,Cl^- \rightleftharpoons Cl_{2(g)} + Br^-$$

but the EMF for the cell is indeed negative, which implies a positive $\Delta G°$, therefore a non-spontaneous reaction. Therefore, the answer is NO!

- Will silver metal (Ag°) react with cupric ions (Cu^{+2})?

    The first thing we should do is look for these substances in Appendix #4. We find the following three reactions (that involve these species and no other ones, other than the metals themselves).

$$Ag^+ + 1\,e^- \rightleftharpoons Ag° \qquad\qquad 0.7994\text{ V}$$

$$Cu^{+2} + 2\,e^- \rightleftharpoons Cu° \qquad\qquad 0.3402\text{ V}$$

$$Cu^{+2} + 1\,e^- \rightleftharpoons Cu^{+1} \qquad\qquad 0.16\text{ V}$$

Chapter 19: Electrochemistry.

Of course, we are trying to combine the substances in bold. It is imperative that one of these substances is in the right side, since we must have an oxidation half reaction to couple with the reduction. This means that the silver reaction must be the anode. That would change the potential to: - 0.7994 V. Independently of which half reaction we choose for the cathode, it is clear that the sum of the two halves will give a negative answer. Therefore, this potential reaction is not spontaneous, and will NOT occur. The best possible scenario is show below.

$$\mathbf{Ag^\circ} \rightleftharpoons \mathbf{Ag^{+1}} + 1\ e^- \qquad -0.7994\ V$$
$$\mathbf{Cu^{+2}} + 2\ e^- \rightleftharpoons \mathbf{Cu^\circ} \qquad 0.3402\ V$$

- If we place a new penny (assume it is pure copper) in a solution of Fe^{+3}, will something happen? If something does happen, tell me what it would be.

    Let's look for all the possible half-reactions from Appendix #4.

$$Cu^{+2} + 2\ e^- \rightleftharpoons \mathbf{Cu^\circ} \qquad 0.3402\ V$$
$$Cu^{+1} + 1\ e^- \rightleftharpoons \mathbf{Cu^\circ} \qquad 0.52\ V$$
$$\mathbf{Fe^{+3}} + 1\ e^- \rightleftharpoons Fe^{+2} \qquad 0.771\ V$$
$$\mathbf{Fe^{+3}} + 3\ e^- \rightleftharpoons Fe^\circ \qquad -0.036\ V$$

Once more, we see that the copper metal is in one side of the reactions, while the ferric ions is in the other side. It bears to repeat that this must always be true since one reaction must be oxidation, while the other must be reduction. The whole question is which should we choose? Remember that we should always follow the path of lowest free energy. Therefore, we should look for the two reactions that will give the most positive EMF. This means reversing the first one (thus forming the least negative number) and keeping the third one (the most positive of the reduction half reactions for the ferric ion).

$$\mathbf{Cu^\circ} \rightleftharpoons Cu^{+2} + 2\ e^- \qquad -0.3402\ V$$
$$\underline{2x\ [\mathbf{Fe^{+3}} + 1\ e^- \rightleftharpoons Fe^{+2}]} \qquad \underline{+0.771\ V}$$
$$Cu^\circ + 2\ Fe^{+3} \rightleftharpoons Cu^{+2} + 2\ Fe^{+2} \qquad +0.431\ V$$

Therefore, a reaction will take place and we will observe that the penny will slowly dissolve into solution (to form cupric ions), while the ferric ions are being converted into ferrous ions.

Chapter 19: Electrochemistry.

- Let's say you lose a 24-karat gold ring (pure gold) and it is exposed to air and rain (that means oxygen gas and water) for a year before you find it. Will the gold have oxidized by that time?

    The two half reactions of interest would be:

    $$Au^{+3} + 3\,e^- \rightleftharpoons Au° \qquad +1.50\text{ V}$$
    $$O_{2(g)} + 4\,H^+ + 4\,e^- \rightleftharpoons 2\,H_2O \qquad +1.23\text{ V}$$

    The interaction of gold with the oxygen gas leads to a negative potential, therefore, this reaction will not occur spontaneously.

    $$Au° \rightleftharpoons Au^{+3} + 3\,e^- \qquad -1.50\text{ V}$$
    $$O_{2(g)} + 4\,H^+ + 4\,e^- \rightleftharpoons 2\,H_2O \qquad +1.23\text{ V}$$

- Can we oxidize nitrate ions any further?

    This question can be answered in two different ways. The simplest (since it involves no critical thinking) is NO because it does not appear in any half reaction in the right side of the Appendix #4. However, this does not tell us why. Let's think about this ion.

    The oxidation number of the nitrogen in the nitrate ion is +5. This means that in this particular form, the nitrogen has lost five valence shell electrons. What is the ground state electronic configuration for the nitrogen atom? It is $2s^2\,2p^3$. That gives the nitrogen a total of five valence shell electrons. In the nitrate ion, the nitrogen has already lost all of its valence shell electrons, therefore, it will NOT lose anymore. Conclusion: it cannot be oxidized any further.

19.10 EMF As a Function of Concentration.

What happens if we do not have a standard cell? Remember that a standard cell has all the species at unit activity. If they are not, we should be able to still determine the amount of work that we can get out of the cell. Now, from thermodynamics,[1] we learned that:

$$\Delta G = \Delta G° + RT \ln Q$$

---
[1] See Section 15.13.

Chapter 19: Electrochemistry.

In electrochemistry we have seen that $\Delta G° = -nF\varepsilon°$. Similarly, $\Delta G = -nF\varepsilon$. Substituting into the equation above, we get:

$$-nF\varepsilon = -nF\varepsilon° + RT \ln Q$$

Dividing by $(-nF)$ we get:

$$\varepsilon = \varepsilon° - (RT/nF) \ln Q$$

and changing to log (base 10), we must multiply by 2.303:

$$\varepsilon = \varepsilon° - 2.303\,(RT/nF) \log Q$$

If we substitute R = 8.314 J/mol K, T = 298 K, and knowing that F = 96,500 C, we get:

$$\varepsilon = \varepsilon° - \frac{0.0592V}{n} \times \log Q$$

We refer to this equation as the **Nernst equation**. This equation will enable us to calculate the EMF of any non-standard cell. Remember that Q is the reaction quotient and in order to determine its value, we need to know the cell reaction (concentration of products over those of the reactants).

- Determine the EMF (ε) for the following cell:

$$Zn° / Zn^{+2}\ (0.100\ M)\ //\ Cu^{+2}\ (0.200\ M)\ /\ Cu°$$

We immediately recognize that this is not a standard cell (since the concentration of at least one of the ions is not unity). Therefore, we need to figure out the EMF (ε) of this cell using the Nernst equation.

$$\varepsilon = \varepsilon° - \frac{0.0592V}{n} \log Q$$

Since we need to find out the numerical value for $\varepsilon°$ anyway, let's do that first. We start this process by writing the two half-reactions (one of oxidation and one of reduction) and get their respective EMF's from the table in Appendix #4. Remember that in the cell notation, the first thing we write is the anode (where oxidation occurs) and then the cathode (where reduction occurs). Therefore, the zinc metal is being oxidized, whereas the cupric ions are being reduced.

Chapter 19: Electrochemistry.

$$Zn° \rightleftharpoons Zn^{+2} + 2\,e^- \qquad\qquad \varepsilon° = -(-0.76\text{ V})$$
$$\underline{2\,e^- + Cu^{+2} \rightleftharpoons Cu° \qquad\qquad \varepsilon° = +0.34\text{ V}}$$
$$Zn° + Cu^{+2} \rightleftharpoons Zn^{+2} + Cu° \qquad\qquad \varepsilon° = +1.10\text{ V}$$

What is Q in the Nernst equation? It is the reaction quotient, which in this case it is simply:[1]

$$Q = \frac{[Zn^{+2}]}{[Cu^{+2}]}$$

Well, the concentrations of these two ions can be found in the cell notation, and we can see from above that there were two electrons transferred from the Zn° to the Cu^{+2} ions, therefore, n = 2. Finally,

$$\varepsilon = +1.10V - \frac{0.0592V}{2}\log(\frac{0.100M}{0.200M})$$

$$\therefore\ \varepsilon° = +1.11\text{ V}\ \checkmark$$

- Calculate the EMF for the following cell:

$$Zn° / Zn^{+2} // Cr^{+3}\ (0.0100\text{ M}) / Cr°$$

Once more, this is not a standard cell so we must use Nernst equation. Therefore, we will calculate $\varepsilon°$ first.

$$[Zn° \rightleftharpoons Zn^{+2} + 2\,e^-]\,x3 \qquad\qquad \varepsilon° = -(-0.76\text{ V})$$
$$\underline{[3\,e^- + Cr^{+3} \rightleftharpoons Cr°]\,x2 \qquad\qquad \varepsilon° = -0.73\text{V}}$$
$$3\,Zn° + 2\,Cr^{+3} \rightleftharpoons 3\,Zn^{+2} + 2\,Cr° \qquad\qquad \varepsilon°_{cell} = +0.03\text{ V}$$

In this case, n = 6, because we transferred six electrons from the Zinc electrode to the chromium electrode, and Q is given by:

$$Q = \frac{[Zn^{+2}]^3}{[Cr^{+2}]^2}$$

The concentrations are read directly from the cell notation, therefore,

---

[1] Remember that solids are not included in the mathematical expression for Q. Zn° and Cu° are the solid metals themselves.

$$\varepsilon = +0.03V - \frac{0.0592V}{6}\log[\frac{1}{(0.0100)^2}]$$

$$\therefore \varepsilon = -0.01 \text{ V}$$

This result is shocking. This is telling us that this is NOT a spontaneous cell, therefore, it is not a galvanic cell. Actually, the galvanic cell would be the reverse reaction. Therefore, the galvanic cell notation should have been:

$$Cr° / Cr^{+3} \text{ (0.0100 M)} // Zn^{+2} / Zn°$$

for which:

$$\varepsilon = +0.01 \text{ V}. \checkmark$$

- Find the EMF for the following cell.

$$Pt° / H_{2(g)} \text{ (0.900 atm)} / H^+ \text{ (pH = 1.00)} // Ag^+ \text{ (0.0100 M)} / Ag°$$

Following the same routine, let's determine the standard EMF for this cell, the cell reaction and the number of electrons transferred from the anode to the cathode.

$H_{2(g)} \rightleftharpoons 2 H^+ + 2 e^-$		$\varepsilon° = 0.00$ V
$[1 e^- + Ag^+ \rightleftharpoons Ag°] \times 2$		$\varepsilon° = +0.80$ V
$H_{2(g)} + 2 Ag^+ \rightleftharpoons 2 H^+ + 2 Ag°$		$\varepsilon° = +0.80$ V

We can see that there are two electrons transferred from the platinum electrode to the silver electrode, therefore, n = 2. Q is given by:

$$Q = \frac{[H^+]^2}{[Ag^+]^2 P_{H_2}}$$

At first glance, it seems weird that we are combining molar concentrations with pressures, but we must remember that these ratios, just like equilibrium constants, are in terms of activities. The activity of a pure solid is equal to one, for an ion in solution it is numerically equal to its molar concentration, and if it is a gas, it is numerically equal to its pressure in atmospheres. The key word here is **numerically equal**, meaning that these are unitless numbers, that happen to be equal to the numerical value of the pressure, if the pressure were expressed in atmospheres, for example.

Substitution of all the values yields:

Chapter 19: Electrochemistry.

$$\varepsilon = +0.80V - \frac{0.0592V}{2} \log \frac{(0.100)^2}{(0.0100)^2 (0.900)}$$

$$\therefore \varepsilon = +0.74 \text{ V} \checkmark$$

19.11 Calculation of Equilibrium Constants.

In order to calculate the numerical value for the equilibrium constant of a redox reaction, we need to know the numerical value for its ΔG°. Why is this necessary? Because there is a relationship between the ΔG° of a reaction and its equilibrium constant. This is given by:

$$\Delta G° = -RT \ln K$$

Earlier in this chapter, we learned that the ΔG° for an electrochemical cell is given by:

$$\Delta G° = -nF\varepsilon°$$

Setting these two equation equal to one another, we van get the following expression.

$$-nF\varepsilon° = -RT \ln K = -2.303 RT \log K$$

$$\log K = \frac{n\varepsilon°}{0.0592V}$$

- Calculate the equilibrium constant for the following reaction.

$$2 \text{ Cr}^{+3} + 3 \text{ Zn}° \rightleftharpoons 3 \text{ Zn}^{+2} + 2 \text{ Cr}°$$

It should be clear that the equilibrium constant can be determined directly from that equation, independent of whether this reaction is standard or not. Therefore, all we need to know is the numerical value for the standard EMF and the number of electrons transferred from the anode to the cathode.

$$[\text{Zn}° \rightleftharpoons \text{Zn}^{+2} + 2e^-] \times 3 \qquad \varepsilon° = -(-0.76 \text{ V})$$

$$\underline{[3e^- + \text{Cr}^{+3} \rightleftharpoons \text{Cr}°] \times 2 \qquad \varepsilon° = -0.73V}$$

$$3 \text{ Zn}° + 2 \text{ Cr}^{+3} \rightleftharpoons 3 \text{ Zn}^{+2} + 2 \text{ Cr}° \qquad \varepsilon° = +0.03 \text{ V}$$

We can see that six electrons are transferred from the zinc electrode to the chromium electrode, therefore, n = 6. Substitution into the equation yields the following results.

$$\log K = 6 (0.03 \text{ V}) / 0.0592 \text{ V}$$

$$\log K = 3.0$$

$$\therefore K = 1,000 \checkmark$$

19.12 <u>Electrolysis.</u>

It is the production of a chemical reaction by means of an electrical current. Such a unit is an electrolytic cell containing two electrodes. These are *non-spontaneous* reactions, therefore we must do work on the system to get the reaction to occur.

It is essential that the ions are free to move in order to conduct electricity. Solid NaCl does not conduct electricity *but* molten NaCl or aqueous NaCl will indeed conduct electricity. Therefore, without ionic mobility, there is no conduction of electricity.

<u>Electrolysis of Molten NaCl[1].</u>

The temperature must be very high to melt the NaCl. In this state we will have mobile Na^+ and Cl^- ions which is the essential prerequisite for conduction of electricity which we just mentioned above. Keep in mind that there is no water in this system, just sodium and chloride ions (at a very high temperature).

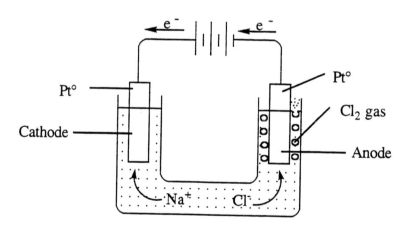

◊ Cathode reaction: $\quad Na^+ + 1\,e^- \rightleftharpoons Na^\circ$

◊ Anode reaction: $\quad 2\,Cl^- \rightleftharpoons Cl_{2(g)} + 2\,e^-$

Obviously, the cell reaction is:

$$2\,Na^+ + 2\,Cl^- \rightleftharpoons 2\,Na^\circ + Cl_{2(g)}.$$

---

[1] This is the principal commercial process for producing metallic sodium.

Chapter 19: Electrochemistry.

Electrolysis of Aqueous NaCl.

Let us follow a very similar example, the electrolysis of **aqueous** NaCl. As always we must first decide if we have mobile ions or not. In an aqueous solution of NaCl we have sodium ions, chloride ions, and lots of water (the solvent). These ions *are free to move*. As soon as we connect the batteries to this cell we find that it takes a lot less batteries[1] connected in series in order to force this reaction to occur. This immediately tells us that the reaction itself must be different.

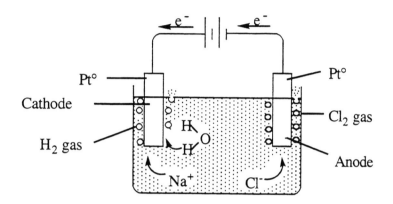

We already know that the Na^+ ions will rush to the cathode (cations are attracted to the cathode) and that the Cl^- ions will rush to the anode (anions are attracted to the anode). However, this time we have a choice: will these ions be discharged or will the water itself be discharged at the electrodes? When we carry out the electrolysis in the laboratory we see that a green gas (Cl_2) is produced at the anode and that a colorless and odorless gas (H_2) is produced at the cathode. Therefore water is the substance being reduced at the cathode and not the Na^+ ions.

Conclusion: It is easier to reduce water than sodium ions.

It is easier to oxidize chloride ions than water.

Obviously, if we are told what the products are at the electrodes, it should be very simple to predict what the half reaction is.

a) When water is reduced it forms H_2 gas and a basic solution:

$$e^- + H_2O \implies \tfrac{1}{2} H_2(g) + OH^-$$

---

[1] Less batteries in series implies we did less work in order to get the reaction to go.

b) When water is oxidized it forms O_2 gas and an acidic solution:

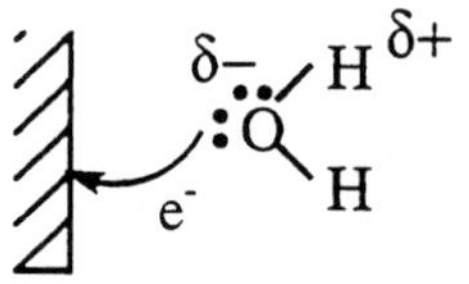

The balanced reactions for these two cases can be seen below.

a) Oxidation of water (acidic).

$$2 H_2O \rightleftharpoons O_{2(g)} + 4 H^+ + 4 e^-$$

b) Reduction of water (basic).

$$2 e^- + 2 H_2O \rightleftharpoons H_{2(g)} + 2 OH^-$$

Let's go back to the reaction we are discussing, the electrolysis of an aqueous sodium chloride solution. Could we have predicted the outcome of this reaction using the table of standard reduction potentials found in Appendix #4? Let's check and find out. We can start with the cathode, that is, we need to make a decision as to whether the sodium ions or the water will be reduced at the cathode.

$$Na^+ + 1 e^- \rightleftharpoons Na° \qquad \varepsilon° = -2.71 \text{ V}$$

$$2 H_2O + 2 e^- \rightleftharpoons H_{2(g)} + 2 OH^- \qquad \varepsilon° = -0.83 \text{ V}$$

Analyzing these two standard EMF's, we can see that it is a lot easier to reduce the water than it is to reduce the sodium ions. Remember that we are supplying the electrical energy necessary for the reaction by connecting a few batteries in series until the potential is more than enough for the reaction to take place. Well, as the supplied voltage increases, it will reach the necessary one for the reduction of water way before the one needed for the reduction of sodium.

How about the anode? We need to decide whether the chloride ions or the water will be oxidized at the anode. In the tables we find the following:

$$2 Cl^- \rightleftharpoons Cl_{2(g)} + 2 e^- \qquad \varepsilon° = -1.36 \text{ V}$$

$$2 H_2O \rightleftharpoons O_{2(g)} + 4 H^+ + 4 e^- \qquad \varepsilon° = -1.23 \text{ V}$$

It seems like we have a discrepancy here. Looking at these numbers, we would expect that it should be easier to oxidize the water, although not by much, than it is to oxidize the chloride ions. However, experimentally we observe that the chloride ions were oxidized preferentially over the water. The reason for this is because it turns out that we need a much higher voltage than expected in order to oxidize the water. The excess voltage needed is referred to as **OVERVOLTAGE**. It is very difficult to try to explain why we have overvoltage for water oxidation but it suffice to say that it is not always simple to predict, especially if the EMF's are close. So, how are we supposed to know exactly what happens at the anode if the EMF's are close? We observe experimentally what is released at the electrodes and then make a clear decision based on the experimental facts. If no information is given, then we will use the table and follow it, knowing full well that there is a chance that we may be incorrect because of overvoltage.

19.13 Faraday's Laws.

1) The mass of a substance produced or consumed at the electrodes is proportional to the quantity of electricity that passes through the cell.

   ◊ 1 Faraday = 1 F = charge carried by 1 mole of electrons.

   $$\therefore 1\,F = 96{,}487\,C = 96{,}500\,C$$

2) The number of faradays that pass through a cell when one mole of a substance is produced or consumed at the electrodes is a <u>whole</u> number.

   Ex. $\quad 2\,H_2O \rightleftharpoons O_{2(g)} + 4\,H^+ + 4\,e^-$

   Therefore, to get <u>one</u> mole of O_2, we must pass <u>4 F</u> of electricity through the cell.

3) The number of Faradays of electricity that pass through the cathode of a cell is identical to the number of Faradays of electricity that pass through the anode of that same cell. It is also true that is we have various cells connected in series, the number of Faradays of electricity that pass through the first cell is equal to the number of Faradays of electricity that pass through each of the cells.

- Consider the electrolysis of aqueous $Cu(NO_3)_2$.

   Let's see if in this case we can predicted the outcome by using the table. Starting with the cathode: we need to make a choice between the cupric ions and the water.

   $$Cu^{+2} + 2\,e^- \rightleftharpoons Cu^\circ \qquad\qquad \varepsilon^\circ = +0.34\,V$$

$$2\ H_2O\ +\ 2\ e^-\ \rightleftharpoons\ H_{2(g)}\ +\ 2\ OH^- \qquad \varepsilon° = -0.83\ V$$

Based on these values, we could predict that the cupric ions should be reduced preferentially over the water. This is exactly what we observe experimentally.

How about the anode? At the anode we need to make a choice between the oxidation of the nitrate ions or that of water. In this case there is no argument whatsoever, since it is impossible to oxidize the nitrate ions any further than they are already. Therefore, the water must be getting oxidized at the anode.

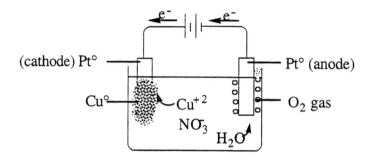

Therefore, the two half reactions are:

$$Cu^{+2}\ +\ 2\ e^-\ \rightleftharpoons\ Cu°$$
$$2\ H_2O\ \rightleftharpoons\ O_{2(g)}\ +\ 4\ H^+\ +\ 4\ e^-$$

and the overall cell reaction (after adding the two half reactions and making sure that the electrons cancel out) is:

$$2\ Cu^{+2}\ +\ 2\ H_2O\ \rightleftharpoons\ O_{2(g)}\ +\ 2\ Cu°\ +\ 4\ H^+$$

a) If a current of 1.50 A is passed through the solution for 2.00 hours, how many grams of Cu° will be deposited at the <u>cathode</u>? Recall that the half reaction is $Cu^{+2}\ +\ 2\ e^-\ \Rightarrow\ Cu°$. This will tell us the relationship between the moles of Cu^{+2} and the number of faradays.

The problems concerning Faraday's laws are easily solved using a gigantic conversion factor. We should always start with the time and convert it into seconds. This is because once we have the time in seconds, we can use the current (in Amps) to change it into coulombs. Once we generate coulombs, we can use the definition of a Faraday (1 F = 96,500 C) to change into Faradays. The whole purpose of getting to the Faradays was to get to the real conversion factor: from faradays into moles. This is the

Chapter 19: Electrochemistry.

conversion factor we get from the balanced half-reaction. In this case, for every mole of copper metal formed, we must have passed two faradays of electricity through the cell. Finally, we can convert from moles to grams using the molecular or atomic weight.

$$2.00 hrs \times \frac{3,600 s}{1 hr} \times \frac{1.50 C}{1 s} \times \frac{1 F}{96,500 C} \times \frac{1 mol}{2 F} \times \frac{63.55 g}{1 mol}$$

$$\therefore g_{Cu°} = 3.56 \text{ g} \quad \checkmark$$

b) How many liters of O_2 gas, measured at STP, will be formed at the anode simultaneously?

The easiest way to do this problem is to figure out how much oxygen gas is formed directly from the amount of Cu° formed (3.56 g, from part **a**). Don't forget that one mole of any gas at STP occupies 22.4 L!

$$3.56 g Cu° \times \frac{1 mol_{Cu}}{63.55 g_{Cu}} \times \frac{1 mol_{O_2}}{2 mol_{Cu}} \times \frac{22.4 L}{1 mol_{O_2}}$$

$$\therefore V_{O2} = 0.627 \text{ L} \quad \checkmark$$

This problem can also be worked out in exactly the same way as we did part **a**, keeping in mind that this time the conversion factor is 4 F will produce one mole of oxygen gas.

$$V_{O_2} = 2.00 hrs \times \frac{3,600 s}{1 hr} \times \frac{1.50 C}{1 s} \times \frac{1 F}{96,500 C} \times \frac{1 mol}{4 F} \times \frac{22.4 L}{1 mol}$$

$$\therefore V_{O2} = 0.627 \text{ L} \quad \checkmark$$

LiCl(s) ΔH°f = -408.61
LiF(s) ΔH°f = -615.97
ClF₃(s) ΔH°f = -163.2

Appendices.

Appendices

Appendix #1.

Thermodynamic Parameters for Some Selected Substances @ 298K and 1 atm

Substance	$\Delta H°_f$, kJ/mol	$\Delta G°_f$, kJ/mol	S°, J/K mol
$Al_{(s)}$	0	0	28.3
$Al_2O_{3(s)}$	-1675.7	-1582.3	50.9
$AlCl_{3(s)}$	-704.2	-628.8	110.7
$Ba_{(s)}$	0	0	67
$BaCO_{3(s)}$	-1219	-1139	112
$BaO_{(s)}$	-582	-552	70
$BaSO_{4(s)}$	-1473.2	-1362.2	132.2
$Be_{(s)}$	0	0	9.5
$BeO_{(s)}$	-599	-569	14
$Br_{2(\ell)}$	0	0	152.2
$Br_{2(g)}$	30.9	3.11	245.46
$Br_{2(aq)}$	-3	4	130
$Br^-_{(aq)}$	-121	-104	82
$HBr_{(g)}$	-36.40	-53.45	198.70
$Cd_{(s)}$	0	0	52
$CdO_{(s)}$	-258	-228	55
$CdS_{(s)}$	-162	-156	65
$CdSO_{4(s)}$	-935	-823	123
$Ca_{(s)}$	0	0	41.42
$CaCO_{3(s)}$	-1206.9	-1128.8	92.9
$CaO_{(s)}$	-635.1	-604.0	39.75
$CaSO_{4(s)}$	-1434.1	-1321.8	106.7
$C_{(s, graphite)}$	0	0	5.74
$C_{(s, diamond)}$	1.895	2.900	2.377
$CO_{(g)}$	-110.5	-137.2	197.7
$CO_{2(g)}$	-393.5	-394.4	213.7
$CH_{4(g)}$	-74.81	-50.72	186.26
$CH_3OH_{(g)}$	-200.66	-161.96	239.81

Substance	$\Delta H°_f$, kJ/mol	$\Delta G°_f$, kJ/mol	$S°$, J/K mol
$CH_3OH_{(\ell)}$	-238.66	-166.27	126.8
$H_2CO_{(g)}$	-116	-110	219
$HCOOH_{(g)}$	-363	-351	249
$HCN_{(g)}$	135.1	125	202
$C_2H_{2(g)}$	226.73	209.20	200.94
$C_2H_{4(g)}$	52.26	68.15	219.56
$CH_3CHO_{(g)}$	-166	-129	250
$CH_3CH_2OH_{(\ell)}$	-278	-175	161
$C_2H_{6(g)}$	-84.7	-32.9	229.5
$C_3H_{6(g)}$	20.9	62.7	266.9
$C_3H_{8(g)}$	-103.8	-23.49	269.9
$CH_3COOH_{(\ell)}$	-484	-389	160
$CCl_{4(\ell)}$	-135.44	-65.21	216.40
$Cl_{2(g)}$	0	0	223.07
$HCl_{(g)}$	-92.307	-95.299	186.91
$Cr_{(s)}$	0	0	23.8
$Cr_2O_{3(s)}$	-1139.6	-1058.2	81.2
$CrO_{3(s)}$	-579	-502	72
$Cu_{(s)}$	0	0	33.15
$CuCO_{3(s)}$	-595	-518	88
$Cu_2O_{(s)}$	-170	-148	93
$CuO_{(s)}$	-157.3	-129.7	42.63
$F_{2(g)}$	0	0	202.79
$HF_{(g)}$	-271.1	-273.2	173.78
$H_{2(g)}$	0	0	130.68
$H·_{(g)}$	217.97	203.25	114.71
$H_2O_{(\ell)}$	-285.830	-237.129	69.91
$H_2O_{(g)}$	-241.818	-228.572	188.825
$I_{2(s)}$	0	0	116.14
$I_{2(g)}$	62.44	19.33	260.7
$Fe_{(s)}$	0	0	27.78
$FeO_{(s)}$	-272	-255	61
$Fe_3O_{4(s)}$ (magnetite)	-1118.4	-1015.4	146.4
$Fe_2O_{3(s)}$ (hematite)	-824.2	-742.2	87.4

Substance	$\Delta H°_f$, kJ/mol	$\Delta G°_f$, kJ/mol	$S°$, J/K mol
$FeS_{(s)}$	-95	-97	67
$FeSO_{4(s)}$	-929	-825	121
$Pb_{(s)}$	0	0	64.81
$PbO_{2(s)}$	-277	-217	69
$PbSO_{4(s)}$	-920	-813	149
$Mg_{(s)}$	0	0	32.68
$MgCO_{3(s)}$	-1095.8	-1012.1	65.7
$MgO_{(s)}$	-601.70	-569.43	26.94
$Mn_{(s)}$	0	0	32
$MnO_{(s)}$	-385	-363	60
$MnO_{2(s)}$	-521	-466	53
$Hg_{(\ell)}$	0	0	76.02
$Hg_2Cl_{2(s)}$	-265	-211	196
$HgCl_{2(s)}$	-230	-184	144
$HgO_{(s)}$	-90	-59	70
$HgS_{(s)}$	-58	-49	78
$Ni_{(s)}$	0	0	29.87
$NiCl_{2(s)}$	-316	-272	107
$NiO_{(s)}$	-239.7	-211.7	37.99
$NiS_{(s)}$	-93	-90	53
$N_{2(g)}$	0	0	191.61
$NH_{3(g)}$	-46.11	-16.45	192.45
$NO_{(g)}$	90.3	86.6	210.8
$NO_{2(g)}$	33.18	51.31	240.06
$N_2O_{(g)}$	82.1	104.2	219.9
$N_2O_{4(g)}$	9.16	97.89	304.29
$N_2O_{4(\ell)}$	-20	97	209
$N_2O_{5(s)}$	-42	134	178
$N_2H_{4(\ell)}$	50.63	149.34	121.21
$HNO_{3(\ell)}$	-174.10	-80.71	155.60
$NH_4ClO_{4(s)}$	-295	-89	186
$NH_4Cl_{(s)}$	-314.43	-202.87	94.6
$O_{2(g)}$	0	0	205.138
$O·_{(g)}$	249.17	231.73	161.06

Substance	ΔH°$_f$, kJ/mol	ΔG°$_f$, kJ/mol	S°, J/K mol
$O_{3(g)}$	142.7	163.2	238.93
$P_{4(s)}$ (white)	0	0	164.36
$P_{4(s)}$ (red)	- 70.4	- 48.4	91.2
$P_{(s)}$ (black)	- 39	- 33	23
$PF_{5(g)}$	- 1578	- 1509	296
$PH_{3(g)}$	5.4	13.4	310.23
$H_3PO_{4(s)}$	- 1279.0	- 1119.1	110.5
$P_4O_{10(s)}$	- 2984	- 2698	229
$K_{(s)}$	0	0	64.18
$KCl_{(s)}$	- 436.75	- 409.14	82.59
$KClO_{3(s)}$	- 397.73	- 296.25	143.1
$KClO_{4(s)}$	- 433	- 304	151
$K_2O_{(s)}$	- 361	- 322	98
$K_2O_{2(s)}$	- 496	- 430	113
$KO_{2(s)}$	- 283	- 238	117
$KOH_{(s)}$	- 424.76	- 379.08	78.9
$SiO_{2(s)}$ (quartz)	- 910.94	- 856.64	41.84
$SiCl_{4(\ell)}$	- 687	- 620	240
$Ag_{(s)}$	0	0	42.55
$AgBr_{(s)}$	- 100	- 97	107
$AgCN_{(s)}$	146	164	84
$AgCl_{(s)}$	- 127.07	- 109.79	96.2
$Ag_2CrO_{4(s)}$	- 712	- 622	217
$AgI_{(s)}$	- 62	- 66	115
$Ag_2O_{(s)}$	- 31	- 11	122
$Ag_2S_{(s)}$	- 32	- 40	146
$Na_{(s)}$	0	0	51.21
$NaBr_{(s)}$	- 361.06	- 348.98	86.82
$Na_2CO_{3(s)}$	- 1130.68	- 1044.44	134.98
$NaHCO_{3(s)}$	- 948	- 852	102
$NaCl_{(s)}$	- 411	- 384	72
$NaH_{(s)}$	- 56	- 33	40
$NaI_{(s)}$	- 288	- 282	91
$NaNO_{3(s)}$	- 467	- 366	116

Appendices.

Substance	$\Delta H°_f$, kJ/mol	$\Delta G°_f$, kJ/mol	$S°$, J/K mol
$Na_2O_{(s)}$	-416	-377	73
$NaOH_{(s)}$	-425.61	-379.49	64.45
$S_{(s)}$ (rhombic)	0	0	31.80
$S_{(s)}$ (monoclinic)	0.3	0.1	33
$SF_{6(g)}$	-1209	-1105	292
$H_2S_{(g)}$	-20.63	-33.56	205.79
$SO_{2(g)}$	-296.83	-300.194	248.22
$SO_{3(g)}$	-395.72	-371.06	256.76
$H_2SO_{4(\ell)}$	-813.989	-690.003	156.904
$Sn_{(s)}$ (white)	0	0	51.55
$Sn_{(s)}$ (gray)	-2.09	0.13	44.14
$TiCl_{4(g)}$	-763.2	-726.7	354.9
$TiO_{2(s)}$	-939.7	-884.5	49.92
$U_{(s)}$	0	0	50
$UF_{6(s)}$	-2137	-2008	228
$UF_{6(g)}$	-2113	-2029	380
$UO_{2(s)}$	-1084	-1029	282
$Xe_{(g)}$	0	0	170
$XeF_{2(g)}$	-108	-48	254
$XeF_{4(s)}$	-251	-121	146
$Zn_{(s)}$	0	0	41.63
$ZnO_{(s)}$	-348.28	-318.30	43.64
$ZnS_{(s)}$ (zinc blende)	-206	-201	58
$ZnSO_{4(s)}$	-983	-874	120

Appendix #2.

Equilibrium Constants for Some Monoatomic Weak Acids

Acid	Formula	Ka for this Acid
Hydrogen Sulfate Ion	HSO_4^{-1}	1.2×10^{-2}
Chlorous Acid	$HClO_2$	1.2×10^{-2}
Monochloro Acetic Acid	$HC_2H_2ClO_2$	1.35×10^{-3}
Hydrofluoric Acid	HF	7.2×10^{-4}
Nitrous Acid	HNO_2	4.0×10^{-4}
Formic Acid	HCO_2H	1.8×10^{-4}
Lactic Acid	$HC_3H_5O_3$	1.38×10^{-4}
Benzoic Acid	$HC_7H_5O_2$	6.4×10^{-5}
Acetic Acid	CH_3COOH	1.8×10^{-5}
Propanoic Acid	$HC_3H_5O_2$	1.3×10^{-5}
Hypochlorous Acid	$HOCl$	3.5×10^{-8}
Hypobromous Acid	$HOBr$	2×10^{-9}
Hydrocyanic Acid	HCN	4.0×10^{-10}
Phenol	HOC_6H_5	1.6×10^{-10}

Equilibrium Constants for Some Weak Bases

Base	Formula	Kb for this Base
Ammonia	NH_3	1.8×10^{-5}
Methylamine	CH_3NH_2	4.38×10^{-4}
Ethylamine	$C_2H_5NH_2$	5.6×10^{-4}
Diethylamine	$(C_2H_5)_2NH$	1.3×10^{-3}
Thiethylamine	$(C_2H_5)_3N$	4.0×10^{-4}
Hydroxylamine	$HONH_2$	1.1×10^{-8}
Hydrazine	H_2N-NH_2	3.0×10^{-6}
Aniline	$C_6H_5NH_2$	3.8×10^{-10}
Pyridine	C_5H_5N	1.7×10^{-9}

Equilibrium Constants for Some Common Polyprotic Acids

Name	Formula	K_{a1}	K_{a2}	K_{a3}
Phosphoric Acid	H_3PO_4	7.5×10^{-3}	6.2×10^{-8}	4.8×10^{-13}
Arsenic Acid	H_3AsO_4	5×10^{-3}	8×10^{-8}	6×10^{-10}
Carbonic Acid	H_2CO_3	4.3×10^{-7}	5.6×10^{-11}	
Sulfuric Acid	H_2SO_4	Very Large	1.2×10^{-2}	
Sulfurous Acid	H_2SO_3	1.5×10^{-2}	1.0×10^{-7}	
Hydrosulfulric Acid	H_2S	1.0×10^{-7}	1.3×10^{-13}	
Oxalic Acid	$H_2C_2O_4$	6.5×10^{-2}	6.1×10^{-5}	
Ascorbic Acid (Vitamin C)	$H_2C_6H_6O_6$	7.9×10^{-5}	1.6×10^{-12}	
Citric Acid	$H_3C_6H_5O_7$	8.4×10^{-4}	1.8×10^{-5}	4.0×10^{-6}

Appendix #3.
Ksp Values for Some Common Ionic Solids @ 25°C.

BaF_2	1.43×10^{-6}
MgF_2	6.41×10^{-9}
PbF_2	2.72×10^{-8}
SrF_2	7.91×10^{-10}
CaF_2	3.96×10^{-11}
$PbCl_2$	1.63×10^{-5}
$AgCl$	1.78×10^{-10}
$PbBr_2$	4.59×10^{-6}
$AgBr$	5.00×10^{-13}
PbI_2	1.43×10^{-8}
AgI	8.31×10^{-17}
$CaSO_4$	9.12×10^{-6}
Ag_2SO_4	1.37×10^{-5}
$SrSO_4$	3.20×10^{-7}
$PbSO_4$	1.60×10^{-8}
$BaSO_4$	1.12×10^{-10}
$BaCrO_4$	1.22×10^{-10}
Ag_2CrO_4	1.11×10^{-12}
$PbCrO_4$	2.80×10^{-13}
Ag_2S	1.60×10^{-49}
HgS	1.61×10^{-54}
PbS	8.02×10^{-28}
$NiCO_3$	1.41×10^{-7}
$CaCO_3$	8.70×10^{-9}
Ag_2CO_3	8.11×10^{-12}
$MgCO_3$	1.01×10^{-15}
$Fe(OH)_3$	4.0×10^{-38}
$Ca(OH)_2$	5.55×10^{-6}
$Fe(OH)_2$	8.02×10^{-16}
$Zn(OH)_2$	1.23×10^{-17}
$Al(OH)_3$	2.1×10^{-32}
Ag_3PO_4	1.80×10^{-18}

Appendix #4.
Standard Reduction Potentials @ 25°C for Some Selected Half-Reactions

Half-Reaction	$\varepsilon°$ in Volts
$F_{2(g)} + 2e^- \rightleftharpoons 2F^-$	2.87
$Co^{+3} + e^- \rightleftharpoons Co^{+2}$	1.952
$H_2O_2 + 2H^+ + 2e^- \rightleftharpoons 2H_2O$	1.78
$Ce^{+4} + e^- \rightleftharpoons Ce^{+3}$	1.70
$PbO_{2(s)} + 4H^+ + SO_4^{-2} + 2e^- \rightleftharpoons PbSO_{4(s)} + 2H_2O$	1.69
$MnO_4^- + 4H^+ + 3e^- \rightleftharpoons MnO_{2(s)} + 2H_2O$	1.68
$2e^- + 2H^+ + IO_4^- \rightleftharpoons IO_3^- + H_2O$	1.60
$MnO_4^- + 8H^+ + 5e^- \rightleftharpoons Mn^{+2} + 4H_2O$	1.51
$Au^{+3} + 3e^- \rightleftharpoons Au°$	1.50
$PbO_{2(s)} + 4H^+ + 2e^- \rightleftharpoons Pb^{+2} + 2H_2O$	1.455
$Cl_{2(g)} + 2e^- \rightleftharpoons 2Cl^-$	1.3583
$Cr_2O_7^{-2} + 14H^+ + 6e^- \rightleftharpoons 2Cr^{+3} + 7H_2O$	1.33
$O_{2(g)} + 4H^+ + 4e^- \rightleftharpoons 2H_2O$	1.23
$MnO_{2(s)} + 4H^+ + 2e^- \rightleftharpoons Mn^{+2} + 2H_2O$	1.224
$IO_3^- + 6H^+ + 5e^- \rightleftharpoons 0.5\,I_{2(s)} + 3H_2O$	1.195
$Br_{2(\ell)} + 2e^- \rightleftharpoons 2Br^-$	1.095
$NO_3^- + 4H^+ + 3e^- \rightleftharpoons NO_{(g)} + 2H_2O$	0.96
$ClO_2 + e^- \rightleftharpoons ClO_2^-$	0.954
$Ag^+ + e^- \rightleftharpoons Ag°$	0.7994
$Fe^{+3} + e^- \rightleftharpoons Fe^{+2}$	0.771
$O_{2(g)} + 2H^+ + 2e^- \rightleftharpoons H_2O_2$	0.68
$I_{2(s)} + 2e^- \rightleftharpoons 2I^-$	0.535
$Cu^+ + e^- \rightleftharpoons Cu°$	0.52
$O_{2(g)} + 2H_2O + 4e^- \rightleftharpoons 4OH^-$	0.40
$Cu^{+2} + 2e^- \rightleftharpoons Cu°$	0.3402
$AgCl_{(s)} + e^- \rightleftharpoons Ag° + Cl^-$	0.222
$SO_4^{-2} + 4H^+ + 2e^- \rightleftharpoons H_2SO_3 + H_2O$	0.20
$Cu^{+2} + e^- \rightleftharpoons Cu^+$	0.16
$2H^+ + 2e^- \rightleftharpoons H_{2(g)}$	0.00
$Fe^{+3} + 3e^- \rightleftharpoons Fe°$	−0.036
$Pb^{+2} + 2e^- \rightleftharpoons Pb°$	−0.13

Half-Reaction	$\varepsilon°$ in Volts
$Sn^{+2} + 2e^- \rightleftharpoons Sn°$	-0.14
$Ni^{+2} + 2e^- \rightleftharpoons Ni°$	-0.23
$Co^{+2} + 2e^- \rightleftharpoons Co°$	-0.277
$PbSO_{4(s)} + 2e^- \rightleftharpoons Pb° + SO_4^{-2}$	-0.35
$Cd^{+2} + 2e^- \rightleftharpoons Cd°$	-0.4026
$Fe^{+2} + 2e^- \rightleftharpoons Fe°$	-0.44
$Cr^{+3} + e^- \rightleftharpoons Cr^{+2}$	-0.50
$Cr^{+3} + 3e^- \rightleftharpoons Cr°$	-0.73
$Zn^{+2} + 2e^- \rightleftharpoons Zn°$	-0.76
$2H_2O + 2e^- \rightleftharpoons H_{2(g)} + 2OH^-$	-0.83
$Mn^{+2} + 2e^- \rightleftharpoons Mn°$	-1.18
$Al^{+3} + 3e^- \rightleftharpoons Al°$	-1.66
$H_{2(g)} + 2e^- \rightleftharpoons 2H^-$	-2.23
$La^{+3} + 3e^- \rightleftharpoons La°$	-2.37
$Na^+ + e^- \rightleftharpoons Na°$	-2.71
$Ca^{+2} + 2e^- \rightleftharpoons Ca°$	-2.76
$Ba^{+2} + 2e^- \rightleftharpoons Ba°$	-2.90
$K^+ + e^- \rightleftharpoons K°$	-2.925
$Li^+ + e^- \rightleftharpoons Li°$	-3.045

$PV = nRT$

$V = \frac{nRT}{P}$ $q_p = \Delta H = n C_p \Delta T$

$W = -nRT \ln \frac{V_2}{V_1}$

$W = P_{ext} \Delta V \times 101.3 \text{ J/L·atm}$

$W = \Delta E - q$

Σ
$(2(66.7)) - 2(86.6)$

$\Delta G° = -39.8$

-39.8

know state function

P1   6 T/F   Like homework questions
11-17
explain why it is false

P2   Theory - 4 processes
table   expansion against a vac
asks for sign of W
⊕ ⊖ or 0

Few essays

either adiabatic $\Delta q = 0$
or (isothermal) non res
reversible $P_{int} = P_{ext} = P_2$
isothermal
isobaric   Multi-step prob on chem formation
Absolute entropy
Given conditions - find other stuff